Performance and Safety Enhancement Strategies in Vehicle Dynamics and Ground Contact

Performance and Safety Enhancement Strategies in Vehicle Dynamics and Ground Contact

Editors

Flavio Farroni
Andrea Genovese
Aleksandr Sakhnevych

MDPI • Basel • Beijing • Wuhan • Barcelona • Belgrade • Manchester • Tokyo • Cluj • Tianjin

Editors
Flavio Farroni
University of Naples Federico II
Italy

Andrea Genovese
University of Naples Federico II
Italy

Aleksandr Sakhnevych
University of Naples Federico II
Italy

Editorial Office
MDPI
St. Alban-Anlage 66
4052 Basel, Switzerland

This is a reprint of articles from the Special Issue published online in the open access journal *Applied Sciences* (ISSN 2076-3417) (available at: https://www.mdpi.com/journal/applsci/special_issues/vehicle_dynamics).

For citation purposes, cite each article independently as indicated on the article page online and as indicated below:

LastName, A.A.; LastName, B.B.; LastName, C.C. Article Title. *Journal Name* **Year**, *Volume Number*, Page Range.

ISBN 978-3-0365-3417-6 (Hbk)
ISBN 978-3-0365-3418-3 (PDF)

Cover image courtesy of Gaia Santamaria Amato

Contents

About the Editors

Flavio Farroni is a Professor of Applied Mechanics at the University of Naples; founder and CEO of its spinoff, MegaRide; and works as an academic advisor in vehicle dynamics with several companies and racing teams. He works on the development of physical models for the optimization of performance in motorsports, in Drivers in the Loop systems and in digital mobility scenarios, cooperating within an international research network. During TireExpo2015, he received the "Young Scientist of the Year" award from MIT Technology Review, was among "Italian Innovators Under 35" of 2018, and as the start-up of MegaRide, won "Tire Technology of the Year" at the 2018 Tire Technology Conference.

Andrea Genovese received an M.S. in Mechanical Engineering (summa cum laude) and PhDs in Industrial Engineering from the University of Naples, in 2011 and 2018, respectively. He is Assistant Professor in Applied Mechanics to Machines at the University of Naples Federico II. His research interests include the design and development of mechatronic systems, smart systems, tribology, non-destructive materials characterization, energy harvesting, vehicle dynamics and vibration control. He is author of more than 30 papers in international journals and refereed conferences. He serves on the Editorial Board of several international scientific journals and conferences.

Aleksandr Sakhnevych received a Ph.D. degree in Mechanical Engineering from the Faculty of Engineering at the Università degli Studi di Napoli, Italy, in 2018. He is Assistant Professor in Applied Mechanics at the University of Naples. His research is centred on vehicle dynamics and on the multi-physical comprehension of tire–road interaction phenomena with the specific purpose of vehicle performance optimization and control. Prof. Sakhnevych has been a member of IAENG since 2015.

Preface to "Performance and Safety Enhancement Strategies in Vehicle Dynamics and Ground Contact"

This book was created with the aim of collecting the latest and most interesting advances in the research domain the guest editors' scientific and technical activities are focused towards.

We hope that it serves as a useful contribution to the state-of-the-art developments in vehicle dynamics, which nowadays concern onboard controls and computational science, attracting experimental and in-simulation papers, focusing on both basic research and perspective applications for future mobility. This book's articles involve research, studies, and projects derived from vehicle dynamics and contact mechanics, with the outlook of embracing activities that aim to enhance vehicle performance in terms of handling, comfort, and adherence and to examine safety optimization in the emerging contexts of smart, connected, and autonomous driving.

Our work on the editing of the volume allowed us to get in touch with brilliant scientists and professionals from the automotive community, providing further value to this challenging research topic.

Our sincere thanks go to all of them for their contributions and to the MDPI Editorial Office, with particular reference to Ms. Shirley Wang, who provided continuous and extensive support.

Flavio Farroni, Andrea Genovese, Aleksandr Sakhnevych
Editors

applied sciences

MDPI

Editorial

Performance and Safety Enhancement Strategies in Vehicle Dynamics and Ground Contact

Flavio Farroni *, Andrea Genovese and Aleksandr Sakhnevych

Department of Industrial Engineering, University of Naples Federico II, 80131 Naples, Italy; andrea.genovese2@unina.it (A.G.); ale.sak@unina.it (A.S.)
* Correspondence: flavio.farroni@unina.it

Recent trends in vehicle engineering prove the great effort that scientists and industries have made in seeking solutions to enhance both the performance and the safety of vehicular systems. Physical models concerning vehicle–ground interaction, control strategies for the vehicle and its subsystems, and new technologies are developing all over the world for this purpose. The published Special Issue contributed to the study of modern vehicle dynamics, attracting recent experimental and in-simulation advances that are the basis for the current technological growth and for future mobility. Such areas involve research, studies, and projects coming from both vehicle dynamics and contact mechanics, with the perspective to embrace activities aiming to enhance vehicle performance in terms of handling, comfort, and adherence and to examine safety optimization also in the emerging contexts of smart, connected, and autonomous driving.

The accepted scientific contributions covered topics concerning new results and studies in the following areas related to the interaction of vehicle dynamics and the ground:

- Physical models concerning tire–road and vehicle–ground interaction: In particular, Ref. [1] refers to new developments in airless (or non-pneumatic) tires, representing a significant perspective in the future evolution of such components. Regarding tires, Ref. [2] proposes strategies to optimize tread wear and minimize the dispersion of rubber particles, properly acting on wheel and suspension setup. Moreover, Refs. [3,4] focus on materials characterization and local contact phenomena, analyzing, respectively, innovative polynomial formulations for the reproduction of viscoelastic compounds' behaviors and the adhesive effects of dimpled textures in contact with flat surfaces. Finally, Ref. [5] proposes artificial neural networks to identify the parameters of Pacejka's Magic Formula tire models, widely adopted in the context of automotive simulations;
- Experimental activities aimed at the investigation and the comprehension of interaction phenomena: Among the published papers, some developed an approach based on the macroscale effects, analyzing the whole vehicle data as proposed in Ref. [6], mainly centered on ride analysis on wavy profiles; In Ref. [7], accounting for suspension sensitivity to road roughness, longitudinal speed, and vehicle segment; and in Ref. [8], switching to the effects on the directional capabilities. Some other authors worked on the microscale, accounting for indentation, friction, and contact mechanics at the ground, as investigated in Ref. [9], relating to gravel surfaces and noise, and in Ref. [4], focusing also on aspects related to adhesive local contact phenomena;
- Control strategies focused on vehicle performance enhancement, in terms of handling/grip, comfort, and safety: In Ref. [10], a safety control strategy is proposed, acting on the steering system and differential, useful for performing emergency maneuvers for obstacle avoidance; In Ref. [11], a central predictive control system is proposed, acting on a non-linear, model-based predictive algorithm; and in Ref. [12], the onboard implementation of friction estimation, in autonomous driving and vehicle following applications, is illustrated. The authors of one of the submitted papers also

Citation: Farroni, F.; Genovese, A.; Sakhnevych, A. Performance and Safety Enhancement Strategies in Vehicle Dynamics and Ground Contact. *Appl. Sci.* **2022**, *12*, 2034. https://doi.org/10.3390/app12042034

Received: 7 February 2022
Accepted: 11 February 2022
Published: 16 February 2022

focused on traffic contexts, in particular reporting a case study involving Duisburg Ring environment [13];
- Innovative technologies to improve the safety and performance of the vehicle and its subsystems, such as adopting active/semiactive suspension in in-wheel architectures, enhancing the roadholding [14] and stability-oriented steering systems in articulated vehicle applications [15];
- Identification of vehicle and tire/wheel model parameters and state with innovative methodologies and algorithms, based on machine learning techniques, as described in Ref. [5];
- Implementation of real-time software, logics, and models in onboard architectures [13] with a main target involving applications oriented towards autonomous driving and connected mobility scenarios;
- Studies and analyses oriented toward the correlation among the factors affecting vehicle consumptions, such as in powertrain architectures in electric mobility described in Ref. [16], or performance and stability, with the target to propose strategies for the minimization of undesired phenomena, as proposed by the authors of the article [2], who focused on tire tread wear;
- Application use cases in scenarios not only concerning car and conventional four-wheeled vehicles or common asphalt roads. The published papers represent advances in vehicle dynamics also involving off-road vehicles, as analyzed in Ref. [9], heavy articulated vehicles [15], or motorcycles, for which [17] proposed a study on their stability, developing an innovative approach based on the so-called screw axis instead of the usual phase plane.

Author Contributions: Conceptualization, F.F.; writing—original draft preparation, F.F.; writing—review and editing, A.S. and A.G.; All authors have read and agreed to the published version of the manuscript.

Funding: This research received no external funding.

Institutional Review Board Statement: Not applicable.

Data Availability Statement: Not applicable.

Conflicts of Interest: The authors declare no conflict of interest.

References

1. Genovese, A.; Garofano, D.; Sakhnevych, A.; Timpone, F.; Farroni, F. Static and Dynamic Analysis of Non-Pneumatic Tires Based on Experimental and Numerical Methods. *Appl. Sci.* **2021**, *11*, 11232. [CrossRef]
2. Papaioannou, G.; Jerrelind, J.; Drugge, L. Multi-Objective Optimisation of Tyre and Suspension Parameters during Cornering for Different Road Roughness Profiles. *Appl. Sci.* **2021**, *11*, 5934. [CrossRef]
3. Tassara, M.; Grigoriadis, K.; Mavros, G. Empirical Models for the Viscoelastic Complex Modulus with an Application to Rubber Friction. *Appl. Sci.* **2021**, *11*, 4831. [CrossRef]
4. Papangelo, A. On the Effect of a Rate-Dependent Work of Adhesion in the Detachment of a Dimpled Surface. *Appl. Sci.* **2021**, *11*, 3107. [CrossRef]
5. Olazagoitia, J.; Perez, J.; Badea, F. Identification of Tire Model Parameters with Artificial Neural Networks. *Appl. Sci.* **2020**, *10*, 9110. [CrossRef]
6. Wedig, W. Speed Oscillations of a Vehicle Rolling on a Wavy Road. *Appl. Sci.* **2021**, *11*, 10431. [CrossRef]
7. Lukoševičius, V.; Makaras, R.; Rutka, A.; Keršys, R.; Dargužis, A.; Skvireckas, R. Investigation of Vehicle Stability with Consideration of Suspension Performance. *Appl. Sci.* **2021**, *11*, 9778. [CrossRef]
8. Lukoševičius, V.; Makaras, R.; Dargužis, A. Assessment of Tire Features for Modeling Vehicle Stability in Case of Vertical Road Excitation. *Appl. Sci.* **2021**, *11*, 6608. [CrossRef]
9. Žuraulis, V.; Sivilevičius, H.; Šabanovič, E.; Ivanov, V.; Skrickij, V. Variability of Gravel Pavement Roughness: An Analysis of the Impact on Vehicle Dynamic Response and Driving Comfort. *Appl. Sci.* **2021**, *11*, 7582. [CrossRef]
10. Lin, Y.; Chen, B. Handling Enhancement of Autonomous Emergency Steering for Reduced Road Friction Using Steering and Differential Braking. *Appl. Sci.* **2021**, *11*, 4891. [CrossRef]
11. Sieberg, P.; Schramm, D. Central Non-Linear Model-Based Predictive Vehicle Dynamics Control. *Appl. Sci.* **2021**, *11*, 4687. [CrossRef]
12. Santini, S.; Albarella, N.; Arricale, V.; Brancati, R.; Sakhnevych, A. On-Board Road Friction Estimation Technique for Autonomous Driving Vehicle-Following Maneuvers. *Appl. Sci.* **2021**, *11*, 2197. [CrossRef]

13. Ma, X.; Hu, X.; Weber, T.; Schramm, D. Traffic Simulation of Future Intelligent Vehicles in Duisburg City Inner Ring. *Appl. Sci.* **2021**, *11*, 29. [CrossRef]
14. Woo, S.; Shin, D. A Double Sky-Hook Algorithm for Improving Road-Holding Property in Semi-Active Suspension Systems for Application to In-Wheel Motor. *Appl. Sci.* **2021**, *11*, 8912. [CrossRef]
15. Lei, T.; Wang, J.; Yao, Z. Modelling and Stability Analysis of Articulated Vehicles. *Appl. Sci.* **2021**, *11*, 3663. [CrossRef]
16. Yildiz, A.; Özel, M. A Comparative Study of Energy Consumption and Recovery of Autonomous Fuel-Cell Hydrogen–Electric Vehicles Using Different Powertrains Based on Regenerative Braking and Electronic Stability Control System. *Appl. Sci.* **2021**, *11*, 2515. [CrossRef]
17. Bova, M.; Massaro, M. A Screw-Axis Approach to the Stability of Two-Wheeled Vehicles. *Appl. Sci.* **2021**, *11*, 7393. [CrossRef]

applied sciences

MDPI

Article

Identification of Tire Model Parameters with Artificial Neural Networks

José Luis Olazagoitia [1,*], Jesus Angel Perez [2] and Francisco Badea [1]

1 Universidad Antonio de Nebrija, 28040 Madrid, Spain; fbadea@nebrija.es
2 Universidad de Oviedo, 33394 Gijón, Spain; perezangel@uniovi.es
* Correspondence: jolazago@nebrija.es; Tel.: +34-914-521-101

Received: 15 November 2020; Accepted: 17 December 2020; Published: 20 December 2020

Abstract: Accurate modeling of tire characteristics is one of the most challenging tasks. Many mathematical models can be used to fit measured data. Identification of the parameters of these models usually relies on least squares optimization techniques. Different researchers have shown that the proper selection of an initial set of parameters is key to obtain a successful fitting. Besides, the mathematical process to identify the right parameters is, in some cases, quite time-consuming and not adequate for fast computing. This paper investigates the possibility of using Artificial Neural Networks (ANN) to reliably identify tire model parameters. In this case, the Pacejka's "Magic Formula" has been chosen for the identification due to its complex mathematical form which, in principle, could result in a more difficult learning than other formulations. The proposed methodology is based on the creation of a sufficiently large training dataset, without errors, by randomly choosing the MF parameters within a range compatible with reality. The results obtained in this paper suggest that the use of ANN to directly identify parameters in tire models for real test data is possible without the need of complicated cost functions, iterative fitting or initial iteration point definition. The errors in the identification are normally very low for every parameter and the fitting problem time is reduced to a few milliseconds for any new given data set, which makes this methodology very appropriate to be used in applications where the computing time needs to be reduced to a minimum.

Keywords: tire model parameters identification; artificial neural networks; curve fitting; Pacejka's magic formula

---

1. Introduction

The identification of parameters in tire models has been addressed through different techniques. It is common practice in the literature to consider the cost function that takes into account only the vertical component of the error between measured data and the mathematical model, assuming indirectly that the independent variable is well known or, at least, that its contribution to the total error is negligible.

Historically, the most used methodology to identify tire model parameters has been the Ordinary Least Squares on the Pacejka's Magic Formula (MF) model [1]. The most widely used methods have been derived from the different alternatives of Newton's method [2]: Gauss–Newton, Levenberg–Marquardt [3], quasi-Newton [4], sequential quadratic programming SQP [5], Nelder–Mead [6], etc. A comparison of various algorithms can be found in [7]. Recently, a Weighted Orthogonal Regression method was presented to account for errors in the slip and the force values [8]. These methods are by their nature iterative and rely on a suitable starting point. In this case, this means that an appropriate initial set of parameters needs to be given to the algorithm to be able to provide a satisfactory solution. This is due to the fact that multiple local minima can be found in the cost function, and these Newtonian methods search for local minima.

An interesting alternative for the identification of these coefficients is to use Genetic Algorithms (GA) and Neural Networks (NN). GA has been found to be good at exploring the parameter space [9,10] and at estimating the MF coefficient, even starting from a bad estimation of the initial parameters. However, GA it is not very suitable for precise exploitation (i.e., identifying the optimum set of parameters). Recently, Talebitooti [11] has combined GA with a local search algorithm, proving that this hybrid two-step method is much more efficient.

ANN has been successfully used along with state observers for dynamic control of cars [12,13]. Palkovic and El-Gindy [14] investigated some applications of the ANN to vehicle dynamics and control and compared it to the MF. They used tire test data to directly train the ANN and the results were promising. However, the starting data for the network training was very limited, as it was based on data from individual tires, of which there was also little data available for the network training.

Regarding tire data obtained directly from the direct interaction of the tire with the road, Farroni [15] proposes a tool called TRICK to estimate the tire interaction forces as a function of slip rates, thus avoiding expensive and time consuming test bed characterizations. Later, Farroni [16] also proposed the use of another tool called TRIP-ID, for the interactive identification of tire parameters based on data from experimental bench data or direct interaction with the road. In the same way, Wang [17] presented the identification of parameters through ANN for a specific tire, based on real test data of a race car tire (Hoosier 18.0 x 6-10R25B) on a test bench for different test conditions.

In this line, in the evaluation of the parameters of Pacejka's MF model in the road operation of tires based on data acquisition, there is the problem of not having continuous data in the whole range and combinations of the slips. Therefore, it would be necessary to combine this identification of partially observable data in the vehicle dynamic system (with noisy and/or incomplete acquired data) with methodologies to reconstruct the complete model, as proposed by [18].

Unlike previous studies into ANN identifying tire parameters in MF, the present study is based on the evaluation of the possibility of creating a widely valid ANN, which can be trained to predict the parameters of Pacejka's MF, for a wide range of real tires. For this, a sufficiently ample database of curves is artificially created. Unlike previous ANN works, limited training data (due to data from only one, or a limited set of tires) is not a problem, since the methodology provides means to create as much training data as is necessary to ensure good network training. Based on this information, the ANN topology is selected and its performance is evaluated.

Therefore, this article presents a novel methodology to create a "universal" ANN-based mathematical model, which can be trained before knowing the specific data obtained from bench tests, or from the acquisition of data of the tire-road contact. Furthermore, this ANN can consider all the specific variations that the Pacejka curve admits, being therefore valid to represent any of the effects (pure-slip, combined-slip, camber angle, pressure, temperature, humidity, etc.) that make such a mathematical model vary. Obviously, since it is based on a mathematical model that may not be perfect (in the sense that it may not be exactly faithful to reality), the ANN obtained will have at least the same errors as the model it represents (in this case Pacejka's MF). The advantage of having a pre-trained network that can represent a wide variety of conditions for different types of tires and immediately identify their Pacejka parameters is evident when you consider that the run time of such a network is only a few milliseconds. This low cost computation allows it to be integrated into real-time vehicle computer models, allowing identification from the data obtained from the vehicle, to have a system dynamic adaptable to the specific conditions and environmental variables of the tire on the vehicle.

The results shown in this paper are part of an ongoing research on the application of different technologies to the identification of vehicle system dynamic parameters [19]. The identification of parameters has also been extended to other mathematically simpler tire models [20] and the identification errors have also been small.

The study presented in this paper is organized as follows. Section 2 presents ANNs in general, their configuration and operation. Section 3 presents the methodology followed in this paper. Section 3.1 deals with the generation of sufficient data (without errors) to train the ANN. Selection of the topology of the ANN, its training and validation, are explained in Section 3.2. In Section 3.3, data (with errors) acquired in bench in tire tests are used to analyze the predictive performance of the network in real situations. Finally, in Section 4, final conclusions are exposed.

2. Artificial Neural Networks (ANNs)

The ANNs can be defined as processing algorithms that infer patterns and relationships between known input and output data, provided sufficient information is given to the system to learn. An ANN is set up mimicking the neural structure of the human brain, where hundreds of thousands of "simple" processor units (neurons) work together by interconnections among them (synapses) to allow learning. In this way, in one type of ANN, the network topology is composed of simple nodes that are organized in layers. The first layer usually has as many nodes as input parameters. The last layer is set up with as many nodes as output parameters (Figure 1).

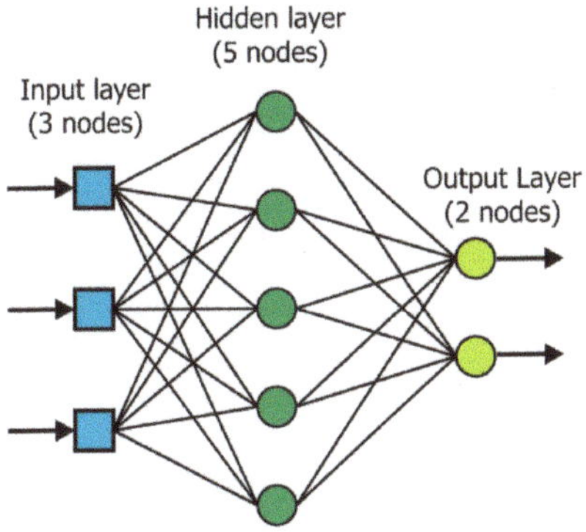

Figure 1. Example of an ANN (3-5-2).

Between these two layers there exist some hidden layers, each of them comprised of a given number of nodes. Nodes from two consecutive layers are usually fully connected between them. Each connection is given a weighting factor that is used to modify the information traveling through it. These connections allow passing information from one node to others connected to it. When the information is passed from left to right, it is called a "feed forward" connection. Nodes between consecutive layers can develop connections among them passing in this way information from one node to the other. Nodes sum up all the weighted information from their input connections in the previous layer and transmit the result to the output connections in the following layers.

To start the training process, after proper randomizing of the weights, the input layer of the ANN gets the input information from the algorithm and passes it on to the following layers. Each node receiving information acts on it and modifies it according to its activation configuration (normally a sigmoid, or similar function). In this way, the information travels through the ANN from the input nodes to the output ones. The training algorithm compares this result to the known solution of the provided training data. If the solution is not ok, then a so-called "backpropagation" algorithm modifies the weights of the connections according to the size of the error difference. Once the weights of all the connections have been updated, all the process is repeated until the answer given by the output layer is satisfactory (or a given stop flag is activated).

This procedure is called "supervised learning", because one knows in advance the correct answer and, if the algorithm fails to obtain it, the ANN weights are corrected in order to get a better solution. Once the ANN has learned from the training data, it can provide solutions for unknown data. Therefore, the ANN is trained with the so-called "training data", and its performance is usually measured by getting an answer from unknown "test data". If the learning of the ANN has been successful, the answer to the "test data" will be satisfactory.

ANNs have already been used in a countless number of applications where analytical solutions do not exist or are very costly to obtain. There are many types of ANN. The configuration of the connections of ANNs is very dependent on the type of problem under study [21–23]. In order to build an efficient ANN, it is necessary to determine its type, topology and learning rule.

In this case, it is desired to know the optimum tire parameters related to the known test data of a particular tire. The input data will be the known test data for certain slips. In this paper, the input data to the network will be the longitudinal force measurements corresponding to slips ranging from −75% to 75%, discretized in steps of 2% for the slip. This gives us 76 entries to the network. These inputs are of a continuous character since the measured longitudinal force can have any continuous real value.

The output of the ANN are the parameters of the chosen tire model. In this case, it has been decided to work with the well-known Magic Formula [1], which is a semi-empirical tire model, and is the most used in the work related to tire models.

In its reduced form, the Magic Formula has six parameters that must be identified to know the whole curve. The output of the ANN will, therefore, have six outputs, one for each parameter. It is important to realize that this model (1) is very nonlinear since it has some nested trigonometric functions, which makes its complexity very high and, as will be seen, will affect the selection of the chosen ANN (see Section 3).

$$F_i = f_y + Fz{\cdot}D \cdot \sin(C \cdot \arctan(B \cdot (x_i + f_x) - E \cdot (B \cdot (x_i + f_x) - \arctan(B \cdot (x_i + f_x))))) \tag{1}$$

The model (1) has the following nomenclature: xi is the longitudinal (or lateral) slip in % and Fi is the longitudinal (or lateral) force, in N. The pairs of values (xi, Fi) are known from tests performed on tire test benches. The remaining variables are the parameters that need to be identified. The parameter *fx* reflects the "horizontal shift" of the curve and *fy* the "vertical shift" that show the deviations in x and y from the curve that does not exactly pass through the coordinate origin. The parameter D is the peak factor, which multiplies to the *Fz* (vertical force on the tire). Both parameters together *Fz·D* stand for the maximum amplitude height of the curve. Parameter C represents the shape factor of the curve and B is the stiffness factor. Finally, the parameter E is the "curvature factor".

Usually, ANNs are used for classification issues or discrete outputs. In this case, all the parameters to be identified are continuous variables, so it is necessary that the ANN support real-valued continuous output variables. The input and output data of the network are known, so a supervised training will be used. The ANN will be based on a Multi-Layer Perceptron (MLP), with nodes totally connected to each other between successive layers (Feed-Forward) [24], activated through the sigmoid function (with continuous derivative) that allows training through the Backpropagation Algorithm [21]. The selection of the chosen ANN topology is detailed in the following section as part of the working methodology.

3. Methodology and Results

In this paper, we intend to study the extent to which ANNs are adequate and could be used to predict the parameters of tire curves such as MF from known data from bench tests.

In order to answer this question, the work has been proposed according to the methodology shown in Figure 2, which is divided into three large blocks: (1) creation of curves to train the ANN; (2) determination of the topology of the ANN and its training; and (3) validation of the ANN against real test bench data.

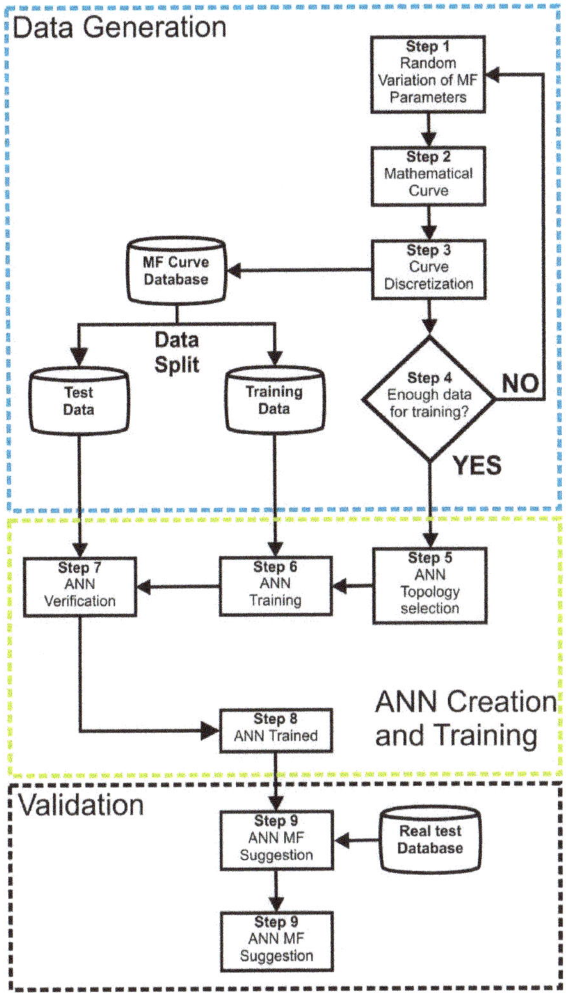

Figure 2. ANN learning and verification flowchart.

In summary, this methodology intends to train the ANN from a sufficiently high number of known tire curves. The main problem is that conducting tests on real tires is very costly and it would not be economically feasible to carry out so many tests to train a universal ANN that can predict the parameters of the MF curve in all possible conditions and situations.

For this reason, the proposed methodology creates synthetic curves based on the use of the MF model. The idea is simple. Knowing the range of values in which the parameters f_x, f_y, B, C, $F_z{\cdot}D$, and E can take in real tire curves, it is possible to generate a sufficiently large set of random combinations. In this way, it is possible to create as many curves as needed so that ANN training can be performed properly.

This is the purpose of the first part (Data Creation) of the methodology proposed in Steps 1 to 4 and explained in Section 3.1. The second part (ANN Creation and Training) studies the topology that the ANN must have to be able to adapt to the nonlinearity of the application (discussed in Section 3.2) and the ability of the trained network to predict the parameters of unknown synthetic curves (Steps 5 to 8). The third part (discussed in Section 3.3) examines the ANN's ability to predict the parameters of unknown curves with measurement errors from actual tests on tire test benches (Steps 9 to 10).

3.1. Data Creation

The creation of synthetic data, from the mathematical model provided by the MF, is carried out in Steps 1 to 4 (Figure 2). First, in Step 1, the parameters of the tire model are varied randomly and independently of each other, in a sufficiently wide range of values to cover the largest possible combination of true tire curves. For this study, the limits of the values that these parameters can take are indicated in Table 1. It is important to note that the synthetic data obtained are free from errors arising from the application in which the tires are used (from test bench or vehicle dynamic system acquisition). However, it must be stressed that mathematical errors always exist in the model (either because the model chosen, in this case Pacejka's model, does not perfectly match reality, or because of mathematical calculation errors, rounding or truncation).

Table 1. Range of values for MF model parameters.

Variable	Minimum	Maximum
x_i	−0.75	+0.75
B	5	30
C	0.5	2
E	−2	0
Fz	1000	5000
D	0.1	2
fx	−0.02	0.02
fy	−300	+300

In Step 2, the mathematical curves are obtained. It can be noted that each set of random parameters produces a very different mathematical curve F_i. For example, in Figure 3, some different curves are shown from Equation (1) for different sets of random parameters.

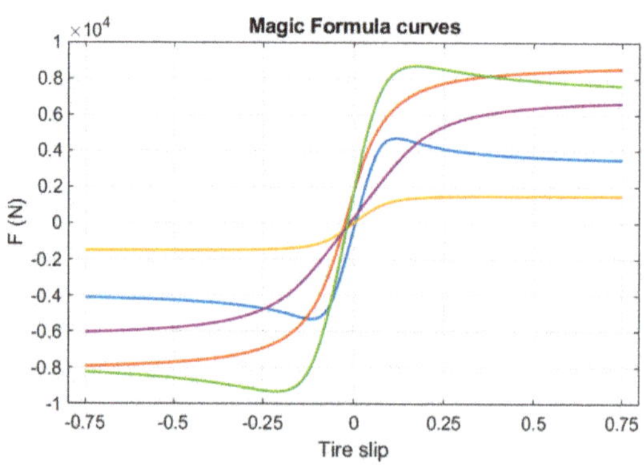

Figure 3. Some MF curves generated by random parameters.

It is important to remark that the range of the parameters has been selected to be sufficiently broad, so that the MF curve can have, mathematically, very different shapes.

In Step 3, the curves are discretized in 2% slip steps in the x_i slip range (e.g., Figure 4). These data points are saved in the training database together with the original set of MF parameters that produced the original curve. This information will be used later in the supervised training of the ANN.

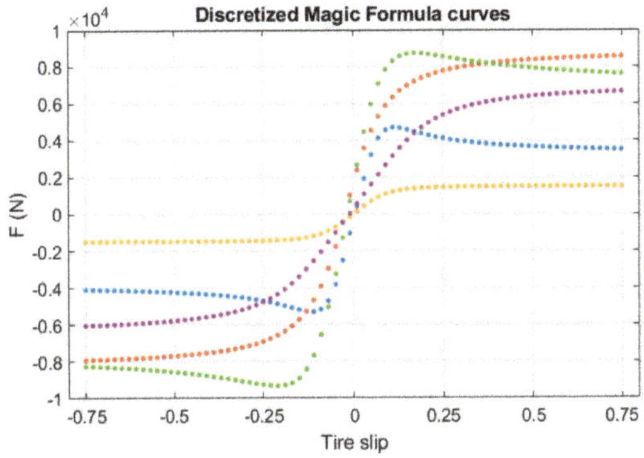

Figure 4. Discretization of MF curves.

It is important to note that the curves generated from the MF are mathematically perfect. That is, as they have been generated mathematically, there is no error in the parameters, the slip or the longitudinal/lateral force. This is very convenient at this point to be able to evaluate the ability of the ANN to represent a sufficiently complicated mathematical model without including other uncertainties or error variables.

One of the objectives of the ANN training is that the network learns to generalize the relationships between the parameters and the actual curve. This can be achieved if the training data is big enough. Step 1 to step 3 are repeated until sufficient data is gathered to train the ANN.

There are no fixed rules to define how many data sets are necessary to train an ANN successfully. This number depends on the complexity and non-linearity of the model under study. In order to be able to determine how big the data size needs to be, during the works in this investigation, the ANN has been trained with sets of a different number of curves and the training error was measured. Figure 5 shows the root-mean-square deviation (RMS Error) at the increasing number of curves in the training set and at a different number of training iterations (epochs). The trend shows that the higher the number of data sets the smaller the training error is. It can also be seen that the error is quite similar at 25k, 50k or 100k curves (with an RMS Error around 0.010 at 8000 epochs). In this study, to make sure a good training is achieved, a data set of 100k curves was selected (Step 4).

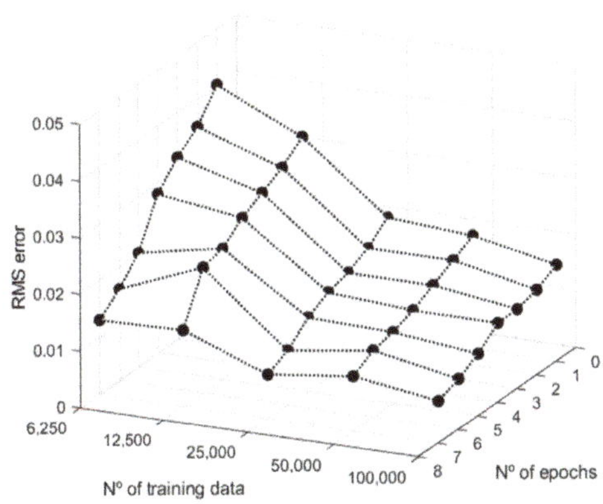

Figure 5. RMS error at different Training Data Sizes.

3.2. Ann Selection, Training and Validation

After the selection of the training data, we can move on to the second part of the methodology (Steps 5 to 8 in Figure 2), which focuses on the determination of the topology of the ANN (Step 5), its training (Step 6) and its verification (Step 7).

In Step 5 topology of the ANN is selected. This topology has to assure full connection between nodes of consecutive layers in the ANN. The selected learning rule type is supervised. Connection weights between neurons are updated in sequential training iterations according to the output targets and the obtained outputs.

For the hidden layers, different architectures may be valid and there is no general rule on how to obtain an optimum one as it depends on the complexity of the problem under study. Most of the literature suggests that from a theoretical point of view it is possible to approximate almost any function with one layer of ANN. If this is correct, it means that a single layer ANN with a sufficient number of hidden neurons can provide a good approximation for most problems. It also means that adding a second or third layer yields little benefit.

Following these recommendations, the possibility of using an ANN with a single hidden layer was explored. Figure 6 shows that more nodes on the hidden layer reduce the training RMS Error. This is true for different training data sets sizes (e.g., 6.25k or 25k). However, the improvement is very small despite continuing to increase the number of nodes in the network. Therefore, for this problem, it does not seem appropriate to use a single hidden layer ANN.

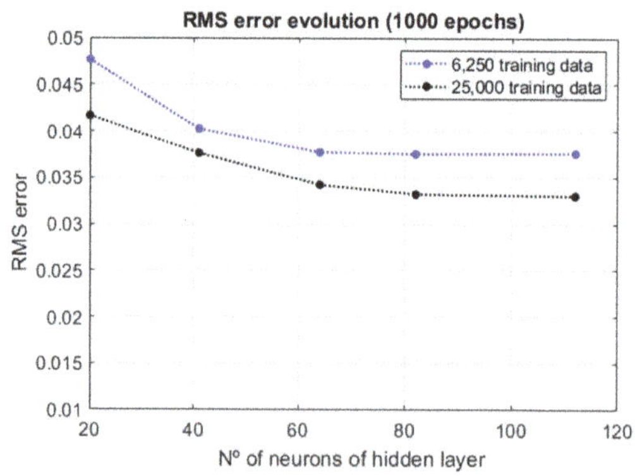

Figure 6. ANN with one hidden layer.

The identification of tire parameters is a highly non-linear problem. For this reason, a study was carried out in order to identify the best network topology that best suits this situation. Several network topologies were tested from one (1H) to five (5H) internal hidden layers with the same number of training curves (6.25k and 25k) running over 1000 iterations (epochs). The distribution of nodes within the network between successive hidden layers was done progressively so that the transition between the 76 inputs and the 6 outputs was "smooth". The distribution of nodes was as follows: 76-41-6 (1H), 76-50-25-6 (2H), 76-50-40-20-6 (3H), 76-60-45-30-15-6 (4H), 76-60-45-35-25-15-6 (5H). Figure 7 shows the results of the network training for each of these networks. The best training performance was obtained for a three hidden layer network with 25k training curves.

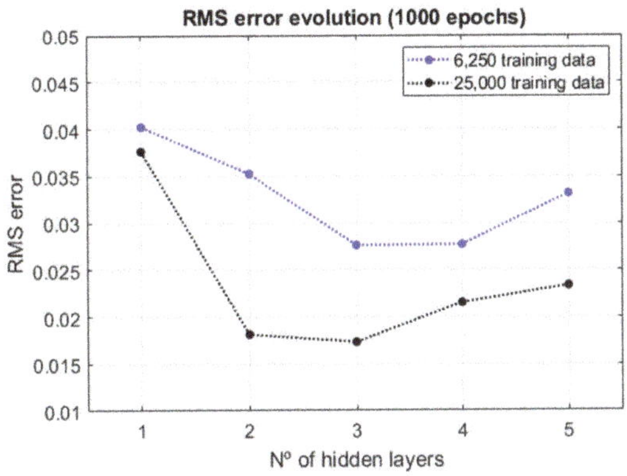

Figure 7. ANN with a different number of hidden layers.

In the course of this research, other less progressive network topologies between input and output for 3H networks were tested, and the results were no better. For this reason, it was decided to use the 3H configuration (76-50-40-20-6) (Figure 8) as a sufficiently good network topology for the problem of the identification of tire parameters. Taking into account that the ANN has a weight to be determined for each connection and that the connections were configured only between adjacent layers, the number of weights to be learned by the ANN can be calculated ($76 \times 50 + 50 \times 40 + 40 \times 20 + 20 \times 6 = 6720$ weights).

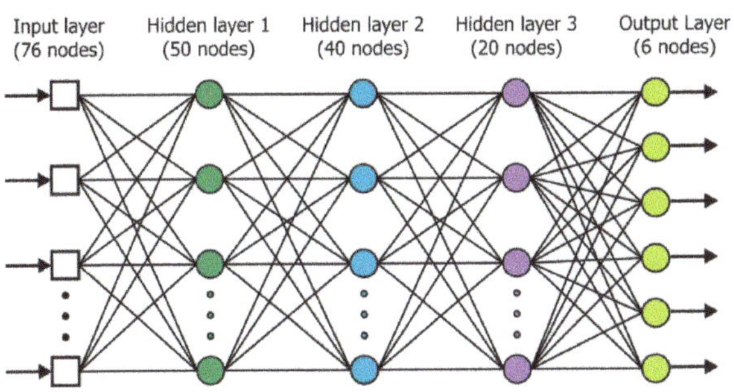

Figure 8. Selected ANN Topology (76-50-40-20-6).

Once the ANN configuration is defined and the training dataset is available, the training can start. Figure 9 shows the details of the activities that are executed in this Step 6. In this case, due to the massive input data chosen (100k curves) for the training, the time to complete it was in the range of a full week with an Intel i5 processor. It is important to emphasize that the objective is to have a tool that can give a precise estimate of the parameters of the tire curve from the inputs provided in a universal way (within the range of parameters defined in Table 1). For this reason, the training time of the ANN is not key, since it will be a one-time calculation. Once the network is trained (Step 8), the ANN can be used as an estimation tool that takes only a few milliseconds for the prediction.

It is usually common to divide the number of available curves to train the network into two subsets. The first is used to train the network (training dataset) and the second is a test dataset. The test dataset is not known by the network since their curves have not been used for its training. The performance of the training (Step 7) can be checked in different ways. For example, in Figure 5 the total RMS Error was calculated at eight different epochs. Figure 10 shows the RMS Error for both the training and testing datasets, which have a continuously decreasing slope. This indicates that no overtraining has been produced on the ANN and that its prediction capabilities could still be improved providing the training algorithm is run longer. After this process, the ANN has been trained (Step 8) and it can be used for prediction.

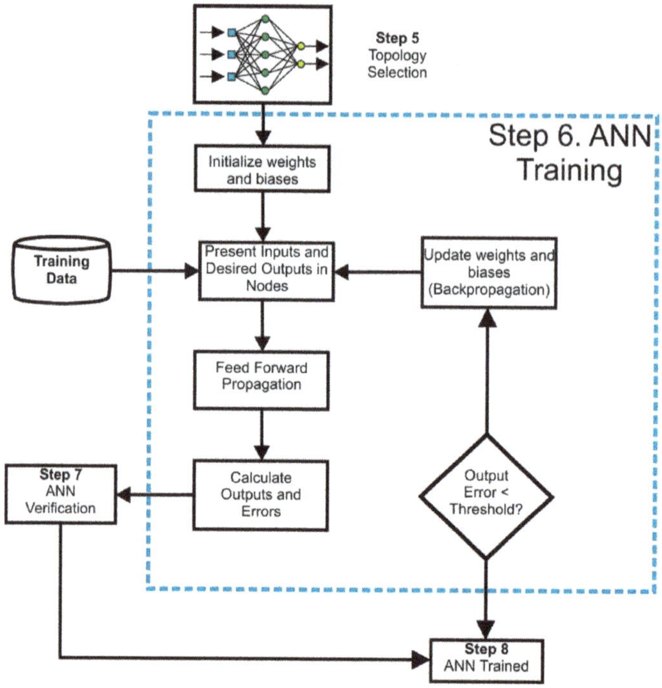

Figure 9. ANN Training Flowchart.

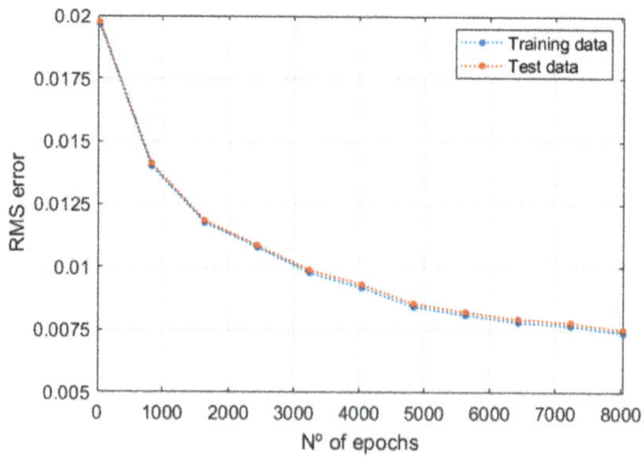

Figure 10. Training and test RMS errors.

In Figure 11, the prediction error frequencies for the 100k validation curves are shown for the different MF parameters, along with the error standard deviation. Additionally, Table 2 classifies the errors frequencies in classes at every 5%.

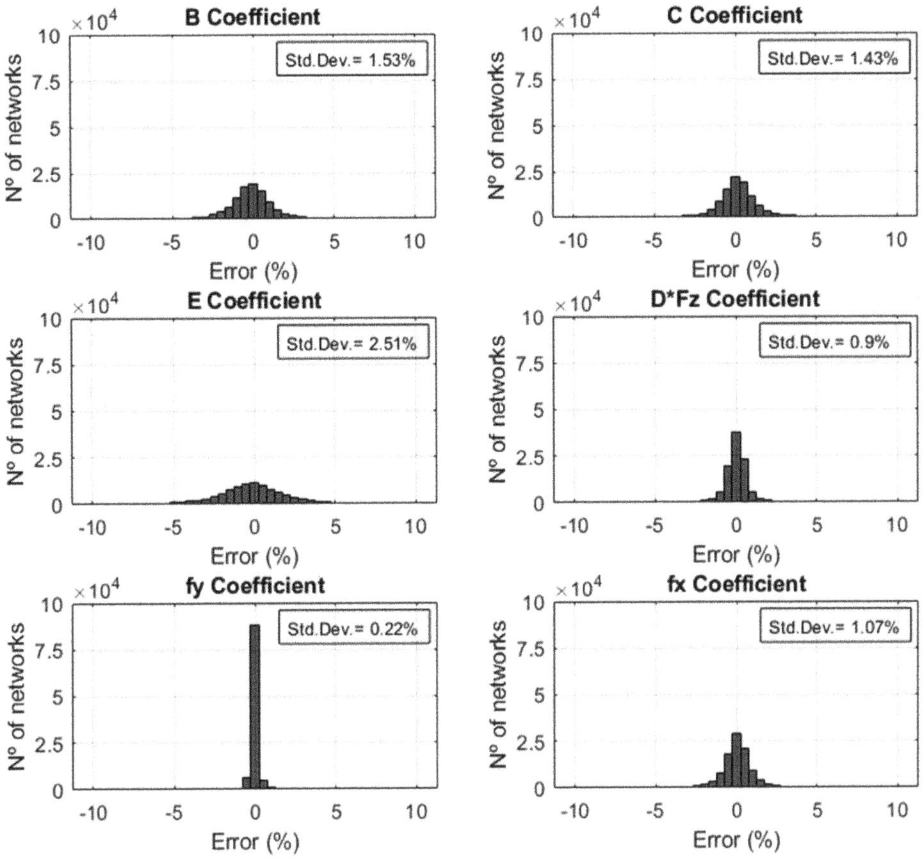

Figure 11. Example of *fx* parameter and ANN error.

Table 2. Range of values for MF model parameters.

Error Range	*B* (%)	*C* (%)	*E* (%)	D^*F_z (%)	f_y (%)	f_x (%)
<5%	98.67	98.81	94.78	99.61	99.99	99.64
5–10%	1.25	1.13	4.71	0.37	0.01	0.35
10–15%	0.06	0.06	0.42	0.02	0	0.01
15–20%	0.02	0.01	0.06	0	0	0
>20%	0	0	0.02	0	0	0

It is clearly noticed that a fairly good performance of the network, where the vast majority of the population (over 98%) can be predicted with and error less than 5%, even for *D*Fz*, *fy*, and *fx* parameters, no error data above this value was registered over the complete validation. The most difficult parameter to be predicted is the parameter *E*, but even in this case, the ANN manages to estimate correctly over 94% of the curves, which can be considered very satisfactory.

3.3. Ann Validation vs. Real Test Data

In the previous section, ANN predictions were validated through "clean" prefabricated data, unknown to the ANN. In this section, we want to take an additional step and analyze the response of the same network when predicting the parameters of real curves obtained in bench tire tests (with different types of errors). The results obtained will serve to draw conclusions about the validity of this methodology against the real problem of identification of parameters of curves from tests (Steps 9 and 10, in Figure 2).

For this purpose, 21 curves from tests measuring the longitudinal force for a 206/65 R15 tire have been used. The curves include seven different slip angles (−8°, −4°, −2°, 0°, 2°, 4° and 8°) in three different normal load (F_z) conditions (2, 3 and 4kN) with a camber of 0. These curves, coming from real tests, have different types of unknown errors. To begin with, the measurement errors are present due to the limited accuracy of the measurement equipment, with their intrinsic uncertainty. On the other hand, it is necessary to take into account the application of slip in steps of 1% between measure and measure, whose uncertainty is not available. In addition, errors and uncertainties can be found regarding the initial condition of the tire or the wear the tire had due to the previous tests or to the roughness of the surface on which the tire runs. Additionally, even environmental variables such as ambient temperature, humidity or wet rolling surface can influence the result of bench acquisitions.

The values of these curves are discretized in the same way as the synthetic curves created in the previous sections so that the ANN can use them directly as input data. In this way, the ANN can give a prediction (Step 9) of the appropriate parameters according to its previous training. As an example, Figure 12 shows the actual data (in dotted red circles) of one of those 21 real curves. The black-discontinued line represents the curve predicted by the ANN.

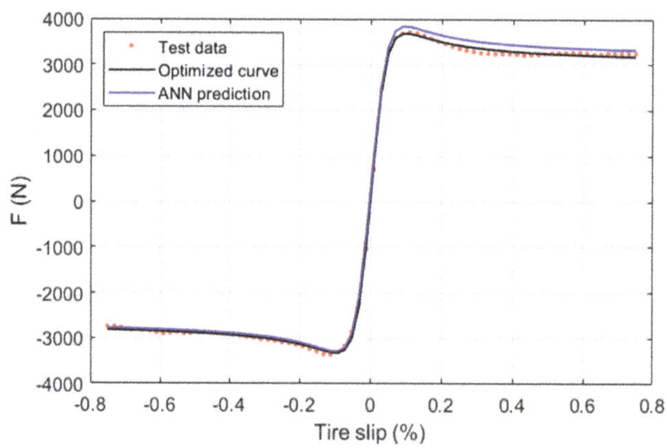

Figure 12. Comparison of real bench data point vs. predicted and optimized curves.

It is important to be able to evaluate the effectiveness of the prediction of Pacjeka's parameters by the ANN. For this reason, it is necessary to compare the results of the ANN with other methods. As seen in the state of the art in the introduction, the identification of parameters can be done through many methods (least squares adjustment, genetic algorithms, particle-swarm optimization, etc.). In this case, for the validation, one of the most effective methods has been used, which, although it is not the fastest in the identification of parameters, it is certainly robust: the least squares adjustment [20]. The method has been applied individually to each of the actual tire test curves. In this sense, in Figure 12, the optimal curve, obtained from a non-linear least squares fitting adjustment using the ANN parameters as starting parameters, is drawn in a blue line. It can be seen that the ANN curve (in black) is quite close to the test data and that the optimal curve (in blue), obtained from this through successive iterations, gets an even better fit to the test data (in red).

In order to quantify the accuracy of the ANN prediction, a comparison has been made between the parameters obtained from the ANN and those obtained from the optimum curves after a subsequent non-linear curve fitting least squares optimization. The results are shown in Figure 13 and Table 3.

From the observation of Figure 13, it is noticed that there is a very good match for B, $D{\cdot}F_z$ and f_x parameters. C and E also show a high similarity in general, but some isolated predictions with higher differences were also observed. It seems that the most difficult parameter to predict in this case corresponds to f_y.

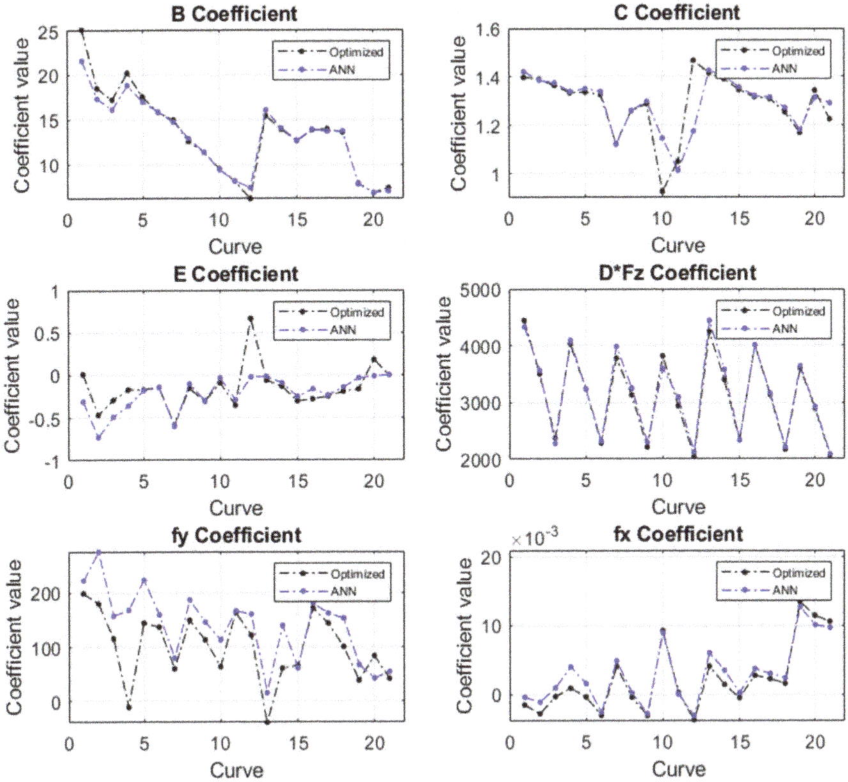

Figure 13. Comparison of real bench data point vs predicted and optimized curves.

Table 3. Range of values for MF model parameters.

Error Range	B (%)	C (%)	E (%)	D^*F_z (%)	f_y (%)	f_x (%)
<5%	85.71	90.48	85.71	100	42.86	90.48
5–10%	9.52	0	9.52	0	38.10	9.52
10–15%	4.76	4.76	0	0	9.52	0
15–20%	0	4.76	4.76	0	4.76	0
>20%	0	0	0	0	4.76	0

The ANN, for an error prediction range below 5%, is able to identify correctly more than 85% of the test curves for the parameters B, C, E, $D{\cdot}F_z$ and f_x. The parameter f_y identifies in that range up to 42.86% of the curves. If the prediction error range is allowed in 10%, then the f_y parameter identifies correctly 81% of the curves. On the other hand, the parameter $D{\cdot}F_z$ is predicted adequately in 100% of cases.

Therefore, it can be concluded that prediction of the real parameters of the real test bench curves is, for most of the parameters, quite good, as the success percentage is over 85% of the cases. It should be noted however that the f_y parameter can only be predicted within the 5% range only in 42.86% of the cases.

Besides, it should be noted that in all the 21 real cases it was possible to carry out the optimum adjustment of the curves using as initial iteration parameters the ones provided by the ANN prediction (Step 10). Therefore, it seems clear that the results predicted by the ANN, despite errors in the input data, can be used, at least as initial iteration parameters for finer posterior curve fitting if needed.

4. Conclusions

This paper has presented a methodology that allows the creation of an ANN that is able to predict with a high precision the optimum parameters of the basic Pacejka's Magic Formula when sufficient input data is provided for the training. This ANN can be used as a virtual brain that allows predicting these parameters as long as they are within the range that was used for their training.

In this study, we managed to train an ANN to predict these parameters from perfect mathematical curves obtained from the random variation of the Pacejka's MF curve parameters. The range of these parameters has been broad enough to cover true values in actual tire curves. The training performed allows for prediction ratios in synthetic curves above 94% for all parameters.

The ANN trained with these mathematical curves has also been used to predict the optimal parameters of real curves obtained in test benches. In this case, the prediction for the majority of parameters is good, obtaining a prediction, within an error range of 5%, of more than 85%. The parameter f_y is the most difficult parameter to predict since only 42.86% of the curves can be predicted within a 5% error range. If the error threshold is raised to 10% then the ANN adequately predicts the f_y parameter for the 80% of the curves.

From the results obtained, the errors identified in the evaluation of curves from real data are limited and, most of the time, very precise. This fact is remarkable if we consider that the ANN was trained for a synthetic data set (with no more error than the Pacejka's model could have in front of reality), and without having previous information or training of real data coming from tire tests in test benches or from direct interaction data in road tire vehicles.

In addition, once the ANN is pre-trained, it can be applied to any tire or operating condition that falls within the training limits of that ANN. Therefore, it can faithfully represent the operating conditions of a tire by immediately identifying the key parameters of Pacejka's MF. The computation time is minimal, making it possible to integrate it into more advanced vehicle models. Furthermore, if the ANN is further combined with a real-time dynamic or HIL (hardware in the loop) systems, providing directly acquired tire data on tire-road interaction, a complete adaptive tire model could be achieved, adapted to each of the four wheels separately and to the individual operating or environmental conditions on each of them. This use is left to future research work.

In conclusion, the trained ANN can be used as a standalone reliable identifier of the parameters of a tire curve or as a convenient initial iteration point generator for further refinement.

Author Contributions: Conceptualization: J.L.O.; data curation: J.L.O., J.A.P. and F.B.; formal analysis: J.L.O.; funding acquisition: J.L.O.; investigation: J.L.O.; methodology: J.L.O. and F.B.; resources: J.L.O.; software: J.L.O.; supervision: J.A.P.; validation: J.L.O. and F.B.; visualization: J.L.O., and J.A.P.; writing—original draft: J.L.O., and F.B.; writing—review and editing: J.L.O. and J.A.P. All authors have read and agreed to the published version of the manuscript.

Funding: The authors thank the Global Nebrija-Santander Chair of Energy Recovery in Surface Transport for their financial support. This work was also supported by the Comunidad de Madrid [grant SEGVAUTO 4.0-CM-P2018EEMT-4362]; and the Agencia Estatal de Investigación [grant RETOS 2018-RTI2018-095923-B-C22].

Conflicts of Interest: The authors declare no conflict of interest.

References

1. Pacejka, H. *Tyre and Vehicle Dynamics*; Butterworth-Heinemann, Ed.; Elsevier: Oxford, UK, 2006; Volume 2, ISBN 980-0-7506-6918-4.
2. Nocedal, J.; Wright, S.J.J. *Numerical Optimization*; Springer: Cham, Switzerland, 1999; Volume 43, ISBN 9780387303031.
3. Marquardt, D.W. An Algorithm for Least-Squares Estimation of Nonlinear Parameters. *J. Soc. Ind. Appl. Math.* **1963**, *11*, 431–441. [CrossRef]
4. van Oosten, J.J.M.; Bakker, E. Determination of Magic Tyre Model Parameters. *Veh. Syst. Dyn.* **1992**, *21*, 19–29. [CrossRef]
5. Sun, W.; Yuan, Y. *Optimization Theory and Methods: Nonlinear Programming*; Springer: New York, NY, USA, 2006; ISBN 9780387249759.
6. McKinnon, K.I.M. Convergence of the Nelde—Mead Simplex Method to a Nonstationary Point. *SIAM J. Optim.* **1998**, *9*, 148–158. [CrossRef]
7. Alagappan, A.V.; Rao, K.V.N.; Kumar, R.K. A comparison of various algorithms to extract Magic Formula tyre model coefficients for vehicle dynamics simulations. *Veh. Syst. Dyn.* **2014**, *53*, 154–178. [CrossRef]
8. Olazagoitia, J.; López, A. Weighted orthogonal distance regression for tire models parameters identification. In Proceedings of the ASME Design Engineering Technical Conference, Boston, MA, USA, 2–5 August 2015; Volume 3.
9. Vetturi, D.; Gadola, M.; Manzo, L.; Faglia, R. Genetic algorithm for tyre model identification in automotive dynamics studies. In Proceedings of the ISATA—International Symposium on Automotive Technology and Automation, Florence, Italy, 3–6 June 1996.
10. Ortiz, A.; Cabrera, J.A.; Castillo, J.; Simón, A. Analysis and evaluation of a tyre model through test data obtained using the IMMa tyre test bench. *Veh. Syst. Dyn.* **2005**, *43*, 241–252. [CrossRef]
11. Talebitooti, R.; Torabi, M. Identification of tire force characteristics using a Hybrid method. *Appl. Soft Comput.* **2016**, *40*, 70–85. [CrossRef]
12. Melzi, S.; Sabbioni, E. On the vehicle sideslip angle estimation through neural networks: Numerical and experimental results. *Mech. Syst. Signal Process.* **2011**, *25*, 2005–2019. [CrossRef]
13. Boada, M.J.L.; Boada, B.L.; Garcia-Pozuelo, D.; Diaz, V. Application of Neural Networks for Estimation of Tyre/Road Forces. In Proceedings of the Volume 10: Mechanical Systems and Control, Parts A and B, Lake Buena Vista, FL, USA, 13–19 November 2009; ASME: New York, NY, USA; pp. 427–433.
14. Palkovich, L.; El-Gindy, M. Neural-Network Representation of Tyre Characteristics—The Neuro-Tyre. *Int. J. Veh. Des.* **1993**, *14*, 463–591.
15. Farroni, F. T.R.I.C.K.-Tire/Road Interaction Characterization & Knowledge—A tool for the evaluation of tire and vehicle performances in outdoor test sessions. *Mech. Syst. Signal Process.* **2016**, *72–73*, 808–831. [CrossRef]
16. Farroni, F.; Lamberti, R.; Mancinelli, N.; Timpone, F. TRIP-ID: A tool for a smart and interactive identification of Magic Formula tyre model parameters from experimental data acquired on track or test rig. *Mech. Syst. Signal Process.* **2018**, *102*, 1–22. [CrossRef]
17. Wang, J.; Liu, Y.; Ding, L.; Li, J.; Gao, H.; Liang, Y.; Sun, T. Neural Network Identification of a Racing Car Tire Model. *J. Eng.* **2018**. [CrossRef]
18. Karimov, A.; Nepomuceno, E.G.; Tutueva, A.; Butusov, D. Algebraic Method for the Reconstruction of Partially Observed Nonlinear Systems Using Differential and Integral Embedding. *Mathematics* **2020**, *8*, 300. [CrossRef]
19. Lopez, A.; Moriano, C.; Olazagoitia, J.L.; Paez, F.J. Fast Computing on Vehicle Dynamics Using Chebyshev Series Expansions. *IEEE/ASME Trans. Mechatron.* **2015**, *20*, 2563–2574. [CrossRef]
20. López, A.; Olazagoitia, J.L.; Moriano, C.; Ortiz, A. Nonlinear optimization of a new polynomial tyre model. *Nonlinear Dyn.* **2014**, *78*, 2941–2958. [CrossRef]
21. Hecht-Nielsen, R. Theory of the backpropagation neural network. In *Neural Networks for Perception*; Academic Press: Cambridge, MA, USA, 1992; pp. 65–93. [CrossRef]
22. Fang, X.; Luo, H.; Tang, J. Structural damage detection using neural network with learning rate improvement. *Comput. Struct.* **2005**, *83*, 2150–2161. [CrossRef]

23. Chidrawar, S.K.; Bhaskarwar, S.; Patre, B.M. Implementation of Neural Network for Generalized Predictive Control: A Comparison between a Newton Raphson and Levenberg Marquardt Implementation. In Proceedings of the 2009 WRI World Congress on Computer Science and Information Engineering, Los Angeles, CA, USA, 31 March–2 April 2009; IEEE: Washington, DC, USA, 2009; Volume 1, pp. 669–673.
24. Lopes, N.; Ribeiro, B. An evaluation of multiple feed-forward networks on GPUs. *Int. J. Neural Syst.* **2011**, *21*, 31–47. [CrossRef] [PubMed]

applied sciences

MDPI

Article

Traffic Simulation of Future Intelligent Vehicles in Duisburg City Inner Ring

Xiaoyi Ma , Xiaowei Hu *, Thomas Weber and Dieter Schramm

Department of Mechatronics, University of Duisburg-Essen, 47057 Duisburg, Germany; xiaoyi.ma@uni-due.de (X.M.); thomas.markus.weber@uni-due.de (T.W.); dieter.schramm@uni-due.de (D.S.)
* Correspondence: xiaowei.hu@uni-due.de

Featured Application: This article can be applied to the simulations of vehicles with different degrees of automation in different city scenarios.

Abstract: Intelligent vehicles gradually enter the vehicular fleet with advanced driver-assistance technologies. Their impact on traffic should, therefore, be considered by transportation decision-makers. This paper examines the effect of vehicles with different levels of automation on traffic flow, such as non-assisted vehicles, vehicles with driver assistance systems, and fully autonomous vehicles. The accuracy of the examined traffic scenario is also an important factor in microscopic traffic simulation. In this paper, the central part of the city of Duisburg, Duisburg's inner ring, is chosen for the traffic scenario. Through the cooperation with local government, official data of Origin/Destination matrices, induction loops, and traffic light plans are provided for this work. Thus, traffic demand from Origin/Destination matrices and induction loops are generated and compared, respectively. Finally, vehicles with different levels of automation are simulated in the Duisburg inner ring scenario.

Keywords: intelligent vehicles; autonomous vehicles; microscopic traffic simulation

Citation: Ma, X.; Hu, X.; Weber, T.; Schramm, D. Traffic Simulation of Future Intelligent Vehicles in Duisburg City Inner Ring. *Appl. Sci.* **2021**, *11*, 29. https://dx.doi.org/10.3390/app11010029

Received: 9 December 2020
Accepted: 21 December 2020
Published: 23 December 2020

1. Introduction

As the urban population grows continuously, traffic congestion, traffic accidents, and other negative impacts of the cities are becoming worse. The United Nations predict that, by 2050, the proportion of the global urban population will increase from 55 to 68% in 2018 [1]. The development of urbanization may increase residents' demand for urban mobility, and emerging technologies including intelligent vehicles may make transportation modes more complicated but have the potential to address some of the problems mentioned. Worldwide, about 1.2 million people die from vehicle-related traffic accidents each year [2]; the resulting medical expenses, legal expenses, property losses, insurance costs, and loss of quality of life exceed 1 trillion US dollars [3]. The majority of traffic accidents can be attributed to human error, such as driving under external influences, drowsiness, or distraction. Automated vehicle systems support or replace human drivers with sensors such as cameras and radars. These devices that cannot be drunk, do not experience burnout, and cannot be distracted, and thus may reduce or even eliminate driver-related errors. Intelligent vehicles may have great significance for the development of public health. In addition to reducing traffic accident deaths, autonomous driving systems can improve people's quality of life [2,4]. For energy consumption, partially automated vehicles might reduce greenhouse gas emissions and energy use by nearly half [5].

Before the realization of fully autonomous vehicles, the development of autonomous driving still faces many challenges [6]. However, a large number of simulation experiments and field trials are helping to promote the development of autonomous driving technology. As of August 2019, as many as 90 cities around the world have implemented pilot projects for autonomous vehicles [7]. So far, investment in the development of autonomous driving technology has reached more than 80 billion US dollars and is expected to show an upward trend [8]. Although some people expect the popularizing of fully autonomous vehicles by 2050 [9], the process of realization is still gradual, from few to many, from simple to complex. In the early stages of adopting autonomous vehicles, ordinary people cannot afford autonomous vehicles [4]. Although prices may decline in the foreseeable future, the cost of autonomous vehicles is expected to be higher than that of non-autonomous vehicles [10]. This has been reflected in the addition of existing vehicles and the addition of certain automatic functions, such as adaptive cruise control, night vision, and pedestrian detection, making vehicles equipped with these systems more expensive than ordinary vehicles. In addition to price factors, many drivers have doubts about the safety of autonomous vehicles. Urban residents and car users of different counties, ages, genders educational backgrounds, and income levels have different views on the current and future safety of autonomous vehicles [11]. Due to the considerations of price and safety, road vehicles will gradually shift from low level to high-level degrees of automation. The choice of different types of vehicle owners and the transformation of urban alternative transportation modes (such as car-sharing and self-driving taxis) will result in a long hybrid period of vehicles of various automation levels. From the perspective of travel demand, almost 85% of urban autonomous car-sharing trips can be covered using vehicles with significantly reduced requirements in the simulation scenario of the year 2035 [12]. An increasing number of partly and fully automated vehicles will likely enter into public roads [13].

Previous works regarding the traffic flow effects of different degrees of automation focus on many different aspects. Zohdy et al. [14] pointed out that modellinlg and simulation is an effective tool in understanding different aspects of interactions between vehicles with different levels of automation in a mixed environment, but no specific models or simulations are implemented. Van Arem et al. [15] introduced a MIXIC (Microscopic Model for Simulation of Intelligent Cruise Control) simulation model for vehicles with CACC (Cooperative Adaptive Cruise Control) and conducted simulations in a highway scenario. They concluded that only a high-CACC-penetration rate (>60%) has benefits on traffic stability while a low penetration rate of CACC (<40%) does not affect traffic flow throughput. Reece et al. [16] introduced a computational driving model called Ulysses for autonomous vehicles and simulated the traffic in a simple intersection scenario. However, no comparisons with vehicles of other degrees of automation were performed. Based on the previous studies, Talebpou et al. [17] introduced a car-following and lane-changing model for autonomous vehicles with limited sensors' range and accuracy. Subsequently, the model was simulated with regular vehicle models in highway segments, in order to prove the benefits of reserved lanes for autonomous vehicles [18]. In our previous work, a vehicle guidance model with a close-to-reality driver model and different levels of vehicle automation was developed, and three models with different levels of automation were simulated at a simple intersection [19]. Previous work in this field has always used two kinds of models in simulation, the non-automated model and the automated model; the intermediate transition phase is not part of the comparison. Simulation scenarios are limited to highway segments or one simple intersection. As an extension of the previous work, three levels of automation of vehicles, representing the automation level of the present, the near, and the more distant future, are simulated in a real highly meshed urban network scenario in this paper.

2. Methodology

2.1. Simulation Scenario

The city of Duisburg is a medium-sized city with approximately 500,000 inhabitants [20]. As a type of medium–large-scale European city, it constitutes the "most important class of cities in Europe in demographic terms" [21]. Compared to large cities, challenges like fewer resources, funding, and organizing capacity are constantly faced by this kind of medium-sized city [21]. However, cities of this kind have the advantage of various transportation networks [22]. For example, Duisburg has great connectivity to the national and European transport network in air, rail, and road aspects. In the transformation process to a smart city, smart mobility including road transportation is an indispensable part [23].

Duisburg's inner ring is located in the heart of Duisburg, as shown in (Figure 1). As the transportation hub of the city, the inner ring is next to Duisburg's Central Railway Station, and all three subway lines pass by this area. The east-west pedestrian street in the ring is not only the largest shopping and entertainment street in Duisburg city, but government agencies (like city hall), the district court, and tax office are located nearby as well. Regarding the future traffic pattern, many researchers have a positive attitude towards car-sharing and estimate that 90% of car-sharing bookings will occur in city centers [24]. Thus, the simulation scenario of Duisburg's inner ring might also be interesting for future car-sharing or autonomous car-sharing planning.

Figure 1. The location of Duisburg inner ring (Source: Here map).

2.2. *Traffic Demand*

With the help of the Duisburg city officials and the WBD (Wirtschaftsbetriebe Duisburg in German), two kinds of data sources are provided: OD (Origin–Destination) matrix data and induction loops data. The OD matrix describes people's movement in a certain area; it is very valuable for modeling transport demand in rural and urban areas [25]. In principle, an OD matrix divides an area into smaller parts and describes the traffic volume among these parts. From an overall perspective, these small parts can be seen as points; the traffic volumes are described from point to point. Induction loop data are recorded by the induction loops installed below the roads. When a vehicle passes by, the metal vehicle cuts magnetic induction lines, and the current changes. Multiple induction loops on the same section can not only detect the number of vehicles but also their speed. The induction loops data are relatively more accurate in a specific road compared to other data sources, but sufficient induction loops layout and reasonable pre-processing of data are the prerequisites for ensuring this accuracy.

Figure 2a shows the partition of the OD matrix of Duisburg's inner ring. There are 18 parts with roads in the inner ring and these parts are shown in five colors. Shown in Figure 2b is the induction loops distribution of this area. As the induction loops here are normally located in the intersections, all the intersections with traffic lights are marked with their intersection numbers. The blue numbers stand for fixed traffic light control intersections, and the red numbers represent dynamic traffic light control intersections. Unfortunately, not all the roads around each intersection have installed induction loops; the green arrows in the figure represent the in-city direction roads equipped with induction loops. In this figure, the bus stations and subway stations are also marked with H and U, respectively. There are, in total, 16 bus lines and 3 subway lines that pass through this area.

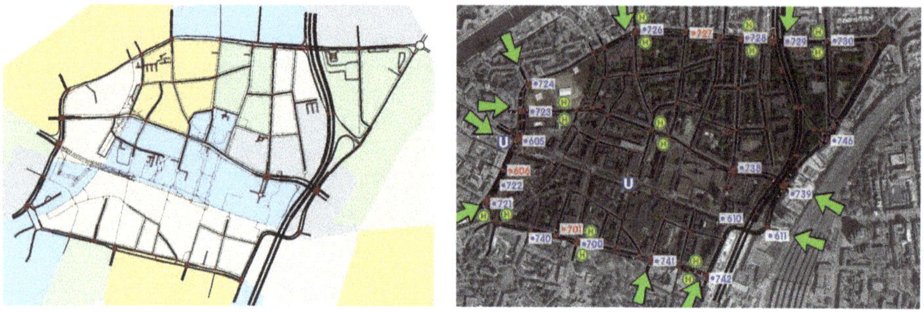

(a) (b)

Figure 2. Two data sources of traffic demand ((**a**) Origin–Destination (OD) matrix, where the distribution of OD areas is shown in different colors; (**b**) induction loops, where the numbers are intersections of the inner ring area).

In this work, both traffic demand data sources are used in generating traffic, and the generated traffic volumes are compared with actual induction loops data. Public transportation, such as bus and subway, is neither included in the OD matrix data source nor the induction loops data. Therefore, in our simulation, the emphasis is placed on passenger cars.

2.3. Vehicle Models with Different Levels of Automation

In this work, a driver–vehicle separate model is used for simulation. An obvious benefit of this kind of model is its flexibility. The driver model can be combined with different vehicle models (fuel, electric, hybrid, etc.), and the vehicle model can also combine with different driver models (autonomous driver, human driver, aggressive driver, etc.). As can be seen in Figure 3, driver model and vehicle model are independent of each other and connected by data transmission. The driver model simulates the reaction of a human driver on the road. Depending on the situation of the ego vehicle, leading vehicle, and other road information, the human driver/driver model controls the vehicle with a suitable gas/brake pedal position. In the car-following models of microscopic traffic simulation, the Krauss model has fewer conflicts with the lane-changing model because of its simplicity [26]. The driver model in this work consists of two parts. The modified Krauss car-following model [27] generates the desired speed of the driver, and the fuzzy control model outputs the gas/brake pedal position depending on the desired speed of the driver and the actual speed of the vehicle. The modified Krauss model can be expressed as:

$$\begin{cases} v(t+t_l) = \max(0, V_n - \epsilon\, a\, \eta) \\ V_n = \min[v_s,\, V_{max}, v_n(t) + a_n t_l] \\ v_s = -\,\tau\, b + \sqrt{(\tau\, b)^2 + v_{n-1}(t)^2 + 2b\, g_n(t)} \\ x(t+1) = x(t) + v_n(t)\, t_l \end{cases} \tag{1}$$

where $v(t+\Delta t)$ represents the speed of the ego vehicle after time Δt, V_n is the desired speed, φ is a random perturbation to allow for deviations from optimal driving, ϵ is the imperfection factor of the driver, a is the acceleration of the ego vehicle, η is a random number between 0 and 1, v_s is the safe speed, V_{max} is the allowed speed of the road, $v_n(t)$ is the speed of the ego vehicle at time t, a_n is the maximum acceleration, t_l is the time step of the simulation, τ is the reaction time of the driver, $v_{n-1}(t)$ is the speed of the leading vehicle, b is the maximum deceleration of the ego vehicle, $g_n(t)$ is the gap between the leader and the follower, $x(t)$ is the position of the ego vehicle at time t, and $x(t+1)$ is the position of the ego vehicle at the next time step.

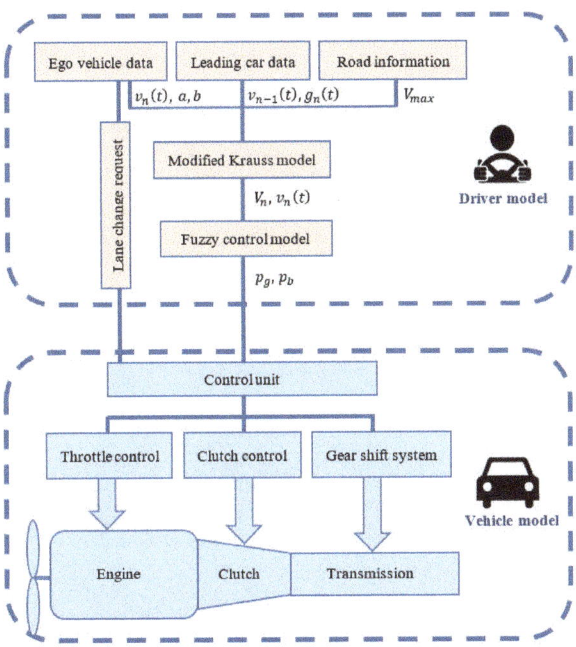

Figure 3. Driver–vehicle model in this work.

The fuzzy control model is established by a driving experiment with 34 human drivers in a driving simulator [19]. Based on the data transferred from the modified Krauss model, the desired speed of the driver V_n and the actual speed of the vehicle $v_n(t)$, the fuzzy control model outputs gas pedal position p_g or brake pedal position p_b with human-like fuzzy rules. The vehicle model considers the characteristic of the vehicle transmission and the driving resistances and is optimized by the data from the driving experiment.

Different levels of automation are reflected in the parameters of the modified Krauss model and fuzzy control model. Three degrees of automation are selected in representing the vehicle automation level of the current stage, as well as in the near future and the far-distant future. Considering the driver assistance functions of high-selling vehicles and the current automation level of vehicles on the road, Level 0 (no automation) is selected for the current stage. Level 2 (partial automation level) and Level 5 (fully autonomous vehicle) are selected for the automation levels in the near future and far-distant future, respectively. In order to effectively distinguish the difference between Level 0 and Level 2, it is assumed that, in the simulation, Level 2 vehicles are in the driving assisted mode. According to the different operators (human or machine) of reaction tasks at different levels, the reaction time τ of the driver is set to 1, 1, 0.5 for Level 0, Level 2, and Level 5, respectively, in the modified Krauss model. The randomness factor r is set to 0.5, 0, 0, respectively, for the three separate levels of automation in the fuzzy control model, in representing the random driving behavior of human driver compared to the machine driver. Under safety considerations, the speed limit of autonomous vehicles is adjusted. High driving speed will lead to a sharp increase in the number and severity of traffic accidents [28]. Therefore, the Level 5 driver models in this work have a lower speed limit compared to the Level 0 and Level 2 vehicles.

3. Establishment of Simulation

In order to simulate the Duisburg inner ring with vehicles of different degrees of automation, an open-source microscopic traffic simulation software SUMO (Simulation of Urban MObility) [29], developed by the German Aerospace Center (DLR), is used in this work. Compared to similar simulation software, SUMO has more possibilities of model extension with less complexity. Software with high complexity would have negative impacts with other models (such as the lane changing model, dynamic assignment model, etc.) which are not easy to be resolved [26]. Hence, the SUMO simulation package is selected in this paper for simulation with both data sources mentioned in 2.2.

3.1. OD Matrix Traffic Demand

In the process of generating traffic demand from the OD matrix, the traffic volume from the divided area (with different colors in Figure 2a) is assigned to roads in the area. As the OD matrix data source used in this work describes the total traffic volume in 24 h, there is no more specific data for the traffic volume per hour. Common daily timelines retrieved from cities in West Germany are used [30]. For passenger cars and trucks, timelines named TGw2_PKW and TGw_LKW are used separately for describing the traffic distribution on a workday. The overall process of generating traffic demand from the OD matrix is listed below:

- **Areas described by the OD matrix transfer into polygons in SUMO:** The areas described by the OD matrix data source are saved in shapefiles (a data format for geometric location information); they are also described by geometric information (longitude and latitude). The most similar data format in SUMO is polygons. The boundary lines can be also described by geometric points in order. In this step, *polyconvert* in SUMO is used to transfer shapefiles into polygons.
- **Polygons to TAZ:** The Traffic Assignment Zone (TAZ) is a district defined in SUMO and the vehicles can be coded to drive from one TAZ to another within a certain period. The TAZ also comprises districts constituted by points; for better visualization, the geometric coordinate format is transferred to the XY coordinate in SUMO. Figure 4 shows the TAZ distribution of the studied area; all the areas in OD matrix are marked with red boxes.
- **TAZ to edges/roads:** To distribute the traffic demand from the whole zone into edges/roads, a python program called *edgesInDistricts.py* is used. Depending on the length of the roads, the traffic demand is distributed with different weights.
- **OD matrix to trips:** After the scope-related work is done, the traffic volume is processed. A function called *od2trips* has been used. The timeline has been used here to assign traffic into hours. Trucks and passenger cars are processed separately.
- **Trips to routes:** At last, through the network file and trip files in the last step, *duarouter* is used to generate the route file. The route file describes when and where each vehicle starts on the map, and through which edges/roads it arrives at the destination.

Figure 4. Traffic Assignment Zone (TAZ) distribution of Duisburg inner ring; each red box shows the roads in one TAZ.

3.2. Induction Loops Traffic Demand

The induction loops data source consists of two parts: position and amount. The position file is in pdf format and the induction loops must be manually added into the network file in SUMO one by one. The amount file is in a machine-readable format. Unlike the method mentioned above, in this section, a fixed route, with the route file and vehicle file, are used. The route file describes the edges/roads a vehicle passes along, and the vehicle file represents the departure time, position, speed, and route of each vehicle. The process of generating traffic demand from induction loop data is listed below:

- **Loop position and type:** There are three types of induction loops in SUMO; "source" loops are the ones where vehicles start, "between" loops are in the middle position, and "sink" loops are the loops where vehicles vanish from the simulation. In this paper, the roads outside the ring in the entering directions are selected to be the source loops. In total, induction loops on 8 roads marked with green arrows in Figure 2b are set as the source loops.
- **Traffic flow:** In order to be recognized by the program, the flow file in SUMO should be in csv format with fixed form. However, the vehicle amount of each induction loop is saved by intersection names. MATLAB is used here to transform the data into the SUMO-required format.
- **Generate route files and vehicles:** Using the network file and flow file, the route file and vehicle file required in the simulation are generated by a sub-program in SUMO called *flowrouter.* In this step, the departure lanes are also modified to avoid conflicts by generating routes.
- **Change source loops position:** The positions of induction loops are always close to an intersection. Sometimes the distance between the induction loops is less than the length of a vehicle. This causes unwanted traffic jams upon vehicle insertion, which in turn affects the entire simulation. Therefore, all the positions of source loops are moved to the side, far from the intersection.

The induction loop data of a coherent week in 2019 are provided. However, some intersections have no data in the given period, some intersections have no installed induction loops, and data from one intersection are not usable due to a malfunction. Therefore, the induction loops used as source loops are adjusted according to the data situation. Finally, the induction loops used for source are marked with green arrows in Figure 2b. In total, 24 induction loops on 12 roads are selected to be the source loops. Each road has 1–3 lanes and each lane has one source induction loop. A total of 76 induction loops with valid data are used as traffic volume sources.

3.3. Verification Data

In order to verify the generated traffic volume from the OD matrix data and induction loops data, 4 verification points (V1–V4) on 4 roads are selected. These verification points are also induction loops near intersections. V1 belongs to intersection 727, in the west to east direction; the road with the induction loops has only one lane. V2 and V3 are in intersection 723; V2 records the vehicles from north to south while V3 records the vehicles from south to north. V2 has 3 lanes; only the two lanes with valid data are selected for the verification. V3 has 3 lanes, all of which are recorded. V4 belongs to intersection 742; the direction is from north to south. There are two lanes of this road section, and both are recorded. In total, 4 verification points with 8 induction loops are selected, covering the north, south, west, and east roads of the inner ring area of the scenario. Due to the data source of the OD matrix being 24 h, the simulated time of the two data sources is also set to 24 h. Furthermore, the verification data are also extracted for 24 h.

4. Results and Discussion

First, the accuracies of the traffic demand generated from different data sources (OD matrix and induction loops) are compared in Section 4.1. With the better data source, simulations of vehicles with different degrees of automation are carried out in Section 4.2. The results and comparison of the simulations are then introduced.

4.1. Comparison of Different Traffic Demand Sources

Figures 5 and 6 show the traffic volume determined with the simulation settings presented in the last section and the average speed at the four verification points. The red lines represent the real data recorded by the induction loops. The blue lines show the simulation results from Section 3.2 and the black lines represent the simulation results from Section 3.1. From left to right, the four figures represent the four verification points from V1 to V4, respectively. In order to reduce the interference of different lane choices, the traffic volume is the sum of different lanes on the same road.

As can be seen in Figure 5, the blue lines are closer to the red lines than the black lines. That means, for the number of vehicles in the simulation, the simulation using induction loops data as the source of traffic demand has better results than the simulation using the OD matrix data. The reason could be that the OD matrix data only contains the trips that start or end in the range of the inner ring, and the trips passing by this area are not included. For V2 and V3, the difference between actual traffic volume and the OD matrix generated traffic volume is more significant than V1 and V4. The reason could be that most vehicles passing intersection 723 do not need to enter the inner ring area. This location of the intersection also proves that it bears more out-of-region traffic.

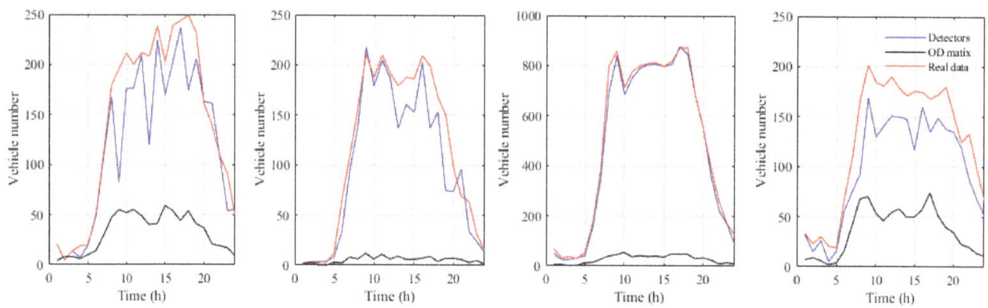

Figure 5. Traffic volume of the verification points. (Red lines, real data; blue lines: simulation of intersection loops data; black lines: simulation of OD matrix data).

For the average velocity comparison in Figure 6, the simulation results using OD matrix data as traffic demand are closer to the real data at 24 h for a day. In contrast, the average speeds of simulation using induction loops data and OD matrix data are similar; sometimes lower than the real data, and sometimes higher. The zero average speed represents that there is no vehicle passing the verification point during this hour. The two simulations use the same traffic scenario and the same network file. Hence, the speed limit of the roads is also the same.

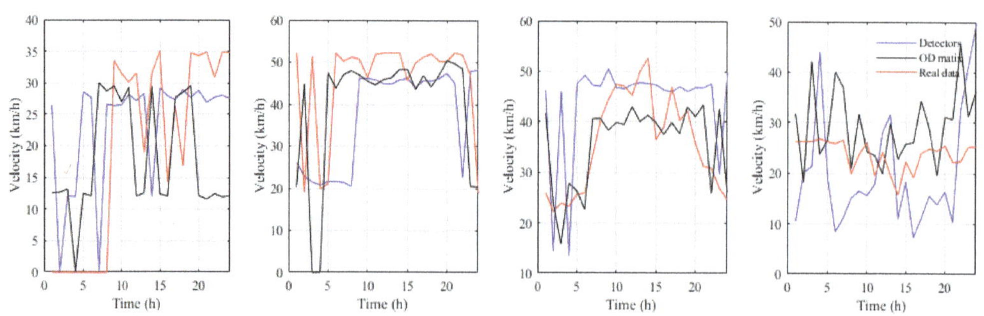

Figure 6. Average velocity of different hours in the verification points. (Red lines, real data; blue lines: simulation of intersection loops data; black lines: simulation of OD matrix data).

The deviation of the detector simulation data and the real data in traffic volume is $\sigma = 37.43$, which is better than the deviation of the OD matrix simulation data and the real data, which is $\sigma = 129.20$. Regarding velocity, the deviation of the detector's data is $\sigma = 7.44$, and therefore slightly better than the deviation of the OD matrix, which is $\sigma = 8.72$. According to the comparison of simulation results, the simulation using induction loops data have a more realistic result than the simulation using the OD matrix in the aspect of traffic volume. The reason might be that the induction loops data record every vehicle passing by, but the OD matrix data only generates the vehicles that depart or end their journey at the selected area. In the next section, the induction loops data are also used as the traffic demand source for the simulation of vehicles with different degrees of automation.

4.2. Comparison of Different Degrees of Automation

To verify the effects of vehicles with different degrees of automation, vehicles of automation Level 0 (no automation), Level 2 (partial automation), and Level 5 (fully autonomous) are simulated in the traffic scenario of Duisburg inner ring. In this simulation, the number of vehicles, vehicle routes, departure time, and departure location on the network are the same, and only the degree of automation is varied. The simulation is for 24 h of a day, and the traffic demand is generated from the close-to-reality results of road detectors. To observe more details in the change of traffic volume, the data collection time cycle of verification detectors is changed to 600 s. The simulation results can be seen in Figures 7 and 8; the average speed of vehicles in the simulation are shown separately (Figure 7) and in comparison (Figure 8). For the same traffic volume input, the vehicles with a higher automation level, represented by the red and blue lines in Figure 8, show a higher average speed than the vehicles without automation, represented by the black lines. In this simulation, the traffic status represents the real traffic situation of the Duisburg inner ring for a certain day. For the current traffic volume, vehicles in all three simulations can finish the driving task with average speeds close to the speed limits of the roads.

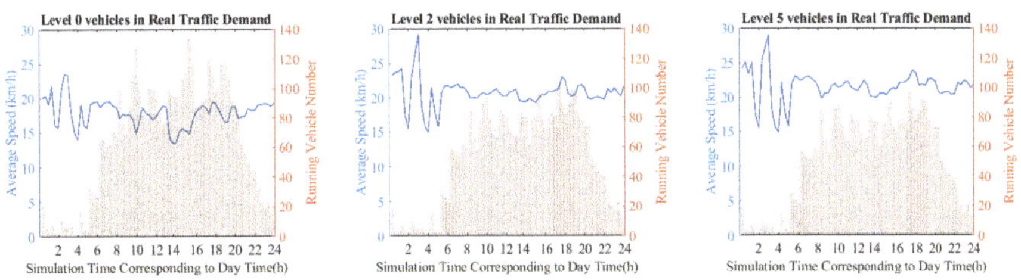

Figure 7. Average speeds of vehicles and running vehicle numbers of simulations with vehicles of different automation levels.

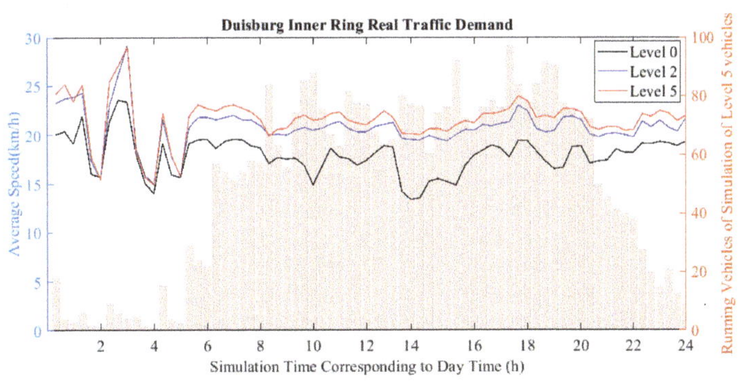

Figure 8. Comparison of average speeds of vehicles with different automation levels in real traffic volume (100% traffic volume).

To create a heavy traffic demand situation, an extra 50% traffic volume is added to the scenario. If traffic conditions continue to deteriorate, there will be more vehicles on the road in the future, whether automated vehicles can improve the congestion problem or not. Figure 9 shows the average speed of vehicles in three different scenarios with Level 0, Level 2, and Level 5 vehicles. For the same heavy traffic volume input, the average speed of vehicles in the simulation with Level 0 vehicles is greatly affected (black line) and, in the other two simulations with vehicles of higher automation levels, the average speed is only slightly lower as in real traffic volume input. The simulation results show that, whether in the actual traffic volume or the possibly more congested future traffic situation, automated vehicles have a positive impact on the traffic flow. Vehicles with higher automation levels can take on more congested traffic conditions.

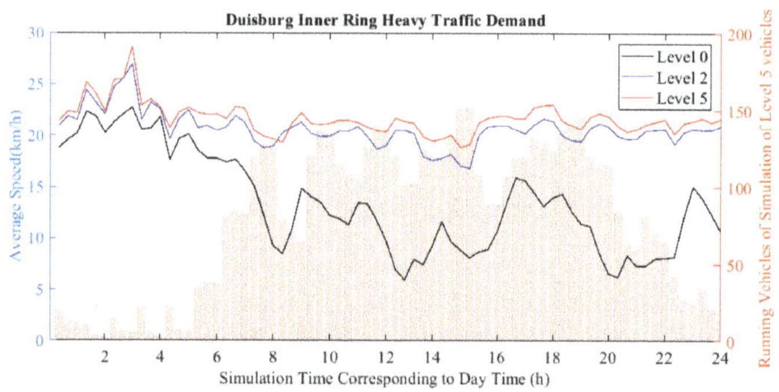

Figure 9. Comparison of the average speed of vehicles with different automation levels in heavy traffic volume (150% traffic volume).

5. Conclusions and Discussions

In this work, a traffic scenario of Duisburg city inner ring has been built, and traffic demand of two different data sources has been generated and compared. In the four selected verification points, the traffic volume generated from induction loop data has a better reproduction (σ =37.43, compared to real data) than the OD matrix data (σ =129.20, compared to real data). Vehicles with different automation levels (Level 0/no automation, Level 2/partial automation, and Level 5/fully autonomous) are simulated continuously in the Duisburg city inner ring scenario for 24 h of a day. Based on the real traffic volume collected in 2019, the vehicles with automation Level 2 and Level 5 have positive effects on traffic. For more congested traffic status, the positive effects of vehicles with higher automation levels are more obvious.

This article uses a real scenario, the inner ring area of the city of Duisburg, as the traffic simulation scenario. Two kinds of data sources are provided for this article, and traffic demand generated from the two kinds of data sources are used separately. In the selected simulation scenario, induction loops data show better accuracy in reproducing the traffic. In this article, vehicles with different degrees of automation are simulated in the scenario. With the same traffic demand, highly automated vehicles can effectively increase average speed. In the case of increased road congestion (150% traffic volume), the increased average speed of automated vehicles is more obvious.

Author Contributions: X.M.: conceptualization, data curation, formal analysis, methodology, software, writing and editing. X.H.: conceptualization, methodology, software, validation, writing and editing, T.W. & D.S.: conceptualization, project administration and supervision, result validation and critical revision. All authors have read and agreed to the published version of the manuscript.

Funding: This research received no external funding.

Institutional Review Board Statement: Not applicable.

Informed Consent Statement: Not applicable.

Data Availability Statement: Restrictions apply to the availability of these data. Data was obtained from Duisburg city officials and the WBD and are available from the authors with the permission of Duisburg city officials and the WBD.

Acknowledgments: We acknowledge support by the Open Access Publication Fund of the University of Duisburg-Essen.

Conflicts of Interest: The authors declare no conflict of interest.

References

1. United Nations. *2018 Revision of World Urbanization Prospects*; United Nations: New York, NY, USA, 2018.
2. Kalra, N.; Groves, D.G. *The Enemy of Good: Estimating the Cost of Waiting for Nearly Perfect Automated Vehicles*; Rand Corporation: Santa Monica, CA, USA, 2017; ISBN 1977400019.
3. Blincoe, L.; Miller, T.R.; Zaloshnja, E.; Lawrence, B.A. The Economic and Societal Impact of Motor Vehicle Crashes, 2010 (Revised). *Ann. Emerg. Med.* **2015**, *66*, 194–196. [CrossRef]
4. Fagnant, D.J.; Kockelman, K. Preparing a nation for autonomous vehicles: Opportunities, barriers and policy recommendations. *Transp. Res. Part A Policy Pract.* **2015**, *77*, 167–181. [CrossRef]
5. Wadud, Z.; MacKenzie, D.; Leiby, P. Help or hindrance? The travel, energy and carbon impacts of highly automated vehicles. *Transp. Res. Part A Policy Pract.* **2016**, *86*, 1–18. [CrossRef]
6. Marshall, A. After peak hype, self-driving cars enter the trough of disillusionment. *Wired*, 29 December 2017.
7. Philanthropies, B. Is your city getting ready for AVs. This is a guide to who's doing what, where, and how. *Hentet* **2018**, *12*, 18.
8. Kerry, C.F.; Karsten, J. *Gauging Investment in Self-Driving Cars*; Retrieved 28 November 2017; Brookings: Washington, DC, USA, 2017.
9. Lanctot, R. *Accelerating the Future: The Economic Impact of the Emerging Passenger Economy*; Strategy Analytics: Boston, MA, USA, 2017; Volume 5, 30p.
10. Nunes, A.; Hernandez, K.D. Autonomous Vehicles and Public Health: High Cost or High Opportunity Cost? *Transp. Res. Part A Policy Pract.* **2019**, *138*, 28–36. [CrossRef]
11. Green, W.H.; Armstrong, R.C.; Ben-Akiva, M.; Heywood, J.; Knittel, C.; Paltsev, S.; Reimer, B.; Vaishnav, C.; Zhao, J.; Gross, E. Insights Into Future Mobility: A Report from the Mobility of the Future Study. 2019. Available online: http://energy.mit.edu/publication/insights-into-future-mobility/ (accessed on 20 December 2020).
12. Peters, P.L.; Kröger, L.; DeMuth, R.; Schramm, D. Derivation of scaled design premises for future vehicle concepts based on a forecast of travel demand using the example of a commercial fully automated on-demand fleet. *Automot. Engine Technol.* **2019**, *4*, 45–61. [CrossRef]
13. Vlakveld, W.P. *Transition of Control in Highly Automated Vehicles: A Literature Review*; SWOV: Hague, The Netherlands, 2016.
14. Zohdy, I.; Kamalanathsharma, R.; Sundararajan, S.; Kandarpa, R. *Automated Vehicles from Modeling to Real World. Road Vehicle Automation 2*; Springer: Berlin, Germany, 2015; pp. 187–191.
15. Van Arem, B.; Van Driel, C.J.G.; Visser, R. The Impact of Cooperative Adaptive Cruise Control on Traffic-Flow Characteristics. *IEEE Trans. Intell. Transp. Syst.* **2006**, *7*, 429–436. [CrossRef]
16. Reece, D.A.; Shafer, S.A. A computational model of driving for autonomous vehicles. *Transp. Res. Part A Policy Pr.* **1993**, *27*, 23–50. [CrossRef]
17. Talebpour, A.; Mahmassani, H.S.; Bustamante, F.E. Modeling Driver Behavior in a Connected Environment: Integrated Microscopic Simulation of Traffic and Mobile Wireless Telecommunication Systems. *Transp. Res. Rec.* **2016**, *2560*, 75–86. [CrossRef]
18. Talebpour, A.; Mahmassani, H.S.; Elfar, A. Investigating the Effects of Reserved Lanes for Autonomous Vehicles on Congestion and Travel Time Reliability. *Transp. Res. Rec.* **2017**, *2622*, 1–12. [CrossRef]
19. Ma, X.; Hu, X.; Schweig, S.; Pragalathan, J.; Schramm, D. A Vehicle Guidance Model with a Close-to-reality Driver Model and Different Levels of Vehicle Automation. *Appl. Sci.* **2021**, *11*, 380. [CrossRef]
20. Organisation for Economic Cooperation and Development. Population by Region—Urban Population by City Size—OECD Data. Available online: https://data.oecd.org/popregion/urban-population-by-city-size.htm (accessed on 18 May 2020).
21. Giffinger, R.; Fertner, C.; Kramar, H.; Meijers, E. City-ranking of European medium-sized cities. *Cent. Reg. Sci. Vienna UT* **2007**, 1–12. Available online: http://www.smart-cities.eu/model.html (accessed on 18 June 2008).
22. Kunzmann, K.R. (Ed.) Medium-Sized Towns, Strategic Planning and Creative Governance. In *Making Strategies in Spatial Planning*; Springer: Berlin, Germany, 2010; pp. 27–45.
23. Bergan, P.; Mölders, A.-M.; Rehring, K.; Ahlemann, F.; Decker, S.; Reining, S. Towards Designing Effective Governance Regimes for Smart City Initiatives: The Case of the City of Duisburg. In Proceedings of the 53rd Hawaii International Conference on System Sciences, Maui, HI, USA, 7–10 January 2020.
24. Alessandrini, A.; Campagna, A.; Delle Site, P.; Filippi, F.; Persia, L. Automated Vehicles and the Rethinking of Mobility and Cities. *Transp. Res. Procedia* **2015**, *5*, 145–160. [CrossRef]
25. Willumsen, L.G. *Estimation of an OD Matrix from Traffic Counts–A Review*; University of Leeds: Leeds, UK, 1978.
26. Bjärkvik, E.; Fürer, F.; Pourabdollah, M.; Lindenberg, B. Simulation and Characterisation of Traffic on Drive Me Route around Gothenburg using SUMO. In Proceedings of the SUMO User Conference 2017, Berlin, Germany, 8–10 May 2017.

27. Krauss, S. *Microscopic Modeling of Traffic Flow. Investigation of Collision Free Vehicle Dynamics*; Als Ms. gedr; Dt. Zentrum für Luft- und Raumfahrt e.V., Abt; Unternehmensorganisation und -Information: Köln, Germany, 1998.
28. Richards, D.C. *Relationship between Speed and Risk of Fatal Injury: Pedestrians and Car Occupants*; Transportation Research Laboratory, Department of Transport: London, UK, 2010.
29. Lopez, P.A.; Behrisch, M.; Bieker-Walz, L.; Erdmann, J.; Flötteröd, Y.-P. Microscopic traffic simulation using sumo. In Proceedings of the 2018 21st International Conference on Intelligent Transportation Systems (ITSC), Maui, HI, USA, 4–7 November 2018.
30. Schmidt, G.; Thomas, B. Hochrechnungsfaktoren für manuelle und automatische Kurzzeitzählungen im Innerortsbereich. In *Forschung Straßenbau Straßenverkehrstechnik*; Bundesministerium für Verkehr: Bonn, Germany, 1996.

applied sciences

MDPI

Article

On-Board Road Friction Estimation Technique for Autonomous Driving Vehicle-Following Maneuvers

Stefania Santini [1], Nicola Albarella [1], Vincenzo Maria Arricale [2], Renato Brancati [2,*] and Aleksandr Sakhnevych [2]

[1] Department of Electrical Engineering and Information Technology, University of Napoli Federico II, 80125 Naples, Italy; stefania.santini@unina.it (S.S.); nicola.albarella@unina.it (N.A.)
[2] Department of Industrial Engineering, University of Napoli Federico II, 80125 Naples, Italy; vincenzomaria.arricale@unina.it (V.M.A.); ale.sak@unina.it (A.S.)
* Correspondence: renato.brancati@unina.it

Abstract: In recent years, autonomous vehicles and advanced driver assistance systems have drawn a great deal of attention from both research and industry, because of their demonstrated benefit in reducing the rate of accidents or, at least, their severity. The main flaw of this system is related to the poor performances in adverse environmental conditions, due to the reduction of friction, which is mainly related to the state of the road. In this paper, a new model-based technique is proposed for real-time road friction estimation in different environmental conditions. The proposed technique is based on both bicycle model to evaluate the state of the vehicle and a tire Magic Formula model based on a slip-slope approach to evaluate the potential friction. The results, in terms of the maximum achievable grip value, have been involved in autonomous driving vehicle-following maneuvers, as well as the operating condition of the vehicle at which such grip value can be reached. The effectiveness of the proposed approach is disclosed via an extensive numerical analysis covering a wide range of environmental, traffic, and vehicle kinematic conditions. Results confirm the ability of the approach to properly automatically adapting the inter-vehicle space gap and to avoiding collisions also in adverse road conditions (e.g., ice, heavy rain).

Keywords: autonomous driving; friction estimate; tire-based control; ADAS; potential friction

Citation: Santini, S.; Albarella, N.; Arricale, V.M.; Brancati, R.; Sakhnevych, A. On-Board Road Friction Estimation Technique for Autonomous Driving Vehicle-Following Maneuvers. *Appl. Sci.* **2021**, *11*, 2197. https://doi.org/10.3390/app11052197

Academic Editor: Flavio Farroni

Received: 30 December 2020
Accepted: 25 February 2021
Published: 3 March 2021

1. Introduction

Rapid economic growth has lead to a considerable expansion of circulating vehicles, especially in big cities [1], exceeding the growth rate of the road infrastructure, and therefore leading to the traffic congestion [2], the growth of risk of accidents and fatalities [3,4], as well as of pollution-linked issues, due to CO_x, NO_x and C_xH_y emissions [5,6]. In recent years, thanks to the availability of continuously improving embedded hardware solutions, there has been a vast increase in the employment of advanced electronic systems to manage both safety and performance driving. Advanced Driver Assistance Systems (ADAS) are currently considered as a valued solution for reducing the main road transport issues (i.e., the congestion, traffic accidents, environmental stress and fuel consumption mentioned above) supporting the driver by informing on, actively assisting in, or taking over part of the driving task [7–9].

In a broader perceptive, by the superposition of sensing, planning, ADAS and control applications, the vehicle is going to become more and more automated in the very next future, thus leading to its full autonomy making the driver simply a passenger [10,11]. Within this technological paradigm, the ability of the vehicles to drive themselves in a safe manner highly depends on their prior capability to understand the external environment and to correctly estimate the vehicle state in all the possible operating and environment conditions [12–14]. It is worth to note that, as stated by SAE International, the difference between a Level 4 and Level 5 autonomous vehicle is the capability of driving itself in any situation, which implies adverse environmental scenarios like heavy rain, snow, or ice on the road surface [15].

Hence, in order to guarantee a greater safety-level with respect to environmental conditions [3,4,16,17], it is necessary to account for their effect since from the very beginning of the ADAS design phase, introducing advanced control strategies that could leverage both real-time measurements, coming from different in-vehicles sensors (camera, radar, lidar and combinations of those via sensor-based fusion techniques [18–21]), and on-board environmental estimation modules. Indeed, the use of only sensors' measurements could be not enough to perceive properly the external environment, since the vehicle control system has also to predict and discern how heavy rain, snow, ice condition or road singularities (e.g., oil stains, puddles, holes, or disconnected cobblestone) could impact on safety, so that the driving policy is to be tuned according to the actual environmental adversities.

Moreover, in extreme scenarios vehicle dynamics may be deeply affected by the non-linearity of tires' dynamic behavior, therefore limiting the maneuverability in terms of both longitudinal and lateral accelerations and significantly reducing drive-ability and steer-ability. Furthermore, during emergency situations, which typically involve abrupt deceleration or steering, the tires can be easily pushed to their unstable dynamic region, thus requiring a specific control policy depending from the current dynamics of the vehicle and its sub-components, that hence have to be estimated at each time instant [22].

To solve the above open issues, in this paper, the authors propose a control architecture responsible for the longitudinal dynamics of the vehicle chassis, composed of two ADAS functionalities—namely Adaptive Cruice Control (ACC) and Autonomous Emergency Braking (AEB)—in addition to the blueAntilock Braking System (ABS), which is road-grip aware in the sense that it is able to properly regulate the vehicle motion on the base of the on-line estimation of the road friction coefficient per single tire, based on the T.R.I.C.K. (Tire Road Interaction Characterization & Knowledge) methodology [23]. In particular, the in-vehicle friction estimation module, starting from the acceleration, angular speed and steering angle data acquirable from widely-adopted sensors, allows us to calculate in run-time the kinematics and the dynamics of all the tires, in terms of interaction slips and forces, respectively. The vehicle is modelled using a totally physical approach, whose parameters (inertia, geometry, etc.) are independent from the external environment, whereas its tires' sub-models response deeply depends on the peculiar asphalt texture characteristics.

The technique has been developed by combining two model-based approaches: a bicycle vehicle model evaluates the state of the vehicle in terms of forces and actual friction coefficient, on the basis of the quantities measured by the sensors installed on board, and the tire Magic Formula (MF)-based model evaluates the potential friction value, achievable on a particular road surface, based on the slip-slope procedure. The estimation of the potential friction procedure relies on the fact that the tires' characteristics have been identified on a reference tarmac texture, called reference road surface, towards which the tires' dynamic response is compared for the same kinematic and dynamic operating conditions.

The methodology, implementable on board within the vehicle Electronic Control Units (ECUs), shows how a potential grip coefficient information, crucial for the design of the innovative control logics, can be adopted in the autonomous driving vehicle systems, allowing the optimization and maximization of the performance, reducing the collisions number and their severity, even in strongly unfavourable and changing environment and road conditions.

The effectiveness of the theoretical approach is validated via a purposely developed co-simulation platform able to emulate both the dynamics of the vehicle under control (or ego vehicle), and the environment, in terms of traffic and road surface conditions. The time step chosen for each iteration of the proposed estimation algorithm is 0.05 s (200 Hz). Results disclose that the developed control strategy is able to achieve higher performance in terms of safety than a commonly adopted ones, demonstrating the potential to decrease the risk of collisions in every studied scenario.

The paper is organized as follows: the Sections 2 and 3 introduce related work and problem statement, respectively; the designed on-board road-grip estimation is represented in the Section 4; the design of the road-grip aware control modules is discussed in the Section 5. Finally, in Sections 6 and 7, the simulation platform and numerical results are shown, respectively.

2. Related Work

ADAS for the safe automatic driving mostly tackles the stabilization of the chassis longitudinal motion and the actuation of the emerging braking via a wide variety of control techniques. Among others, Model Predictive Control (MPC) has been effectively used in [24–28] in order to synthesize an ACC system. Regarding the AEB, typically event-based controllers have been realized through a continuous evaluation of the braking distance [29,30] or the collision time [31]. Alternative formulations can be found in [32], where the authors classify the collision risk upon the definition of potential fields, or in [33] where an impedance controller is synthesized, thus resulting in a time based controller. Some ADAS combine ACC and AEB strategies for multiple driving situations: in [34] a Linear Quadratic Regulator (LQR) works jointly with a Time-To-Collision (TTC)-based logic and in [35] a Proportional-Integral-Derivative (PID)-based velocity control embeds a continuous time collision avoidance mechanism with the aim of reducing excessive jerk. More complex architectures can be found in [36,37], where nonlinear MPC and reinforcement learning formulations have been designed for safely steering the longitudinal vehicle's dynamics.

Attempts to account for the road conditions into specific ADAS driving features have only recently been developed; research of the tire-road friction estimation is a topic that has been extensively addressed and the study is continuing to this day. According to [38], it is possible to divide the different approaches on friction estimation into two main groups: experiment-based and model-based approaches.

The experimental based methods use additional sensor measurement as optical or acoustic sensors and cameras to evaluate the friction based on the fact that wet asphalt is dark grey with a higher clarity of texture than dry asphalt [39]. The disadvantages associated with this category lie in the high frequency of these sensors get dirty, and therefore distort the results. In addition, the vehicles are generally not equipped with the sensors mentioned above and are difficult to maintain.

With regard to the model-based group, the friction information is evaluated thanks to the mathematical models describing the vehicle system and its subsystems, starting from the information, measured by the sensors installed on the vehicle. Such methodology has demonstrated to be able to evaluate the actual grip in the most environmental condition, but not the potential grip. In [40], the authors experimentally evaluate a set of parameters, as peak friction, interaction shape and curvature factors, for different road environmental condition (dry, snow, ice) to estimate the tire stiffness. Then a run-time switch selects the set of parameters in memory corresponding to the current stiffness of the tire, leading to evaluate the potential grip.

The limit of this approach is dictated by the number of parameters to be stored in the memory, able to describe the different asphalt conditions [41]. Differently from the [40], the friction peak value has been researched imposing relatively large magnitudes of braking/accelerating or steer inputs to achieve sufficient variations in tires' dynamic responses. To this purpose, a different speed control logics have been developed for the front and rear wheels in order to identify the stiffness and the tire road friction coefficient without severely influencing vehicle forward speed. However, these maneuvers may not be practical in every vehicle operating condition, as in [42], in which the tire-road friction estimator has been activated when the vehicle reached constant speeds. The latest methodology belonging to model-based approaches is the slip-slope [43], based on the assumption that in small slip ranges the correlation between slip and μ could be represented by a linear function, and at higher values of slip ratio the normalized longitudinal interaction force is assumed to saturate. The potential friction coefficient can be then evaluated starting from knowledge of the slope of the tire-road interaction curve even from low slip values, obtainable during not particularly aggressive driving conditions, employing linear regression models.

3. Problem Statement and Control Architecture

Consider a front-wheel driven vehicle where the propulsion is obtained through an electric engine. Moreover assume the vehicle is equipped with proprioceptive sensors for the measurement of its state variables (e.g., chassis velocity, acceleration, yaw rate, and tires' angular velocities), as well as with exteroceptive sensors (e.g., radar, camera, lidar, or a combination of these) for the sensing of the external environment and for the mapping of external obstacles (details on sensing technologies can be found in [44] and the corresponding references).

The aim of this work is to describe a methodology capable to perform the autonomous vehicle-following process in a safe, controlled and comfortable manner even in poor weather conditions, like ice, snow and heavy rain, starting from the information available thanks to a computationally cost effective model-based tire-road friction coefficient technique. The data from proprioceptive sensors are collected in run-time, processed with the physical model-based estimator and, then, employed in loop with a vehicle control logic. From the point of view of the control, the objective is to develop grip-aware functionalities in order to improve driving performance and safety, starting from the strategies for ACC, AEB and ABS longitudinal maneuvers, leveraging the on-board estimation of the road conditions.

To achieve the above mentioned tracking capability, the ACC system has not only to safely adjust the ego-vehicle speed to approach the velocity of the leading vehicle, but it has also to keep the vehicle spacing to an expected value d_{des} that must be adaptable on the base of the estimated road grip, as:

$$d(t) \to d_{des}(t,\mu), \tag{1}$$

$$\Delta v(t) \to 0, \tag{2}$$

where $v(t)$ is the ego vehicle velocity measured on-board by proprioceptive sensors, while $d(t)$ is the distance between the ego vehicle and the leading one and $\Delta v(t)$ is the relative velocity w.r.t. the leading v_{lead}, computed leveraging the on-board exteroceptive sensors. Here, the grip-aware desired space gap d_{des} can be set according to the following the headway time rule [45]:

$$d_{des}(t,\mu) = d_0 + \tau_H(\mu)v(t), \tag{3}$$

where d_0 is the constant spacing at standstill and $\tau_H(\mu)$ is the headway time to be properly adapted on the base of the road friction coefficient to be on-line estimated.

In order to further reduce the risks of crashes, the ACC works jointly with the AEB that, sharing the same on-board sensors, continuously monitors the area in front of the car, automatically detects a risk and hence activates the vehicle braking system (via the ABS, Anti Brake-locking System) decelerating the vehicle with the purpose of avoiding or mitigating a possible collision. It follows that, unlike the ACC, the AEB is activated only when a collision index highlights the possibility of a crash. Here, we exploit the well-known TTC index [46] and the AEB is hence activated if its value is under some threshold depending on the estimate road conditions, as:

$$TTC = -\frac{d(t)}{\Delta v(t)} < TTC_{th}(\mu), \quad \text{being } \Delta v(t) < 0. \tag{4}$$

When the emergency braking is requested by the AEB, the maximum torque is applied to the wheels via the ABS control chain. The wheel actuation systems decreases the longitudinal slip value, thus generating a braking force on the chassis. However, if the slip ratio is below the optimal value, depending on the actual road condition, the dynamics could become unstable with a consequent lock of the wheels. It follows that an efficient control strategy for automatic safe braking during emergency has to adapt the optimal slip value on the base of the estimation of the grip in order to enhance the performance of the ABS, and thus of the overall vehicle.

The above grip-aware ACC, AEB and ABS functionalities have been embedded into the on-board control architecture depicted in Figure 1. The on-line road-grip estimate module implemented on-board firstly calculates the actual friction conditions, estimating the tire-road interaction kinematics and dynamics, and then it provides the actual and the potential friction coefficients per each tire. This estimate, indicated in what follows as $\widehat{\mu}$, is hence exploited to control the longitudinal dynamics of the ego vehicles via grip-aware ACC, AEB and ABS controllers. Note that a supervisor (the so called decision-making unit in Figure 1) is responsible of classifying the specific driving conditions and of choosing the required driving functionality accordingly [47].

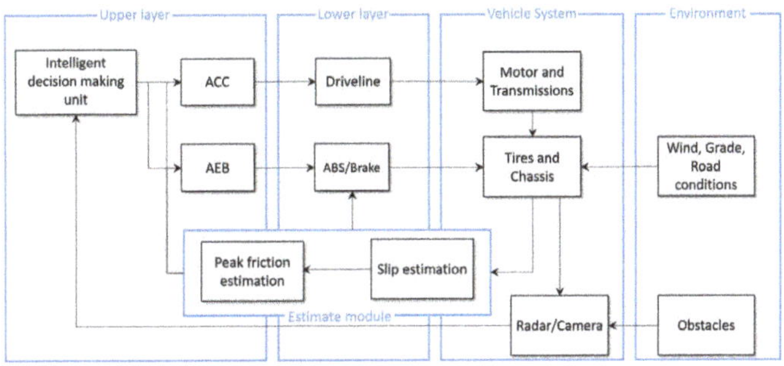

Figure 1. On-board Control Architecture.

4. In-Vehicle Road-Grip Estimation

4.1. From Vehicle Sensors to Tires' State

In recent years the number of sensors installed on vehicles has increased exponentially, facilitating the modelization of the entire system towards the target to consider the standard-instrumented vehicle as a mobile laboratory. Indeed, starting from the acquisition of the physical signals coming from all the sensors installed, employing currently widely-available and affordable mobile calculators, it is possible to properly process the time-evolving dynamic quantities with the aim to feed the real-time state estimators directly on-board. Furthermore, starting from the global quantities referring to the vehicle total behavior, it is currently possible to evaluate even the kinematic and dynamic states of its sub-components, as tires. The developed algorithm, based on the T.R.I.C.K. methodology described in [23], allows us to evaluate in a specifically dedicated on-board module the fundamental kinematic and dynamic quantities for the tire characterization in real time, starting from the experimental signals available within the vehicle CAN bus (Controller Area Network) and s-motion measurement or, as the case in exam, employing a specific set of sensors pre-configured on the vehicle. Such methodology also allows us to evaluate the potential of an estimation process in terms of tire interaction curves, such as in [48].

The originally designed model, described in [23], referred to a quadricycle vehicle fully described from the dynamic point of view. Since the study under analysis aims at simulating the emergency braking manoeuvres et similia, involving only the vehicle longitudinal dynamics, the model can be simplified considering its plane of symmetry **xz** (ISO reference system). Taking into account the above hypothesis, the vehicle can be represented as a bicycle model, whose constitutive equations are described by 3 degrees of freedom within the reference plane **xy**. The above assumption allows us to reduce also the analytical computational cost linked to the model state evaluation per step, as well as the number of parameters to be identified in order to physically reproduce the model dynamics concerning the longitudinal maneuvers, object of investigation. The simplified vehicle model, able to evaluate the kinematics and the dynamics at each axle, feeds the specifically designed logic of the control system providing both actual and potential friction coefficient in run-time.

To perform the analyses, the following modelling and environment assumptions have been considered:

- The road is modelled completely flat with eventual banking and local geometrical effects (i.e., potholes, kerbs, micro- and macro- roughness) absent.
- The tire is modelled only in terms of its kinematic-dynamic transfer function without taking into account its eventual transient dynamics. Furthermore, the multi-physical effects, as thermal or wear abrasive and degradation influences, have not been taken into account at the current stage.
- Since the vehicle is involved in analyses concerning only the longitudinal dynamic maneuvers and considering the vehicle body symmetry hypotheses, the steering angle signal is assumed to be always zero and, therefore, it is not employed within the modelling and the estimation of the vehicle state.
- The vehicle is described only in terms of its intrinsic global geometric and mass-inertia parameters. The longitudinal load transfer is considered taking into account the position of the vehicle body centre of gravity.
- The vertical load distribution on each axle is evaluated starting from the static load data, load transfers due to the geometric position of the vehicle body centre of gravity within the **xz** plane and the aerodynamic force. The estimation of the tangential interaction forces, due an intrinsic non-linearity of each tire system, need an additional convergence algorithm for a correct partition of the global longitudinal force, located at the centre of gravity, into its two contributes based on the front and on the rear axles. Indeed, starting from the vertical loads calculated at each axle the convergence algorithm evaluates the above longitudinal forces, consistent with the vertical loads applied, the kinematics evaluated and the intrinsic dynamic characteristics of a pre-calibrated tire (neglecting the tires' transient behavior at the current stage).
- The suspensions and steering system kinematics and compliances have been taking into account by acquiring the invariable KC curves by means of physical bench testor as an output of simulations performed by means of a multibody model.

The inputs of the T.R.I.C.K.-based methodology, optimized for the longitudinal vehicle dynamics estimation, comprise the following signals acquired thanks to the sensors acquired and processed directly on-board:

- Wheels' angular velocity (rad/s).
- Longitudinal velocity evaluated at the vehicle's centre of gravity (m/s).
- Longitudinal acceleration evaluated at the vehicle's centre of gravity (m/s^2).
- Throttle position (%).
- Braking position (%).

The model outputs, referring to the axle kinematic and dynamic quantities as well as to the additional, are reported below:

- Axles' slip ratio (-).
- Axles' vertical interaction force (N).
- Axles' longitudinal interaction force (N).
- Axles' actual friction coefficient (-).

Since the double track model, i.e. since the dynamics of the vehicle axle, and analyzing a longitudinal maneuvers, the assumption in [49] related to consider the left and right gear ratio of the steering system almost equal, small steering angles and negligible of lateral load transfer and the body roll effect are accepted. Due to the vehicle body symmetry hypotheses made to develop a single track model, the forces acting on the tire have been considered equal. Therefore, the forces acting on the single tires of a single track model are equal to the forces of the entire axle. The vehicle model and the reference system considered are shown in Figure 2.

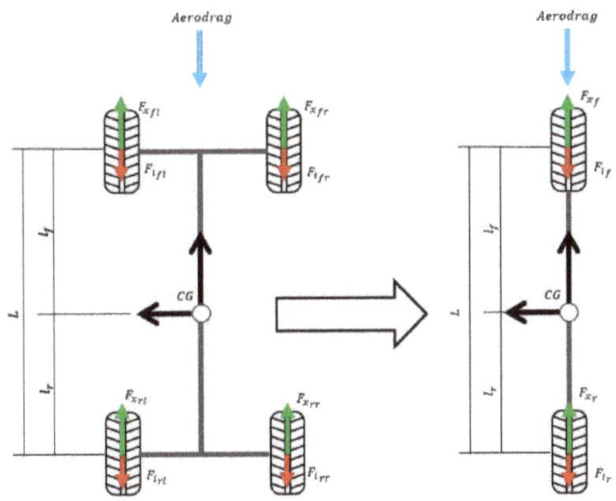

Figure 2. Friction Estimator: Vehicle model and reference system in which the **z** axis is perpendicular to the road equivalent plane **xy**.

In order to evaluate the vertical forces, the loads acting on axles in a stationary condition ($v = 0$ and $a = 0$), called "static loads", W_f and W_r, have to be evaluated. Such values depend on the position of the vehicle body centre of gravity:

$$W_f = \frac{mgl_r}{L}, \tag{5}$$

$$W_r = \frac{mgl_f}{L}. \tag{6}$$

For the longitudinal load equation [49], the load transfer are:

$$\Delta F_z = \frac{mha_x}{l}. \tag{7}$$

The aerodynamic downforces are expressed by following equation:

$$Fz_{aero_i} = \frac{1}{2}\rho A_v v^2 C_{z_i}, \tag{8}$$

with $i = [1,2]$ are defined the axles (respectively front and rear).

Therefore, the axles vertical loads result equal to:

$$F_{z_i} = -(W_i - \Delta F_z + Fz_{aero_i}). \tag{9}$$

The effect due to the inertia resistance of the axles is equal to:

$$F_{inertia_i} = \frac{I_w \dot{\Omega}_i}{R_{r_i}}. \tag{10}$$

The longitudinal interaction forces can be estimated starting from the information regarding the velocity estimated at the vehicle centre of gravity, acquirable by means of specific sensors or employing a model-based technique, taking into account the vehicle kinematics and the vertical load estimated at each wheel hub [50]. Therefore, in order to obtain the axle forces, the kinematic and load vehicle state estimator provides the accurate vehicle speed v_x, the longitudinal acceleration a_x, the wheel speed Ω, the inclination angle (IA) and the normal load F_z.

To this purpose, the global longitudinal dynamic equilibrium of the vehicle has been implemented considering axles' longitudinal forces as given by sums of singular tires' forces, distributed equally between the left and the right side:

$$F_x = F_{x_l} + F_{x_r}, \tag{11}$$

in which the contribution of the single tire, for symmetry hypothesis, is assumed to be equal to:

$$F_{x_i} = \frac{F_x}{2}. \tag{12}$$

The longitudinal interaction forces is a non-linear function of longitudinal acceleration, normal loads, inclination angle, longitudinal speed evaluated at the contact point, wheel speed and longitudinal spindle velocity:

$$F_{x_{i,j}} = f(a_x, F_{z_i}, v_{xCP_{i,j}}, IA_{i,j}, \Omega_{i,j}, v_{x_{spindle_{i,j}}}). \tag{13}$$

The $v_{x_{CP}}$ has been evaluated as:

$$v_{xCP_{i,j}} = R_{i,j}\Omega_{i,j}, \tag{14}$$

with $R_{i,j}$ has been assumed the rolling radius as:

$$R_{i,j} = f(F_{z_{i,j}}, IA_{i,j}, \Omega_{i,j}). \tag{15}$$

Finally, the slip ratio (λ) is:

$$\lambda_{i,j} = \frac{v_{xCP_{i,j}} - v_{x_{spindle_{i,j}}}}{v_{x_{spindle_{i,j}}}}. \tag{16}$$

4.2. On-Board Estimation of Actual and Potential Friction

It is common knowledge that the tribological characteristics of an asphalt can vary significantly depending on the distributed uniform dry, wet, snow or icy conditions (linked to meteorological aspects), or on the presence of the eventual local singularities as oil spots, puddles, kerbs or potholes (linked to local maintenance conditions of the road surface). In order to guarantee the optimum employment of the advanced functionalities of the autonomous driving logic, besides the information concerning the actual friction condition of the road surface, it is even more important providing the potential friction coefficient and the kinematic conditions, in terms of the tire-road interaction slip ratio quantity, it could be reached applying the external inputs, as throttle or braking pedals. The tire model parameters, employed within the estimation of the potential friction coefficient, depend on the parameters characterized and identified on the road characteristics where the experimental activities took part. Starting from the pre-calibrated set parameters of the tire model, depending, in its turn, on the peculiar dynamic set of equation chosen to describe the tire dynamics, and on the actual grip quantity obtainable from the vehicle state information, the potential friction coefficient achievable by each tire is evaluated. The potential friction quantity is assumed reachable varying only the slip ratio quantity (i.e., relative velocity within the tire-road interface) with all other operating conditions remaining the same (wheel alignment, vertical load and wheel spindle longitudinal speed). There are different approaches to tire modelling in the literature, which can be both physical and empirical. Several authors refer to the tire modelling using the Finite Element Method (FEM), adopted to evaluate static characteristics or to the multi-body tire approaches, as [51–53], commonly adopted to study dynamic phenomena on uneven surfaces. Although, the above modelling techniques should be evaluated carefully to the purpose of their employment within the embedded on-board control electronics due to their particularly significant computational cost. It becomes, therefore, necessary the adoption of simpler modelling approaches, as semi-empirical and analytical models [54], whose computational cost is compliant with the capabilities of the modern on-board systems.

The typical tire characteristics curve is described in Figure 3, where three different regions of tire working range are represented.

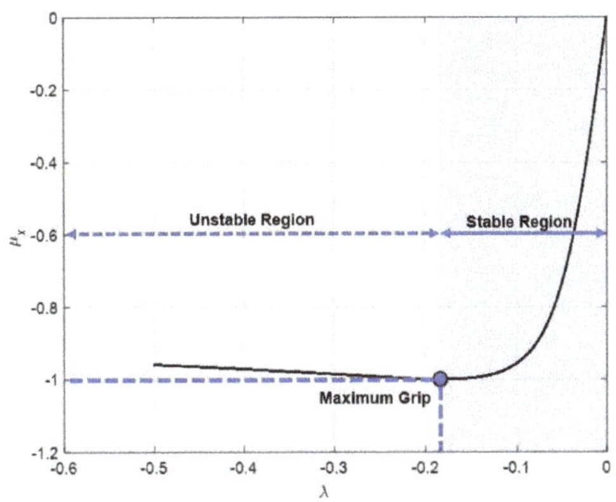

Figure 3. Tire characteristics curve and potential friction coefficient.

The ratio between longitudinal and vertical forces gives the instantaneous friction coefficient, i.e., the actual run-time coefficient between road surface and tire, expressed as follows:

$$\mu_{x_{i,actual}} = \frac{F_{x_i}}{F_{z_i}}. \tag{17}$$

The actual friction coefficient μ_x depends both on the condition of the asphalt and on the peculiar operating conditions the tire is stressed with (i.e., vertical load, wheel alignment, slip ratio, longitudinal speed). Therefore, each tire operating point, describable by the actual friction coefficient μ_x and the corresponding slip ratio λ, can be represented in Figure 4, the point 1.

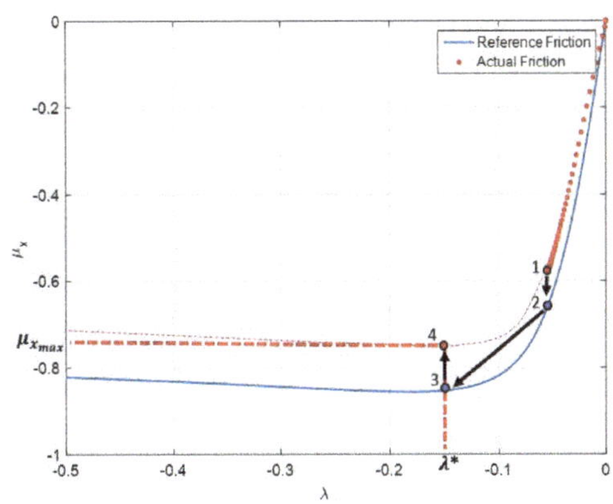

Figure 4. Procedure to evaluate potential friction coefficient.

The eventual changes in terms of friction coefficient within the tire-road interface are assumed to be referred only to the road surface, since the tire has been especially pre-calibrated on a reference asphalt surface. Assuming a linear behaviour of the tire in the typical working conditions of the vehicle, a linear proportion between the reference tire-road and the actual tire-road friction coefficients can be assumed. Starting from the actual friction coefficient quantity, obtained in particular working conditions of the vehicle and therefore of the tire, and from the model parameters already able to properly represent the tire dynamics in run-time, the potential friction is evaluated in the following steps, represented in Figure 4:

- Once the actual friction coefficient (point 1) has been calculated (17), the equivalent grip for the reference tire-road (point 2) can be evaluated:

$$\mu_{x_{i,refRoad}} = \frac{F_{x_{i,refRoad}}}{F_{z_i}}, \tag{18}$$

- Furthermore, starting from the tire model parameters calibrated on a reference road surface, the model is able to provide a valuable output in terms of the maximum longitudinal force, achievable for the same conditions of vertical load, wheel alignment and vehicle longitudinal speed, at the optimal value λ^* of the slip ratio (point 3 in Figure 4):

$$\mu^{max}_{x_{i,refRoad}} = \frac{F^{max}_{x_{i,refRoad}}}{F_{z_i}}, \tag{19}$$

- The potential friction coefficient (point 4) is obtainable, using the proportionality criterion already adopted for the point 2, assuming the linearity of the tire behavior within the working conditions of the vehicle, as follows:

$$\widehat{\mu}_{x_i} = \frac{\frac{F_{x_i}}{F_{z_i}}}{\frac{F_{x_{i,refRoad}}}{F_{z_i}}} \mu^{max}_{x_{i,refRoad}}. \tag{20}$$

In Figure 5, the overall architecture of the developed model is shown. In particular, starting from the sensor-acquired input channels (on the left), the kinematic and load estimator calculate the vehicle state up to the kinematics on the wheel hubs. Then the tire model evaluates the state at the tire-road interface, and, using the above information, the actual and potential friction estimator module.

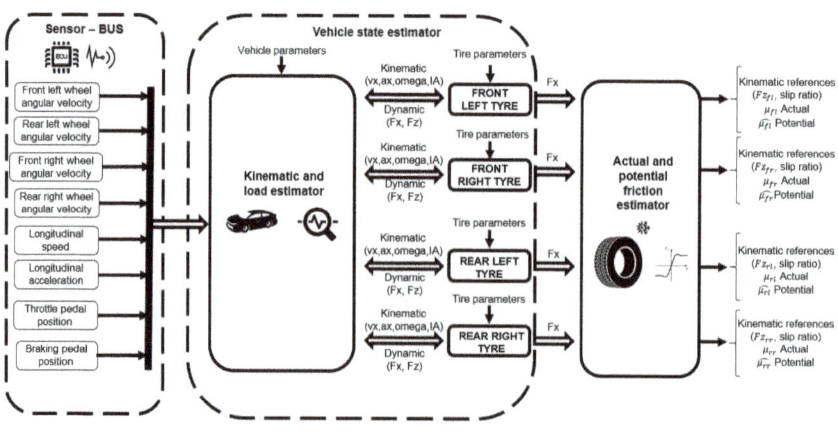

Figure 5. Architecture of the vehicle state estimation system.

5. Design of the Road-Grip Aware Control Modules

The problem stated in Section 3 is here solved by designing road-grip aware driving functionalities for the four-wheel electric vehicle, leveraging the in-vehicle road grip estimate as in Equation (20). Specifically $\widehat{\mu}$ has been chosen as $\widehat{\mu}_{x_i}$ with $i = 1$ if only the front wheel drive is used, and as a weighted sum of $\widehat{\mu}_{x_i}$ with with $i = 1, 2$, if both axles are used for actuation.

5.1. Predictive ACC Design

The ACC is responsible for longitudinal tracking in the autonomous vehicle-following process, so that the vehicle velocity is regulated to a desired speed, while maintaining a safety distance from the preceding vehicle, often named as leading vehicle in the technical literature. The controller is hierarchical and it is composed of a double feedback layer. Namely, the upper-control layer, acts as a reference governor generating the appropriate acceleration profile to be tracked, while also complying additional constraints related to driving comfort and energy consumption. The lower level is responsible for commanding the actuators and, hence, its robust design depends on the specific vehicle configuration.

Here, we focus on the design of the upper layer controller generating reference trajectories able to also improve driving safety leveraging the on-line prediction of road conditions. The predicted ACC is designed following the Model Predictive Control (MPC) approach allowing the continuous constrained optimization of the vehicle longitudinal dynamics. In contrast to the LQR, the MPC solves the problem over a finite time window, or prediction horizon, to make it tractable online. The optimization generates a sequence of control inputs to be imposed over the control horizon, but, according to the receding horizon principle, only the first element of the sequence is effectively applied to the plant. New inputs are received at the following time intervals and the procedure is iteratively repeated.

In order to design the controller, let us define a control oriented mode according the vehicle-following paradigm [55]:

$$\dot{d}(t) = \Delta v(t), \tag{21}$$

$$\Delta\dot{v}(t) = a_{lead}(t) - a(t), \tag{22}$$

where $d(t)$ is the distance between the leading vehicle and the ego vehicle, while $\Delta v(t) = v - v_{lead}$ is their relative velocity (while a_{lead} is the leading acceleration) and the velocity of the chassis of the ego vehicle $v(t)$ undergoes the following longitudinal dynamics [55]:

$$\dot{v}(t) = a(t), \tag{23}$$

$$\dot{a}(t) = \frac{1}{\tau}(-a(t) + u(t)), \tag{24}$$

where $a(t)$ is the actual vehicle acceleration and τ the driveline time constant and $u(t)$ the acceleration input.

Define now the distance error with respect to the desired space gap as $e(t) = d(t) - d_{des}(t)$, where the spacing policy $d_{des}(t)$ is computed as in Equation (3) with the head-way time being the following piece-wise function of the road grip:

$$\tau_H = \begin{cases} \widetilde{\tau}_H/0.2 & \widehat{\mu} \leq 0.2 \\ \widetilde{\tau}_H/\widehat{\mu} & 0.2 < \widehat{\mu} \leq 1 \\ \widetilde{\tau}_H & \widehat{\mu} > 1 \end{cases} \tag{25}$$

where $\widehat{\mu}$ is the estimated maximum available grip and $\widetilde{\tau}_H$ is the constant headway for an ideal dry road [28]. Note that Equation (25) ensures that the safety distance increases as the peak road friction decreases.

Let the state vector as $x(t) = [d(t)\ \Delta v(t)\ v(t)\ a(t)]^T \in \mathbb{R}^4$, the output vector as $y(t) = [e(t)\ \Delta v(t)\ v(t)\ a(t)]^T \in \mathbb{R}^4$ and $w(t) \in \mathbb{R}$ as the leader acceleration, i.e., $w(t) = a_{lead}(t)$. The system in Equations (21) and (23) can be easily recast in the following state space representation

$$\dot{x}(t) = Ax(t) + Bu(t) + Ww(t), \tag{26}$$

$$y(t) = Cx(t) - Z, \tag{27}$$

being

$$A = \begin{bmatrix} 0 & 1 & 0 & 0 \\ 0 & 0 & 0 & -1 \\ 0 & 0 & 0 & 1 \\ 0 & 0 & 0 & -\frac{1}{\tau} \end{bmatrix}, B = \begin{bmatrix} 0 \\ 0 \\ 0 \\ \frac{1}{\tau} \end{bmatrix}, W = \begin{bmatrix} 0 \\ 1 \\ 0 \\ 0 \end{bmatrix}, C = \begin{bmatrix} 1 & -\tau_H & 0 & 0 \\ 0 & 1 & 0 & 0 \\ 0 & 0 & 1 & 0 \\ 0 & 0 & 0 & 1 \end{bmatrix}, Z = \begin{bmatrix} d_0 \\ 0 \\ 0 \\ 0 \end{bmatrix}. \tag{28}$$

However, in order to synthesize the MPC controller, Equation (26) are discretized with a fixed sample time T_s leveraging the zero-order-hold method, thus yielding:

$$x(k+1) = Ax(k) + Bu(k) + Ww(k), \tag{29}$$

$$y(k) = Cx(k) - Z, \tag{30}$$

where, with an abuse of notation, the discrete-time system matrices have been labelled as the ones of the continuous-time model. Moreover, by augmenting the state vector as $[x(k)\ u(k)]^T$, we can resort to the following off-set free formulation as:

$$x(k+1) = \begin{bmatrix} A & B \\ 0 & 1 \end{bmatrix} x(k) + \begin{bmatrix} B \\ 1 \end{bmatrix} \Delta u(k) + \begin{bmatrix} W \\ 0 \end{bmatrix} w(k), \tag{31}$$

$$y(k) = \begin{bmatrix} C & 0_{4\times 1} \end{bmatrix} x(k) - Z. \tag{32}$$

Note that the above formulation is really beneficial since the increment $(\Delta u(k)/T_s)$ is the chassis jerk in discrete time, which is the crucial index for the driving comfort.

The ACC problem in Equation (1) is solved when system in Equation (31) is regulated to the origin while fulfilling at the same time additional tracking, comfort, consumption and safety constraints for all times k. In this perspective, the cost function embeds three different indexes, namely tracking capability, energy consumption and driving comfort. Tracking capability measures the performances in terms of distance and velocity errors, driving comfort is evaluated in terms of acceleration and jerk, finally the acceleration command is chosen as the performance index for energy consumption, yielding the following cost:

$$J_1(y(k), \Delta u(k)) = q_1|d(k) - d_{des}(k, \hat{\mu})|^2 + q_2|\Delta v(k)|^2, +q_3|a(k)|^2 + q_4|u(k)|^2 + r|\Delta u(k)|^2 \tag{33}$$

which can be recasted in matrix form as:

$$J(y(k), \Delta \mathcal{U}) = \sum_{i=0}^{H_p-1} y_{k+i|k}^T Q y_{k+i|k} + \Delta u_{k+i}^T R \Delta u_{k+i}, \tag{34}$$

where $Q = \text{diag}\{q_1, q_2, q_3, q_4\}$ is a positive definite diagonal matrix, $R \in \mathbb{R}^+$, H_p and H_c are the prediction and control horizon, respectively, $\Delta \mathcal{U} \triangleq [\Delta u_k, \dots, \Delta u_{k+H_c-1}]$ is the control sequence during the horizon H_c, while $y_{k+i|k}$ is the predicted output vector at time $k+i$, obtained by applying the input sequence starting from the state $x(k)$.

With respect to safety, road-grip constraints are introduced for the desired and actual acceleration, as:

$$u_{min}(\hat{\mu}) \leq u(k) \leq u_{max}(\hat{\mu}), \tag{35}$$
$$a_{min}(\hat{\mu}) \leq a(k) \leq a_{max}(\hat{\mu}), \tag{36}$$

with $a_{max}(\hat{\mu}) = u_{max}(\hat{\mu}) = \min(2, \hat{\mu}g)$, $a_{min}(\hat{\mu}) = u_{min}(\hat{\mu}) = \max(-4, -\hat{\mu}g)$, where $\hat{\mu}$ is the estimated maximum available grip and g the acceleration of gravity. The saturation values (i.e., 2 and -4) are chosen as upper and lower limit, related to the ideal value for the grip set as 1 [28].

The constraints on the spacing and the maximum velocity are given as:

$$d_{min} \leq d(k) \leq d_{max}, \tag{37}$$
$$v(k) \leq v_{max}, \tag{38}$$

where d_{min} is set to the standstill value d_0 (see Equation (3)) and v_{max} is the maximum admissible speed depending on the legal requirements on the specific traveled road (urban, extra-urban, etc.). Note that this information can be acquired from a map-based on board service leveraging the GPS (Global Positioning System).

Finally, additional constraints on the control input are defined for further improving the driving comfort as:

$$\Delta u_{min} \leq \Delta u(k) \leq \Delta u_{max}. \tag{39}$$

Given the above definitions, the constrained optimization problem to be on-line solved at each time step can be written as:

$$\min_{\Delta \mathcal{U}} \quad J(y(k), \Delta \mathcal{U}) \tag{40}$$
$$\text{subject to} \quad x(k+1) = Ax(k) + B\Delta u(k) + Ww(k) \tag{41}$$
$$y(k) = Cx(k) - Z \tag{42}$$
$$d_{min} \leq \Delta d(k) \leq d_{max} \tag{43}$$
$$v(k) \leq v_{max} \tag{44}$$
$$a_{min} \leq a(k) \leq a_{max} \tag{45}$$
$$u_{min} \leq u(k) \leq u_{max} \tag{46}$$
$$\Delta u_{min} \leq \Delta u(k) \leq \Delta u_{max} \tag{47}$$

Evaluating the output prediction as $y_{k+i|k} = CA^i x_k + \sum_{j=0}^{i-1} CA^j B u_{k+i-1-j} - Z$ and substituting it into Equation (34), it is possible to obtain the cost J as a function of the only control sequence. Hence, after some algebraic manipulations, the cost in Equation (34) can be recasted in the following compact form:

$$J(\Delta \mathcal{U}) = \Delta \mathcal{U}^T F \Delta \mathcal{U} + \Lambda \Delta \mathcal{U}, \tag{48}$$

where $F = \bar{R} + \bar{D}^T \bar{Q} \bar{D}$ and $\Lambda = 2(x(k)^T \bar{C}^T + W^T(k)\bar{E}^T - \bar{Z}^T)\bar{Q}\bar{D}$,
being $\bar{R} = \text{diag}\{R, R, \ldots, R\} \in \mathbb{R}^{H_c}$, $\bar{Q} = \text{diag}\{Q, Q, \ldots, Q\} \in \mathbb{R}^{5H_p \times 5H_p}$,
$\bar{Z} = [Z^T, Z^T, \ldots, Z^T]^T \in \mathbb{R}^{4H_p \times 1}$, $W(k) = [w(k)\; w(k) \ldots w(k)]^T \in \mathbb{R}^{H_p}$,
$\bar{C} = [(CA)^T\; (CA^2)^T \ldots (CA^{H_p})^T]^T$

$$\bar{D} = \begin{bmatrix} CB & 0 & \dots & 0 \\ CAB & CB & \dots & 0 \\ \vdots & \vdots & \ddots & \vdots \\ CA^{H_p-1}B & CA^{H_p-2}B & \dots & CA^{H_p-H_c}B \end{bmatrix}, \quad (49)$$

$$\bar{E} = \begin{bmatrix} CW & 0 & \dots & 0 \\ CAW & CW & \dots & 0 \\ \vdots & \vdots & \ddots & \vdots \\ CA^{H_p-1}W & CA^{H_p-2}W & \dots & CA^{H_p-H_c}W \end{bmatrix}. \quad (50)$$

The constraints Equations (43)–(47) can be be also recast in a compact matrix form as:

$$\bar{H}\mathcal{X} \leq \bar{S}, \quad (51)$$

$$\Delta\mathcal{U} \leq U_{max}, \quad (52)$$

$$-\Delta\mathcal{U} \leq -U_{min}, \quad (53)$$

where $U_{max} = \Delta u_{max} \cdot 1_{H_c \times 1}$, $U_{min} = \Delta u_{min} \cdot 1_{H_c \times 1}$, $\bar{H} = I_{H_p \times H_p} \otimes [H - H]^T$, $\bar{S} = 1_{H_p \times 1} \otimes S$, being $S = [d_{max}\ v_{max}\ a_{max}\ u_{max}\ \ -d_{min}\ 0\ -a_{min}\ \ -u_{min}]^T$, and

$$H = \begin{bmatrix} 1 & 0 & 0 & 0 & 0 \\ 0 & 0 & 1 & 0 & 0 \\ 0 & 0 & 0 & 1 & 0 \\ 0 & 0 & 0 & 0 & 1 \end{bmatrix}. \quad (54)$$

Hence, substituting the state prediction $x_{k+i|k} = A^i x_k + \sum_{j=0}^{i-1} A^j B u_{k+i-1-j}$ in Equation (5 we obtain:

$$\bar{H}\bar{B}\Delta\mathcal{U} \leq \bar{S} - \bar{H}\bar{A}x_k - \bar{H}\bar{E}W(k), \quad (55)$$

$$\Delta\mathcal{U} \leq U_{max}, \quad (56)$$

$$-\Delta\mathcal{U} \leq -U_{min}. \quad (57)$$

From Equations (48)–(55) we finally formulate the linearly constrained quadratic optimization problem in a compact form as:

$$\min_{\Delta\mathcal{U}} \quad J(\Delta\mathcal{U}) \quad (58)$$

$$\text{subject to} \quad G\Delta\mathcal{U} \leq \Gamma \quad (59)$$

where $G = [(\bar{H}\bar{B})^T\ I_{H_c} - I_{H_c}]^T$ and $\Gamma = [(\bar{S} - \bar{H}\bar{A}x_k - \bar{H}\bar{E}W(k))^T\ U_{max}^T\ U_{min}^T]^T$.

One last consideration must be made about the feasibility of problem Equation (58). Indeed, due to the fact that d_{des} depends on the estimated grip ratio, at some point in time the constraint in Equation (43) could be violated, thus leading to the unfeasibility of the problem. This violation can actually be tolerated for small periods of time, so the it is treated in practice as a soft constraint [56], i.e., it is added to the cost function with a slack variable. When the constraint is not violated the slack variable is null and the original problem is obtained.

5.2. Autonomous Emergency Brake

Road accidents and fatalities statistics are reported annually, showing the relation between accidents and drivers behaviour [3,4]. Moreover, the authors in [57] showed that the collision risk increases with the degradation of road conditions. The Autonomous Emergency Brake is one of the most effective driving functionalities for collision prevention and social cost lowering linked to accidents. Nonetheless, EuroNCAP tests are being carried on roads with friction peaks of at least 0.9, even if in real situations a lower value reduces the safeness and the robustness of the whole system.

In this perspective, the aim of the grip-aware AEB system proposed by the authors is to identify the collision risk depending on the actual road conditions and, hence, to take control of the brakes to avoid possible accidents or at least to reduce their severity. Here, we base the decision-making of AEB according to the Time-To-Collision (TTC) in Equation (4), where the detection threshold depends on the estimated road-grip as:

$$TTC_{th}(t,\widehat{\mu}) = \frac{v(t)}{\widehat{\mu} a_{brk}}, \tag{60}$$

where a_{brk} is the deceleration value commanded to the ABS in case of emergency, i.e., $9.8\,\mathrm{m/s^2}$.

5.3. Anti-Lock Braking System

Once an emergency braking is commanded from the AEB, the ABS has to drive the brake system preventing wheels from locking during the hard braking maneuver. Here we propose a Sliding Mode (SMC) ABS controller that leverages the on-line estimation of the road-grip in order to provide a safe braking automatic maneuver for a vehicle-following process also in the presence of hard rainy or icy pavement. This choice is due to SMC's enhanced stability performances with respect to classical control architectures [58] (e.g., proportional action). In particular it can be shown that matched disturbances (uncertainties entering the system through the same channel as the control) are rejected, at least below the actuation limits, moreover due to the controller nonlinear nature, larger stability margins can be achieved.

First, let us define a control-oriented model, i.e., the quarter car model, in which we neglect the lateral and yaw motion of the wheel, thus obtaining a model dealing with the wheel rotational dynamics and longitudinal vehicle dynamics. The rotational dynamics of the wheel is described by

$$I_w \dot{\Omega} = -T_b - R_r F_x, \tag{61}$$

where I_w is the moment of inertia about the wheel axis of rotation, Ω is the angular velocity, T_b is the braking torque, R_r is the wheel rolling radius and F_l is the force produced by the friction reaction. The longitudinal vehicle dynamics are simply modeled as

$$m\dot{v} = -F_x. \tag{62}$$

where m is the vehicle mass.

The control goal is to yield λ to a reference value λ^* during braking [59]. To this aim we define the following sliding surface

$$\sigma(t) = \lambda(t) - \lambda^\star, \tag{63}$$

where $\lambda^\star$ is the optimal slip obtained from the friction estimator (see Section 4) and λ is the longitudinal slip with dynamics as:

$$\dot{\lambda} = -\frac{1}{v}\left(\frac{1-\lambda}{m} + \frac{R_r^2}{I_w}\right)F_x + \frac{R_r}{vI_w}T_b. \tag{64}$$

Due to the inertia differences between wheel and vehicle, we can consider the velocity v as slowly varying, thus reducing Equation (64) to a single-input single-output system, where the control law can be defined as:

$$u(t) = T_b = u_c(t) + u_{sw}(t), \tag{65}$$

where u_c is the continuous term, or *equivalent control* [58], and u_{sw} the discontinuous term. The equivalent control input is is responsible for keeping the trajectories on σ, i.e.,

$$\dot{\sigma} = 0 \Rightarrow u_c = (\frac{(1-\lambda)I_w}{mR_r} + R_r)F_z\mu_{x_{i,actual}}, \tag{66}$$

where the force $F_x = F_z\mu_{x_{i,actual}}$, where F_z is the tire vertical load, and $\mu_{x_{i,actual}}$ the instantaneous friction value provided by the estimation module; note that the subscript $i = [1, 2]$ identifies front and rear tire depending from the axle involved (see Equation (17)).

Closed-loop stability can be easily proven by considering the following Lyapunov function $V(\lambda) = \frac{1}{2}\sigma^2$ and its derivative $\dot{V}(\lambda) = \sigma\dot{\sigma}$. Substituting Equations (65) and (66) into the expression of $\dot{V}$, we obtain

$$\dot{V}(\lambda) = \sigma\dot{\sigma} = \sigma(\frac{R_r}{vI_w}u_{sw}). \tag{67}$$

Hence, selecting $u_{sw} = -\frac{vI_w}{R_r}\eta sgn(\sigma)$ it follows that

$$\dot{V}(\lambda) = -\eta\sigma sgn(\sigma) = -\eta|\sigma| < 0, \tag{68}$$

where $\eta > 0$. In so doing, the surface σ is attractive and the closed-loop is asymptotically stable.

Note that, in order to avoid the well-known chattering problem of sliding mode controllers, for its practical implementation the sign function in Equation (68) has been substituted by the hyperbolic tangent function. Furthermore, since controllability is lost when the vehicle speed is approaching zero (see Equation (64)), following a common practice for implementing the ABS, the controller is disabled at the very low velocities.

6. Co-Simulation Platform

The design for improved solutions of safety related features has been significantly eased thank to the usage of appropriate simulation platforms, enabling engineers to design, test and validate the control architectures through models in a singular platform, and therefore reducing the development cost and the time to market.

Here, we propose a co-simulation platform for Model-In-the-Loop (MIL), where autonomous vehicle has been tested in a realistic traffic scenario. This co-simulation environment, represented in Figure 6, has been built leveraging the following four main components :

- MATLAB/Simulink platform, a widely used framework to model dynamical systems and to design control architectures. Indeed, through an easy to use of its graphical interface, it is possible to develop controllers according to the well-known Model-based Control Design approach. The vehicle dynamics model, implemented in the MATLAB/Simulink environment and employed for the evaluation of the control logic performance in vehicle-following maneuvers, is an efficient 15 degrees of freedom lumped-parameter full vehicle model (LPFVM), described in [60] with a MF-based tire model [54]. The LPFVM is based on a set of Ordinary Differential Equations (ODEs) governing the dynamic equilibrium of the vehicle chassis, including three translational and three rotational equilibrium conditions, and of each wheel, comprising a translational equilibrium along the vertical direction and a rotational equilibrium around the spindle axis. Furthermore, the braking actuation is achieved through a standard hydraulic system made of a master cylinder, a reservoir, a pump and two valves for each wheels which are used to build braking pressure. Details on its model, and the related parameters, can be found in [61] and references therein.
- SUMO (Simulation of Urban MObility), an open-source road traffic simulation package, enabling the user to model entities such as vehicles, traffic lights, road networks, vehicle routing. Each entity is simulated microscopically, meaning that it is possible to

control each of them singularly, while the whole scenario is emulated by its internal engine built upon realistic driving models.
- Friction estimator module, allowing the on-board estimation of the current tire-road interaction state and the potential friction value.

In particular, the Simulink environment has been adopted to describe a highly detailed dynamical behavior of the autonomous vehicle under control, while SUMO emulates the traffic scenario and the road network, where the actual road grip in different scenario can change to mimic the effect of different environmental conditions to be studied.

The interaction between the different modules for the co-simulation is allowed by *Traci*, an integration tool provided by SUMO. The library *TraCI4Matlab* has been employed to couple the vehicle, the road and the SUMO environment model in Simulink.

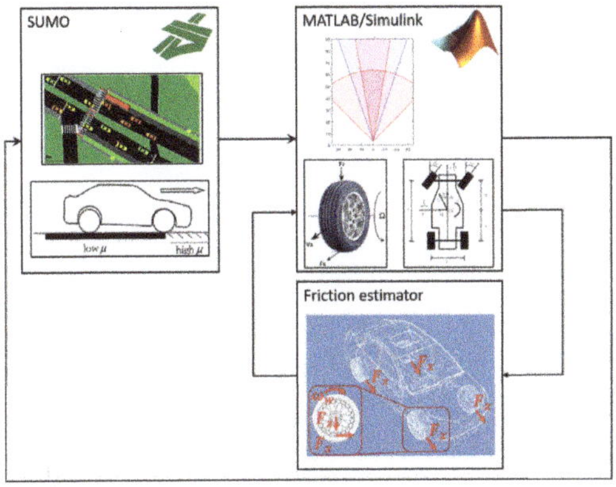

Figure 6. Co-simulation Platform.

7. Performance Analysis

The co-simulation platform described in Section 6 has been exploited to assess the effectiveness of the proposed grip-aware functionalities.

The illustrative results, reported in the following, refer to a vehicle-following process along a typical motorway where the ego vehicle moves with an initial velocity of $v(0) = 30$ m/s, having an initial space gap $d(0) = 90$ m from its predecessor (leader) that moves with an initial speed of $v_{lead}(0) = 20$ m/s. The leader is a human-driven vehicle emulated through SUMO, whose realistic velocity profile accounts for both speed limits and driver imperfection parameters (details on how to model the human-drivers via SUMO can be found in [62] and references therein). Further characteristics parameters of the road scenario, as well as the ones for the controller, are summarized in Table 1.

Table 1. Ego vehicle parameters used for simulation purposes and control tuning parameters.

Parameter	Description	Value
m	vehicle mass	1521 [kg]
I_z	chassis moment of inertia	2875 [kg · m^2]
l_f	distance c.o.g.-front axle	1.2 [m]
l_r	distance c.o.g.-rear axle	1.6 [m]
C_x	longitudinal drag coefficient	0.28
I_w	wheel moment of inertia	1 [kg · m^2]
R_r	wheel radius	0.315 [m]
h	height c.o.g.	0.54 [m]
τ	driveline constant	0.05 [s]
$\tilde{\tau}_H$	headway time	1.1 [s]
d_0	minimum spacing	2 [m]
T_s	ACC sampling time	0.1 [s]
H_p	ACC prediction horizon	15
H_c	ACC control horizon	15
Δu_{min}	ACC minimum control	−0.1 [m/s^2]
Δu_{max}	ACC maximum control	0.1 [m/s^2]
q_1	ACC spacing tracking weight	2
q_2	ACC velocity tracking weight	5
q_3	ACC acceleration weight	20
q_4	ACC control effort weight	20
r	ACC incremental control effort weight	20

The first exemplar driving scenario refers to vehicles moving in the presence of heavy rain, with actual road grip $\mu = 0.5$. Due to the presence of an obstacle, at the time instant $t = 150$ s the leading vehicle performs a sudden hard-brake inducing the maximum deceleration allowed by the road grip, i.e., μg . Results in Figure 7 show how leveraging the on line estimation of the actual road condition (reported in Figure 7c), the ego vehicle is able to safely perform the velocity tracking while always preserving the desired safe space gap about $d_{des}(t, \hat{\mu})$, depending on the grip estimate.

In addition, it is worth to note that the emergency brake is safely performed and vehicles correctly reach the required standstill distance when they finally stops without colliding. According to the theoretical derivation, the Predictive ACC also guarantees both acceleration and jerk of the ego vehicle fulfill the comfort constraints until the leading vehicle performs the hard brake at time instant $t = 150$ s (see Figure 8). Indeed, from this time instant the ACC tries to handle this hazardous braking maneuver, but the necessity of hard deceleration leads to the activation of the AEB, which hence commands the maximum braking torque to be imposed, obviously ignoring the comfort constraints which have less priority w.r.t. the safety.

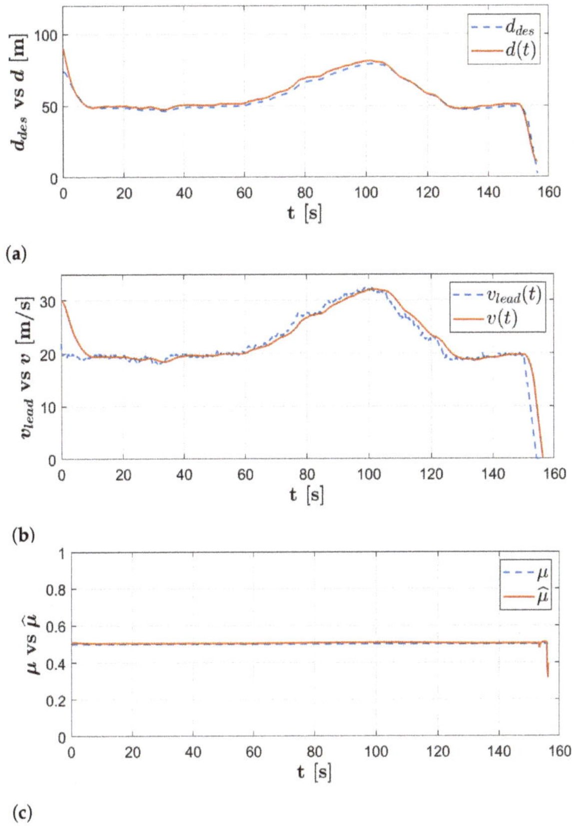

Figure 7. Road-Grip aware Driving Functionalities: vehicle-following and hard emergency braking. (**a**) Time-history of the current distance gap, d, and of the desired one, d_{des}. (**b**) Time-history of the ego velocity v and leader velocity v_{lead}. (**c**) On-board road-grip estimate, $\mu(t)$ vs. $\widehat{\mu}$.

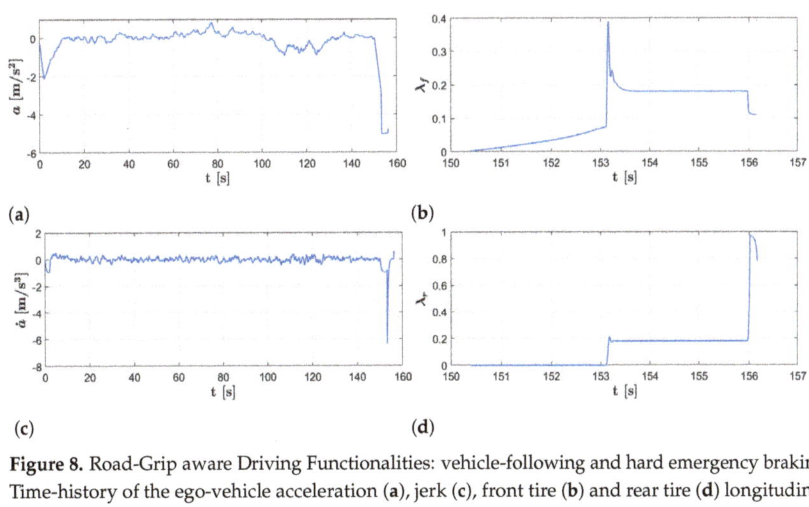

Figure 8. Road-Grip aware Driving Functionalities: vehicle-following and hard emergency braking. Time-history of the ego-vehicle acceleration (**a**), jerk (**c**), front tire (**b**) and rear tire (**d**) longitudinal slip ratios.

In so doing, the collision is safely avoided (see Figure 7), but higher acceleration and jerk can be appreciated during the braking until the stop, as shown in Figure 8. It is also worth to note that, when the emergency braking maneuver is commanded from AEB, than the ABS is responsible ensuring that the longitudinal slips of the tires are regulated to the optimal reference value λ^* returned by the friction estimator module as described in Section 4 (see Figure 8b–d).

In order to clearly appreciate the enhancement of the here proposed grip-aware driving functionalities with respect to classical ACC, AEB and ABS strategies, the above maneuver has been repeated without leveraging the knowledge of the actual road-grip.

Results in Figure 9 disclose that in this case the ACC is still capable of tracking the velocity references, while ensuring a desired gap that obviously depends only from the actual vehicle velocity, namely $d_{des}(t) = d_0 + \tilde{\tau}_H v(t)$. However, when the leading vehicle performs the emergency hard brake, the safety distance results to be too small, the AEB is activated too late and it is impossible to avoid the collision that, hence, occurs at the time instant $t = 154$ s with a velocity of $v \simeq 10$ m/s = 36 km/h.

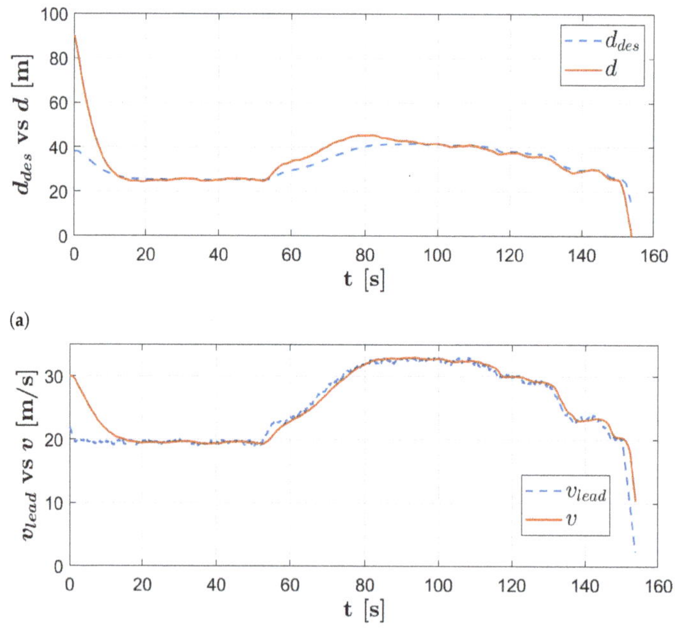

Figure 9. Driving Functionalities without on-board road-grip estimate: vehicle-following and hard emergency braking. (**a**) Time-history of the current distance gap, d , and of the desired one, d_{des}. (**b**) Time-history of the ego velocity v and leader velocity v_{lead}.

A further investigation of the achievable performance has been performed in the case when vehicles are moving in variable environmental conditions, i.e., the actual grip changes in time due to different climatic condition that have to be faced during travelling. Specifically, vehicle drives from dry asphalt to wet road, i.e., the actual maximum road grip starts from $\mu = 1$ and than decreases, into two steps, until $\mu = 0.5$ (see Figure 10c). The initial dynamic condition of the vehicles match the ones chosen in the previous driving scenario.

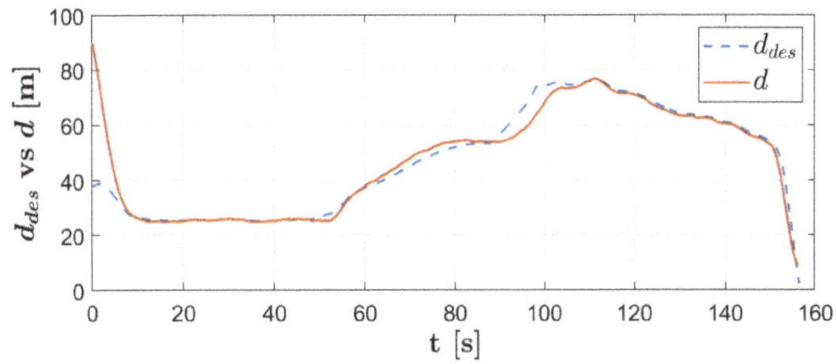

(a)

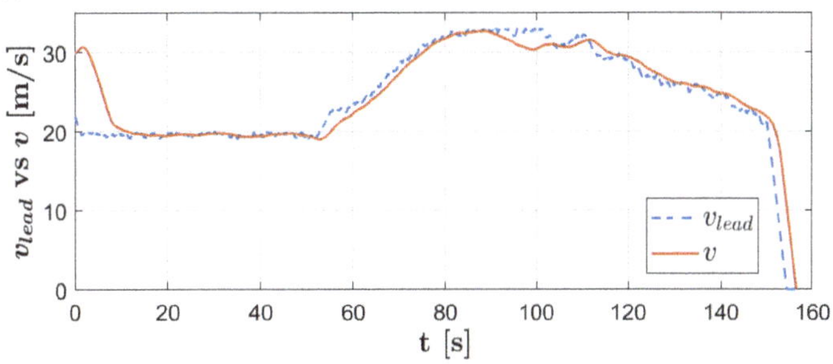

(b)

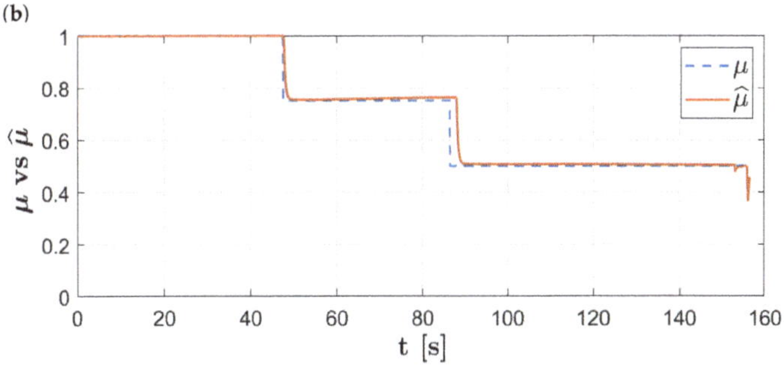

(c)

Figure 10. Road-Grip aware Driving Functionalities in case of varying μ: vehicle-following and hard emergency braking. (**a**) Time-history of the current distance gap, d, and of the desired one, d_{des}. (**b**) Time-history of the ego velocity v and leader velocity v_{lead}. (**c**) On-board road-grip estimate, $\mu(t)$ vs. $\widehat{\mu}$.

Results depicted in Figure 10 show how the on-line road-grip estimate is performed with good precision (always below 1% at steady state). Furthermore, as the road grip decreases the safe distance is correctly adapted in order to provide a safer spacing with respect to the current adhesion (see Figure 10a) and the predictive ACC correctly tracks the reference values without any constraints violation. As in the previous driving scenario, at the time instant a $t = 150$ s an emergency situation emerges inducing the hard braking maneuver. Also in this case the combination of the grip-aware AEB and ABS is able of ensuring a safe stopping without collision.

Final exemplar results refer to a typical Stop & Go scenario where continuous smooth accelerations and decelerations occur due to traffic congestion. In order to better assess the collision risk and the safety margins during traffic jam, we leverage the following well-known non-dimensional collision-index $\gamma(t)$ [34]:

$$\gamma(t) = \frac{d(t) - d_{br}}{d_w - d_{br}}, \tag{69}$$

where $d(t)$ is the actual distance between the vehicles, d_{br} is the breaking critical distance and d_w is the warning critical distance. Note that the above index witnesses the possibility of an incoming crash due to the current driving situation. Specifically, when it is positive and greater than the unity a safe situation is detected, while, if it is below the unity, a possible dangerous scenario is signalized.

Results depicted in Figure 11 clearly confirm that, leveraging the road-aware driving control architecture, the safety index $\gamma(t)$ never goes below the unity, while, on the other hand, if the estimate of the road-grip is not exploited for the automated driving functionalities the collision index alerts for possible dangerous situations during the deceleration phases (see Figure 11), reducing the vehicle safety.

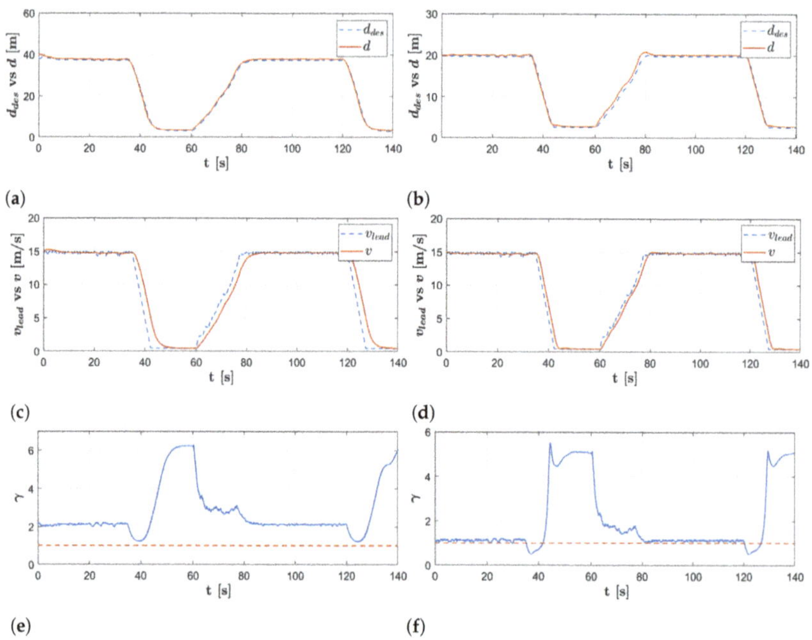

Figure 11. Road-Grip aware Driving Functionalities during Stop&Go. Time-history of the current distance gap, d, and of the desired one, d_{des} (**a**) ego velocity v and leader velocity v_{lead} (**c**) safety index, γ (**e**) with road grip adaptation. Time-history of the current distance gap, d, and of the desired one, d_{des} (**b**) ego velocity v and leader velocity v_{lead} (**d**) safety index, γ (**f**) without road grip adaptation.

The effectiveness of the approach w.r.t. safety is finally summarized in Table 2. Here, results clearly disclose that, adapting in real-time the driving policy to the current road-grip, the overall automated driving performance can be enhanced with the minimal values assumed by the safety indexes [34] (i.e., time-to-collision, relative distance w.r.t. the predecessor and collision index γ as in Equation (69)) comparable with the ones required in the case of ideal road conditions.

Table 2. Summary of the minimal values assumed by the safety indexes in every described scenario.

Scenario	min TTC	min λ	min d(t)
Vehicle following with estimation	2.02	−0.15	10.30
Vehicle following w/o estimation	0	−1.08	0
Stop&Go with estimation	2.73	1.22	3.20
Stop&Go w/o estimation	2.13	0.48	2.70

8. Conclusions

This paper is focused on the development of a new control architecture for vehicles, based on the estimation of the maximum achievable road friction coefficient.

The proposed technique for real-time road friction estimation in different environmental conditions is based on both bicycle model to evaluate the state of the vehicle and a tire Magic Formula model based on a slip-slope approach. The introduction of the grip value in ADAS application, in particular for the longitudinal dynamics of the vehicle chassis, composed of Adaptive Cruise Control (ACC) and Autonomous Emergency Brake (AEB), and the Antilock Braking System (ABS), allows the vehicle to work at the maximum performance in all operating conditions.

In this study, the technique has been performed on a simulation platform, consisting of different components working simultaneously. A vehicle dynamics model, implemented in the MAT-LAB/Simulink environment, has been employed for the evaluation of the control logic performance in vehicle-following maneuvers. In addition, a SUMO environment has been implemented to realize different traffic scenarios and environmental conditions. Finally, the Friction algorithm has been used to perform the estimation of the road coefficient.

The simulation has been carried out in different environmental and vehicle kinematic conditions. According to the results, the control system has been improved by the involvement of the current and potential friction coefficient evaluated in run-time. The improvement basically results from the development of a control system that is able to avoid collisions in any environmental condition. The potential grip has been therefore demonstrated to be crucial for the autonomous driving systems.

Future developments will also comprehend the lateral dynamics. Several authors have already tackled the problem describing different techniques and modelling approaches [63–65]. The authors plan also to include the impact of the road bank angle and slope, the tire combined interaction characteristics, as well as, the variations of the vehicle dynamic behaviour due to the tire intrinsic multi-physics (i.e., wear and temperature effects) [66,67].

Author Contributions: Conceptualization and supervision, S.S. and A.S.; methodology, S. S., V.M.A. and A.S.; software and investigation, N.A. and V.M.A.; formal analysis, S.S.; resources and visualization, R.B.; data curation, N.A. All authors have read and agreed to the published version of the manuscript.

Funding: This research received no external funding.

Institutional Review Board Statement: Not applicable.

Informed Consent Statement: Informed consent was obtained from all subjects involved in the study.

Data Availability Statement: No new data were created or analyzed in this study. Data sharing is not applicable to this article.

Conflicts of Interest: The authors declare no conflict of interest.

Abbreviations

The following abbreviations are used in this manuscript:

$ADAS$	Advanced Driver Assistance System
MPC	Model Predictive Control
ACC	Adaprive Cruise Control
AEB	Autonomous Emergency Braking
LQR	Linear Quadratic Regulator
TTC	Time To Collision
PID	Proportional-Integral-Derivative
SAE	Society of Automotive Engineers
ABS	Anti-lock Braking System
$TRICK$	Tire Road Interaction Characterization and Knowledge
MF	Magic Formula
$ECUs$	Electronic Control Units
CAN	Controller Area Network
GPS	Global Positioning System
SMC	Sliding Mode Control
MIL	Model In the Loop
$SUMO$	Simulation of Urban MObility
$LPFVM$	Lumped-Parameter Full Vehicle Model
FEM	Finite element Methods
$c.o.g.$	Center of Gravity
v	longitudinal velocity evaluated at the c.o.g.
a	longitudinal acceleration
Ω	wheel speed
IA	inclination angle
$v_{x_{CP}}$	longitudinal velocity evaluated at the contact point
$v_{x_{spindle}}$	spindle velocity
m	vehicle mass
g	acceleration of gravity
l_f	distance c.o.g. - front axle
l_r	distance c.o.g. - rear axle
L	wheelbase
ρ	air density
h	height c.o.g.
C_x	longitudinal drag coefficient
C_z	lift coefficient
I_ω	moment of inertia about the wheel axis of rotation
I_z	wheel moment of inertia
R_r	rolling radius
A_v	master section
W_f	static load - front axle
W_r	static load - rear axle
ΔF_z	load transfer
Fz_{aero_i}	aerodynamic down force
$F_{i,inertia}$	inertial force
F_{x_i}	longitudinal force
$F_{x_{i,refRoad}}$	longitudinal force evaluated on a reference road surface
F_{z_i}	normal force
$\mu_{x_{i,actual}}$	actual friction coefficient estimated
$\mu_{x_{i,refRoad}}$	actual friction coefficient estimated on a reference road surface
$\widehat{\mu}_{x_i}$	potential friction coefficient estimated
λ	slip ratio
e	distance error to the desired d_{des}
T_b	braking torque

d	distance lead to ego vehicle
d_{des}	ACC desired distance
d_0	minimum spacing
τ_H	headway time
a_{lead}	leading vehicle acceleration
v_{lead}	leading vehicle velocity
τ	driveline time constant
u	control input
J	ACC cost function
Q	ACC output weight
r	ACC incremental control effort weight
H_c	ACC control horizon
H_p	ACC prediction horizon
T_s	ACC sampling time
a_{brk}	AEB deceleration command
σ	sliding surface
η	switching control gain
d_ω	warning critical distance
d_{br}	braking critical distance
γ	safety index

References

1. Urmson, C.; Anhalt, J.; Bagnell, D.; Baker, C.; Bittner, R.; Clark, M.; Dolan, J.; Duggins, D.; Galatali, T.; Geyer, C.; et al. Autonomous driving in urban environments: Boss and the urban challenge. *J. Field Robot.* **2008**, *25*, 425–466. [CrossRef]
2. INRIX. Congestion Costs Each American 97 h, $ 1,348 A Year. Available online: https://inrix.com/press-releases/scorecard-2018-us/#:~:text=Americans%20lost%20an%20average%20of,the%20top%2060%20urban%20areas (accessed on 7 June 2018).
3. USDOT. Early Estimate of Motor Vehicle Traffic Fatalities for the First 9 Months of 2019. Available online: https://crashstats.nhtsa.dot.gov/Api/Public/ViewPublication/812874#:~:text=A%20statistical%20projection%20of%20traffic,as%20shown%20in%20Table%201. (accessed on 3 September 2019).
4. WHO. *Global Status Report on Road Safety*; WHO: Geneva, Switzerland, 2018.
5. Matthaios, V.N.; Kramer, L.J.; Crilley, L.R.; Sommariva, R.; Pope, F.D.; Bloss, W.J. Quantification of within-vehicle exposure to NOx and particles: Variation with outside air quality, route choice and ventilation options. *Atmos. Environ.* **2020**, *240*, 117810. [CrossRef]
6. Wallington, T.J.; Sullivan, J.L.; Hurley, M.D. Emissions of CO2, CO, NOx, HC, PM, HFC-134a, N2O and CH4 from the global light duty vehicle fleet. *Meteorol. Z.* **2008**, *17*, 109–116. [CrossRef]
7. Walta, L.; Marchau, V.; Brookhuis, K. Stakeholder preferences of advanced driver assistance systems (ADAS)—A literature review. In Proceedings of the 13th World Congress and Exhibition on Intelligent Transport Systems and Service, London, UK, 8–12 October 2006.
8. Tsourveloudis, N.C.; Valavanis, K.P.; Hebert, T. Autonomous vehicle navigation utilizing electrostatic potential fields and fuzzy logic. *IEEE Transact. Robot. Automat.* **2001**, *17*, 490–497. [CrossRef]
9. Levinson, J.; Askeland, J.; Becker, J.; Dolson, J.; Held, D.; Kammel, S.; Kolter, J.Z.; Langer, D.; Pink, O.; Pratt, V.; et al. Towards fully autonomous driving: Systems and algorithms. In Proceedings of the 2011 IEEE Intelligent Vehicles Symposium (IV), Baden-Baden, Germany, 5–9 June 2011; pp. 163–168.
10. Minguez, J.; Lamiraux, F.; Laumond, J.P. Motion planning and obstacle avoidance. In *Springer Handbook of Robotics*; Springer: Berlin, Germany, 2016; pp. 1177–1202.
11. Paden, B.; Čáp, M.; Yong, S.Z.; Yershov, D.; Frazzoli, E. A survey of motion planning and control techniques for self-driving urban vehicles. *IEEE Transact. Intell. Veh.* **2016**, *1*, 33–55. [CrossRef]
12. Antonov, S.; Fehn, A.; Kugi, A. Unscented Kalman filter for vehicle state estimation. *Veh. Syst. Dynam.* **2011**, *49*, 1497–1520. [CrossRef]
13. Wenzel, T.A.; Burnham, K.; Blundell, M.; Williams, R. Dual extended Kalman filter for vehicle state and parameter estimation. *Veh. Syst. Dynam.* **2006**, *44*, 153–171. [CrossRef]
14. Liu, W.; He, H.; Sun, F. Vehicle state estimation based on minimum model error criterion combining with extended Kalman filter. *J. Franklin Inst.* **2016**, *353*, 834–856. [CrossRef]
15. Collado, J.M.; Hilario, C.; De la Escalera, A.; Armingol, J.M. Model based vehicle detection for intelligent vehicles. In Proceedings of the IEEE Intelligent Vehicles Symposium, Parma, Italy, 14–17 June 2004; pp. 572–577.
16. Hojjati-Emami, K.; Dhillon, B.; Jenab, K. Reliability prediction for the vehicles equipped with advanced driver assistance systems (ADAS) and passive safety systems (PSS). *Int. J. Ind. Eng. Comput.* **2012**, *3*, 731–742. [CrossRef]
17. Cafiso, S.; Di Graziano, A. Evaluation of the effectiveness of ADAS in reducing multi-vehicle collisions. *Int. J. Heavy Veh. Syst.* **2012**, *19*, 188–206. [CrossRef]

18. Park, S.J.; Kim, T.Y.; Kang, S.M.; Koo, K.H. A novel signal processing technique for vehicle detection radar. In Proceedings of the IEEE MTT-S International Microwave Symposium Digest, Philadelphia, PA, USA, 8–13 June 2003; Volume 1, pp. 607–610.
19. Varghese, J.Z.; Boone, R.G. Overview of autonomous vehicle sensors and systems. In Proceedings of the International Conference on Operations Excellence and Service Engineering, Orlnado, FL, USA, 10–11 September 2015; pp. 178–191.
20. Dickson, M.A.; Noguchi, N.; Zhang, Q.; Reid, J.F.; Will, J.D. Sensor-Fusion Navigator for Automated Guidance of Off-Road Vehicles. U.S. Patent 6,445,983, 3 September 2002.
21. Kocić, J.; Jovičić, N.; Drndarević, V. Sensors and sensor fusion in autonomous vehicles. In Proceedings of the 2018 26th Telecommunications Forum (TELFOR), Belgrade, Serbia, 20–21 November 2018; pp. 420–425.
22. Ndoye, M.; Totten, V.F.; Krogmeier, J.V.; Bullock, D.M. Sensing and signal processing for vehicle reidentification and travel time estimation. *IEEE Transact. Intell. Transport. Syst.* **2010**, *12*, 119–131. [CrossRef]
23. Farroni, F. T.R.I.C.K.-Tire/Road Interaction Characterization & Knowledge—A tool for the evaluation of tire and vehicle performances in outdoor test sessions. *Mech. Syst. Signal Proces.* **2016**, *72*, 808–831. [CrossRef]
24. Weißmann, A.; Görges, D.; Lin, X. Energy-optimal adaptive cruise control combining model predictive control and dynamic programming. *Control Eng. Pract.* **2018**, *72*, 125–137. [CrossRef]
25. Weißmann, A.; Görges, D.; Lin, X. Energy-optimal adaptive cruise control based on model predictive control. *IFAC-Papers OnLine* **2017**, *50*, 12563–12568. [CrossRef]
26. Sun, C.; Chu, L.; Guo, J.; Shi, D.; Li, T.; Jiang, Y. Research on adaptive cruise control strategy of pure electric vehicle with braking energy recovery. *Adv. Mech. Eng.* **2017**, *9*, 1687814017734994. [CrossRef]
27. Zhang, S.; Luo, Y.; Li, K.; Li, V. Real-Time Energy-Efficient Control for Fully Electric Vehicles Based on an Explicit Model Predictive Control Method. *IEEE Transact. Veh. Technol.* **2018**, *67*, 4693–4701. [CrossRef]
28. Zhang, S.; Zhuan, X. Model-Predictive Optimization for Pure Electric Vehicle during a Vehicle-Following Process. *Math. Probl. Eng.* **2019**, *2019*, 5219867. [CrossRef]
29. Lee, D.; Kim, S.; Kim, C.; Huh, K. Development of an autonomous braking system using the predicted stopping distance. *Int. J. Automot. Technol.* **2014**, *15*, 341–346. [CrossRef]
30. Xiong, H.; Ling, Z.; Yue, R.; Yinong, L.; Zhenfei, Z.; Yusheng, L.; Qiang, Z.; Zhoubing, X. Research on control strategy of automatic emergency brake system based on Prescan. In Proceedings of the IET International Conference on Intelligent and Connected Vehicles (ICV 2016), Chongqing, China, 22–23 September 2016; pp. 1–6.
31. Naseralavi, S.; Nadimi, N.; Saffarzadeh, M.; Mamdoohi, A.R. A general formulation for time-to-collision safety indicator. *Proc. ICE Transp.* **2013**, *166*, 294–304.10.1680/tran.11.00031. [CrossRef]
32. Abdul Hamid, U.Z.; Ahmad Zakuan, F.R.; Zulkepli, K.; Azmi, M.Z.; Zamzuri, H.; Abdul Rahman, M.A.; Zakaria, M. Autonomous emergency braking system with potential field risk assessment for frontal collision mitigation. In Proceedings of the 2017 IEEE Conference on Systems, Process and Control (ICSPC), Malacca, Malaysia, 15–17 December 2017; pp. 71–76.
33. Lee, I.H.; Luan, B.C. *Design of Autonomous Emergency Braking System Based on Impedance Control for 3-Car Driving Scenario*; SAE Technical Paper; SAE International: Warrendale, PE, USA, 2016. [CrossRef]
34. Moon, S.; Moon, I.; Yi, K. Design, tuning, and evaluation of a full-range adaptive cruise control system with collision avoidance. *Control Eng. Pract.* **2009**, *17*, 442–455. [CrossRef]
35. Mullakkal B.; Wang, M.; Arem, B.; Happee, R. Design and Analysis of Full Range Adaptive Cruise Control with Integrated Collision Avoidance Strategy. In Proceedings of the 2016 IEEE 19th International Conference on Intelligent Transportation Systems (ITSC), Rio de Janeiro, Brazil, 1–4 November 2016. [CrossRef]
36. Shakouri, P.; Ordys, A.; Askari, M. Adaptive cruise control with stop&go function using the state-dependent nonlinear model predictive control approach. *ISA Transact.* **2012**, *51*, 622–631. [CrossRef]
37. Zhao, D.; Hu, Z.; Xia, Z.; Alippi, C.; Zhu, Y.; Wang, D. Full-range adaptive cruise control based on supervised adaptive dynamic programming. *Neurocomputing* **2014**, *125*, 57–67. [CrossRef]
38. Khaleghian, S.; Emami, A.; Taheri, S. A technical survey on tire-road friction estimation. *Friction* **2017**, *5*, 123–146. [CrossRef]
39. Leng, B.; Jin, D.; Xiong, L.; Yang, X.; Yu, Z. Estimation of tire-road peak adhesion coefficient for intelligent electric vehicles based on camera and tire dynamics information fusion. *Mech. Syst. Signal Proces.* **2021**, *150*, 107275. [CrossRef]
40. Berntorp, K.; Quirynen, R.; Di Cairano, S. Friction Adaptive Vehicle Control. U.S. Patent App. 16/299,285, 17 September 2020.
41. Hu, J.; Rakheja, S.; Zhang, Y. Tire-Road Friction Coefficient Estimation under Constant Vehicle Speed Control. *IFAC-Papers OnLine* **2019**, *52*, 136–141. [CrossRef]
42. Chen, Y.; Wang, J. Adaptive vehicle speed control with input injections for longitudinal motion independent road frictional condition estimation. *IEEE Transact. Vech. Technol.* **2011**, *60*, 839–848. [CrossRef]
43. Rajamani, R.; Piyabongkarn, N.; Lew, J.; Yi, K.; Phanomchoeng, G. Tire-road friction-coefficient estimation. *IEEE Control Syst. Mag.* **2010**, *30*, 54–69.
44. Ilas, C. Electronic sensing technologies for autonomous ground vehicles: A review. In Proceedings of the 2013 8th International Symposium on Advanced Topics in Electrical Engineering (ATEE), Bucharest, Romania, 23–25 May 2013; pp. 1–6.
45. Winsum, W.V.; Heino, A. Choice of time-headway in car-following and the role of time-to-collision information in braking. *Ergonomics* **1996**, *39*, 579–592. [CrossRef]
46. Hayward, J.C. *Near Miss Determination through Use of a Scale of Danger*; Sponsored by Committee on Effectiveness of Operational Measures: Basel, Switzerland, 1972.

47. Zhenhai, G.; Jun, W.; Hongyu, H.; Wei, Y.; Dazhi, W.; Lin, W. Multi-argument control mode switching strategy for adaptive cruise control system. *Procedia Eng.* **2016**, *137*, 581–589. [CrossRef]
48. De Martino, M.; Farroni, F.; Pasquino, N.; Sakhnevych, A.; Timpone, F. Real-time estimation of the vehicle sideslip angle through regression based on principal component analysis and neural networks. In Proceedings of the 2017 IEEE International Systems Engineering Symposium (ISSE), Vienna, Austria, 11–13 October 2017; pp. 1–6.
49. Guiggiani, M. *The Science of Vehicle Dynamics: Handling, Braking, and Ride of Road and Race Cars;* Springer Science & Business Media: Berlin, Germany, 2014. [CrossRef]
50. Dell'Annunziata, G.N.; Lenzo, B.; Farroni, F.; Sakhnevych, A.; Timpone, F. A New Approach for Estimating Tire-Road Longitudinal Forces for a Race Car. In *IFToMM World Congress on Mechanism and Machine Science;* Springer: Berlin, Germany, 2019; pp. 3601–3610.
51. Romano, L.; Sakhnevych, A.; Strano, S.; Timpone, F. A hybrid tyre model for in-plane dynamics. *Vehic. Syst. Dynam.* **2020**, *58*, 1123–1145. [CrossRef]
52. Gipser, M. Ftire: A physically based application-oriented tyre model for use with detailed mbs and finite-element suspension models. *Vehic. Syst. Dynam.* **2005**, *43*, 76–91. [CrossRef]
53. Gallrein, A.; Bäcker, M. Cdtire: A tire model for comfort and durability applications. *Vehic. Syst. Dynam.* **2007**, *45*, 69–77. [CrossRef]
54. Pacejka, H. *Tire and Vehicle Dynamics;* Elsevier: Amsterdam, The Netherlands, 2005.
55. Rajamani, R. *Vehicle Dynamics and Control;* Springer Science & Business Media: New York, NY, USA, 2011.
56. Eben Li, S.; Li, K.; Wang, J. Economy-oriented vehicle adaptive cruise control with coordinating multiple objectives function. *Vehic. Syst. Dynam.* **2013**, *51*, 1–17. [CrossRef]
57. Edwards, J.B. The Relationship Between Road Accident Severity and Recorded Weather. *J. Safety Res.* **1998**, *29*, 249–262. [CrossRef]
58. Utkin, V.; Guldner, J.; Shijun, M. *Sliding Mode Control in Electro-Mechanical Systems;* CRC Press: Boca Raton, FL, USA, 1999; Volume 34.
59. Pasillas-Lépine, W.; Loría, A.; Gerard, M. Design and experimental validation of a nonlinear wheel slip control algorithm. *Automatica* **2012**, *48*, 1852–1859. [CrossRef]
60. Perrelli, M.; Farroni, F.; Timpone, F.; Mundo, D. Analysis of Tire Temperature Influence on Vehicle Dynamic Behaviour Using a 15 DOF Lumped-Parameter Full-Car Model. In *International Conference on Robotics in Alpe-Adria Danube Region;* Springer: Berlin, Germany, 2020; pp. 266–274.
61. Vázquez, I.; Galicia, M.I.; Sánchez, J.D.; Loukianov, A.G.; Kruchinin, P.A. Integral Nested Sliding Mode Control for Antilock Brake System*. *IFAC Proc. Vol.* **2010**, *43*, 49–54. [CrossRef]
62. Song, J.; Wu, Y.; Xu, Z.; Lin, X. Research on car-following model based on SUMO. In Proceedings of the 7th IEEE/International Conference on Advanced Infocomm Technology, 1, Fuzhou, China, 4–16 November 2014; pp. 47–55.
63. Pomponi, C.; Scalzi, S.; Pasquale, L.; Verrelli, C.; Marino, R. Automatic motor speed reference generators for cruise and lateral control of electric vehicles with in-wheel motors. *Control Eng. Practice* **2018**, *79*, 126–143. [CrossRef]
64. Zhang, H.; Zhang, X.; Wang, J. Robust gain-scheduling energy-to-peak control of vehicle lateral dynamics stabilisation. *Vehic. Syst. Dynam.* **2014**, *52*, 309–340. [CrossRef]
65. Fergani, S.; Menhour, L.; Sename, O.; Dugard, L.; D'Andréa-Novel, B. Integrated vehicle control through the coordination of longitudinal/lateral and vertical dynamics controllers: Flatness and LPV/-based design. *Int. J. Robust Nonlinear Control* **2017**, *27*, 4992–5007. [CrossRef]
66. Farroni, F.; Russo, M.; Sakhnevych, A.; Timpone, F. TRT EVO: Advances in real-time thermodynamic tire modeling for vehicle dynamics simulations. *Proc. Inst. Mech. Eng. Part D J. Automob. Eng.* **2019**, *233*, 121–135. [CrossRef]
67. Farroni, F.; Sakhnevych, A.; Timpone, F. Physical modelling of tire wear for the analysis of the influence of thermal and frictional effects on vehicle performance. *Proc. Inst. Mech. Eng. Part L: J. Mater. Design Appl.* **2017**, *231*, 151–161. [CrossRef]

applied sciences

MDPI

Article

A Comparative Study of Energy Consumption and Recovery of Autonomous Fuel-Cell Hydrogen–Electric Vehicles Using Different Powertrains Based on Regenerative Braking and Electronic Stability Control System

Ahmet Yildiz [1,*] and Mert Ali Özel [2]

[1] Automotive Engineering Department, Engineering Faculty, Bursa Uludağ University, Görükle, Bursa 16240, Turkey

[2] Mechanical Engineering Department, Engineering Faculty, Bursa Uludağ University, Görükle, Bursa 16240, Turkey; mertaliozel@uludag.edu.tr

* Correspondence: ahmetyildiz@uludag.edu.tr

Citation: Yildiz, A.; Özel, M.A. A Comparative Study of Energy Consumption and Recovery of Autonomous Fuel-Cell Hydrogen–Electric Vehicles Using Different Powertrains Based on Regenerative Braking and Electronic Stability Control System. *Appl. Sci.* **2021**, *11*, 2515. https://doi.org/10.3390/app11062515

Academic Editor: Flavio Farroni

Received: 29 January 2021
Accepted: 25 February 2021
Published: 11 March 2021

Abstract: Today, with the increasing transition to electric vehicles (EVs), the design of highly energy-efficient vehicle architectures has taken precedence for many car manufacturers. To this end, the energy consumption and recovery rates of different powertrain vehicle architectures need to be investigated comprehensively. In this study, six different powertrain architectures—four independent in-wheel motors with regenerative electronic stability control (RESC) and without an RESC, one-stage gear (1G) transmission, two-stage gear (2G) transmission, continuously variable transmission (CVT) and downsized electric motor with CVT—were mathematically modeled and analyzed under real road conditions using nonlinear models of an autonomous hydrogen fuel-cell electric vehicle (HFCEV). The aims of this paper were twofold: first, to compare the energy consumption performance of powertrain architectures by analyzing the effects of the regenerative electronic stability control (RESC) system, and secondly, to investigate the usability of a downsized electrical motor for an HFCEV. For this purpose, all the numerical simulations were conducted for the well-known FTP75 and NEDC urban drive cycles. The obtained results demonstrate that the minimum energy consumption can be achieved by a 2G-based powertrain using the same motor; however, when an RESC system is used, the energy recovery/consumption rate can be increased. Moreover, the results of the article show that it is possible to use a downsized electric motor due to the CVT, and this powertrain significantly reduces the energy consumption of the HFCEV as compared to all the other systems. The results of this paper present highly significant implications for automotive manufacturers for designing and developing a cleaner electrical vehicle energy consumption and recovery system.

Keywords: energy consumption and recovery; transmission layouts; fuel-cell electric vehicles

1. Introduction

In the automotive industry, new vehicular technologies such as battery electric vehicles (BEVs), hybrid electric vehicles (HEVs), and hydrogen fuel-cell electric vehicles (HFCEVs) offer critically important environmentally friendly solutions for achieving zero net carbon emissions [1–6]. In this regard, HFCEVs have gained special attention owing to their greater charge time–range ratios as compared to BEVs [7]. However, HFCEVs are costly, primarily due to the high production cost of hydrogen fuel. Additionally, the HFCEV demonstrates lower efficiency than the BEV [8,9]. Nevertheless, many manufacturers foresee a reduction in hydrogen fuel costs as cheaper production methods are explored using investments from the solar and wind renewable energy sectors [10–12]. The general structure of an HFCEV consists of six major components: a hydrogen storage tank, an air intake system, a fuel-cell system, power electronics, a battery or capacitor, and an electric motor. All these major components directly affect the overall efficiency of the vehicle. The common energy sources available for use in electric vehicles are lithium-ion batteries, supercapacitors, and hydrogen fuel cells. For creating an efficient energy system, each energy source needs to be examined separately, along with studies focusing on hybrid systems that utilize more than one energy source. Hybrid systems offer the advantage of avoiding disadvantages associated with each energy system [13–16]. In general, hydrogen-powered electric vehicles require an auxiliary energy storage system in the vehicle during energy recovery. Therefore, electric vehicles with hydrogen fuel cells require additional batteries or supercapacitors. A supercapacitor acts as a buffer against sudden charging currents, maintaining the overall efficiency, which makes it a more viable choice for use in vehicles to achieve energy recovery as compared to batteries [17,18].

In an electrical vehicle, the electrical motor is generally connected to the wheels through a single-stage transmission path. This simple and relatively economical design is called a two-wheel-drive layout. However, different powertrain architectures are also available, such as an in-wheel motor and four-wheel-drive layouts with multistage gearbox systems. The mechanical layout design is a critically important factor in assuring optimal energy consumption and recovery rates for the electric motor. It also directly determines other parameters such as the size of the electrical motor, power of the fuel cell, and tank capacity, etc. A two-wheel-drive design also allows the construction of a more efficient, lighter, cheaper, and lower-volume electric motor due to the reduction in the rotor size and stator geometries of the electric motor. For analyzing mechanical layouts based on energy consumption and recovery performance, numerous studies have been carried out [19–22]. Bottiglione et al. [23] created an energy consumption model of an electric vehicle for comparing one-stage (1G) and two-stage (2G) gear drives, half and full toroidal continuously variable transmissions (CVTs), and two different types of infinitely variable transmissions (IVTs). In the research, the energy efficiencies of the electric motor and transmission of six different transmission types of EVs were calculated. In another piece of research, Sluis et al. [24] described the advantages of using a single-loop set belt CVT as compared to using 1G, 2G automatic, and 2G double-clutch transmission based on a comparison of their values of energy consumption, cost, and performance in a downsized electric motor. Furthermore, other studies have focused on innovating shifting strategies in electric vehicles. In particular, efforts have been made to improve energy consumption and recovery to ensure a smooth drive during shifting. Shifting innovations are also used in some clutches used in transfer elements. Particularly, a double clutch enhances the power connection between a motor and powertrain during shifting [25–27].

Among the different transmission systems, the continuously variable transmission (CVT) system can increase the efficiency of a powertrain system significantly, as it can continuously change the speed ratio and allows the electric motor to operate around the most efficient regions [28–31]. Therefore, a CVT system improves the overall performance and motor efficiency of an electric vehicle, whilst also helping to downsize the electric motor and other energy sources [32–38]. Another major layout for electric vehicles is an independent in-wheel motor with/without a planetary gear. This system offers some key advantages, one of which is greater vehicle control, owing to its all-wheel design, which can be controlled with high precision and speed. Another advantage is the increase in the internal volume of a vehicle due to the removal of the transmission system, which also prevents the power loss of transmission systems [39,40]. Conversely, this type of electric motor increases the mass of the wheel group, thus increasing the unsprung mass, which in turn affects the vehicle drive. The increased mass of the wheel group reduces the fatigue strength of the elements in the suspension system [41]. Nevertheless, the aforementioned advantages of an in-wheel motor, such as regenerative electronic stability control (RESC), have opened up avenues for further exploration. To this end, research by Najjari et al. [42] proposed a strategy for managing the energy and directional stability of four in-wheel-driven electric vehicles. Another piece of research by Xu et al. [43] modeled four in-wheel hub motors by examining factors such as the regenerative braking performance and its effect on energy recovery for the front and rear wheels. In the study, although the in-wheel motor was used as a generator, its performance in influencing the braking conditions and energy consumption rate need to be further examined based on its comparison with the other types of powertrain architecture.

To date, no research has focused on drawing a comparison between the energy consumption performance of an HFCEV equipped with in-wheel-driven motors and that of an electrical motor utilizing CVT and other powertrain layouts based on RESC technology. In addition, a comprehensive comparison has not been performed using a downsized electrical motor. Thus, this study investigated the energy consumption rate and recovery performance of an autonomous HFCEV equipped with six different powertrain architectures: four independent in-wheel motors using RESC and without RESC, one-stage gear (1G) transmission, two-stage gear (2G) transmission, CVT, and a downsized electric motor utilizing CVT. In the mathematical model, the HFCEV was modeled, and simulations were performed for FTP75 and NEDC urban drive cycles. Additionally, the total energy consumption of the HFCEV with different powertrains was compared. The results demonstrate that the minimum energy consumption can be obtained with 2G transmission for the same motor. In this regard, a regenerative electronic stability control system can increase the energy recovery/ consumption rate of the HFCEV by at least 3%. Furthermore, the results illustrate that it is possible to reduce the size of the electric motor by using CVT, as the required maximum peak torque can be reduced. The simulations were repeated using a downsized electric motor, and the value for the best energy consumption was obtained owing to the light components of the system.

2. Developed Simulation Model for Different Powertrain Layouts

In this section, the details of the developed simulation model are presented. Mathematical formulations were developed based on the block diagram that is shown in Figure 1. The model included the drive cycles, an RESC controller, dynamic equations for the vehicle, transmission models, electrical motor maps, and formulations of the subsystems. It should be noted that the RESC controller was only available in the in-wheel motor driven powertrain architecture, and that the developed model was found to be suitable for all the different types of powertrain architectures used for HFCEVs.

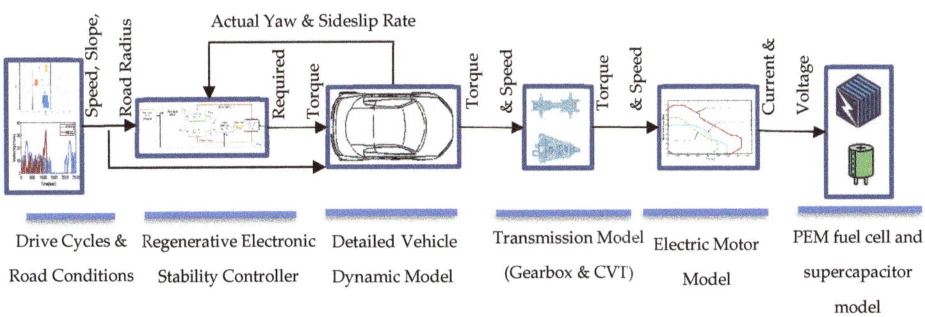

Figure 1. Block diagram of the developed model.

Based on Figure 1, firstly, the necessary powers and torques for the FTP75 and NEDC drive cycles were calculated, and then, these values were used in the electric motor model. For all types of electric motors, the efficiency was determined from real-life working conditions, and the current and voltage were calculated based on the necessary torque with respect to the angular velocity. After defining the current and voltage of the electrical motor, the energy consumption and recovery were determined based on the proton exchange membrane PEM fuel cell and supercapacitor model. At the end of the calculation, the values for the other parameters, such as the energy consumption, energy recovery, and operating points on the efficiency maps, were obtained for all the types of powertrain layouts for HFCEVs.

In Figure 2, different powertrain layouts and an energy supply diagram for the HFCEV are presented for drawing comparisons between their energy consumption and recovery rates. As shown in Figure 2a, there are four main layouts. The first type of vehicle is equipped with four permanent magnets synchronous with the in-wheel electric motor. In the second, third, and fourth models, there are two main central electric motors, and they are coupled with different types of transmission systems: one-stage (1G) and two-stage gear ratios (2G), continuously variable transmission (CVT), and downsized electric motors with CVT. The speed ratios of the one-stage and two-stage automatic gearboxes were defined by considering the optimum operating conditions of the electric motor with regard to the efficiency map of an electric motor (in the Appendix A). Additionally, the efficiency values of the transmission, system were also considered. Furthermore, an energy supply diagram for the HFCEV is illustrated in Figure 2b. The main energy source of the vehicle is defined as a PEM fuel cell, and its output directly connects to 255 V DC–DC boost converters. A supercapacitor is an auxiliary energy source and serves as the energy recovery unit of the vehicle.

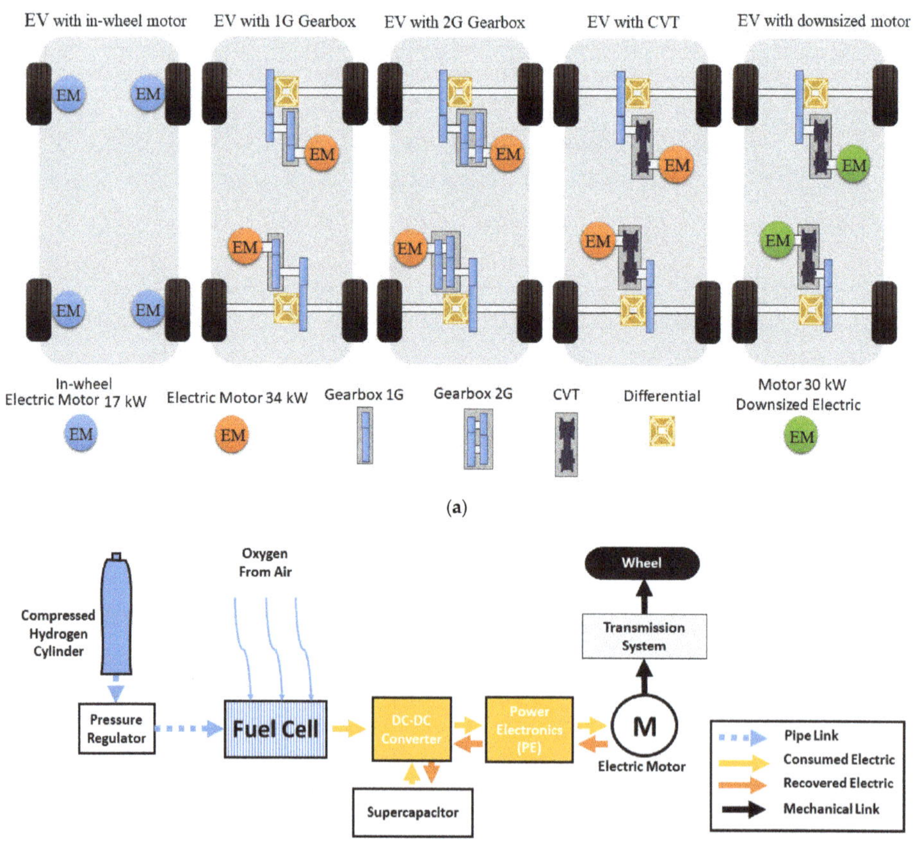

Figure 2. (**a**) Vehicle architectures electric vehicles (EV); (**b**) energy supply diagram of hydrogen fuel-cell electric vehicle (HFCEV).

The case study vehicle was an urban autonomous passenger car. As shown in Figure 3, the selected vehicle model was shown considering its steering angle, suspension, calculations of resistance forces F_l (aerodynamic drag, tire rolling resistance, and slope), and efficiencies of the mechanical and electrical components.

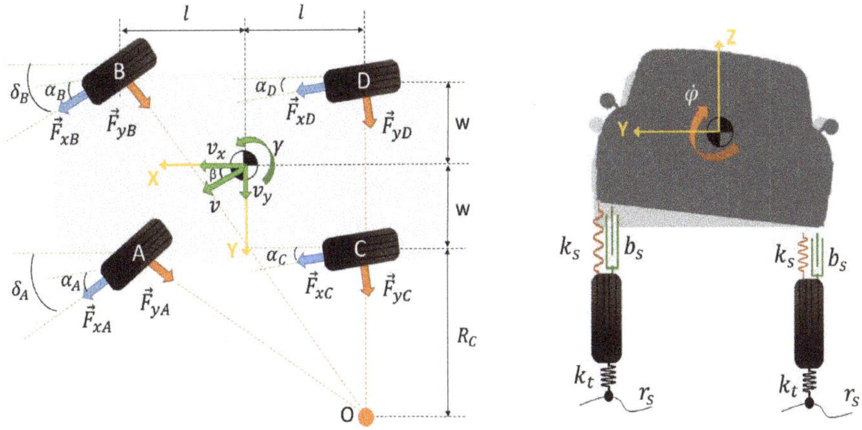

(**a**) (**b**)

Figure 3. Vehicle dynamic model: (**a**) Top view; (**b**) Back view.

The sum of the load forces that were used are given as follows:

$$F_l = \rho c_d A_f \frac{v^2}{2} + m_t g \sin(\beta) + m_t g \left(f_0 + f_1 v^2\right) \tag{1}$$

Here,v, m_t, β, A_f, c_d, and ρ indicate the velocity of the vehicle, the total mass of the vehicle, the slope angle of the road, the frontal area, the drag coefficient, and the air density, respectively. Note that, total mass of the vehicle is sum of the vehicle mass without transmission layout m_v given in Table 1. f_0 and f_1 are the coefficients of the rolling resistance constants with values that equal 0.015 and 0.0067, respectively.

Table 1. Specifications of the selected autonomous electric vehicle.

Drag coefficient	c_d	0.32	Half length of vehicle	l	1.413 [m]
Frontal area	A_f	1.4 [m^2]	Half width of vehicle	w	0.652 [m]
Vehicle mass inertia at x axis	I_{xx}	750 [kg/m^2]	Wheel radius	R_w	0.325 [m]
Rolling stiffness coefficient	K_{roll}	85,450 [Nm/rad]	Air density	ρ	1.225 [kg/m^2]
Height of the rolling axis	h_{roll}	0.27 [m]	Rolling damping coefficient	C_{roll}	4542 [Nms/rad]
Vehicle mass (without powertrain and motor)	m_v	2300 [kg]	Vehicle mass inertia at z axis	I_{zz}	2707 [kg/m^2]

In this study, an autonomous vehicle was chosen to subtract the consideration of driver mistakes and steering response delays in the vehicle. Autonomous vehicles can be used to calculate the curvature of the road in different ways. The referenced vehicle detects the lanes using the image processing sliding window method. This method calculates the radius of the road, R_r, given in Equation (2), with respect to the referenced double-lane-change maneuver. The purpose of this maneuver is to achieve a lane change by varying the steering angle in one direction (left) and then the opposite direction, simultaneously, as the vehicle encounters an obstacle.

$$R_r = \frac{\left[1 + \left(\frac{dy}{dx}\right)^2\right]^{3/2}}{\left|\frac{d^2y}{dx^2}\right|} \tag{2}$$

Here, x and y are coordinates of the roads when looking from top view. The steering angles of the left and right front wheels, δ_A and δ_B, respectively, are represented based on the well-known Ackermann principle for left turning as follows:

$$\delta_A = \arctan\left(\frac{2l}{\sqrt{R_r{}^2 - l^2} - w}\right) \tag{3}$$

$$\delta_B = \arctan\left(\frac{2l}{\sqrt{R_r{}^2 - l^2} + w}\right) \tag{4}$$

where l and w indicate the half length and half width of the vehicle, respectively.

The longitudinal λ_A and lateral slips α_A were determined for front left tire A and for the others were calculated using similar equations. During driving situations, the longitudinal and lateral slip angles for the left front tire were calculated as follows:

$$\lambda_A = \begin{cases} \frac{V_x - R_w\omega_A}{V_x}, & brake\ situation \\ \frac{-V_x + R_w\omega_A}{V_x}, & drive\ situation \end{cases} \tag{5}$$

$$\alpha_A = \delta_A - \arctan\left(\frac{V_y + \gamma l}{V_x - \gamma w}\right) \tag{6}$$

In the above equations, V_x and V_y represent the longitudinal and lateral velocities, R_w denotes the radius of the wheel, ω_A indicates the angular speed of the front left wheel, and γ represents the yaw rate of the vehicle. In addition, for the calculation of the lateral and longitudinal forces, denoted as F_y and F_x respectively, the Pacjeka model was selected. The Magic Formula of Pacjeka is presented in Equations (7) and (8), respectively:

$$F_{yA}(\lambda_A, \alpha_A, \mu_A, F_{zA}) = \frac{\sigma_{yA}}{\sigma_{tA}} F_{yA0}(\alpha_A, \mu_A, F_{zA}) \tag{7}$$

$$F_{xA}(\lambda_A, \alpha_A, \mu_A, F_{zA}) = \frac{\sigma_{xA}}{\sigma_{tA}} F_{xA0}(\alpha_A, \mu_A, F_{zA}) \tag{8}$$

All the parameters of these two equations $(\lambda_A, \alpha_A, \mu_A, F_{zA})$ are explained in the Appendix A. The equations of the motion of the vehicle are listed as follows for the longitudinal, lateral, yaw, and roll motions, respectively:

$$\dot{V}_x = V_y\gamma + \frac{(F_{xA}cos\delta_A + F_{xB}cos\delta_B + F_{xC} + F_{xD})}{m_t} - \frac{F_l + F_{yA}sin\delta_A + F_{yB}sin\delta_B}{m_t} - h_{roll}\gamma\dot{\varphi} \tag{9}$$

$$\dot{V}_y = -V_x\gamma + \frac{(F_{yA}cos\delta_A + F_{yB}cos\delta_B + F_{yC} + F_{yD}) + F_{xA}sin\delta_A + F_{xB}sin\delta_B}{m_t} - h_{roll}\ddot{\varphi}\cos(\varphi) \tag{10}$$

$$\begin{aligned}\dot{\gamma} = \tfrac{1}{I_{zz}}[&l(F_{xA}sin\delta_A + F_{yA}cos\delta_A + F_{xB}sin\delta_B + F_{yB}cos\delta_B - F_{yD} - F_{yC}) + \\ &w(-F_{xA}cos\delta_A + F_{yA}sin\delta_A + F_{xB}cos\delta_B - F_{yB}sin\delta_B + F_{xD} - F_{xC})]\end{aligned} \tag{11}$$

$$\ddot{\varphi} = \frac{1}{I_{xx}}\left[m_s h_{roll}\left(\dot{V}_x + gsin\varphi\right) - K_{roll}\varphi - C_{roll}\dot{\varphi}\right] \tag{12}$$

Here, F_{xA}, F_{xB}, F_{xC}, and F_{xD} denote the longitudinal force of each tire; F_{yA}, F_{yB}, F_{yC}, and F_{yD} are the lateral force of each tire, while I_{zz} and I_{xx} represent the mass inertias for the z- and x-axis, respectively. The term φ represents the rolling angle, and h_{roll} is the height of the rolling axis. m_s is sprung mass and it is found subtract of unsprung mass m_u (41 kg for in wheel-motor, 23 kg for others) from total mass m_t.

3. Regenerative Electronic Strategy and Stability Control Strategy

The main objective of the yaw rate controller design was to compare the vehicle yaw rate and other stability-related states with the desired states obtained using drive commands of the autonomous vehicles. The desired state parameters -yaw rate and side slip angle- were calculated using a two-DoF vehicle single-track model. The two DoF equations of motion are expressed below.

$$\dot{v}_{y_d} = \frac{1}{m_t}[\left(C_f + C_r\right)\frac{v_{y_d}}{v_{x_d}} + \left(lC_f - lC_r - m_t v_{x_d}^2\right)\frac{\gamma_d}{v_{x_d}} - C_f\delta] \quad (13)$$

$$\dot{\gamma}_d = \frac{1}{I_{zz}}[\left(2lC_f - 2lC_r\right)\frac{v_{y_d}}{v_{x_d}} + \left((2l)^2C_f + (2l)^2C_r\right)\frac{\gamma_d}{v_{x_d}} - 2lC_f\delta\,] \quad (14)$$

Here, γ_d desired yaw rate, $\dot{v}_{y_d}$ desired lateral acceleration, $\dot{v}_{x_d}$ desired longitudinal acceleration, C_f and C_r indicate the lateral stiffnesses of the front and rear wheels given as in Table A1, respectively, δ denotes the steering angle of the front tire. Figure 4 shows the yaw rate controller scheme. In this scheme, input parameters such as the vehicle speed, steering angle, and brake situation are defined from the autonomous vehicle with reference to the drive cycle and road profile. The referenced model determined the desired yaw rate and slip angle based on the outlined input parameters. A PID controller calculated the desired moment around the z-axis, $M_{z,des}$, with respect to the vehicle reference coordinate using the error value between the actual yaw rate, which was calculated using the detailed vehicle model, and the desired yaw rate. Additionally, the desired torque value, T_{si}, for each tire was calculated using the value from the wheel slip controller. These value input parameters of the torque distribution algorithm produced values that were the same as those obtained using the desired wheel slip torque, which was calculated from the wheel slip PID controller and drive force at the x-axis defined from the vehicle based on the drive cycle. The torque distribution algorithm requires the input of the torque value for the control of the yaw rate of the vehicle. A parallel PID structure was chosen, as the parallel form allows the complete decoupling of proportional, integral, and derivative actions and also decreases the time for finding the optimum PID coefficients.

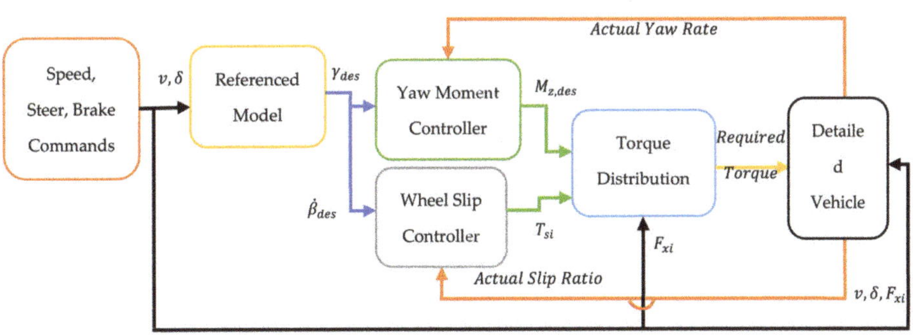

Figure 4. Yaw rate controller scheme.

In electric vehicles, three different regenerative braking strategies are used: series braking with an optimal feel, series braking with optimal energy recovery, and parallel braking [44]. It should be noted that the regenerative braking strategy of HFCEVs has to meet two main requirements. Firstly, the total required braking force has to be distributed and the kinetic energy of the vehicle has to be recovered as much as possible. Secondly, the total requiring braking force on the rear and front axles needs to be distributed to achieve steady-state braking performance. Based on these requirements, in this study, the method of series braking with optimal energy recovery was chosen as in [44]. The main working principle of this strategy is based on this approach: when the vehicle is braked with a deceleration rate which is smaller than road friction, the braking force can be varied within a certain range. In this scenario, the sum of the rear and front braking forces equals the inertia force of the vehicle, and it is assumed that this principle is satisfied in the referenced FTP75 and NEDC drive cycles.

4. Model of Electric Motors and Concurrent Efficiency Maps

All the selected electric motors were a DC permanent magnet (PM) type, and their characteristics are expressed in the Appendix A section. For the in-wheel electric motor architecture, each of the in-wheel motor powers equaled 17 kW, and the total power of the vehicle was determined as 68 kW. On the other hand, for the architectures of 1G, 2G, and CVT, the motors had a separate power of 34 kW, and the total powers of the vehicles were the same for all the architectures. However, the power of the downsized electric motor was 30 kW, and the total power of the vehicle was 60 kW, less than that of the other architectures. The equations of the consumed and recovered currents are expressed in Equation (15), and the energy consumption of the fuel cells was activated during the driving case, while the state of charge of the supercapacitor was run during braking conditions.

$$I = \begin{cases} \frac{T\omega_{EM}}{uV_C}, & driving\ situation \\ \frac{uT\omega_{EM}}{V_C}, & brake\ situation \end{cases} \quad (15)$$

Here, the term I represents the consumed and recovered current, T is the motor shaft torque, ω_{EM} represents the angular speed of the electric motor, u denotes the efficiency of the electric motor with respect to the torque and angular speed, and V_C indicates the boost converter voltage. The efficiency of the electric motor is expressed as follows:

$$\mu = 100\frac{T\omega_{EM}}{T\omega_{EM} + P_0 + k_e T^2 + k_w {\omega_{EM}}^2} \quad (16)$$

where P_0 is the fixed power loss, k_e represents the torque-dependent electrical loss coefficient, and k_w is the speed-dependent iron loss coefficient.

$$T_{peak} = k_T D_{EM}^2 L_{EM} \quad (17)$$

The peak torque value can be found using Equation (17), where k_T is described as a torque constant, D_{EM} is the diameter of the motor, and L_{EM} denotes the length of the motor.

Note that, the motor shaft torque T and angular speed ω_{EM} are exposed to gear ratios and their efficiencies given in Table 2. The efficiency map of CVT is adopted from [24].

Table 2. The specifications of the powertrain layouts.

Differential Ratio	τ_{diff}	1:4
Differential efficiency	η_{diff}	0.9
Gear ratio of 1G	τ_{1G}	1:3.625
Gear efficiency of 1G	τ_{1G}	0.92
First gear ratio of 2G	τ_{2G1}	1:2.724
First gear efficiency of 2G	η_{2G1}	0.92
Second gear ratio	τ_{2G2}	1:1.251
Second gear efficiency of 2G	η_{2G2}	0.92
Minimum CVT speed ratio	$\tau_{CVT,min}$	0.4
Maximum CVT speed ratio	$\tau_{CVT,max}$	2.5
Maximum CVT efficiency	$\eta_{CVT,max}$	0.96
Minimum CVT efficiency	$\eta_{CVT,min}$	0.89

5. PEM Fuel Cell and Supercapacitor Model

The referenced HFCEV had a 68 kW PEM fuel cell (PFC) and 7 kWh maximum stored energy supercapacitor as the two main energy sources. PEM fuel was considered the main energy source for the calculation of the energy consumption and supercapacitor for the energy recovery of the HFCEV. The characteristics of the fuel cell are presented in the Appendix A. The stack of PFCs had 255 cells. Each cell considered had the same efficiency. Equation (18) represents the Nernst equation for the voltage of one cell without energy losses.

$$V_{Nernst} = V_0 + \frac{RT_{FC}}{2F} \ln\left(\frac{P_{H_2}{P_{O_2}}^{0.5}}{P_{H_2O}}\right) \tag{18}$$

Here, V_0 denotes an open-circuit voltage with a value equal to 1.2 V; T_{FC} indicates the temperature of the PFC with a constant value of 50 °C. R is the universal gas constant; F is the Faraday constant; P_{H_2}, P_{O_2}, and P_{H_2O} are the pressures of hydrogen gas, oxygen, and water vapor, respectively. The operating pressure of the fuel-cell system was set at a constant 1.5 bar. Equations (19)–(21) illustrate the voltage losses of the fuel cell in the order of the activation losses, V_{act}; concentration losses, V_{con}; and ohmic losses, V_{ohm}, respectively. To simplify, the activation losses and concentration losses have been defined in the form of an empirical equation as a function of the current and temperature.

$$V_{act} = \frac{RT_{FC}}{n\alpha F} \ln(\frac{I}{I_0}) \tag{19}$$

$$V_{con} = 1.1 \times 10^{-4} - 12 \times 10^{-6}(T_{FC} - 273)e^{(0.008)I} \tag{20}$$

$$V_{\text{ohm}} = \frac{t_m}{\sigma} I \tag{21}$$

where I_0 represents the referenced current, α is the transfer coefficient, t_m denotes the membrane thickness with a value equal to 0.02 mm, and σ denotes the electrical conductivity. The total voltage of the PFC V_{FC} is presented in Equation (22) considering the voltage losses. Here, N represents the stack number, and the fuel-cell unit had 255 cells.

$$V_{FC} = [V_{Nernst} - V_{act} - V_{con} - V_{ohm}]N \tag{22}$$

The fuel consumption of the PEM fuel cell, HC, was calculated using the linear function in Equation (23) based on the characteristics of the PFC.

$$HC = k_{FC} V_{FC} I \tag{23}$$

Here, k_{FC} is the fuel consumption rate with respect to the power of the fuel cell, with a value equal to 0.01. By using Equation (24), it is possible to calculate the value of the charge voltage, V_{sp}, of the supercapacitor with respect to the recovered current during braking.

$$\frac{dV_{sp}}{dt} = \frac{1}{C}\frac{dQ}{dt} = \frac{1}{C_{sp}}I \tag{24}$$

Here, C_{sp} represents the rated capacitance. The state of charge (SoC) rate of the supercapacitor was used for drawing a comparison of the energy recovery rates of the different architectures, according to Equation (25).

$$SoC = \%\frac{{V_{sp}}^2}{{V_{sp,maks}}^2}100 \tag{25}$$

6. Results

In this section, the simulation results for the 14-degree-of-freedom (DoF) autonomous HFCEV model based on the regenerative electronic stability control strategy are demonstrated. To this end, the comparisons of the energy consumption and recovery performance of the six powertrain layouts, and the operation points of each electrical motor regarding the efficiency maps are illustrated. In addition, the vehicle handling performance of the RESC for four in-wheel electric motor architectures is presented. The characteristic profiles of the velocity, acceleration/deceleration, and power requirements for the FTP75 and NEDC drive cycles are shown in Figure 5.

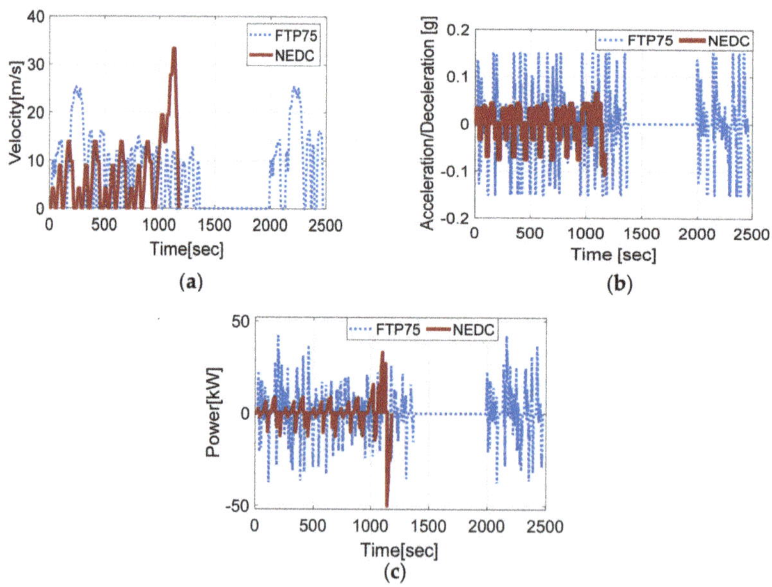

Figure 5. Characteristic profiles of the velocity (**a**), acceleration/deceleration (**b**), and power requirements (**c**) of the FTP75 and NEDC.

Simulations of the RESC system were performed using 18 and 11 km-long double-lane-change maneuver paths for the FTP75 and NEDC urban drive cycles, respectively. Figure 6a shows the yaw rate performance, considering the double-lane-change maneuver, for comparing the RESC performance with the desired yaw rate and uncontrolled HFCEV. In addition, Figure 6b shows the sideslip angle performance with respect to the same maneuver for in-wheel layout with RESC. The obtained yaw rate with the RESC was within 7% of the desired value, while it is evident that the vehicle sideslip angle can be reduced. It can be concluded from this figure that the performance of the autonomous vehicle is satisfactory and the yaw rate is within 12% of the findings in [45]. After creating the illustrations of the RESC steering performance, the total and instant energy recovery performance of the rear in-wheel motors was calculated based on the double-lane-change maneuver using the RESC control strategy, as shown in Figure 7.

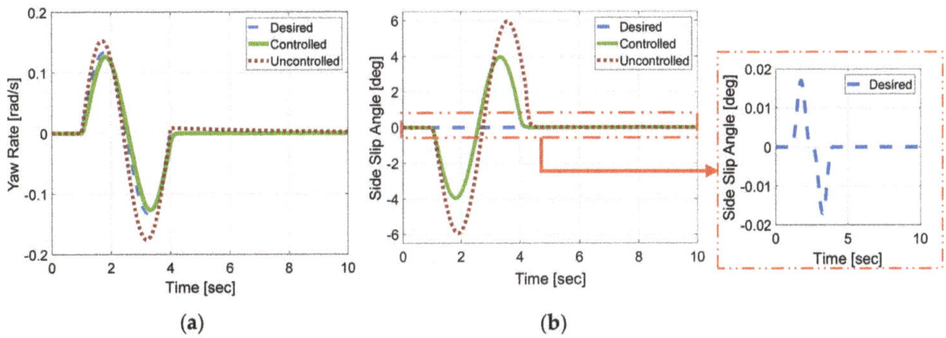

Figure 6. Comparison of regenerative electronic stability control (RESC), (**a**) with respect to yaw rates and (**b**) from sideslip angle for in-wheel layout with RESC.

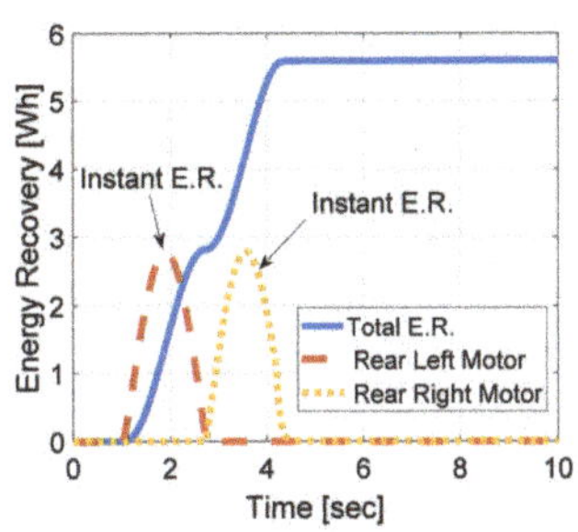

Figure 7. Total and instant energy recovery during regenerative stability control.

In order to determine the operation points of the electric motors in the efficiency map, the simulation results were plotted for FTP75 and NEDC, and these are illustrated in Figure 8. While the motor mode working points yielded approximately the same findings with and without RESC, different operation points were obtained for the generator mode due to the regenerative braking effect.

In Figure 9, the comparisons of the operation points of the electric motors for one-stage (1G), two-stage (2G), and CVT transmissions are presented. As shown in Figure 9a,b, the operation points shifted to the more efficient region, and the required torque was decreased significantly by using a transmission system. This achievement was due to a more flexible speed ratio, which allowed vehicles to run the electrical motors in more efficient operating regions.

Moreover, the mass, volume, and cost of the electric motors can also be reduced due to the decrease in the required peak torque. If the size of the electric motor is decreased, many more advantages can also be achieved, such as lower power losses and energy consumption, less cells in fuel cell stack, and lower total masses of the vehicles. Thus, in the study, the masses of the powertrain architecture were taken into account. A mass comparison between the vehicle architectures is given in Figure 10. It can be seen from this figure that the in-wheel motor layout has the lowest vehicle mass in comparison to the other designs; however, this architecture requires four independent motors, which increase the overall cost. On the other hand, the downsized electric motor layouts decrease the number of motors used and power electronic masses as well. Thus, the downsized layout is lighter than the two-stage transmission layout. Note that, two downsized electric motors were used and each of them was saved almost 10 kg mass. Additionally, since the power requirement of the fuel cell is decreased by using a downsized electric motor, it is also possible to have a lighter weight for the fuel cell. The simulations were repeated using the downsized electric motor, and the findings are illustrated in Figure 11.

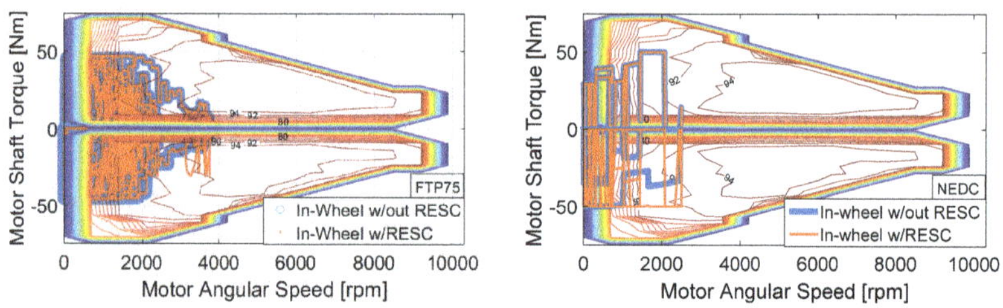

(**a**) (**b**)

Figure 8. In-wheel motor with and without regenerative electronic stability control (RESC) for (**a**) the FTP75 and (**b**) NEDC.

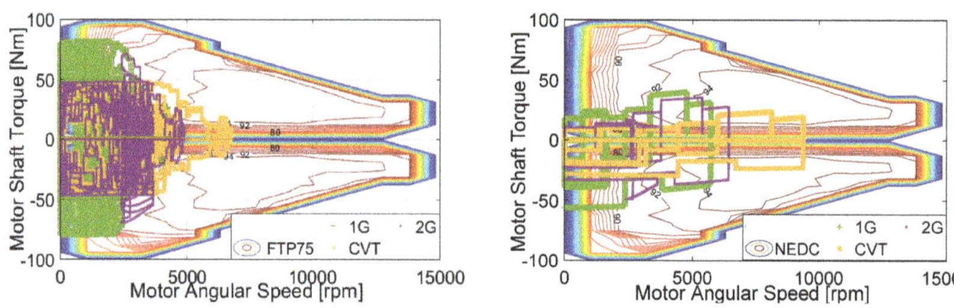

(**a**) (**b**)

Figure 9. Operation points of the EV equipped with one-stage (1G), two-stage (2G), and CVT transmission for (**a**) the FTP75 and (**b**) NEDC.

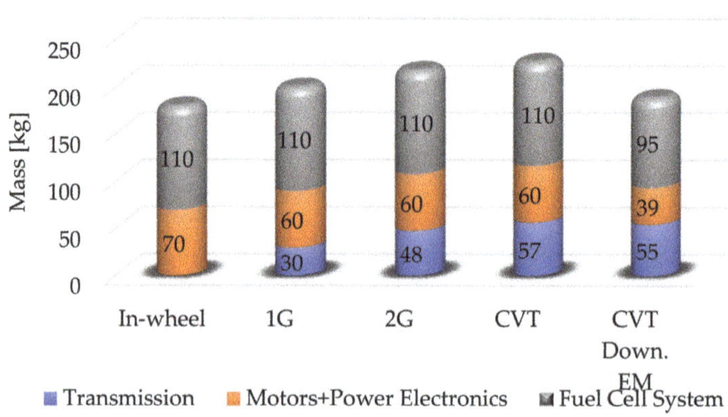

Figure 10. Comparison of the main subsystem masses.

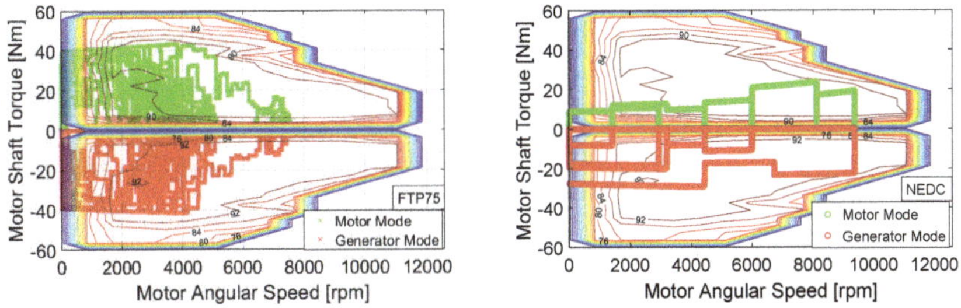

(a) (b)

Figure 11. Downsized CVT operation points for (**a**) the FTP75 and (**b**) NEDC.

Figure 12 illustrates the energy consumption performance of all six different types of EV for the NEDC and FTP75 drive cycles. It is evident that the consumption of in-wheel motors with RESC is less compared to that for the same construction without RESC; this is because energy is recovered in the RESC mode by the in-wheel motor. Moreover, the hydrogen fuel consumption rates with the CVT are better than with the one-stage and worse than with the two-stage gearbox, because a little extra energy is needed to operate the CVT. However, it should be noted that the downsized electric motor architecture shows the lowest consumption rates out of all the layouts owing to the CVT transmission, which allows the motor to work at the most efficient points. The state of charge (SoC) rates of the supercapacitor with respect to the NEDC and FTP75 drive cycles were also compared, and the results are presented in Figure 13. Based on these results, the downsized electric motor architecture recovers more energy than the other constructions.

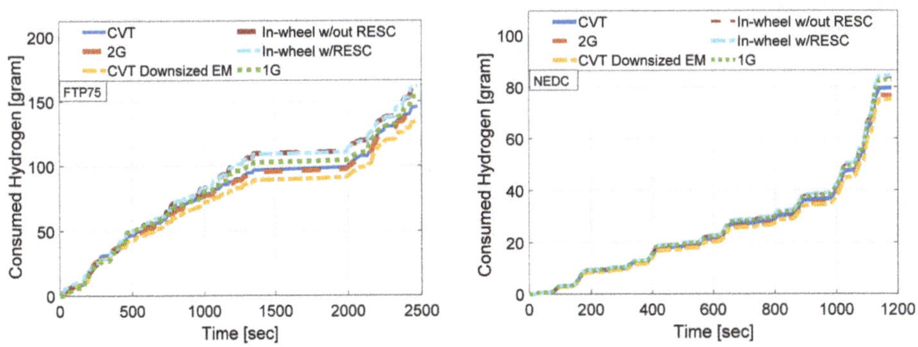

(a) (b)

Figure 12. Comparison of the consumed hydrogen: (**a**) FTP75 drive cycle; (**b**) NEDC drive cycle.

The energy consumption and recovery rates per 100 km for all the different transmission units were also calculated, and these results are presented in Figures 14 and 15, respectively. It can be seen that the downsized electric motor using a CVT yielded the lowest energy consumption rate and highest energy recovery rate. This is because, as emphasized earlier, the CVT allows the electric motor to operate in a more efficient zone. Additionally, it is noteworthy that an in-wheel motor architecture produces no transmission losses and that using an RESC system can increase the energy recovery percentage significantly.

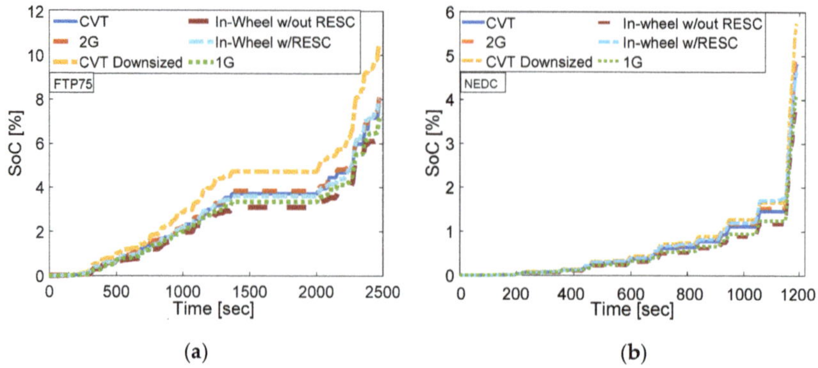

Figure 13. Comparison of energy recovery: (**a**) FTP75 drive cycle; (**b**) NEDC drive cycle.

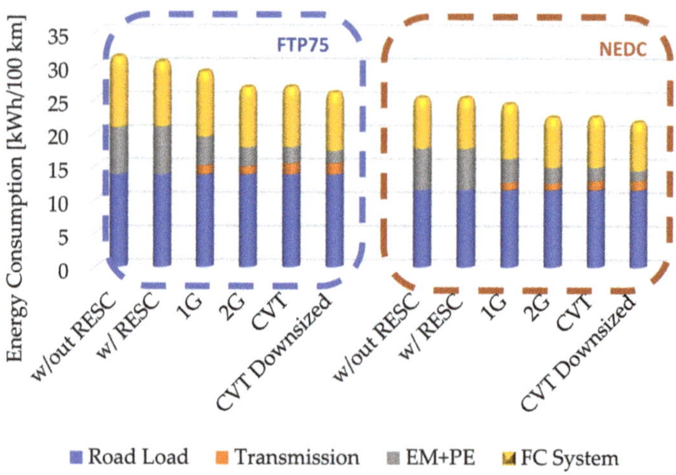

Figure 14. Energy consumption rates per 100 km for FTP75 and NEDC.

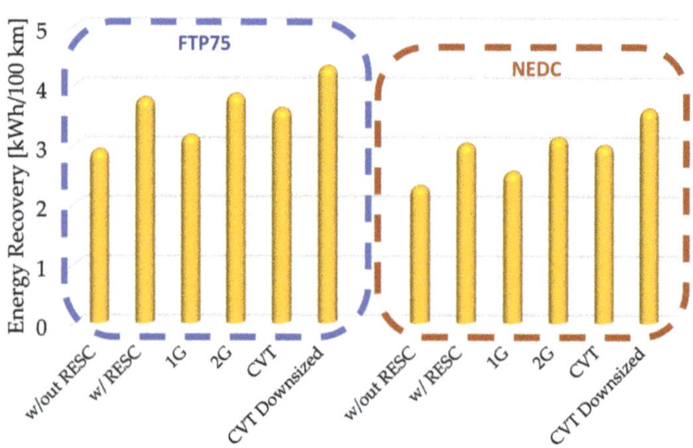

Figure 15. Energy recovery rates per 100 km for FTP75 and NEDC.

Figure 16 illustrates the energy recovery/consumption rates, which are a valuable consideration in designing an efficient vehicle layout for each powertrain architecture. It is evident from this figure that the level of the energy recovery/consumption rate can significantly increase when using a downsized CVT. Additionally, it is apparent that the energy recovery/consumption rates increased with the use of an RESC system within an in-wheel-driven motor layout.

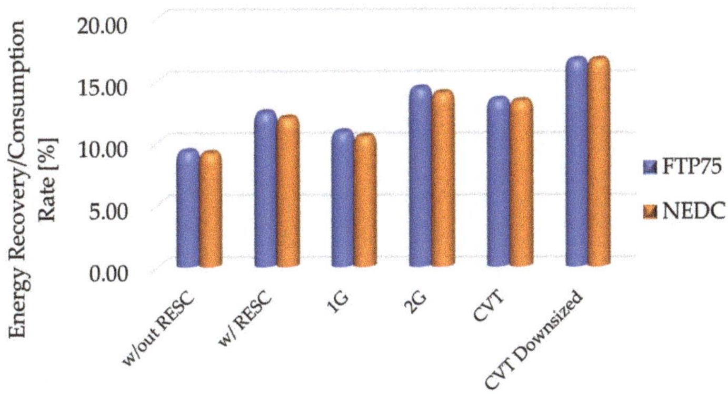

Figure 16. Comparison of energy recovery/consumption rates.

7. Conclusions

In this study, the energy consumption and recovery performance of six different powertrain architectures, including transmission and nontransmission layouts, were analyzed for HFCEVs. The stability control performance of RESC was tested with respect to the desired yaw rate and sideslip angle of the reference model, and the results were satisfactory. The conclusions of the study can be summarized as follows:

- A transmission system can reduce the energy consumption of HFCEVs tremendously as compared to other powertrain layouts, as it provides the optimum operating conditions for electric motors. If high torque capacity is required at low velocities of the vehicle, an automated transmission system must be used for improved energy consumption.
- An in-wheel motor layout is able to recover more energy if an RESC system is added to it. Overall, it spends less energy, except in the case of a downsized electric motor and two-stage automatic transmission. In addition, if the vehicle is used on a curved road, the in-wheel architecture with an RESC system can save a significant amount of energy.
- Using a CVT transmission, the required peak torque of the electric motor can be decreased; therefore, it is possible to downsize the electric motor. Downsizing helps to achieve a reduction in the weight of not only the motor but also its driver and other power units, which results in the lowest energy consumption rate as compared to the other layouts.

The simulation results can be used for designing different types of vehicles, such as electric buses and trucks, as the findings will be easy to implement. Moreover, the simulation results can be further tested in future studies for exploring more innovations.

Author Contributions: Conceptualization, A.Y.; methodology, A.Y.; software, M.A.Ö.; validation, A.Y. and M.A.Ö.; formal analysis, A.Y. and M.A.Ö.; investigation, A.Y. and M.A.Ö.; resources, A.Y. and M.A.Ö.; data curation, A.Y. and M.A.Ö.; writing—original draft preparation, A.Y. and M.A.Ö.; writing—review and editing, A.Y.; visualization, A.Y. and M.A.Ö.; supervision, A.Y.; project administration, A.Y. and M.A.Ö.; funding acquisition, A.Y. and M.A.Ö. Both authors have read and agreed to the published version of the manuscript.

Funding: This research received no external funding.

Conflicts of Interest: The authors declare no conflict of interest.

Appendix A

Appendix A.1. Parameter of Magic Formula for a Lateral and Longitudinal Slip

$$\sigma_{xA} = \frac{\lambda_A}{1+\lambda_A},\ \sigma_{yA} = \frac{tan\alpha_A}{1+\lambda_A},\ \sigma_{tA} = \sqrt{\sigma_{xA}^2 + \sigma_{yA}^2}$$
$$F_{xA}(\alpha_A, \mu_A, F_{zA}) = D^* \sin[C_x^* \arctan(B_x^*(1-E)\lambda_i + E_x \arctan(B_x^* \lambda_A)]$$
$$F_{yA}(\alpha_A, \mu_A, F_{zA}) = D^* \sin[C_y^* \arctan\left(B_y^*(1-E)(\alpha_A + \Delta S_{hA}) + E_y \arctan\left(B_y^* \alpha_A\right)\right] + \Delta s_{vA}$$
$$B_x^* = B_x(2-\mu_A),\ C_x = 1.65,\ E_x = \frac{E}{B_x}$$
$$B_y^* = B_y(2-\mu_A),\ C_y = 1.65,\ E_y = \frac{E}{B_y}$$
$$D^* = \mu_A\left(a_1 F_{zA}^2\right) + a_2 F_{zA},\ B_x = \left(\frac{a_3 F_{zA}^2 + a_4 F_{zA}}{C_x D^* e^{a_5 F_{zA}}}\right),\ B_y = (1 - a_{12}|\xi|)\left(\frac{a_3 \sin(a_4 \arctan(a_5 F_{zA}))}{C_y^* D^*}\right)$$
$$E = a_6 F_{zA}^2 + a_7 F_{zA} + a_8$$
$$\Delta S_{hA} = \left(a_{10} F_{zA}^2 + a_{10} F_{zA}\right)\xi$$
$$\Delta S_{vA} = a_9 \xi$$

Here, B^*, C^*, D^*, and E denote the tire stiffness, shape, peak, and curvature factors, respectively, whereas ξ represents the wheel camber angle. In addition, ΔS_h and ΔS_v indicate the horizontal and vertical shifts, and μ signifies the road coefficient of friction. Equations are determined for front left tire and for others can calculated similar way.

Using Newton's law, the mathematical equations was derived for the vertical motions of suspension system as indicated below:

$$m_s \ddot{s}_{z,i} = \Delta N_{f,i} - b_{s,i}(\dot{s}_{z,i} - \dot{u}_{z,i}) - k_{s,i}(s_{z,i} - u_{z,i})$$
$$m_u \ddot{u}_{z,i} = b_{s,i}(\dot{s}_{z,i} - \dot{u}_{z,i}) + k_{s,i}(s_{z,i} - u_{z,i}) - k_{t,i}(u_{z,i} - r_{s,i})$$

where m_s and m_u denote the quarter of the sprung and unsprung masses of the vehicle, respectively; $s_{z,i}$, $u_{z,i}$ and $r_{s,i}$ show the vertical displacements of sprung mass, unsprung mass and road disturbance for each tire, respectively. $k_{s,i}$, $b_{s,i}$ and $k_{t,i}$ represent the stiffness of spring, damping coefficient and tire stiffness, respectively. Dynamic forces for each tires, $\Delta N_{f,i}$, which occurs during acceleration and deceleration situations is presented as follow:

$$\Delta N_{f,i} = m_s g \mp \frac{m_t \dot{V}_x h_g}{4l} \mp \frac{m_t \dot{V}_y h_g}{4w}$$

where $\dot{V}_x$ and $\dot{V}_y$ represent the longitudinal and lateral accelerations, and h_g, l and w indicate the height of center of gravity, half length and half width of the vehicle, respecively. Sign i is indicated the related tire; A is left front, B is right front, C is left rear and D is right rear tire.

Table A1. The values of the vehicle parameters.

Hydrogen Pressure	P_{H_2}	**1.5 [bar]**	**Oxygen Pressure**	P_{O_2}	**1.5 [bar]**
Front tire lateral stiffness	C_f	41,000 [N/rad]	Rear tire lateral stiffness	C_r	40,000 [N/rad]
Damping coefficient	b_s	3500 [Ns/m]	Stiffness of tire	k_t	193,000 [N/m]
Stiffness of spring	k_s	35,000 [N/m]	Fixed power loss	P_0	0.9 [kW]

Appendix A.2. Efficiency Map for the Electric Motors

Figure A1a presents the efficiency map for the electric motor for 1G, 2G, and CVT. Figure A1b illustrates the efficiency map for the in-wheel electric motor for the in-wheel motor architecture with RESC and without RESC. In addition, Figure A1c provides the efficiency map for the downsized electric motor for CVT. Finally, the comparison of the power cycles of the downsized electric motors is depicted in Figure A1d.

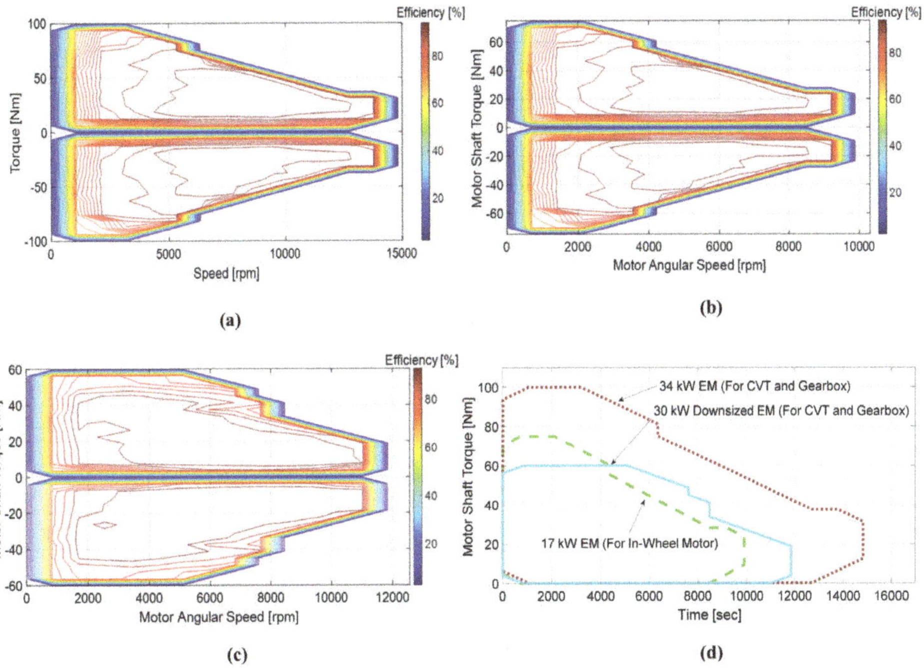

Figure A1. (**a**) Efficiency of electric motor for 1G, 2G, and CVT; (**b**) Efficiency of in-wheel motor with RESC and without RESC; (**c**) Efficiency map for downsized electric motor for CVT; (**d**) Power cycle comparison for the used electric motors.

References

1. Kim, I.; Kim, J.; Lee, J. Dynamic analysis of well-to-wheel electric and hydrogen vehicles greenhouse gas emissions: Focusing on consumer preferences and power mix changes in South Korea. *Appl. Energy* **2020**, *260*, 114281. [CrossRef]
2. Wang, Q.; Xue, M.; Lin, B.; Lei, Z.; Zhang, Z. Well to wheel analysis of energy consumption, greenhouse gas and air pollutants emissions of hydrogen fuel cell vehicle in China. *J. Clean. Prod.* **2020**, *275*, 123061. [CrossRef]
3. Lathia, R.; Dobariya, K.; Patel, A. Hydrogen Fuel Cells for Road Vehicles. *J. Clean. Prod.* **2017**, *141*, 462. [CrossRef]
4. Ahmadi, P. Environmental impacts and behavioral drivers of deep decarbonization for transportation through electric vehicles. *J. Clean. Prod.* **2019**, *225*, 1209–1219. [CrossRef]
5. Xiong, H.; Liu, H.; Zhang, R.; Yu, L.; Zong, Z.; Zhang, M.; Li, Z. An energy matching method for battery electric vehicle and hydrogen fuel cell vehicle based on source energy consumption rate. *Int. J. Hydrogen Energy* **2019**, *44–56*, 29733–29742. [CrossRef]
6. Bartolozzi, I.; Rizzi, F.; Frey, M. Comparison between hydrogen and electric vehicles by life cycle assessment: A case study in Tuscany, Italy. *Appl. Energy* **2013**, *101*, 103–111. [CrossRef]
7. Grüger, F.; Dylewski, L.; Robinius, M.; Stolten, D. Carsharing with fuel cell vehicles: Sizing hydrogen refueling stations based on refueling behavior. *Appl. Energy* **2018**, *228*, 1540–1549. [CrossRef]
8. Wen, C.; Rogie, B.; Kaern, M.R.; Rothuizen, E. A first study of the potential of integrating an ejector in hydrogen fuelling stations for fuelling high pressure hydrogen vehicles. *Appl. Energy* **2020**, *260*, 113958. [CrossRef]
9. Cao, S.; Alanne, K. The techno-economic analysis of a hybrid zero-emission building system integrated with a commercial-scale zero-emission hydrogen vehicle. *Appl. Energy* **2018**, *211*, 639–661. [CrossRef]
10. Nagasawa, K.; Davidson, F.T.; Lloyd AC: Webber, M.E. Impacts of renewable hydrogen production from wind energy in electricity markets on potential hydrogen demand for light-duty vehicles. *Appl. Energy* **2019**, *235*, 1001–1016. [CrossRef]

11. Hoskins, A.L.; Milican, S.L.; Czernik, C.E.; Alshankiti, I.; Netter, J.C.; Wendelin, T.J.; Musgrave, C.B.; Weimer, A.W. Continuous on-sun solar thermochemical hydrogen production via an isothermal redox cycle. *Appl. Energy* **2019**, *249*, 368–376. [CrossRef]
12. Mastropasqua, L.; Pecenati, I.; Giostri, A.; Campanari, S. Solar hydrogen production: Techno-economic analysis of a parabolic dish-supported high-temperature electrolysis system. *Appl. Energy* **2020**, *261*, 114392. [CrossRef]
13. Eckert, J.; Silva, L.; Santiciolli, F.; Costa, E.; Corrêa, F.; Dedini, F. Energy storage and control optimization for an electric vehicle. *Int. J. Energy Storage* **2018**, *42*, 3506–3523. [CrossRef]
14. Hu, J.; Jiang, X.; Zheng, L. Design and analysis of hybrid electric vehicle powertrain configurations considering energy transformation. *Int. J. Energy Storage* **2018**, *42*, 4719–4729. [CrossRef]
15. Yang, X.; Taenaka, B.; Miller, T.; Snyder, K. Modeling validation of key life test for hybrid electric vehicle batteries. *Int. J. Energy Storage* **2010**, *34*, 171–181. [CrossRef]
16. Tie, S.F.; Tan, C.W. A review of energy sources and energy management system in electric vehicles. *Renew. Sustain. Energy Rev.* **2013**, *20*, 82–102. [CrossRef]
17. Changizian, S.; Ahmadi, P.; Raeesi, M.; Javani, N. Performance optimization of hybrid hydrogen fuel-cell electric vehicles in real driving cycles. *Int. J. Hydrogen Energy* **2020**, *0360*, 3199. [CrossRef]
18. Alamili, A.; Xue, Y.; Anayi, F. An experimental and analytical study of the ultra-capacitor storage unit used in regenerative braking systems. *Energy Procedia* **2019**, *159*, 376–381. [CrossRef]
19. Kaya, K.; Hames, Y. Two new control strategies: For hydrogen fuel saving and extending the life cycle in the hydrogen fuel cell vehicles. *Int. J. Hydrogen Energy* **2019**, *44*, 18967–18980. [CrossRef]
20. Tanç, B.; Arat, H.T.; Conker, Ç.; Baltacioğlu, E.; Aydin, K. Energy distribution analyses of an additional traction battery on hydrogen fuel cell hybrid electric vehicle. *Int. J. Hydrogen Energy* **2019**, *45*, 26344–26356. [CrossRef]
21. Ferreira, F.J.T.E.; Almeida, A.T. Induction motor downsizing as a low-cost strategy to save energy. *J. Clean. Prod.* **2012**, *24*, 117–131. [CrossRef]
22. De Pinto, S.; Camocardi, P.; Chatzikomis, C.; Sorniotti, A.; Bottiglione, F.; Mantriota, G.; Perlo, P. On the Comparison of 2- and 4-Wheel-Drive Electric Vehicle Layouts with Central Motors and Single- and 2-Speed Transmission Systems. *Energies* **2020**, *13*, 3328. [CrossRef]
23. Bottiglione, F.; De Pinto, S.; Mantriota, G.; Sorniotti, A. Energy consumption of a battery electric vehicle with infinitely variable transmission. *Energies* **2014**, *7*, 8317–8337. [CrossRef]
24. Sluis, F.; Romers, L.; Spijk GHupkes, I. CVT, promising solutions for electrification. *SAE Int. Conf. Pap.* **2019**, *1*, 0359. [CrossRef]
25. Liang, J.; Walker, P.D.; Ruan, J.; Yang, H.; Wu, J. Gear shift and brake distribution control for regenerative braking in electric vehicles with dual clutch transmission. *Mech. Mach. Theory* **2019**, *133*, 1–22. [CrossRef]
26. Mo, W.; Walker, P.D.; Fang, Y.; Wu, J.; Ruan, J.; Zhang, N. A novel shift control concept for multi-speed electric vehicles. *Mech. Syst. Signal Process.* **2018**, *112*, 171–193. [CrossRef]
27. Zhang, L.; Ca, X. Control strategy of regenetrative braking system in electric vehicles. *Energy Procedia* **2018**, *152*, 496–501. [CrossRef]
28. Ruan, J.; Walker, P.; Zhang, N. A comparative study energy consumption and costs of battery electric vehicle transmissions. *Appl. Energy* **2016**, *165*, 119–134. [CrossRef]
29. Yildiz, A.; Kopmaz, O.; Cetin, S.T. Dynamic modeling and analysis of a four-bar mechanism coupled with a CVT for obtaining variable input speeds. *J. Mech. Sci. Technol.* **2015**, *29*, 1001–1006. [CrossRef]
30. Yildiz, A.; Piccininni, A.; Bottiglione, F.; Carbone, G. Modeling chain continuously variable transmission for direct implementation in transmission control. *Mech. Mach. Theory* **2016**, *105*, 428–440. [CrossRef]
31. Yildiz, A.; Bottiglione, F.; Carbone, G. An experimental study on the shifting dynamics of the chain CVT. *Int. Conf. Adv. Mech. Robot. Eng.* **2015**. [CrossRef]
32. Yildiz, A.; Kopmaz, O. Control-oriented modelling with experimental verification and design of the appropriate gains of a PI speed ratio controller of chain CVTs. *Stroj. Vestn. J. Mech. Eng.* **2017**, *63*, 374–382. [CrossRef]
33. Yildiz, A.; Bottiglione, F.; Piccininni, A.; Kopmaz, O.; Carbone, G. Experimental validation of the Carbone–Mangialardi–Mantriota model of continuously variable transmission. *Proc. Inst. Mech. Eng. Part D J. Automob. Eng.* **2018**, *232*. [CrossRef]
34. Yildiz, A.; Kopmaz, O. Dynamic analysis of a mechanical press equipped with a half-toroidal continuously variable transmission. *Int. J. Mater. Product Technol.* **2015**, *50*, 22–36. [CrossRef]
35. Arango, I.; Muñoz Alzate, S. Numerical Design Method for CVT Supported in Standard Variable Speed Rubber V-Belts. *Appl. Sci.* **2020**, *10*, 6238. [CrossRef]
36. Wang, Z.; Cai, Y.; Zeng, Y.; Yu, J. Multi-Objective Optimization for Plug-In 4WD Hybrid Electric Vehicle Powertrain. *Appl. Sci.* **2019**, *9*, 4068. [CrossRef]
37. Li, H.; Zhou, Y.; Xiong, H.; Fu, B.; Huang, Z. Real-Time Control Strategy for CVT-Based Hybrid Electric Vehicles Considering Drivability Constraints. *Appl. Sci.* **2019**, *9*, 2074. [CrossRef]
38. Mingyao, Y.; Datong, Q.; Xingyu, Z.; Sen, Z.; Yuping, Z. Integrated optimal control of transmission ratio and power split ratio for a CVT-based plug-in hybrid electric vehicle. *Mech. Mach. Theory* **2019**, *136*, 52–71. [CrossRef]
39. Jain, M.; Williamson, S.S. Suitability analysis of in-wheel motor direct drives for electric and hybrid electric vehicles. *IEEE Electr. Power Energy Conf.* **2019**, *988*, 1–5. [CrossRef]

40. Long, G.; Ding, F.; Zhang, N.; Zhang, J.; Qin, A. Regenerative active suspension system with residual energy for in-wheel motor driven electric vehicle. *Appl. Energy* **2020**, *260*, 114180. [CrossRef]
41. Kulkarni, A.; Ranjha, S.A.; Kapoor, A. Fatigue analysis of a suspension for an in-wheel electric vehicle. *Eng. Fail. Anal.* **2016**, *68*, 150–158. [CrossRef]
42. Najjari, B.; Mirzae, M.; Tahouni, A. Constrained stability control with optimal power management strategy for in-wheel electric vehicles. *J. Multi-Body Dyn.* **2019**, 1–19. [CrossRef]
43. Xu, W.; Chen, H.; Zhao, H.; Ren, B. Torque optimization control for electric vehicles with four in-wheel motors equipped with regenerative braking system. *Mechatronics* **2019**, *57*, 95–108. [CrossRef]
44. Ehsani, M.; Gao, Y.; Gay, S.E.; Emadi, A. *Modern Electric, Hybrid Electric, and Fuel Cell Vehicles Fundamentals, Theory, and Design*; CRC Press: Boca Raton, FL, USA, 2005; ISBN 0-8493-3154-4.
45. Tekin, G.; Unlusoy, Y.S. Design and simulation of an integrated active yaw control system for road vehicles. *Int. J. Veh. Des.* **2010**, 52. [CrossRef]

applied sciences

MDPI

Article

On the Effect of a Rate-Dependent Work of Adhesion in the Detachment of a Dimpled Surface

Antonio Papangelo [1,2]

[1] Politecnico di BARI, Department of Mechanics, Mathematics and Management, Via Orabona 4, 70125 Bari, Italy; antonio.papangelo@poliba.it or antonio.papangelo@tuhh.de

[2] Department of Mechanical Engineering, Hamburg University of Technology, Am Schwarzenberg Campus 1, 21073 Hamburg, Germany

Abstract: Patterned surfaces have proven to be a valuable design to enhance adhesion, increasing hysteresis and the detachment stress at pull-off. To obtain high adhesive performance, soft materials are commonly, used, which easily conform to the countersurface, such as soft polymers and elastomers. Such materials are viscoelastic; i.e., they show rate-dependent properties. Here, the detachment of two half spaces is studied, one being flat and the other having a dimple in the limit of short range adhesion and a power law rate-dependent work of adhesion, as observed by several authors. Literature results have suggested that the dimpled surface would show pressure-sensitive adhesion, showing two possible adhered states, one weak, in partial contact, and one strong when full contact is achieved. By accounting for a power law rate-dependent work of adhesion, the "weak state" may be much stronger than it was in the purely elastic case, and hence the interface may be much more tough to separate. We study the pull-off detachment stress of the dimpled surface, showing that it weakly depends on the preload, but it is strongly affected by the dimensionless unloading rate. Finally, possible implications of the presented results in the detachment of soft materials from rough substrates are discussed.

Keywords: adhesion enhancement; dimple model; patterned surfaces; viscoelasticity; enhancement

Citation: Papangelo, A. On the Effect of a Rate-Dependent Work of Adhesion in the Detachment of a Dimpled Surface. *Appl. Sci.* **2021**, *11*, 3107. https://doi.org/10.3390/app11073107

Academic Editor: Alberto Campagnolo

Received: 12 March 2021
Accepted: 29 March 2021
Published: 31 March 2021

1. Introduction

Tribology is a very active field of research of utmost importance in several engineering applications, ranging from automotive [1] to aerospace [2] and bio-engineering [3]. In automotive, for example, the role of adhesion and viscoelasticity is crucial in determining tires' adherence and performance [4]. Adhesion due to van der Waals interactions is commonly exploited in nature, in which very efficient strategies have been developed to adhere to almost any kind of surface [5,6]. Lizardw, geckos and insects in general in most cases outperform the most advanced human designed adhesive technologies, and the topic still has far to go. Geckos for example have developed a multiscale hierarchical structure, so that the "macroscopic" foot splits in several lamellae, which branch in the setae and finally in fine spatulae of the size of nanometers [7,8]. Inspired by nature, several adhesive strategies have been pursued, among the others, that of fabricating patterned surfaces [9] present in mushroom pillars [10] or depressions ("dimples") [9] (see Figure 1a) and have been shown to be able to reach far higher adhesive stress with respect to the smooth case. Nevertheless, simple criteria to discern sticky from unsticky surfaces are still unavailable [11–13], except maybe the criterion introduced by Dalquist at 3M, who suggested that a strong adhesive tape should have a Young's modulus lower than 0.3 MPa to be able to conform well to the harder counterpart [14], which, only recently, has received some support from a theoretical perspective [11,12].

Over the years, several authors have conceived of contact mechanics models to ascertain how the enhancement is originated and possibly indicate routes for further development or better performance [15–17]. A very elegant model for the detachment of a halfspace with a dimple from a flat substrate was proposed by McMeeking et al. [17], who developed a contact mechanics model in the limit of short range adhesion (the so-called "JKR limit" from Johnson, Kendall, and Roberts' seminal paper [18]) and showed that the dimple surface behaved as a pressure-sensitive adhesive: detaching the dimple from its equilibrium position would lead to "weak adhesion" in partial contact, but upon application of a compressive pressure, a full-contact state would be achieved that would require a (theoretically) infinite tensile traction to be detached ("strong adhesion"). In this respect, the finding recalls the seminal work of Johnson [19], who considered the contact of a halfplane with sinusoidal waviness. Johnson had to postulate the presence of a flaw at the interface (air trapping, contaminants, fine scale roughness...), which would guarantee the possibility to separate the two surfaces. Zhou et al. [15] have applied the McMeeking et al. [17] model to study the adhesion capabilities of both cockroach pads (*Nauphoeta cinerea*) and dock beetles pads (*Gastrophysa viridula*), showing that they are strongly influenced by the geometric features of the interfacial pattern, and Cañas et al. [20] adopted the McMeeking et al. [17] model to fit experimental results of adhesion of biomimetic polydimethylsiloxane (PDMS) surfaces patterned with pillars with mushroom-shaped tips.

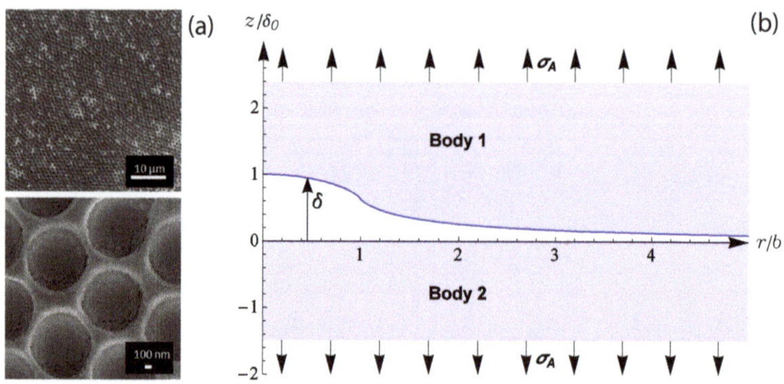

Figure 1. (**a**) SEM images of a surface pattered with dimples (adapted from [9]). (**b**) Geometry of the dimple surface considered in this study. The model is axysimmetric.

Recently, Papangelo and Ciavarella [21] generalized the "dimple" model of McMeeking et al. [17] by using a cohesive Maugis–Dugdale model, which introduces two new parameters, i.e., the theoretical strength of the material and the range of interaction of the adhesive tractions, and allowed investigation of the detachment performance of the dimple ranging from soft to hard materials as a function of a dimensionless parameter that is similar to that introduced by Tabor for the sphere [21,22]. It was shown that the dimple adhesive performance, particularly the "strong adhesive state", becomes rapidly degraded, moving towards the rigid limit; hence for the dimple to effectively work as a pressure-sensitive mechanism, soft materials should be employed.

The above studies have considered the work of adhesion (the work needed to separate two flat surfaces from their equilibrium position up to infinity) as a constant that depends on the contact pair. Nevertheless, it is well-known that soft materials show rate-dependent properties. In particular, the dependence of the work of adhesion w on the velocity of the peeling v_p (the velocity at the crack tip) is well documented both experimentally [23,24] and theoretically [25,26]. Although with some variants in the nomenclature, most of the authors agree that the dependence of the work of adhesion on the peeling velocity is well captured by a power law form (Reference [27])

$$w = w_0 \left[1 + \left(\frac{v_p}{V_0} \right)^n \right] \tag{1}$$

where w_0 is the adiabatic (or thermodynamic) work of adhesion for vanishing peeling velocity, while V_0 and n are material properties with V_0 a reference velocity, and n an exponent usually in a range from 0.1 to 0.9 for realistic materials. Here, we revise the model of McMeeking et al. [17] by accounting for the effect of the rate at which the external load is applied. It will be shown that the dimple still behaves as a pressure-sensitive adhesive, but the "weak" adhesive state is strongly influenced by the unloading rate so that the pull-off stress in partial contact and the interfacial toughness can largely increase for high unloading velocity. The enhancement effect is greatly dependent on the exponent n, larger exponents leading to larger strengthening of the interface, but it is only marginally affected by the level of preload.

2. Griffith Equilibrium Solution for a Dimple

2.1. Elastic Problem

Here, the model introduced by McMeeking et al. [17] for the detachment of two elastic half-spaces is summarized. The geometry consists of two elastic bodies (see Figure 1b), with Young's modulus and Poisson's ratio $\{E_1, \nu_1\}$ and $\{E_2, \nu_2\}$, one of those flat while the other with an axysimmetric dimple, whose geometry is defined by (see Figure 1b)

$$\begin{cases} \delta = \frac{2}{\pi}\delta_0 \varepsilon\left(\frac{r}{b}\right), & \frac{r}{b} \leq 1 \\ \delta = \frac{2r}{\pi b}\delta_0\left[\varepsilon\left(\frac{b}{r}\right) - \left(1 - \frac{b^2}{r^2}\right)\kappa\left(\frac{b}{r}\right)\right], & \frac{r}{b} > 1 \end{cases} \tag{2}$$

where r is the radial coordinate, δ_0 is the dimple amplitude, b is the dimple radius, $\kappa(\theta)$ and $\varepsilon(\theta)$ are, respectively, the complete elliptical integral of first and second kind with argument θ. Johnson [28] has shown that a uniform axisymmetric pressure of magnitude p applied over a region $r < b$ on the surface of a half-space produces surface displacements as in Equation (2); hence, to make the two half-spaces conform to each other, one would need within the dimple ($r \leq b$) a tensile traction T equal to [28]

$$T = \frac{E^* \delta_0}{2b}, \qquad r \leq b \tag{3}$$

where $\frac{1}{E^*} = \frac{1-\nu_1^2}{E_1} + \frac{1-\nu_2^2}{E_2}$ is the composite elastic modulus. Hence, if a remote stress σ_A (>0, when tensile) is applied, the stress σ_{zz} at the interface of the two half-spaces will be

$$\begin{cases} \sigma_{zz} = \sigma_A + \frac{E^*\delta_0}{2b}, & \frac{r}{b} \leq 1 \\ \sigma_{zz} = \sigma_A, & \frac{r}{b} > 1 \end{cases} \tag{4}$$

The combination of the remote stress applied and the inner (constant) stress within the dimple can be easily studied in the framework of the Linear Elastic Fracture Mechanics (LEFM) theory, as the dimple behaves as an axisymmetric crack under internal pressure $p(r)$ [29]. Let us define, for a crack of radius c, the auxiliary function ([29], 3.114a) $g(x) = \int_0^x \frac{sp(s)ds}{\sqrt{x^2-s^2}}$

$$\begin{cases} g(x) = (\sigma_A + T)x, & x < b \\ g(x) = \sigma_A x + T\left(x - \sqrt{x^2 - b^2}\right), & x > b \end{cases} \tag{5}$$

Then, Equation (3.117) in [29] gives the stress intensity factor as $K_I = \frac{2}{\sqrt{\pi c}} g(c)$, or equivalently, the energy release per unit area $G = K_I^2/2E^*$

$$\begin{cases} G = \frac{2c}{\pi E^*}\left(\sigma_A + \frac{E^*\delta_0}{2b}\right)^2, & \frac{c}{b} \leq 1 \\ G = \frac{2c}{\pi E^*}\left[\sigma_A + \frac{E^*\delta_0}{2b}\left(1 - \sqrt{1 - \left(\frac{b}{c}\right)^2}\right)\right]^2 & \frac{c}{b} > 1 \end{cases} \tag{6}$$

Hence McMeeking et al. [17] apply the classical Griffith energetic argument that the energy release per unit area G should be equal to the surface energy w_0 of the contact pair (the "toughness"), leading to their LEFM model [17]. The equilibrium curves that link the remote tension to the crack radius are written in dimensionless form as

$$\begin{cases} \widehat{\sigma}_A = -1 + \alpha_d \sqrt{\frac{1}{\widehat{c}}}, & \widehat{c} \leq 1 \\ \widehat{\sigma}_A = -1 + \alpha_d \sqrt{\frac{1}{\widehat{c}}} + \sqrt{1 - \left(\frac{1}{\widehat{c}}\right)^2}, & \widehat{c} > 1 \end{cases} \tag{7}$$

where we have introduced the following dimensionless parameters

$$\alpha_d^2 = \frac{2\pi w_0 b}{E^* \delta_0^2}, \qquad \widehat{\sigma}_A = \frac{\sigma_A}{T}, \qquad \widehat{c} = \frac{c}{b}, \tag{8}$$

Notice that α_d is analogous to the parameter that Johnson defined for the sinusoid [28] as it is proportional to the ratio between the adiabatic work of adhesion and the strain elastic energy to flatten the dimple. Figure 2 shows the equilibrium solutions for $\alpha_d = [0.1, 0.5, 0.9]$, where we used a solid (dashed) line to identify the stable (unstable) branches of the curve. For the curve $\alpha_d = 0.5$, we have labelled some noteworthy points. The system is in equilibrium under vanishing remote stress $\widehat{\sigma}_A = 0$ at the point "C", which corresponds also to a certain radius (not null) of the crack $\widehat{c}$. If a tensile stress is applied, the crack radius increases and detachment takes place at point "B" in partial contact (Reference [21] reports $\widehat{c}$ at pull-off), which will be referred to as the "weak" adhered state. If, from the equilibrium position "C", a compressive pressure is applied below $\widehat{\sigma}_{A,C} = -1 + \alpha_d$, the dimple jumps from point "A" to a full contact state (the branch below $\widehat{c} = 1$ is unstable); hence, the crack shrinks, and theoretically an infinite tensile pressure will be needed to separate the two bodies. In practice, there may be flaws, contaminants, air trapping or simply the two surfaces will be detached when the theoretical strength is reached (see [21] for a detailed analysis).

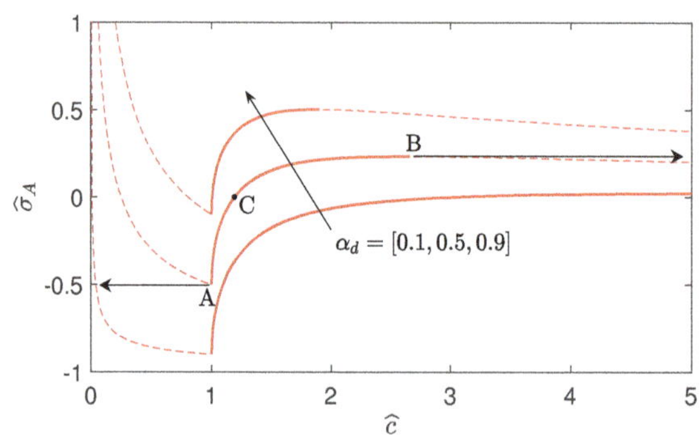

Figure 2. Remote stress $\widehat{\sigma}_A$ versus the crack radius $\widehat{c}$ in the McMeeking et al. [17] elastic model for $\alpha_d = [0.1, 0.5, 0.9]$. For the curve $\alpha_d = 0.5$, point "C" determines the equilibrium position ($\widehat{\sigma}_A = 0$), point "B" the "weak" adhesive state, while point A is the point where the jump to full contact happens.

2.2. *Effect of a Rate-Dependent Work of Adhesion*

We have introduced above the McMeeking et al. [17] LEFM elastic model for the detachment a dimpled half-space from a flat substrate. Nevertheless, soft matter has viscoelastic behavior; hence, we here extend the McMeeking et al. [17] model to account for a rate-dependent work of adhesion. According to Equation (1), the effective work of adhesion is a power-law function of the peeling velocity $v_p = dc/dt$ (t is the time), i.e., the velocity at the crack tip, which can be written as

$$v_p = \frac{dc}{dt} = \pm \dot{\sigma}_A \frac{dc}{d\sigma_A} \tag{9}$$

where $\dot{\sigma}_A = d\sigma_A/dt$ [Pa/s] is the externally imposed unloading rate and the "+" ("−") sign holds when the crack advances $(dc/dt > 0)$ or retracts $(dc/dt < 0)$. Hence, using (1), (6) and (9) and imposing the Griffith equilibrium concept, one obtains

$$\begin{cases} w_0\left[1+\left(-\frac{\dot{\sigma}_A}{V_0}\frac{dc}{d\sigma_A}\right)^n\right] = \frac{2c}{\pi E^*}\left(\sigma_A + \frac{E^*\delta_0}{2b}\right)^2, & \frac{c}{b} \le 1 \\ w_0\left[1+\left(\frac{\dot{\sigma}_A}{V_0}\frac{dc}{d\sigma_A}\right)^n\right] = \frac{2c}{\pi E^*}\left[\sigma_A + \frac{E^*\delta_0}{2b}\left(1-\sqrt{1-\left(\frac{b}{c}\right)^2}\right)\right]^2, & \frac{c}{b} > 1 \end{cases} \tag{10}$$

which in dimensionless notation reads

$$\begin{cases} \frac{d\hat{c}}{d\hat{\sigma}_A} = -\frac{1}{\dot{\hat{\sigma}}_A}\left[\frac{\hat{c}}{\alpha_d^2}(\hat{\sigma}_A+1)^2 - 1\right]^{1/n}, & \hat{c} \le 1 \\ \frac{d\hat{c}}{d\hat{\sigma}_A} = \frac{1}{\dot{\hat{\sigma}}_A}\left\{\frac{\hat{c}}{\alpha_d^2}\left[\hat{\sigma}_A + \left(1-\sqrt{1-\frac{1}{\hat{c}^2}}\right)\right]^2 - 1\right\}^{1/n}, & \hat{c} > 1 \end{cases} \tag{11}$$

where the dimensionless unloading rate $\dot{\hat{\sigma}}_A = \frac{\dot{\sigma}_A}{TV_0/b}$ has been introduced. The ordinary differential equations in (11) are of first order and can be easily solved numerically using as a starting point a solution of the elastic problem in Equation (7). Clearly, physically admissible starting points are those on the stable branch AB in Figure 2; hence, as we shall see later, a rate-dependent work of adhesion will not affect the "strong" adhesive state, but only the "weak" one. In the next section we shall investigate the effects of the various parameters $\left\{n, \dot{\hat{\sigma}}_A, \alpha_d\right\}$ on the detachment stress in partial contact ("weak" state).

3. Results

3.1. Detachment Curves

Let us investigate first the effect that a rate-dependent work of adhesion has on the detachment curve of the dimple by solving Equation (11). Figure 3 shows the unloading curves for the rate-dependent model (Equation (11), black dot-dashed curves) starting from different initial conditions $\hat{c}_0$ on the branch AB (Figure 2) for $\alpha_d = 0.5$, $n = 0.5$ and $\dot{\hat{\sigma}}_A = 1$. The red curve (solid for the stable, dashed for the unstable branches) shows the Griffith elastic reference solution [17]. Clearly, the rate-dependence of the work of adhesion has a toughening effect (dot-dashed black curves). Figure 4a shows the dimensionless peeling velocity $\hat{v}_p = v_p/V_0$ as it changes during the detachment process, while panel (b) shows the corresponding dimensionless work of adhesion $\hat{w} = w/w_0$, while the crack propagates. The peeling process starts with a vanishing velocity; i.e., instantaneously, the remote stress increases without change of the crack radius $\hat{c}$, and this can be easily checked if one substitutes a solution of the Griffith elastic model (Equation (7)) into the rate-dependent model (Equation (11)). By further unloading the dimple surface $\left(\dot{\hat{\sigma}}_A > 0\right)$, the peeling velocity $\hat{v}_p$ increases with a power law behavior, and so does the interfacial toughness. Notice that the slope of the $\hat{\sigma}_A(\hat{c})$ curve is inversely proportional to the peeling velocity; hence, the latter decays with a power law. We define the pull-off detachment stress as $\hat{\sigma}_{A,pull} = \lim_{\hat{c}\to+\infty}\hat{\sigma}_A$. Furthermore, Figure 3 shows the crack retraction curve, when the dimple is forced into the unstable regime ($\hat{c} \le 1$, blue dot-dashed curve). Apart from a different transient behavior, similarly to the elastic solution, the rate-dependent solution will jump into a full contact strong adhesive state. Figure 3 shows that the effect of the initial preload on the detachment stress $\hat{\sigma}_{A,pull}$ is marginal, particularly in light of the dependencies on the parameters $\left\{n, \dot{\hat{\sigma}}_A, \alpha_d\right\}$ that we will show in the following. From now on, we will consider as initial condition only the more "likely" situation of unloading the dimple from its equilibrium position $\hat{\sigma}_A = 0$, which belongs to the stable branch AB for $\alpha_d \leq 1$ (otherwise for $\alpha_d > 1$, the dimple naturally jumps into the full-contact strong adhesive state).

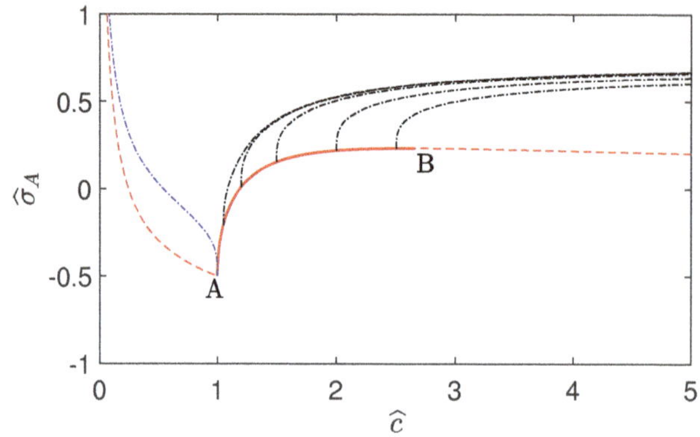

Figure 3. Unloading curves of the dimpled surface starting from different initial conditions $\widehat{c}_0 = [1.05, 1.2, 1.5, 2, 2.5]$ for $\alpha_d = 0.5$, $n = 0.5$ and $\dot{\widehat{\sigma}}_A = 1$. Red lines for the Griffith elastic solution, dot-dashed black lines for the rate-dependent model unloading curves.

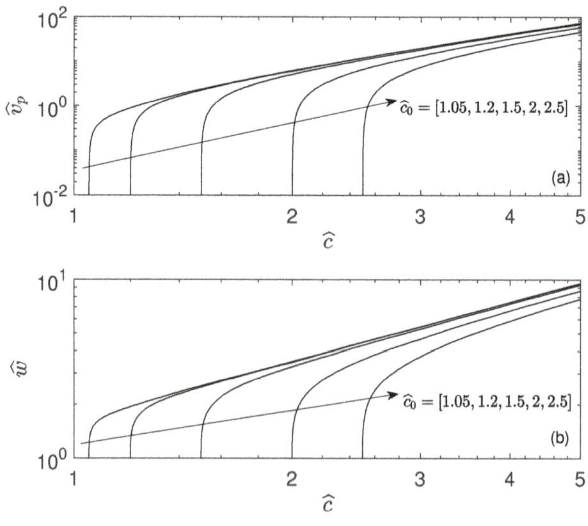

Figure 4. (**a**) Dimensionless peeling velocity and (**b**) corresponding apparent work of adhesion $\widehat{w} = w/w_0$ for the unloading curves in Figure 3.

Let us look at the influence of the exponent n and of the unloading rate $\dot{\widehat{\sigma}}_A$ on the dimple detachment curves. Figure 5a shows that starting from the dimple equilibrium position "C" for $\alpha_d = 0.5$, $n = 0.5$ and $\dot{\widehat{\sigma}}_A = 10^{[-3,0,1,2]}$, where black dot-dashed curves indicate the rate-dependent solutions, while the red curves show the reference elastic solution. Figure 5a shows that the unloading rate has a strong influence on the detachment stress $\widehat{\sigma}_{A,pull}$ and that the elastic solution is approached only for the very low unloading rate $\dot{\widehat{\sigma}}_A = 10^{-3}$. Notice that, in this particular case, for the reference elastic solution, we have $\widehat{\sigma}_A|_{\max} \approx 0.23$,, while in the a rate-dependent case $\widehat{\sigma}_{A,pull} \approx 1.7$ at $\dot{\widehat{\sigma}}_A = 10^2$, which gives a detachment stress increased by a factor $\simeq 7$. Finally panel (b) shows that for the same unloading rate $\dot{\hat{\sigma}}_A = 1$, the pull-off stress $\hat{\sigma}_{A,pull}$ increases with the exponent $n = [0.1, 0.3, 0.5, 0.7, 0.9]$ as for the increasing of the effective surface energy w (see Equation (1)).

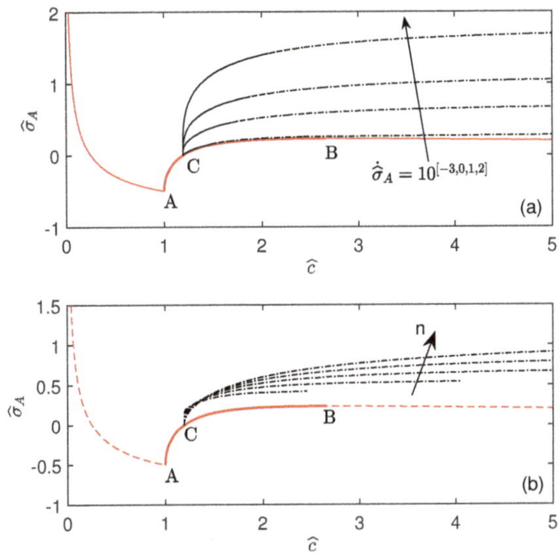

Figure 5. (**a**) Unloading curves of the dimpled surface starting from $\hat{\sigma}_A = 0$ for $\alpha_d = 0.5$, $n = 0.5$ and $\dot{\hat{\sigma}}_A = 10^{[-3,0,1,2]}$. Red lines for the Griffith elastic solution, dot-dashed black lines for the a rate-dependent unloading curves. (**b**) As panel (**a**) but for $\dot{\hat{\sigma}}_A = 1$ and $n = [0.1, 0.3, 0.5, 0.7, 0.9]$.

3.2. Pull-Off Detachment Stress

We have investigated the effect of the unloading rate $\dot{\hat{\sigma}}_A$ and of the exponent n on the detachment curves of a dimple from an elastic substrate with rate-dependent work of adhesion. Here we give a closer look at the dependence of the pull-off stress $\hat{\sigma}_{A,pull}$ on the unloading rate $\dot{\hat{\sigma}}_A$ by varying α_d and n. Figure 6a shows the pull-off stress $\hat{\sigma}_{A,pull}$ as a function of the dimensionless unloading rate $\dot{\hat{\sigma}}_A$ for $\alpha_d = [0.2, 0.4, 0.6, 0.8]$ and $n = 0.5$. Panel (a) shows that the the pull-off stress $\hat{\sigma}_{A,pull}$ approaches the elastic value (obtained from Equation (7)) only at very low unloading rate $\dot{\hat{\sigma}}_A \approx 10^{-4}$, while it increases considerably when $\dot{\hat{\sigma}}_A$ is increased, or by increasing α_d. In the regime of high unloading rate, the pull-off stress dependence on $\dot{\hat{\sigma}}_A$ is well captured by a power law behavior, while that on α_d is linear. For $\dot{\hat{\sigma}}_A \gtrsim 10^{-2}$ and $n = 0.5$, a very good fit of the pull-off stress is given by the power law

$$\hat{\sigma}_{A,pull} = (1.32\alpha_d + 0.065)\dot{\hat{\sigma}}_A^{0.2} \tag{12}$$

whose predictions are given by dot-dashed black lines (almost indistinguishable from the solid curves obtained numerically). Panel (b) shows that for a fixed $\alpha_d = 0.5$, the pull-off stress increases with the exponent n. At low $\dot{\hat{\sigma}}_A$, all the curves start from the same elastic solution (black dashed lines) as the pull-off stress depends only on α_d and then increases for larger unloading rates with a power law behavior whose slope is very well fitted by a power law $\widehat{\sigma}_{A,pull} \propto \dot{\widehat{\sigma}}_A^{0.3n+0.05}$. (For $\dot{\widehat{\sigma}}_A \gtrsim 10^{-2}$, an estimate of the pull-off stress as a function of α_d and n can be obtained by $\widehat{\sigma}_{A,pull} = 2n(1.32\alpha_d + 0.065)\dot{\widehat{\sigma}}_A^{0.3n+0.05}$, where we found this to be more accurate close to $n \approx 0.5$ and $\alpha_d = 0.5$.) Notice that typically $0 < n < 1$; hence, the increase of the pull-off stress with the unloading rate is sublinear.

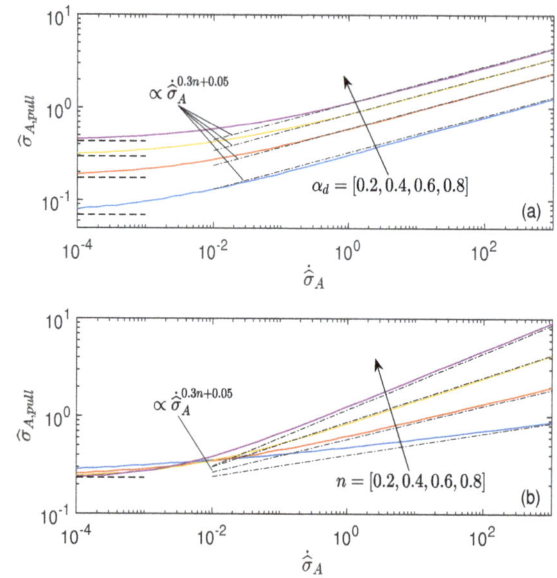

Figure 6. (**a**) Pull-off detachment stress $\widehat{\sigma}_{A,pull}$ as a function of the dimensionless unloading rate $\dot{\widehat{\sigma}}_A$ for $\alpha_d = [0.2, 0.4, 0.6, 0.8]$ and $n = 0.5$. (**b**) Pull-off detachment stress $\widehat{\sigma}_{A,pull}$ as a function of the dimensionless unloading rate $\dot{\widehat{\sigma}}_A$ for $n = [0.2, 0.4, 0.6, 0.8]$, $\alpha_d = 0.5$. For all the curves, the initial condition is $\widehat{\sigma}_A = 0$. In both panels, the black dashed lines indicate the Griffith equilibrium solution from Equation (7). Black dot-dashed lines serve as a guide to the eye.

4. Conclusions

In this work, the detachment of a surface with an axisymmetric dimple from an elastic substrate with rate-dependent work of adhesion has been studied in the limit of short-range adhesion. Previous elastic model with constant work of adhesion has shown that the dimpled surface has two adhered states, one "strong" in full contact, one "weak" in partial contact. The rate dependence of the work of adhesion has been accounted for by adopting a power law dependence (with exponent n) of the effective work of adhesion on the pealing velocity as it is commonly observed in experiments [27]. Notice that, in this analysis, we have neglected bulk effects (we used the relaxed modulus of the viscoelastic material $E_0 = E_{\omega=0}$, with ω being the excitation frequency of the material); hence, the results presented should be representative for not too high peeling velocity. It has been shown that a rate-dependent work of adhesion does not affect the "strong" adhesive state, but it strongly influences the detachment pull-off stress in partial contact, resulting in a strong toughening of the interface. In this respect, we have shown that for a given unloading rate, after a transient, the peeling velocity increases with a power law behavior during the detachment process, and so does the effective work of adhesion. Furthermore, it has been shown that the pull-off stress (in partial contact) does not depend much on the particular initial condition chosen to unload the dimple. Instead, the pull-off stress depends on three dimensionless parameters; i.e., it increases by increasing the unloading rate, the exponent "n" or the dimensionless parameter α_d. Clearly, for a real viscoelastic material, the interfacial toughening will not increase to infinity but will be limited by the ratio $E_{\omega=\infty}/E_{\omega=0}$, which is known to vary anyway by orders of magnitude with the excitation frequency [26]. In this respect, for a real material, the curves in Figure 4 will be bounded. In conclusion, the presented results suggest that viscoelasticity may play a strong role in determining the apparent interfacial toughness in patterned interfaces. In particular, when soft rough substrates are considered, it may be very difficult to recognize other sources of toughening, such as those due to the increase of contact area due to roughness [30] or due the local jump instabilities as in the Guduru problem [31], unless an extremely low unloading rate is adopted as indeed recently attempted by Dalvi and coauthors [32].

Funding: A.P. acknowledges the support by the Italian Ministry of Education, University and Research under the Programme Department of Excellence Legge 232/2016 (Grant No. CUP-D94I 18000260001). A.P. is thankful to the DFG (German Research Foundation) for funding the project PA 3303/1-1. A.P. acknowledges support from "PON Ricerca e Innovazione 2014-2020-Azione I.2"–D.D. n. 407, 27/02/2018, bando AIM (Grant No. AIM1895471). AP acknowledges support for the Open Access fees by Hamburg University of Technology (TUHH) in the funding programme Open Access Publishing.

Institutional Review Board Statement: Not applicable.

Informed Consent Statement: Not applicable.

Data Availability Statement: Not applicable.

Conflicts of Interest: The author declares no conflict of interest.

References

1. Hong, S.; Erdogan, G.; Hedrick, K.; Borrelli, F. Tyre–road friction coefficient estimation based on tyre sensors and lateral tyre deflection: Modelling, simulations and experiments. *Veh. Syst. Dyn.* **2013**, *51*, 627–647.
2. Yu, P.; Zhang, D.; Ma, Y.; Hong, J. Dynamic modeling and vibration characteristics analysis of the aero-engine dual-rotor system with Fan blade out. *Mech. Syst. Signal Process.* **2018**, *106*, 158–175. [CrossRef]
3. Li, X.; Tao, D.; Lu, H.; Bai, P.; Liu, Z.; Ma, L.; Tian, Y. Recent developments in gecko-inspired dry adhesive surfaces from fabrication to application. *Surf. Topogr. Metrol. Prop.* **2019**, *7*, 023001. [CrossRef]
4. Genovese, A.; Farroni, F.; Papangelo, A.; Ciavarella, M. A discussion on present theories of rubber friction, with particular reference to different possible choices of arbitrary roughness cutoff parameters. *Lubricants* **2019**, *7*, 85. [CrossRef]
5. Ciavarella, M.; Joe, J.; Papangelo, A.; Barber, J.R. The role of adhesion in contact mechanics. *J. R. Soc.* **2019**, *16*, 20180738. [CrossRef]
6. Kamperman, M.; Kroner, E.; del Campo, A.; McMeeking, R.M.; Arzt, E. Functional adhesive surfaces with "gecko" effect: The concept of contact splitting. *Adv. Mater.* **2010**, *12*, 335–348. [CrossRef] [PubMed]
7. Huber, G.; Gorb, S.; Hosoda, N.; Spolenak, R.; Arzt, E. Influence of surface roughness on gecko adhesion. *Acta Biomater* **2007**, *3*, 607–610. [CrossRef]
8. Pugno, N.M.; Lepore, E. Observation of optimal gecko's adhesion on nanorough surfaces. *Biosystems* **2008**, *94*, 218–222. [CrossRef]
9. Akerboom, S.; Appel, J.; Labonte, D.; Federle, W.; Sprakel, J.; Kamperman, M. Enhanced adhesion of bioinspired nanopatterned elastomers via colloidal surface assembly. *J. R. Interface* **2015**, *12*, 20141061. [CrossRef] [PubMed]
10. Paretkar, D.; Kamperman, M.; Martina, D.; Zhao, J.; Creton, C.; Lindner, A.; Arzt, E. Preload-responsive adhesion: Effects of aspect ratio, tip shape and alignment. *J. R. Soc.* **2013**, *10*, 20130171. [CrossRef] [PubMed]
11. Violano, G.; Afferrante, L.; Papangelo, A.; Ciavarella, M. On stickiness of multiscale randomly rough surfaces. *J. Adhesion* 2019, 1–19. [CrossRef]
12. Ciavarella, M. Universal features in "stickiness" criteria for soft adhesion with rough surfaces. *Tribol. Int.* **2020**, *146*, 106031. [CrossRef]
13. Tiwari, A.; Wang, J.; Persson, B.N.J. Adhesion paradox: Why adhesion is usually not observed for macroscopic solids. *Phys. Rev. E* **2020**, *102*, 042803. [CrossRef]
14. Dahlquist, C.A. *Treatise on Adhesion and Adhesives*; Patrick, R.L., Ed.; Marcel Dekker: New York, NY, USA, 1969; p. 244. [CrossRef] [PubMed]
15. Zhou, Y.; Robinson, A.; Steiner, U.; Federle, W. Insect adhesion on rough surfaces: Analysis of adhesive contact of smooth and hairy pads on transparent microstructured substrates. *J. R. Interface* **2014**, *11*, 20140499.
16. Kern, M.D.; Qi, Y.; Long, R.; Rentschler, M.E. Characterizing adhesion between a micropatterned surface and a soft synthetic tissue. *Langmuir* **2017**, *33*, 854–864. [CrossRef]
17. McMeeking, R.M.; Ma, L.; Arzt, E. Bi-Stable Adhesion of a Surface with a Dimple. *Adv. Eng. Mater.* **2010**, *12*, 389–397. [CrossRef]
18. Johnson, K.L.; Kendall, K.; Roberts, A.D. Surface energy and the contact of elastic solids. *Proc. R. Soc. Lond. A* **1971**, *324*, 1558. [CrossRef]

19. Johnson, K.L. The adhesion of two elastic bodies with slightly wavy surfaces. *Int. Solids Struct.* **1995**, *32*, 423–430.
20. Cañas, N.; Kamperman, M.; Völker, B.; Kroner, E.; McMeeking, R.M.; Arzt, E. Effect of nano-and micro-roughness on adhesion of bioinspired micropatterned surfaces. *Acta Biomater.* **2012**, *8*, 282–288. [CrossRef]
21. Papangelo, A.; Ciavarella, M. A Maugis–Dugdale cohesive solution for adhesion of a surface with a dimple. *J. R. Soc. Interface* **2017**, *14*, 20160996. [CrossRef]
22. Tabor, D. Surface forces and surface interactions. *J. Colloid Interface Sci.* **1977**, *58*, 2. [CrossRef]
23. Waters, J.F.; Guduru, P.R. Mode-mixity-dependent adhesive contact of a sphere on a plane surface. *Proc. R. A Math. Phys. Eng. Sci* **2010**, *466*, 1303–1325. [CrossRef]
24. Barquins, M.; Maugis, D.; Blouet, J.; Courtel, R. Contact area of a ball rolling on an adhesive viscoelastic material. *Wear* **1978**, *51* 375–384. [CrossRef]
25. Greenwood, J.A.; Johnson, K.L. The mechanics of adhesion of viscoelastic solids. *Philos. Mag. A* **1981**, *43*, 697–711. [CrossRef]
26. Persson, B.N.J.; Brener, E.A. Crack propagation in viscoelastic solids. *Phys. Rev. E* **2005**, *71*, 036123. [CrossRef]
27. Gent, A.N.; Schultz, J. Effect of wetting liquids on the strength of adhesion of viscoelastic material. *J. Adhes.* **1972**, *3*, 281–294 [CrossRef] [PubMed]
28. Johnson, K.L. *Contact Mechanics;* Cambridge University Press: Cambridge, UK, 1985. [CrossRef]
29. Maugis, D. *Contact, Adhesion and Rupture of Elastic Solids (Volume 130);* Springer: New York, NY, USA, 2000.
30. Persson, B.N.J.; Tosatti, E. The effect of surface roughness on the adhesion of elastic solids. *J. Chem. Phys.* **2001**, *115*, 5597–5610.
31. Guduru, P.R. Detachment of a rigid solid from an elastic wavy surface: Theory. *J. Mech. Phys. Solids* **2007**, *55*, 473–488 [CrossRef]
32. Dalvi, S.; Gujrati, A.; Khanal, S.R.; Pastewka, L.; Dhinojwala, A.; Jacobs, T.D. Linking energy loss in soft adhesion to surface roughness. *Proc. Natl. Acad. Sci. USA* **2019**, *116*, 25484–25490. [CrossRef]

applied sciences

MDPI

Article

Modelling and Stability Analysis of Articulated Vehicles

Tianlong Lei, Jixin Wang and Zongwei Yao *

Key Laboratory of CNC Equipment Reliability, Ministry of Education, School of Mechanical and Aerospace Engineering, Jilin University, Changchun 130022, China; leitl19@mails.jlu.edu.cn (T.L.); jxwang@jlu.edu.cn (J.W.)
* Correspondence: yzw@jlu.edu.cn

Abstract: This study constructs a nonlinear dynamic model of articulated vehicles and a model of hydraulic steering system. The equations of state required for nonlinear vehicle dynamics models, stability analysis models, and corresponding eigenvalue analysis are obtained by constructing Newtonian mechanical equilibrium equations. The objective and subjective causes of the snake oscillation and relevant indicators for evaluating snake instability are analysed using several vehicle state parameters. The influencing factors of vehicle stability and specific action mechanism of the corresponding factors are analysed by combining the eigenvalue method with multiple vehicle state parameters. The centre of mass position and hydraulic system have a more substantial influence on the stability of vehicles than the other parameters. Vehicles can be in a complex state of snaking and deviating. Different eigenvalues have varying effects on different forms of instability. The critical velocity of the linear stability analysis model obtained through the eigenvalue method is relatively lower than the critical velocity of the nonlinear model.

Keywords: articulated vehicles; stability analysis; nonlinear dynamic model; snake instability; eigenvalue analysis

Citation: Lei, T.; Wang, J.; Yao, Z. Modelling and Stability Analysis of Articulated Vehicles. *Appl. Sci.* **2021**, *11*, 3663. https://doi.org/10.3390/app11083663

Academic Editor: Flavio Farroni

Received: 29 March 2021
Accepted: 14 April 2021
Published: 19 April 2021

1. Introduction

Articulated vehicles with small steering radius and positive maneuverability are extensively used in mining, construction, forestry, emergency rescue, and other fields [1,2]. Articulated vehicles consist of two separate front and rear vehicles and an articulated steering device connecting the both vehicles. This particular form of construction and steering results in underdeveloped stability, specifically at high speed, and the snaking instability phenomenon will occur [1,3]. This situation increases the operating burden and danger for drivers and limits the speed and efficiency of articulated vehicles. Therefore, the factors influencing the stability and snaking instability of articulated vehicles should be studied.

Numerous studies examined the instability of articulated vehicles, and two research methods for the stability analysis of these vehicles are currently available. The first approach assumes that vehicles travel at a constant speed, obtains the corresponding characteristic equations by linearising the vehicle dynamics model and analyses the vehicle state using the eigenvalue method [4–6]. The other approach uses numerical integration methods to obtain responses to some arbitrary inputs [7,8]. The majority of the related studies have used the first method for the corresponding analysis of articulated vehicles. School and Klein [9] studied the steering system, simulation model and stability criteria to gain an improved understanding of the closed loop stability characteristics. Horton and Crolla [5,10] firstly invoked the eigenvalue method to judge the stability of articulated vehicles. Snaking critical velocity is obtained when the real part of the eigenvalue is positive. He analysed the correspondence between articulated vehicle jack-knife and snaking. Azad [4,11] provided an overview of lateral stability analysis. The influence of some vehicle structure parameters on the stability of articulated vehicles is analysed. Gao [12,13] studied the critical speed of articulated vehicles' hydraulic steering systems with different hydraulic moduli of elasticity. Xu [3] studied the snaking mode under different conditions by the articulated angle. Dudziński and Skurjat [14] analysed the effects of air content in the fluid and vehicle drive form on the articulated angle. Eigenvalue analysis is limited to linear or linearised models and is considered valid near the point of linearization [15]. Rehnberg [6] built a scaled test vehicle model, and verified the practicality of the stability analysis method based on the linearised kinetic model. Stiffer steering leads to high frequency of snaking, and flexible suspension negatively affects snaking stability [16]. Liu [8] constructed a nonlinear vehicle dynamics model and used an integration method to derive an analytical periodic solution of the system in the neighborhood of the critical speed. Mu [17] showed that elastic torsional suspension can improve the roll stability of vehicles, but the directional stability at high speed of vehicles with suspension is poorer than that of vehicles without suspension. Lopatka and Muszynski [18] analysed the limited perception of driver, which has insensitivity zones of the lateral and angular locations. Drivers rely heavily on rotational motion and their senses to prevent vehicles from snaking; hence, drivers have a significant impact on vehicle stability.

The current study constructs a nonlinear articulated vehicle dynamics model and hydraulic steering system model. The nonlinear dynamics model, stability analysis models and equations of state are obtained using Newtonian mechanical equations relative to the previous generalised force approach, specifically by deriving the Lagrange equations. The characteristics of snaking are investigated, and the causes of snaking are analysed using multiple vehicle status parameters. The factors influencing the stability of articulated vehicles are analyzed, and the influence of some new factors are specifically studied. Vehicle instability can be in the snaking and deviating complex compound state. A better analysis of vehicle stability can be conducted by combining eigenvalue curves with vehicle status parameters compared with mere eigenvalue curves.

2. Modelling of Non-Linear Systems for Articulated Vehicles

Articulated vehicle structures are relatively complex, and different simplification methods for establishing articulated vehicle dynamics models can be broadly divided into two-dimensional planar models [10,19] and three-dimensional spatial models [20,21]. Two-dimensional planar models study the lateral, longitudinal and transverse motions of vehicles on a plane. Tire forces are obtained through the vertical distribution of gravity, axial load transfer and relevant experienced or semi-experienced tire models. Three-dimensional spatial models involve substantially complex effects, such as roll motion and suspension, and the models can directly calculate tire forces based on tire vertical displacement and associated tire models. The vehicle coordinate system can be divided into a one-coordinate [19,20] and two-coordinate [22,23] systems. The one-coordinate system often takes the articulated centre as the coordinate origin. The relevant research parameters can be obtained by dereferencing the excursions of the rotational inertia and dynamic balance equations of the front and rear vehicles. The two-coordinate system often considers the front and rear plasmas as the coordinate origin. The relevant research parameters are obtained through coordinate transformation or constructing kinematic relationships between the front and rear vehicles. One of the mathematical means for modelling global dynamics is to construct generalised coordinates and generalised forces, and derive the Lagrange equations [4,24]. The other method uses Newtonian mechanics to construct mechanical equilibrium equations. According to dissimilar research questions, the reasonable simplification of models and selection of appropriate modelling methods can improve efficiency and obtain superior results.

The vehicle dynamics model is the research foundation of vehicle handling and stability characteristics. However, owing to the special steering mechanism of articulated vehicles, front and rear vehicles coupling, non-linear influence of tires, axle load transfer and other factors, accurate articulated vehicle nonlinear system models should be established, including vehicle dynamics, tire and hydraulic steering models.

2.1. Vehicle Dynamics Analysis and Modelling

The articulated vehicle dynamics model is simplified as follows based on this study's research objectives.

(1) The front and rear body centres are located on the longitudinal central axis, and the vehicle is symmetrical with respect to the longitudinal central axis.
(2) The influence of the tire camber angle and return torque on wheel dynamics are disregarded.
(3) Air resistance is disregarded, and the road surface is flat and two-dimensional.

The articulated vehicle dynamics model is shown in Figure 1. The force analysis of the articulated vehicle is performed to obtain the equations of the equilibrium for the lateral, longitudinal and transverse swings of the front and rear vehicles.

$$\begin{cases} m_1(\dot{v}_{x1} - v_{y1}\omega_1) = F_{x1} + F_{x2} + F_{Ox1} \\ m_1(\dot{v}_{y1} - v_{x1}\omega_1) = F_{y1} + F_{y2} - F_{Oy1} \\ I_1\dot{w}_1 = M_{O1} + (F_{x1} - F_{x2})B + (F_{y1} + F_{y2})L_1 + F_{Oy1}L_2 \end{cases} \tag{1}$$

$$\begin{cases} m_2(\dot{v}_{x2} - v_{y1}\omega_2) = F_{x3} + F_{x4} - F_{Ox2}\cos\beta - F_{Oy2}\sin\beta \\ m_2(\dot{v}_{y2} - v_{x2}\omega_2) = F_{y3} + F_{y4} - F_{Ox2}\sin\beta + F_{Oy2}\cos\beta \\ I_2\dot{w}_2 = -M_{O2} + (F_{x3} - F_{x4})B - (F_{y3} + F_{y4})L_3 + (F_{Oy2}\cos\beta - F_{Ox2}\sin\beta)L_4 \end{cases} \tag{2}$$

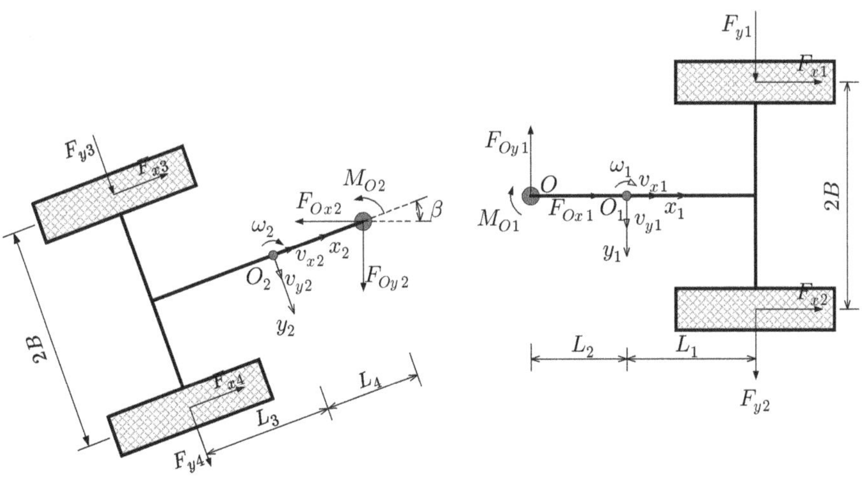

Figure 1. Articulated vehicle dynamics model.

Based on the structural and kinematic relationship of the front and rear vehicles, the relationships between the front and rear vehicles are as follows:

$$\begin{cases} v_{x2} = v_{x1}\cos\beta - (v_{y1} - L_2\omega_1)\sin\beta \\ v_{y2} = v_{x1}\sin\beta + (v_{y1} - L_2\omega_1)\cos\beta - L_4\omega_2 \end{cases} \tag{3}$$

$$\begin{cases} \dot{v}_{x2} = \dot{v}_{x1}\cos\beta - (\dot{v}_{y1} - L_2\dot{\omega}_1)\sin\beta - (v_{y1}\cos\beta + v_{x1}\sin\beta - L_2\omega_1\cos\beta)(\omega_1 - \omega_2) \\ \dot{v}_{y2} = \dot{v}_{x1}\sin\beta + (\dot{v}_{y1} - L_2\dot{\omega}_1)\cos\beta - L_4\dot{\omega}_2 + (v_{x1}\cos\beta - v_{y1}\sin\beta + L_2\omega_1\sin\beta)(\omega_1 - \omega_2) \end{cases} \tag{4}$$

$$\dot{\beta} = \omega_1 - \omega_2 \tag{5}$$

Accordingly, the vehicle dynamics model can be obtained by combining Equations (1) to (5).

$$\begin{cases} (m_1+m_2)\dot{v}_{x1} = (m_1+m_2)\omega_1 v_{y1} + m_2 L_4 \dot{\omega}_2 \sin\beta - m_2 L_2 \omega_1^2 - m_2 L_4 \omega_2^2 \cos\beta \\ \qquad + F_{x1} + F_{x2} + (F_{x3}+F_{x4})\cos\beta + (F_{y3}+F_{y4})\sin\beta \\ (m_1+m_2)\dot{v}_{y1} = m_2 L_2 \dot{\omega}_1 - (m_1+m_2)\omega_1 v_{x1} + m_2 L_4 \dot{\omega}_2 \cos\beta + m_2 L_4 \omega_2^2 \sin\beta \\ \qquad + F_{y1} + F_{y2} + (F_{y3}+F_{y4})\cos\beta - (F_{x3}+F_{x4})\sin\beta \\ I_1\dot{\omega}_1 = M_{O1} + (F_{x1}-F_{x2})B + (F_{y1}+F_{y2})(L_1+L_2) - m_1(\dot{v}_{y1}+v_{x1}\omega_1)L_2 \\ I_2\dot{\omega}_2 = -M_{O2} + (F_{x3}-F_{x4})B - (F_{y3}+F_{y4})L_3 + (F_{y1}+F_{y2})L_4\cos\beta + (F_{x1}+F_{x2})L_4\sin\beta \\ \qquad - m_1(\dot{v}_{y1}+v_{x1}\omega_1)L_4\cos\beta - m_1(\dot{v}_{x1}-v_{y1}\omega_1)L_4\sin\beta \end{cases} \quad (6)$$

An analysis of snaking indicates that articulated vehicles move in a straight line. Given that lateral and longitudinal accelerations cause axial load transfer, the transferred load should be distributed in addition to self-gravity. The articulated vehicle vertical force analysis is shown in Figure 2, and the vertical force of each tire is shown in Equation (7).

$$\begin{cases} F_{z1} = \frac{(m_1+m_2)g}{4}(2k_1+2k_2-1) - \frac{k_2(m_1+m_2)a_x'h}{2c} + \frac{(m_1+m_2)a_y'h}{4B} \\ F_{z2} = \frac{(m_1+m_2)g}{4}(1-2k_1+2k_2) - \frac{k_2(m_1+m_2)a_x'h}{2c} - \frac{(m_1+m_2)a_y'h}{4B} \\ F_{z3} = \frac{(m_1+m_2)g}{4}(1+2k_1-2k_2) + \frac{k_2(m_1+m_2)a_x'h}{2c} + \frac{(m_1+m_2)a_y'h}{4B} \\ F_{z4} = \frac{(m_1+m_2)g}{4}(3-2k_1+2k_2) + \frac{k_2(m_1+m_2)a_x'h}{2c} - \frac{(m_1+m_2)a_y'h}{4B} \end{cases} \quad (7)$$

where $a_{x1} = \dot{v}_{x1} - v_{y1}\omega_1$, $a_{y1} = \dot{v}_{y1} - v_{x1}\omega_1$, $a_{x2} = \dot{v}_{x2} - v_{y2}\omega_2$, $a_{y2} = \dot{v}_{y2} - v_{x2}\omega_2$, $a_x' = \frac{m_1 a_{x1} + m_2 a_{x2}\cos\beta - m_2 a_{y2}\sin\beta}{m_1+m_2}$, $a_y' = \frac{m_1 a_{y1} + m_2 a_{y2}\cos\beta - m_2 a_{x2}\sin\beta}{m_1+m_2}$, $k_1 = \frac{a}{2B}$, and $k_2 = \frac{c}{L_1+L_2+L_3+L_4}$.

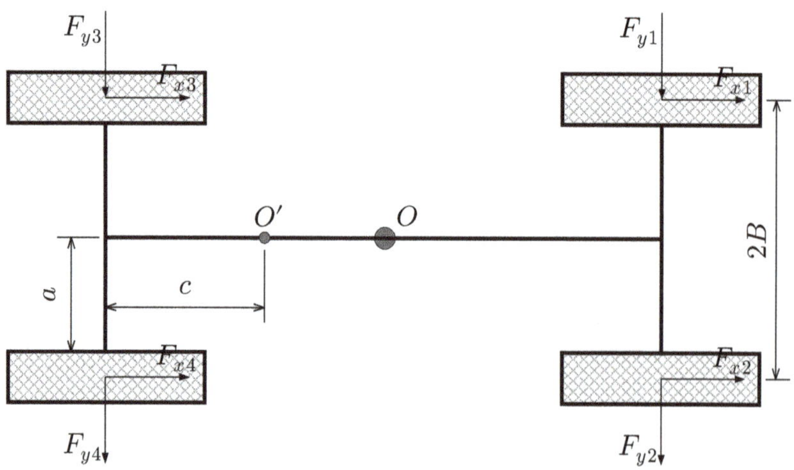

Figure 2. Articulated vehicle vertical force analysis.

The lateral and longitudinal linear velocities at the centre of each wheel are shown in Equations (8) and (9), respectively.

$$\begin{cases} v_{tx1} = v_{x1} + B\omega_1 \\ v_{tx2} = v_{x1} - B\omega_1 \\ v_{tx3} = v_{x2} + B\omega_2 \\ v_{tx4} = v_{x2} - B\omega_2 \end{cases} \tag{8}$$

$$\begin{cases} v_{ty1,2} = v_{y1} + L_1\omega_1 \\ v_{ty3,4} = v_{y2} - L_3\omega_2 \end{cases} \tag{9}$$

The angular acceleration, tire sideslip angle and longitudinal slip rate are shown in Equations (10)–(12), respectively.

$$\dot{\omega}_{ti} = \frac{T_{ti} - (F_{xi} + F_{zi}f)reff}{I_{ti}} \tag{10}$$

$$\alpha_{ti} = -\arctan\frac{v_{tyi}}{v_{txi}} \tag{11}$$

$$S_{ti} = \frac{v_{txi} - reff\omega_{ti}}{reff\omega_{ti}} \tag{12}$$

2.2. Tire Models

External forces on vehicles are mainly applied by the ground to the vehicle body through the tires. Thus, the tire model affects the vehicle dynamics, and is often a non-linear model. This study uses the Dugoff tire model, and the tire model is complementary to the elastic basis analysis model developed by Fiala and the lateral force-vertical force synthesis model of Pacejka and Sharp. The tire model provides a method to calculate forces under the combined lateral and longitudinal forces [18]. The most important feature is that it is fast and requires only a few parameters for calculation. The lateral and longitudinal forces of tires can be obtained according to the vertical force, tire sideslip angle and longitudinal slip rate. The model equations are as follows:

$$\begin{cases} F_x = C_\sigma \dfrac{S_{ti}}{1+S_{ti}} f(\lambda) \\ F_y = C_\alpha \dfrac{\tan\alpha}{1+S_{ti}} f(\lambda) \end{cases} \tag{13}$$

where $\lambda = \frac{\mu F_z(1+S_{ti})}{2\left[(C_\sigma S_{ti})^2 + (C_\sigma \tan\alpha)^2\right]^{1/2}}$, $f(\lambda) = \begin{cases} (2-\lambda)\lambda & \lambda < 1 \\ 1 & \lambda \geq 1 \end{cases}$. Substituting Equations (7) to (12) in Equation (13) obtains the tire lateral and longitudinal forces for each wheel.

2.3. Hydraulic Steering System Model

The hydraulic steering system model has a significant impact on the stability of articulated vehicles. When articulated vehicles travel in a straight line or at a fixed radius, the inlet and outlet ports of the steering valve are closed, and the hydraulic steering system is equivalent to a torsion spring acting at the steering hinge point and connecting the front and rear bodies [12]. The hydraulic steering model is shown in Figure 3.

The lengths of the left and right steering cylinders and the lengths of the two cylinders when $\beta = 0$ are as follows:

$$\begin{cases} L_1L_2 = \sqrt{R^2 + r^2 - 2rR\cos(\theta+\beta)} \\ R_1R_2 = \sqrt{R^2 + r^2 - 2rR\cos(\theta-\beta)} \\ L = \sqrt{R^2 + r^2 - 2rR\cos\theta} \end{cases} \tag{14}$$

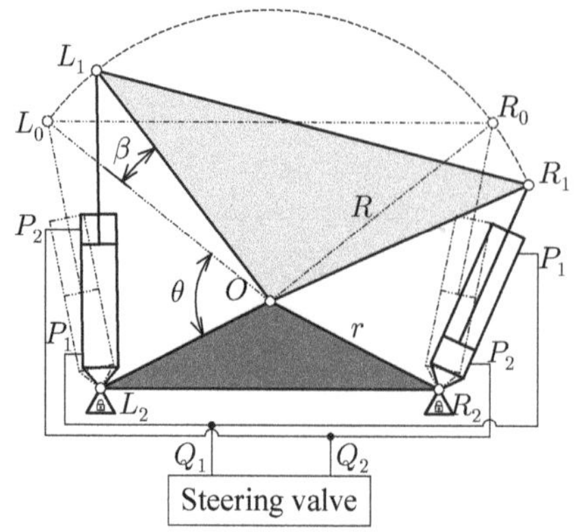

Figure 3. Hydraulic steering system model.

The force arms corresponding to the thrust of the hydraulic cylinders on both sides are as follows:

$$\begin{cases} h_1 = \dfrac{Rr\sin(\theta+\beta)}{L_1L_2} \\ h_2 = \dfrac{Rr\sin(\theta-\beta)}{R_1R_2} \end{cases} \quad (15)$$

$$\begin{cases} V_{c1} = V_0 + A_1(L_1L_2 - L) + A_2(L - R_1R_2) \\ V_{c2} = V_0 + A_2(L - L_1L_2) + A_1(R_1R_2 - L) \end{cases} \quad (16)$$

where V_{c1}, V_{c2} is the fluid volume of the P_1, P_2 chamber, V_0 is the volume of the fluid in the initial position, A_1 and A_2 are the cross-sectional area of the rod less and rod cavities, respectively, Q_1 is the amount of oil supplied to the steering cylinder and Q_2 is the amount of oil discharged from the steering cylinder.

$$\begin{cases} Q_1 = (A_1h_1 + A_2h_2)\dot{\beta} + \dfrac{V_{c1}}{K_c}\dfrac{\mathrm{d}P_1}{\mathrm{d}x} \\ Q_2 = (A_1h_1 + A_2h_2)\dot{\beta} - \dfrac{V_{c2}}{K_c}\dfrac{\mathrm{d}P_2}{\mathrm{d}x} \end{cases} \quad (17)$$

When the articulated vehicle is travelling straight or at a fixed radius, $Q_1 = Q_2 = 0$:

$$\begin{cases} \dfrac{\mathrm{d}P_1}{\mathrm{d}x} = -\dfrac{K_c}{V_{c1}}(A_1h_1 + A_2h_2)\dot{\beta} \\ \dfrac{\mathrm{d}P_2}{\mathrm{d}x} = \dfrac{K_c}{V_{c2}}(A_1h_1 + A_2h_2)\dot{\beta} \end{cases} \quad (18)$$

The left and right hydraulic cylinder forces are as follows:

$$\begin{cases} F_L = P_1A_1 - P_2A_2 \\ F_R = P_2A_1 - P_1A_2 \end{cases} \quad (19)$$

The hydraulic steering system has a torque on the articulation points of the articulated vehicles:

$$M_{O1} = -M_{O2} = -(F_Lh_1 - F_Rh_2) \quad (20)$$

2.4. Model Simulation

On the basis of the nonlinear dynamics mathematical model of the previously described articulated vehicle, the articulated vehicle dynamics simulation model was built in the MATLAB/Simulink simulation environment. The simulation model includes the vehicle dynamics, tire, driver, hydraulic steering system and vertical force calculation models. The parameters in the model are shown in Table 1.

Table 1. Parameters of the vehicle dynamics model.

Parameters	Values	Parameters	Values
m_1	6980 kg	reff	0.875 m
m_2	9767 kg	C_σ	2×10^5 N/m
L_1	0.25 m	C_α	5×10^4 N/rad
L_2	1.35 m	R	0.45 m
L_3	0.2 m	r	0.55 m
L_4	1.459 m	μ	0.8
B	1.15 m	A_1	0.002826 m^2
I_1	32,977 kgm^2	A_2	0.002112 m^2
I_2	13,228 kgm^2	V_0	0.0015 m^3
I_z	117 kgm^2	K_C	7×10^8 Pa

To verify the snaking, the vehicle travels at an initial speed of 5 m/s (disregarding rolling friction) and exerts a pulse moment interference on the articulated points. The parameters of each vehicle are in a state of oscillatory convergence with decreasing amplitude, as shown in Figure 4. Owing to interference, the ground exerts lateral forces on the tires, thereby generating lateral and angular accelerations about the z-axis in different directions and amplitudes for the front and rear vehicles. The front and rear vehicles produce lateral and angular velocities about the z-axis, and the vehicle's swing angle changes accordingly, thereby resulting in the snaking phenomenon. Owing to vehicle structural parameters, ground and speed, the swing angle is in a state of oscillatory convergence, while the snaking oscillation is in a steady state. Ideally, if the friction between the ground and tires is zero, then the entire snaking oscillation system of the vehicle is oscillating without damping, and the swing angle should be an equal amplitude oscillation. If the snaking is unstable under the influence of vehicle parameters, ground, speed and other factors, then the swing angle and corresponding parameters will oscillate and diverge.

Numerous reasons can be cited for snaking, the most important of which is the compressibility of the hydraulic fluid in the hydraulic steering system, thereby results in a certain degree of stiffness of the hydraulic steering system. Other vehicle parameters also affect the stability of vehicles, such as the position of the centre of gravity, tire stiffness, hydraulic system characteristics and other factors. Ideally, when the road is completely flat and in the same condition, the lateral and longitudinal forces on the ground are identical for each tire, thereby substantially producing snaking. However, the fact is that roads will not be in ideal condition, the forces acting on the wheels will be different and non-structural terrain will increase the occurrence of snaking. When drivers operate vehicles under the road surface or other external disturbance, they will have a poor driving experience and misuse the steering system to suppress the effect of the swing or deviation. However, this situation alters the steady state and increases snaking, thereby creating a vicious cycle.

To investigate the effects of velocity variation and different loads on the swing angle, several sets of simulations were performed. As shown in Figure 4e, the vehicles are all in steady state at three different velocities of 2 m/s, 5 m/s and 10 m/s. As the vehicle velocity increases, the convergence rate of the swing angle gradually decreases which means the vehicle stability tends to decrease as the velocity increases. The vehicles are also in steady state at three different loads of 20%, 60% and 100% load factor of the front vehicle, as shown in Figure 4f, the amplitude of the swing angle decreases with the increase of the load factor. The oscillation frequency of the swing angle is lower at high load factor area than that at low load factor area. Due to the influence of vehicle weight and the moment of inertia, the relationship between the swing angle and load factor is not linear. Overall, the increase in vehicle load factor is beneficial for vehicle stability.

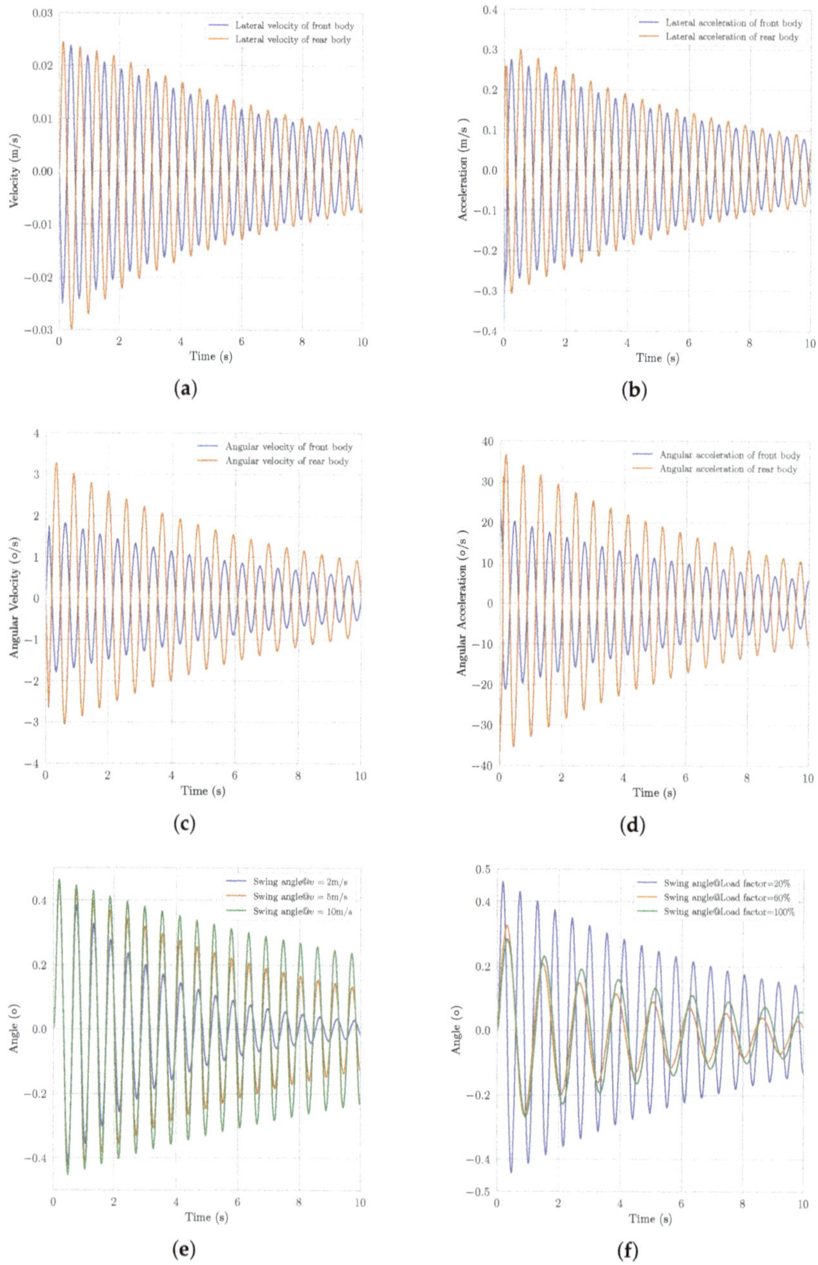

Figure 4. Simulation results: (**a**) lateral velocity, (**b**) lateral acceleration, (**c**) angular velocity, (**d**) angular acceleration, (**e**) variation of the swing angle regarding the vehicle velocity, and (**f**) variation of the swing angle regarding the load factor.

The verification and analysis of the snaking phenomenon of the articulated vehicle dynamics model shows the reliability of the entire vehicle dynamics model. Accordingly, stability and snaking influence factor analyses can be performed.

3. Stability Analysis Model for Articulated Vehicles

On the basis of Lyapunov stability analysis theory, the eigenvalue method is used to analyse the relationship between vehicle parameters and stability. For the analysis, the kinetic model developed in the previous section is linearised. The forward velocity is assumed to be constant. In the event of snaking in articulated vehicles, the longitudinal force is relatively small compared with the lateral force. Hence, the longitudinal force can be disregarded. Moreover, the hydraulic steering system can be equivalent to a torsion spring, as shown in Figure 5. It should be noted that we have enlarged the articulated angle in the figure to facilitate interpretation, i.e., the actual β is not as large as shown in the figure.

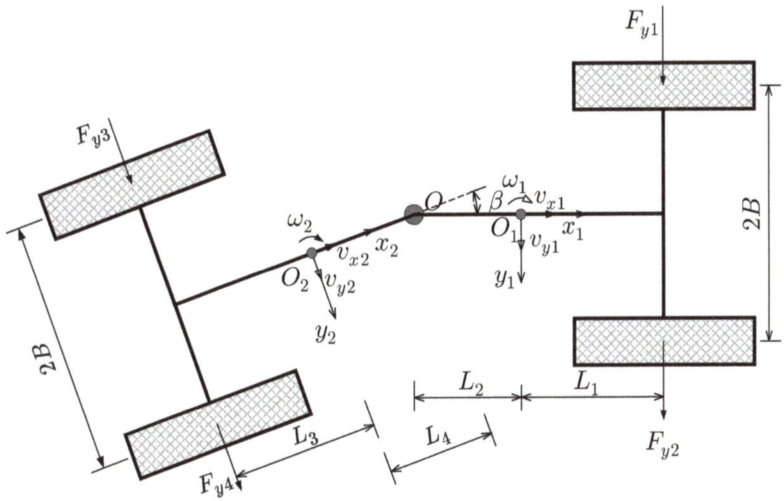

Figure 5. Simplified model of vehicle dynamics.

The vehicle dynamics equation can be expressed as follows:

$$\begin{cases} m_1(\dot{v}_{y1} + v_{x1}\omega_1) + m_2(\dot{v}_{y2} + v_{x2}\omega_2)\cos\beta = F_{y1} + F_{y2} + (F_{y3} + F_{y4})\cos\beta \\ \left(I_1 + m_1 L_2^2\right)\dot{\omega}_1 = M + (F_{y1} + F_{y2})(L_1 + L_2) \\ \left(I_2 + m_2 L_4^2\right)\dot{\omega}_2 = -M - (F_{y3} + F_{y4})(L_3 + L_4) \\ \dot{\beta} = \omega_1 - \omega_2 \end{cases} \tag{21}$$

Since β is considerably small that it is acceptable to assume $\cos\beta \approx 1$, then the first equation in Equation (21) can be simplified to

$$m_1(\dot{v}_{y1} + v_{x1}\omega_1) + m_2(\dot{v}_{y2} + v_{x2}\omega_2) = F_{y1} + F_{y2} + F_{y3} + F_{y4}$$

The front and rear vehicle kinematics relationship is as follows:

$$\begin{cases} v_{ty1,2} = v_{y1} + L_1\omega_1 \\ v_{ty3,4} = v_{y2} - L_3\omega_2 = v_{y1} - (L_2 + L_3 + L_4)\omega_1 + (L_3 + L_4)\dot{\beta} + v_{x1}\dot{\beta} \end{cases} \tag{22}$$

Tire forces can be expressed as follows:

$$\{F_y = C_\alpha \alpha \tag{23}$$

Hydraulic systems can be equated as follows:

$$\left\{ M = \frac{2K_c}{V_0}(A_1h_1 + A_2h_2)^2\beta \right. \tag{24}$$

In conjunction with Equations (21) to (24), the system dynamics differential equation group is obtained with the equation of state as follows:

$$\begin{bmatrix} \dot{v}_{y1} \\ \dot{\omega}_1 \\ \ddot{\beta} \\ \dot{\beta} \end{bmatrix} = \begin{bmatrix} a_{11} & a_{12} & a_{13} & a_{14} \\ a_{21} & a_{22} & a_{23} & a_{24} \\ a_{31} & a_{32} & a_{33} & a_{34} \\ a_{41} & a_{42} & a_{43} & a_{44} \end{bmatrix} \begin{bmatrix} v_{y1} \\ \omega_1 \\ \dot{\beta} \\ \beta \end{bmatrix} \tag{25}$$

where $a_{11} = \frac{2(C_{\alpha1}+C_{\alpha2})}{(m_1+m_2)v_{x1}} - \frac{2m_2L_4(L_3+L_4)C_{\alpha2}}{(I_2+m_2L_4^2)(m_1+m_2)v_{x1}} + \frac{2m_2L_2(L_1+L_2)C_{\alpha1}}{(I_1+m_1L_2^2)(m_1+m_2)v_{x1}}$, $a_{12} = \frac{2L_1C_{\alpha1}-2(L_2+L_3+L_4)C_{\alpha2}}{(m_1+m_2)v_{x1}} + \frac{2m_2L_4(L_3+L_4)(L_2+L_3+L_4)C_{\alpha2}}{(I_2+m_2L_4^2)(m_1+m_2)v_{x1}} + \frac{2m_2L_2(L_1^2+L_1L_2)C_{\alpha1}}{(I_2+m_1L_2^2)(m_1+m_2)v_{x1}} - v_{x1}$, $a_{13} = \frac{2(L_3+L_4)C_{\alpha2}}{(m_1+m_2)v_{x1}} - \frac{2m_2L_4(L_3+L_4)^2C_{\alpha2}}{(I_2+m_2L_4^2)(m_1+m_2)v_{x1}}$, $a_{14} = \frac{2C_{\alpha2}}{(m_1+m_2)} + \frac{m_2L_4K_R-2m_2L_4C_{\alpha2}(L_3+L_4)}{(I_2+m_2L_4^2)(m_1+m_2)} - \frac{m_2L_2K_R}{(m_1+m_2)(I_1+m_1L_2^2)}$, $a_{21} = \frac{2C_{\alpha1}(L_1+L_2)}{(I_1+m_1L_2^2)v_{x1}}$, $a_{22} = \frac{2C_{\alpha1}(L_1^2+L_1L_2)}{(I_1+m_1L_2^2)v_{x1}}$, $a_{23} = 0$, $a_{24} = \frac{-K_R}{I_1+m_1L_2^2}$, $a_{31} = \frac{2C_{\alpha1}(L_1+L_2)}{(I_1+m_1L_2^2)v_{x1}} + \frac{2C_{\alpha2}(L_3+L_4)}{(I_2+m_2L_4^2)v_{x1}}$, $a_{32} = \frac{2C_{\alpha1}(L_1^2+L_1L_2)}{(I_1+m_1L_2^2)v_{x1}} - \frac{2C_{\alpha2}(L_3+L_4)(L_2+L_3+L_4)}{(I_2+m_2L_4^2)v_{x1}}$, $a_{33} = \frac{2(L_3+L_4)^2C_{\alpha2}}{(I_2+m_2L_4^2)v_{x1}}$, $a_{34} = \frac{2C_{\alpha2}(L_3+L_4)-K_R}{I_2+m_2L_4^2} - \frac{K_R}{I_1+m_1L_2^2}$, $a_{41} = 0$, $a_{42} = 0$, $a_{43} = 1$, $a_{44} = 0$.

According to the Lyapunov stability conditions, the stability of articulated vehicles is related to the real part of each eigenvalue of the equation of state. When the real part of each eigenvalue of the equation of state is negative/positive, the vehicles are in a stable/unstable state.

4. Analysis of the Stability Influencing Factors

4.1. Effects of the Centre of Mass Position

Owing to the wide range of engineering applications of articulated vehicles, different loads and working devices can change the position of the centre of mass. In changing the distance from the front and rear mass centre positions to the articulation point, stability analysis theory is used to study the effects of the centre of mass position. The distance between the front and rear mass centres was changed from $L_2 = 1.35$ m, $L_4 = 1.459$ m to $L_2 = 1.85$ m, $L_4 = 1.859$ m and $L_2 = 1.05$ m, $L_4 = 1.159$ m. The eigenvalues and swing angle are shown in Figure 6.

When $L_2 = 1.35$ m and $L_4 = 1.459$ m, the real parts of the eigenvalues of λ_1 and λ_2 are positive, the corresponding oscillation of the swing angle is diverging, the swing angle is increasing and the vehicle is unstable. The other two conditions have negative eigenvalues, corresponding to the convergence of the swing angle, but the $L_2 = 1.05$ m, $L_4 = 1.159$ m corresponding eigenvalues are significantly smaller, and the swing angle converges rapidly. This result means that decreasing the distance between the centre of mass position and articulation point is beneficial for increasing vehicle stability. When the centre of mass position exceeds the position of the wheel axle, the parameter has a significant impact on the stability of vehicles (i.e., may cause instability). Therefore, when designing an articulated vehicle structure, the centre of mass should be positioned as far inside the wheel axle as possible, and the distance between the centre of mass position and articulation point should be as short as possible.

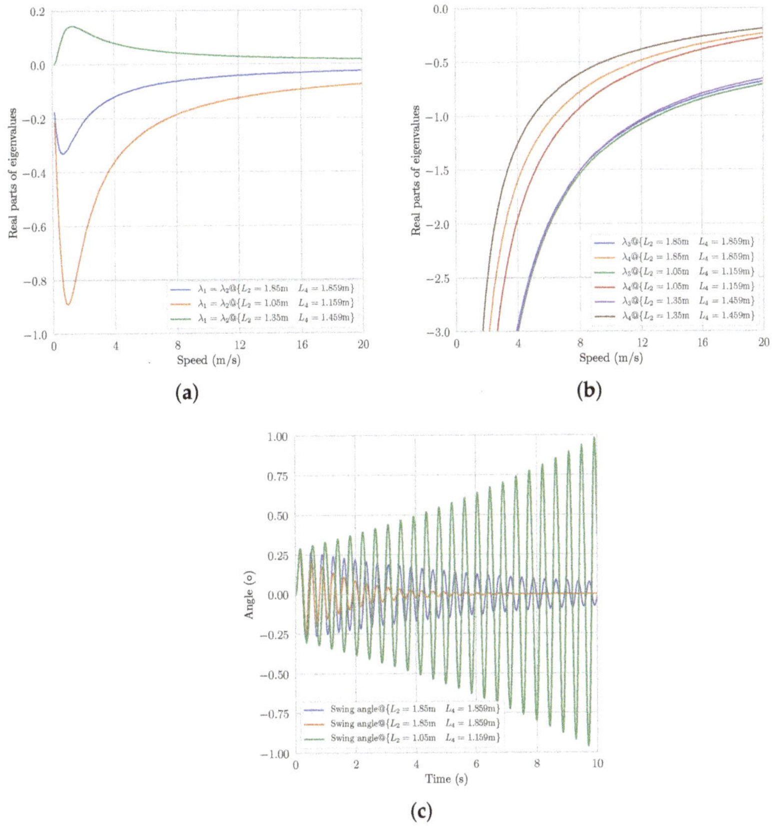

Figure 6. Variation of the real parts of the eigenvalues and swing angle with the center of mass: (**a**) real parts of λ_1 and λ_2, (**b**) real parts of λ_3 and λ_4, and (**c**) swing angle of the vehicle.

4.2. Effects of Torsional Stiffness

Engineering hydraulic oil modulus of elasticity can assume a fixed value, but leakage from hydraulic steering systems, a certain amount of air in the fluid, hydraulic pipes and other influencing factors, the hydraulic oil composite modulus of elasticity of the articulated vehicle steering system will be substantially reduced compared with the ideal situation. The stability of articulated vehicles under different elastic moduli of 0.1 K_C, 0.2 K_C, 0.5 K_C, 1 K_C, and the eigenvalues and swing angles are shown in Figure 7. The eigenvalues λ_1, λ_2 are negative, while the two conditions with the lower modulus of elasticity at higher velocity of the eigenvalues λ_3, λ_4 are positive. The analysis of the results indicate that the higher the hydraulic oil composite modulus of elasticity, the greater the rigidity of the hydraulic steering system, the more stable the vehicles and the faster the convergence of the swing angle. However, the amplitude of the swing angle is not linearly related to the elastic modulus, i.e., the higher the elastic modulus, the higher the oscillation frequency.

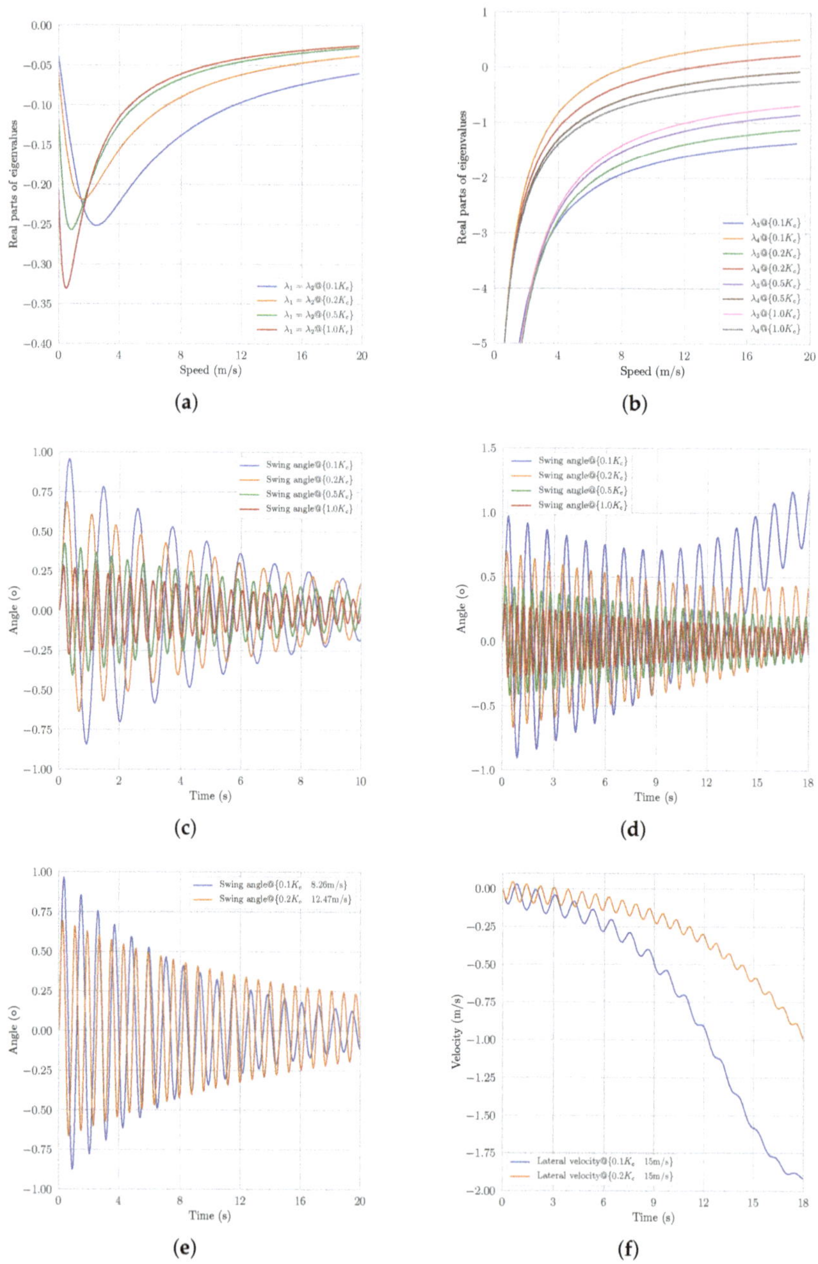

Figure 7. Variation of the real parts of the eigenvalues, swing angles and lateral velocity with the torsional stiffness: (**a**) real parts of λ_1 and λ_2, (**b**) real parts of λ_3 and λ_4, (**c**) swing angle of the vehicle at 5 m/s, (**d**) swing angle of the vehicle at 15 m/s, (**e**) swing angle of the vehicle at critical velocity, and (**f**) lateral velocity.

Kinetic simulations of two lower elastic moduli at the respective critical velocities are performed separately, and the respective swing angles are in a steady state. When the velocity exceeds the critical speed of a certain range, the vehicle bending angle does not oscillate with equal amplitude in the centre of line zero, but gradually deviates from the state of the centre line of the oscillation with a type of amplitude gradually decreasing, and the lateral velocities of the vehicle gradually increase. At the time the vehicle is in a relatively deviating stage with snaking, the smaller the modulus of elasticity, the faster the deviation. The stability analysis model differs from vehicle nonlinear dynamics model owing to linearisation. According to the analytical curves, the critical velocity obtained by the linearised model is relatively lower than the critical velocity obtained by the nonlinear dynamics model.

4.3. Effects of Mass

Different loads on vehicles result in the front and rear masses being in a varying state. The stability characteristics of the front and rear vehicles under 1.5 times the original mass are studied, and the eigenvalues are shown in Figure 8. All eigenvalues are negative, the articulated vehicle is in a stable state and changing the mass will not have a substantial impact on stability in a certain range of vehicle parameters. According to the change in eigenvalue curves, increasing the mass of the front vehicle will increase the stability of the articulated vehicle, while increasing the mass of the rear vehicle will reduce the stability of the articulated vehicle volume.

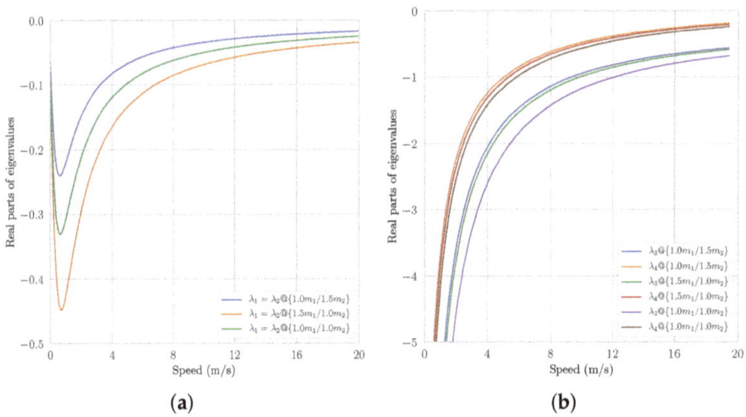

Figure 8. Variation of the real parts of the eigenvalues with the vehicle mass: (**a**) real parts of λ_1 and λ_2 and (**b**) real parts of λ_3 and λ_4.

4.4. Effects of the Moment of Inertia

The stability characteristics of the front and rear vehicles under 1.5 times the original rotational inertia are studied separately, and the eigenvalues are shown in Figure 9. All eigenvalues are negative and the articulated vehicle is in a stable state. Similar to the effects of mass, stable vehicle parameters indicate that changing the moment of inertia within a certain range will not have a significant effect on stability. According to the eigenvalue curves, increasing the moments of inertia of the front and rear vehicles increase and decrease, respectively, the stability of the articulated vehicle.

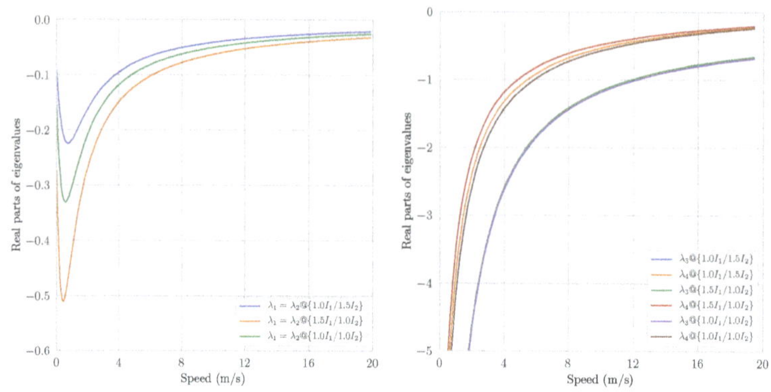

(a) (b)

Figure 9. Variation of the real parts of the eigenvalues with the moment of inertia: (**a**) real parts of λ_1 and λ_2 and (**b**) real parts of λ_3 and λ_4.

4.5. Effects of Tire Cornering Stiffness

Tyres are the component of the vehicle that directly contacts the ground and their condition has a crucial impact on the stability of the vehicle. In addition, pressure is one of the key parameters of a tyre. It is difficult or even impossible to establish a universal model to accurately calculate the relationship between tire pressure and cornering stiffness for different tires. Several related studies [25] showed that the relationship between the cornering stiffness and the tire pressure is approximately linear under heavy load. As articulated vehicles are mostly heavy construction vehicles which means the tires are always under heavy load, so the cornering stiffness is very likely to increases as the tire pressure increase linearly. Therefore, the effect of tires pressure on stability can be analysed implicitly by analysing the effect of cornering stiffness on stability.

The stability characteristics of articulated vehicles under different tire cornering stiffness conditions are studied, and the values are shown in Figure 10. All eigenvalues are negative, and the articulated vehicle is in a stable state. According to the eigenvalue curve, increasing tire stiffness will increase the stability of the vehicle. Therefore, in terms of stability, tires with high rigidity should be chosen as much as possible when designing vehicle matching.

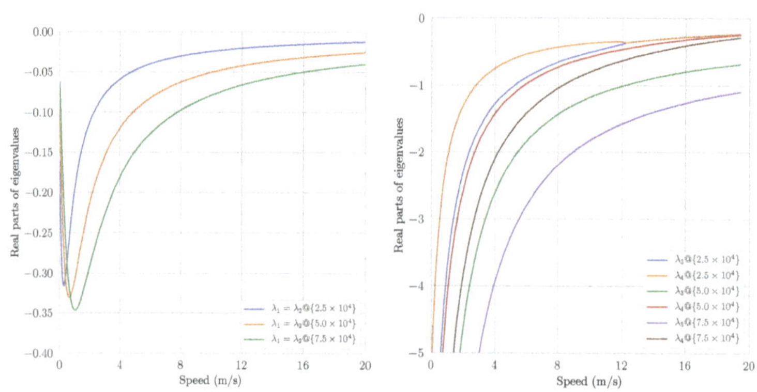

(a) (b)

Figure 10. Variation of the real parts of the eigenvalues with the tire stiffness: (**a**) real parts of λ_1 and λ_2 and (**b**) real parts of λ_3 and λ_4.

4.6. Effects of the Hydraulic Cylinder Force Arm

The position of the hinge point of the hydraulic cylinder affects the stiffness of the hydraulic system and also has an impact on the stability of articulated vehicles. The stability characteristics of articulated vehicles with different force arms are studied, and the values are shown in Figure 11. λ_1 and λ_2 are negative, while λ_3 and λ_4 are positive at high velocity for the small force arms. The analysis of the results indicates that the larger the force arm, the more rigid the hydraulic steering system and the more stable the articulated vehicles. Moreover, articulated vehicles with small force arm will seemingly deviate with high speed, thereby indicating that the vehicle is in an unstable state. Therefore, the force arm should be designed as large as possible to increase the stability.

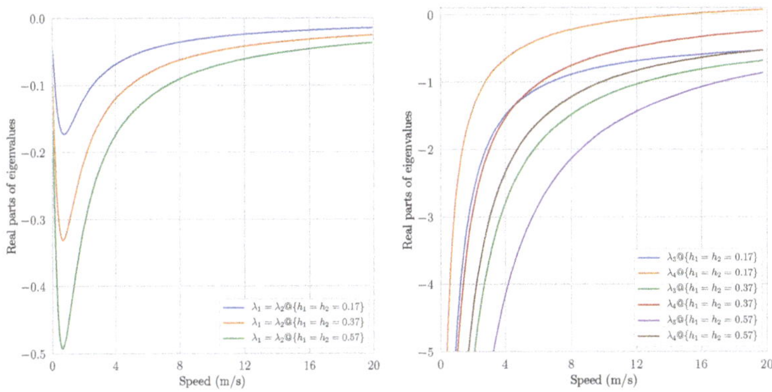

(**a**) (**b**)

Figure 11. Variation of the real parts of the eigenvalues with the hydraulic cylinder force arm: (**a**) real parts of λ_1 and λ_2 and (**b**) real parts of λ_3 and λ_4.

5. Discussion

A nonlinear articulated vehicle dynamics model and hydraulic steering system model are established in this research. By studying the curve characteristics of the swing angle, lateral velocity, lateral acceleration and angular velocity and angular acceleration parameters in the snaking phenomenon, the evaluation and analysis indexes of the snaking are determined. The causes of the snaking phenomenon are analysed from the perspective of mechanical principles and subjective and objective factors (e.g., drivers and environment).

5.1. Conclusions

A linearised stability analysis model was established to analyse the factors influencing the stability of articulated vehicles. The centre of mass position, modulus of elasticity of the hydraulic fluid, tire stiffness, mass, rotational inertia and hydraulic cylinder arm have an impact on vehicle stability. An effective method for determining vehicle stability was established by combining eigenvalue curves with vehicle condition. Stability analysis indicates that the stability of articulated vehicles decreases with increasing speed regardless of the state of these vehicles. Among the vehicle parameters, centre of mass position and hydraulic system have a more substantial influence on the stability of articulated vehicles than the other parameters. The real parts of λ_1 and λ_2 have a substantial influence on snaking instability, while the real parts of λ_1 and λ_2 have a significantly deviating instability influence. Articulated vehicles are in unstable and compound state of snaking and deviating. The critical velocity obtained by the linearised model is relatively lower than the critical velocity obtained by the nonlinear dynamics model. The eigenvalue curve can be used to identify and compare the state of vehicles. An effective stability analysis method is formed to provide the basis for vehicle structural design and vehicle parameter matching and control methods in optimising vehicle stability.

5.2. Limitation and Outlook

Owing to several limitations, no experiments were carried out in this paper. Considering the safety factor, in the subsequent research, this study will focus on the development and improvement of the unmanned articulated vehicle, and carry out experiments to verify and correct the theory and simulation when the vehicle satisfies the experimental requirement.

Since the paper focuses on the transverse swing characteristics, the vehicle dynamics model is established as a planar model, ignoring the three degrees of freedom of z, roll and pitch, and simplifying some conditions; also, the stability analysis method used in this paper linearises the equation of state, which has little effect when the swing angle is small, but as the swing angle increases, the non-linear characteristics of the vehicle will increase and linearisation is no longer appropriate. These issues can have a considerable negligence on the accuracy and credibility of the analysis results and will be the focus of our subsequent research.

Author Contributions: Conceptualization, Z.Y. and J.W.; methodology, T.L. and Z.Y.; software, T.L.; validation, T.L. and Z.Y.; formal analysis, T.L.; investigation, J.W. and Z.Y.; resources, J.W. and Z.Y.; data curation, T.L.; writing—original draft preparation, T.L.; writing—review and editing, Z.Y.; visualization, Z.Y.; supervision, J.W.; project administration, J.W.; funding acquisition, J.W. and Z.Y. All authors have read and agreed to the published version of the manuscript.

Funding: This research was funded by the National Natural Science Foundation of China grant number 51875239 and 51875232, and the National Key Research and Development Program of China grant number 2016YFC0802904.

Institutional Review Board Statement: Not applicable.

Informed Consent Statement: Not applicable.

Conflicts of Interest: The authors declare no conflict of interest. The funders had no role in the design of the study; in the collection, analyses, or interpretation of data; in the writing of the manuscript, or in the decision to publish the results.

Nomenclature

c	Distance from the centre of mass of the whole vehicle to the rear axle
C_α	Cornering stiffness
C_σ	Longitudinal tire stiffness
F_{ox1}	Longitudinal force of the steering mechanism on the front vehicle
F_{ox2}	Longitudinal force of the steering mechanism on the rear vehicle
F_{oy1}	Lateral force of the steering mechanism on the front vehicle
F_{oy2}	Lateral force of the steering mechanism on the rear vehicle
F_{xj}	Longitudinal tire force (j = 1,2,3,4)
F_{yj}	Lateral tire force (j = 1,2,3,4)
F_{zj}	Vertical tire force (j = 1,2,3,4)
I_1	Vehicle rotational inertia about the z-axis of the front vehicle
I_2	Vehicle rotational inertia about the z-axis of the rear vehicle
L_1	Distance from the centre of the front vehicle gravity to the front axles
L_2	Distance from the articulated point to the centre of the front vehicle gravity
L_3	Distance from the centre of the rear vehicle gravity to the rear axles
L_4	Distance from the articulated point to the centre of the rear vehicle gravity
M_{O1}	Torque of the steering mechanism on the front vehicle
M_{O2}	Torque of the steering mechanism on the rear vehicle

m_1	Mass of the front vehicle
m_2	Mass of the rear vehicle
$O_1x_1y_1$	Front vehicle coordinate system
$O_2x_2y_2$	Rear vehicle coordinate system
R	Distance between the hinge points of the hydraulic cylinder rod and articulated point
r	Distance between the hinge points of the hydraulic cylinder seat and articulated point
v_{x1}	Longitudinal velocity of the front vehicle
v_{x2}	Longitudinal velocity of the rear vehicle
v_{y1}	Lateral velocity of the front vehicle
v_{y2}	Lateral velocity of the rear vehicle
ω_1	Angular velocity about the z-axis of the front vehicle
ω_2	Angular velocity about the z-axis of the rear vehicle
β	Swing angle
θ	Initial angle of the hydraulic cylinder
μ	Friction coefficient

References

1. Azad, N.L.; Khajepour, A.; Mcphee, J. Robust state feedback stabilization of articulated steer vehicles. *Veh. Syst. Dyn.* **2007**, *45*, 249–275. [CrossRef]
2. Li, X.; Chen, W.; Xu, Q. A Novel Dynamic Measurement System for Evaluating the Braking Coordination of Articulated Vehicles. *J. Sens.* **2016**, *2016*, 1–10. [CrossRef]
3. Xu, T.; Ji, X.; Liu, Y.; Liu, Y. Differential Drive Based Yaw Stabilization Using MPC for Distributed-Drive Articulated Heavy Vehicle. *IEEE Access* **2020**, *8*, 104052–104062. [CrossRef]
4. Azad, N.L.; Khajepour, A.; McPhee, J. Effects of locking differentials on the snaking behaviour of articulated steer vehicles. *Int. J. Veh. Syst. Model. Test.* **2007**, *2*, 101. [CrossRef]
5. Crolla, D.A. The steering behaviour of articulated body steer vehicles. In *Road Vehicle Handling, I Mech E Conference Publications 1983–1985. Sponsored by Automobile Division of the Institution of Mechanical Engineers under patronage of Federation Internationale des Societies d'Ingenieurs des Techniques de l'Automobile (FISITA) he*; Number C123/83; MIRA: Nuneaton, Warwickshire, UK, 1983; pp. 139–146.
6. Rehnberg, A.; Edren, J.; Eriksson, M.; Drugge, L.; Trigell, A.S. Scale model investigation of the snaking and folding stability of an articulated frame steer vehicle. *Int. J. Veh. Syst. Model. Test.* **2011**, *6*, 126. [CrossRef]
7. Vlk, F. Lateral stability of articulated buses. *Int. J. Veh. Des.* **1988**, *9*, 35–51.
8. Liu, Z.; Hu, K.; Chung, K. Nonlinear analysis of a closed-loop tractor-semitrailer vehicle system with time delay. *Mech. Syst. Signal Process.* **2016**, *76–77*, 696–711. [CrossRef]
9. Scholl, R.D.; Klein, R.E. *Stability Analysis of an Articulated Vehicle Steering System*; SAE Technical Paper; SAE: Warrendale, PA, USA, 1971.
10. Horton, D.N.L.; Crolla, D.A. Theoretical Analysis of the Steering Behaviour of Articulated Frame Steer Vehicles. *Veh. Syst. Dyn.* **1986**, *15*, 211–234. [CrossRef]
11. Azad, N.L.; Khajepour, A.; McPhee, J. A survey of stability enhancement strategies for articulated steer vehicles. *Int. J. Heavy Veh. Syst.* **2009**, *16*, 26. [CrossRef]
12. Gao, Y.; Shen, Y.; Yang, Y.; Zhang, W.; Güvenç, L. Modelling, verification and analysis of articulated steer vehicles and a new way to eliminate jack-knife and snaking behaviour. *Int. J. Heavy Veh. Syst.* **2019**, *26*, 375. [CrossRef]
13. Gao, Y.; Shen, Y.; Xu, T.; Zhang, W.; Guvenc, L. Oscillatory Yaw Motion Control for Hydraulic Power Steering Articulated Vehicles Considering the Influence of Varying Bulk Modulus. *IEEE Trans. Control. Syst. Technol.* **2019**, *27*, 1284–1292. [CrossRef]
14. Dudziński, P.; Skurjat, A. Directional dynamics problems of an articulated frame steer wheeled vehicles. *J. Kones* **2012**, *19*, 89–98. [CrossRef]
15. Islam, M.M.; He, Y.; Zhu, S.; Wang, Q. A comparative study of multi-trailer articulated heavy-vehicle models. *Proc. Inst. Mech. Eng. Part D J. Automob. Eng.* **2015**, *229*, 1200–1228. [CrossRef]
16. Rehnberg, A.; Drugge, L.; Trigell, A.S. Snaking stability of articulated frame steer vehicles with axle suspension. *Int. J. Heavy Veh. Syst.* **2010**, *17*, 119. [CrossRef]
17. Chai, M.; Rakheja, S.; Shangguan, W.B. Relative ride performance analysis of a torsio-elastic suspension applied to front, rear and both axles of an off-road vehicle. *Int. J. Heavy Veh. Syst.* **2019**, *26*, 765. [CrossRef]
18. Lopatka, M.; Muszynski, T. Research of the snaking phenomenon to improve directional stability of remote controlled articulated wheel tool-carrier. In *ISARC2003: The future site: Proceedings of the 20th International Symposium on Automation and Robotics in Construction, Eindhoven, The Netherlands, 21–24 September 2003*; Technische Universiteit Eindhoven: Eindhoven, The Netherlands, 2003; p. 95.

19. Gao, G.; Wang, J.; Ma, T.; Liu, W.; Lei, T. Multistage Estimators for the Distributed Drive Articulated Steering Vehicle. *Math. Probl. Eng.* **2020**, *2020*, 1–16.
20. Li, X.; Wang, G.; Yao, Z.; Yang, Y. Research on lateral stability and rollover mechanism of articulated wheel loader. *Math. Comput. Model. Dyn. Syst.* **2014**, *20*, 248–263. [CrossRef]
21. Pazooki, A.; Rakheja, S.; Cao, D. A three-dimensional model of an articulated frame-steer vehicle for coupled ride and handling dynamic analyses. *Int. J. Veh. Perform.* **2014**, *1*, 264. [CrossRef]
22. Yin, Y.; Rakheja, S.; Yang, J.; Boileau, P. Effect of articulated frame steering on the transient yaw responses of the vehicle. *Proc. Inst. Mech. Eng. Part D J. Automob. Eng.* **2018**, *232*, 384–399. [CrossRef]
23. Pazooki, A.; Rakheja, S.; Cao, D. Kineto-dynamic directional response analysis of an articulated frame steer vehicle. *Int. J. Veh. Des.* **2014**, *65*, 1. [CrossRef]
24. Yao, Z.; Wang, G.; Guo, R.; Li, X. Theory and experimental research on six-track steering vehicles. *Veh. Syst. Dyn.* **2013**, *51*, 218–235.
25. Yang, C.; Xu, N.; Guo, K. *Incorporating Inflation Pressure into UniTire Model for Pure Cornering*; Technical Report; SAE Technical Paper; SAE: Warrendale, PA, USA, 2016.

applied sciences

MDPI

Article

Central Non-Linear Model-Based Predictive Vehicle Dynamics Control

Philipp Maximilian Sieberg * and Dieter Schramm

Chair of Mechatronics, Faculty of Engineering, University of Duisburg-Essen, 47057 Duisburg, Germany; dieter.schramm@uni-due.de

* Correspondence: philipp.sieberg@uni-due.de; Tel.: +49-203-379-1862

Featured Application: This contribution presents a central predictive control of the vehicle dynamics regarding the roll, self-steering and pitch behavior.

Abstract: Considering automated driving, vehicle dynamics control systems are also a crucial aspect. Vehicle dynamics control systems serve as an important influence factor on safety and ride comfort. By reducing the driver's responsibility through partially or fully automated driving functions, the occupants' perception of safety and ride comfort changes. Both aspects are focused even more and have to be enhanced. In general, research on vehicle dynamics control systems is a field that has already been well researched. With regard to the mentioned aspects, however, a central control structure features sufficient potential by exploiting synergies. Furthermore, a predictive mode of operation can contribute to achieve these objectives, since the vehicle can act in a predictive manner instead of merely reacting. Consequently, this contribution presents a central predictive control system by means of a non-linear model-based predictive control algorithm. In this context, roll, self-steering and pitch behavior are considered as control objectives. The active roll stabilization demonstrates an excellent control quality with a root mean squared error of 7.6953×10^{-3} rad averaged over both validation maneuvers. Compared to a vehicle utilizing a conventional control approach combined with a skyhook damping, pitching movements are reduced by 19.75%. Furthermore, an understeering behavior is maintained, which corresponds to the self-steering behavior of the passive vehicle. In general, the central predictive control, thus, increases both ride comfort and safety in a holistic way.

Keywords: central control; non-linear model-based predictive control; pitch behavior; predictive control; roll behavior; self-steering behavior; vehicle dynamics

Citation: Sieberg, P.M.; Schramm, D. Central Non-Linear Model-Based Predictive Vehicle Dynamics Control. *Appl. Sci.* **2021**, *11*, 4687. https://doi.org/10.3390/app11104687

Academic Editors: Flavio Farroni, Andrea Genovese and Aleksandr Sakhnevych

Received: 24 April 2021
Accepted: 17 May 2021
Published: 20 May 2021

1. Introduction

Two major driving factors in vehicle development are increasing the safety and enhancing the ride comfort of the vehicle [1]. Especially in the context of automated driving, where the driver becomes a passenger, the perception of ride comfort changes significantly and at the same time gains in importance [2]. Moreover, a predictive mode of operation of the vehicle is beneficial. The implementation of a central predictive control of the vehicle dynamics addresses the objectives of increasing safety and ride comfort. The central control structure exploits synergies in terms of the control quality [3]. An overview of the state of the art for centralized integrated vehicle dynamics control systems is given in [4]. Furthermore, a predictive mode of operation allows the vehicle to act in a predictive way instead of just reacting [5]. In addition to classical approaches to control vehicle dynamics such as skyhook damping [6] and model-based control algorithms [7], the utilization of artificial intelligence is gaining increased attention. This mainly includes reinforcement learning [8], fuzzy inference systems [9] as well as deterministic artificial intelligence [10]. Due to the current regulation for the use of artificial intelligence in vehicles, a model-based approach is considered in this contribution.

For this purpose, a non-linear model-based predictive control algorithm implements the central predictive control. The control objectives pursued are an active roll stabilization, the manipulation of the self-steering behavior as well as the reduction of pitching movements. Mathematical models are used to predict the system behavior as a function of the manipulated variables [11]. Subsequently, the predicted system behavior is adapted to the desired system behavior in the form of reference trajectories. Furthermore, the manipulated variables can also be taken into account in the cost function to be minimized, so that the energy requirement within the central predictive vehicle dynamics control can likewise be reduced. A further advantage of the model-based predictive control is that it can also consider constraints on the manipulated variables. This enables actuator limits to be respected already during the optimization within the model-based predictive control algorithm [12]. As a result, this class of algorithms exhibits an excellent control quality. In [13], a model-based predictive control is used to stabilize a vehicle at its vehicle dynamics limits. Due to limitations of the side-slip angle and the yaw rate within the control algorithm the safety is enhanced. [14] apply a model-based predictive control algorithm to reduce vertical vehicle body motions. The vehicle is equipped with active suspension elements. Compared to a passive vehicle the ride comfort is significantly increased. A model-based predictive control algorithm is used in [15] to control the semi-active suspensions of a vehicle. This algorithm is validated against classical control approaches such as the skyhook damping presented in [16] and a clipped control strategy presented in [17]. For various road excitations, the model-based predictive control algorithm outperforms the classical control approaches.

This contribution is organized as follows: Section 2 presents the simulation framework, which is used to develop and validate the central predictive vehicle dynamics control. Section 3 introduces the central non-linear model-based predictive control algorithm with respect to the control objectives of roll, self-steering and pitch behavior. The central predictive control is then validated in Section 4. The contribution concludes in Section 5 with a summary as well as an outlook on future research tasks.

2. Simulation Framework

A simulation framework is used to implement the central predictive vehicle dynamics control and its validation. This framework is based on a co-simulation between IPG CarMaker and MATLAB & Simulink. Figure 1 illustrates the simulation framework. The multi-body simulation within the software IPG CarMaker is used for a realistic simulation of the vehicle. In addition to this realistic representation of the vehicle and the vehicle dynamics, IPG CarMaker also offers the possibility to edit and simulate the environment as well as driver models. In the context of the contribution, a vehicle of the sport utility vehicle class, a Lexus RX400h, is utilized. Due to the heightened center of gravity, this class of vehicle features higher tendencies towards movements in terms of rolling and pitching, which ultimately presents a more challenging task for the vehicle dynamics control. In order to accomplish the control and to achieve the control objectives, the vehicle is equipped with active stabilizers and semi-active dampers. The sensor equipment of the vehicle in IPG CarMaker is based on a minimalistic configuration. Available measured quantities are the longitudinal acceleration a_x, the lateral acceleration a_y, the yaw rate $\dot{\psi}$, the steering wheel angle δ_{SW}, the velocity v and the wheel speeds n_{ij}. Further fixed parameters of the vehicle are listed in Table 1.

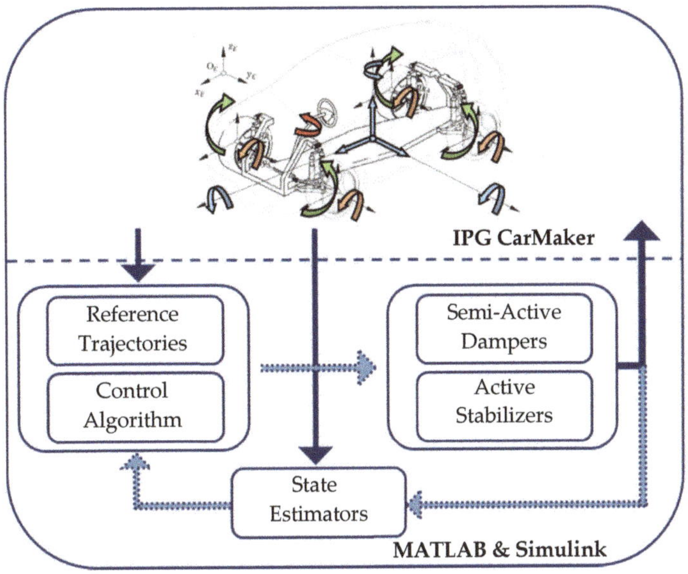

Figure 1. Simulation Framework, © 2021 IEEE. Reprinted, with permission, from [18].

Table 1. Vehicle Parameters.

Parameter	Value	Unit
Vehicle Body Mass	1820	kg
Track Width	1.538	m
Wheelbase	2.75	m
Tires	235/55R18	-
Distance of the Center of Gravity to the Front Axle	1.343	m
Distance of the Center of Gravity to the Rear Axle	1.407	m
Height of the Center of Gravity	0.682	m
Height of the Center of Pitching	0.3257	m
Height of the Center of Rolling	0.2826	m
Moment of Inertia about the Lateral Axis	2654	kg m^2
Moment of Inertia about the Longitudinal Axis	760	kg m^2
Moment of Inertia about the Vertical Axis	2774	kg m^2

The implementation of all algorithms is done in MATLAB & Simulink. In addition to the central predictive control, this also includes the generation of reference trajectories representing the control targets, the simulation of the actuators with regard to a realistic mapping, as well as the implementation of state estimators, which estimate the states necessary for the control not determined by sensors. Examples for the implementation of these state estimators are presented in [19–21].

In the following, the focus is on the central non-linear model-based predictive control algorithm.

3. Central Predictive Control

The steps of prediction and subsequent optimization characterize the central predictive control based on the non-linear model-based predictive control algorithm [22]. Within this contribution, the integrated model-based predictive control presented in [5] is extended and elaborated with respect to influencing the self-steering behavior. In this context, the control of roll behavior features the highest priority. Influencing the self-steering behavior and reducing pitching movements are subordinate control objectives.

3.1. Prediction

Theoretical modeling is used to generate the prediction models of the vehicle dynamics as a function of the manipulated variables. As a result, three interrelated prediction models are determined, which are presented individually. The split is made in relation to the control objectives of influencing the roll, the self-steering and the pitch behavior. In this context, u_1 and u_2 represent the manipulated variables of the counter roll torques at the front and rear axles, respectively. The variable u_3 is the damping factor of the semi-active damper at the front left, u_4 the damping factor of the semi-active damper at the front right, u_5 the damping factor of the semi-active damper at the rear left and u_6 the damping factor of the semi-active damper at the rear right.

3.1.1. Roll Behavior

To build the prediction model for the roll behavior, the vehicle body is cut free in the $y - z$ plane. Subsequently, the principle of angular momentum is set up around the vehicle's roll center. The resulting equation can be transformed according to the roll acceleration $\ddot{\varphi}(k)$ at a certain time step k:

$$\begin{aligned}
&\tfrac{1}{J_{xx}}[h_{\mathrm{GR}} m a_y \cos\varphi(k) + h_{\mathrm{GR}} m g \sin\varphi(k) - u_1(k) - u_2(k) \\
&-2\left(s_{\mathrm{S,f}}^2 c_{\mathrm{S,f}} + s_{\mathrm{S,r}}^2 c_{\mathrm{S,r}}\right)\sin\varphi(k) \\
&-\left((u_3(k)+u_4(k))s_{\mathrm{D,f}}^2\right)\dot{\varphi}(k)\cos\varphi(k) \\
&-\left((u_5(k)+u_6(k))s_{\mathrm{D,r}}^2\right)\dot{\varphi}(k)\cos\varphi(k)] \\
&= \ddot{\varphi}(k).
\end{aligned} \tag{1}$$

Here, J_{xx} represents the moment of inertia about the x-axis, h_{GR} the distance between the center of gravity and the roll center and m the mass of the vehicle body. The external input variables are the lateral acceleration a_y and the gravitational acceleration g. In addition to the external input variables, the chassis elements also have an effect on the roll motion. Apart from the active stabilizers and the semi-active dampers, the vehicle is equipped with passive springs. These passive springs are characterized by the spring stiffnesses $c_{\mathrm{S},i}$. Furthermore, $s_{\mathrm{S},i}$ and $s_{\mathrm{D},i}$ indicate the distances of the springs and dampers from the vehicle's center plane. The index i indicates which vehicle axle is concerned.

Using the scheme of the semi-implicit Euler method [23], the roll rate $\dot{\varphi}(k+1)$ and the roll angle $\varphi(k+1)$ can be predicted as a function of the manipulated variables, starting from the roll acceleration $\ddot{\varphi}(k)$:

$$\dot{\varphi}(k+1) = \dot{\varphi}(k) + \ddot{\varphi}(k)t_{\mathrm{S}}, \tag{2}$$

$$\varphi(k+1) = \varphi(k) + \dot{\varphi}(k+1)t_{\mathrm{S}}. \tag{3}$$

Here, t_{S} denotes the fixed step size.

3.1.2. Self-Steering Behavior

The basis for the prediction of the self-steering behavior is the single-track model [24]. Here, the wheels of an axle are virtually combined for modeling. The single-track model can be used to describe and predict the self-steering behavior and, thus, the response of the vehicle to steering movements [25]. Within the control system, the self-steering gradient SSG is used as the characteristic variable:

$$SSG(k) = \frac{(\alpha_{\mathrm{f}}(k) - \alpha_{\mathrm{r}}(k))}{a_y} \tag{4}$$

The variables α_{f} and α_{r} are the slip angles at the front and rear axles, respectively. The slip angles are dependent of the yaw rate $\dot{\psi}$, the velocity v and the side-slip angle β. In addition, the steering angle δ affects the slip angle at the front axle α_{f}. The parameters

l_f and l_r represent the distance from the center of gravity to the front axle and the rear axle, respectively.

$$\alpha_\mathrm{f}(k) = \delta - \arctan\left(\frac{l_\mathrm{f}\dot{\psi}(k) + v\sin\beta(k)}{v\cos\beta(k)}\right) \tag{5}$$

$$\alpha_\mathrm{r}(k) = -\arctan\left(\frac{-l_\mathrm{r}\dot{\psi}(k) + v\sin\beta(k)}{v\cos\beta(k)}\right) \tag{6}$$

Whereas the steering angle and the velocity are kept constant within the prediction, the yaw rate and the side-slip angle are predicted. For this purpose, both Newton's principle in the lateral direction and the principle of angular momentum in the $x-y$ plane are applied:

$$\frac{(\cos\delta(F_{y,\mathrm{fl}}(k) + F_{y,\mathrm{fr}}(k)) + (F_{y,\mathrm{rl}}(k) + F_{y,\mathrm{rr}}(k)))}{mv\cos\beta(k)} - \dot{\psi}(k) = \dot{\beta}(k), \tag{7}$$

$$\frac{1}{J_{zz}}(l_\mathrm{f}\cos\delta(F_{y,\mathrm{fl}}(k) + F_{y,\mathrm{fr}}(k)) - l_\mathrm{r}(F_{y,\mathrm{rl}}(k) + F_{y,\mathrm{rr}}(k))) = \ddot{\psi}(k). \tag{8}$$

These two equations are solved for $\dot{\beta}$ and $\ddot{\psi}$, respectively. $F_{y,\mathrm{fl}}$, $F_{y,\mathrm{fr}}$, $F_{y,\mathrm{rl}}$ and $F_{y,\mathrm{rr}}$ represent the lateral forces at the tire front left, front right, rear left and rear right, respectively. The moment of inertia about the vertical axis is denoted as J_{zz}. By applying the scheme of the explicit Euler integration method [26], the time derivative of the side-slip angle $\dot{\beta}(k)$ and the yaw acceleration $\ddot{\psi}(k)$ are then used to predict the side-slip angle $\beta(k+1)$ and the yaw rate $\dot{\psi}(k+1)$, respectively:

$$\beta(k+1) = \beta(k) + \dot{\beta}(k)t_\mathrm{S}, \tag{9}$$

$$\dot{\psi}(k+1) = \dot{\psi}(k) + \ddot{\psi}(k)t_\mathrm{S}. \tag{10}$$

The influence of the chassis elements and, thus, the actuators on the self-steering behavior is exerted indirectly via the lateral tire forces $F_{y,ij}$. Here, the index j denotes the vehicle side. The lateral tire forces $F_{y,ij}$ correspond to the product of the slip angles α_i and the respective cornering stiffnesses $c_{\alpha,ij}$:

$$F_{y,ij}(k) = c_{\alpha,ij}(k)\alpha_i(k). \tag{11}$$

The cornering stiffness $c_{\alpha,ij}$ depends on the current wheel load $F_{z,ij}$. The dependency features a degressive characteristic. This characteristic is illustrated in Figure 2.

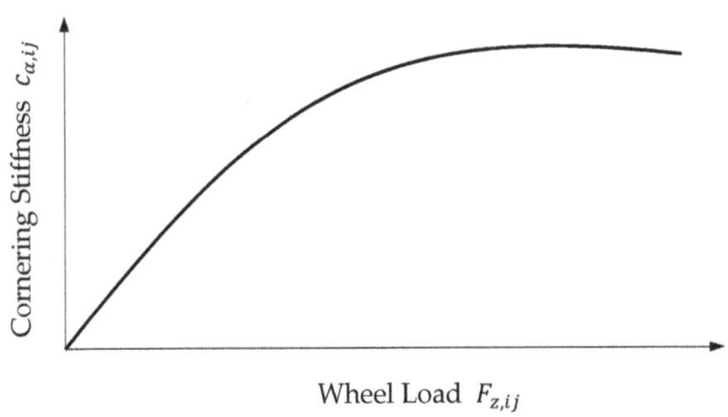

Figure 2. Degressive Characteristic of the Cornering Stiffness Regarding the Wheel Load.

This relationship is modeled using a semi-empirical approach according to [27]. The degressive characteristic is described by a mathematical model, which is parameterized by empirical measurements:

$$c_{\alpha,ij}(k) = c_1 c_2 F_{z0,ij} \sin\left(2\arctan\left(\frac{F_{z,ij}(k)}{c_2 F_{z0,ij}}\right)\right). \quad (12)$$

The basis for the parameter identification is the tire model used in IPG CarMaker. This results in the parameters c_1 and c_2 of the semi-empirical tire model. Furthermore, $F_{z0,ij}$ is the nominal wheel load, which is present under static conditions. The influence of the actuators and, thus, of the control is taken into account via the wheel loads $F_{z,ij}$. The forces of the chassis elements are determined as a function of the roll behavior. The parameter $s_{\mathrm{St},i}$ indicates the distance of the stabilizer force application point from the vehicle's center plane:

$$F_{\mathrm{z,fl}}(k) = \frac{l_\mathrm{r}}{l_\mathrm{f}+l_\mathrm{r}}\left(0.5mg - u_3(k)s_{\mathrm{D,f}}\,\dot{\varphi}(k)\cos\varphi(k) - s_{\mathrm{S,f}}c_{\mathrm{S,f}}\sin\varphi(k) - \frac{1}{2s_{\mathrm{St,f}}}u_1(k)\right), \quad (13)$$

$$F_{\mathrm{z,fr}}(k) = \frac{l_\mathrm{r}}{l_\mathrm{f}+l_\mathrm{r}}\left(0.5mg + u_4(k)s_{\mathrm{D,f}}\,\dot{\varphi}(k)\cos\varphi(k) + s_{\mathrm{S,f}}c_{\mathrm{S,f}}\sin\varphi(k) + \frac{1}{2s_{\mathrm{St,f}}}u_1(k)\right), \quad (14)$$

$$F_{\mathrm{z,rl}}(k) = \frac{l_\mathrm{f}}{l_\mathrm{f}+l_\mathrm{r}}\left(0.5mg - u_5(k)s_{\mathrm{D,r}}\,\dot{\varphi}(k)\cos\varphi(k) - s_{\mathrm{S,r}}c_{\mathrm{S,r}}\sin\varphi(k) - \frac{1}{2s_{\mathrm{St,r}}}u_2(k)\right), \quad (15)$$

$$F_{\mathrm{z,rr}}(k) = \frac{l_\mathrm{f}}{l_\mathrm{f}+l_\mathrm{r}}\left(0.5mg + u_6(k)s_{\mathrm{D,r}}\,\dot{\varphi}(k)\cos\varphi(k) + s_{\mathrm{S,r}}c_{\mathrm{S,r}}\sin\varphi(k) + \frac{1}{2s_{\mathrm{St,r}}}u_2(k)\right). \quad (16)$$

3.1.3. Pitch Behavior

In order to obtain the prediction model for the pitch behavior, the vehicle body is cut free in the $x - z$ plane. Subsequently, the principle of angular momentum is set up around the vehicle's pitch center. The transformation of the resulting equation to the pitch acceleration $\ddot{\theta}(k)$ yields:

$$\begin{aligned} &\tfrac{1}{J_{yy}} lbrack h_{GP} m a_x \cos\theta(k) + h_{GP} mg \sin\theta(k) - 2\left(l_{S,f}^2 c_{S,f} + l_{S,r}^2 c_{S,r}\right)\sin\theta(k) \\ &-u_1(k)\tfrac{l_{St,f}}{s_{St,f}} - u_2(k)\tfrac{l_{St,r}}{s_{St,r}} \\ &-\left((u_3(k)+u_4(k))l_{D,f}^2\right)\dot{\theta}(k)\cos\theta(k) \\ &-\left((u_5(k)+u_6(k))l_{D,r}^2\right)\dot{\theta}(k)\cos\theta(k)] = \ddot{\theta}(k). \end{aligned} \quad (17)$$

J_{yy} denotes the moment of inertia about the lateral axis of the vehicle. The distance between the center of gravity and the pitch center is defined by h_{GP}. The parameters $l_{\mathrm{S},i}$, $l_{\mathrm{D},i}$ and $l_{\mathrm{St},i}$ represent the distances between the center of gravity plane and the force application points of the springs, dampers and stabilizers, respectively. Based on the pitch acceleration $\ddot{\theta}(k)$ at time k, the pitch rate $\dot{\theta}(k+1)$ and the pitch angle $\theta(k+1)$ for the time $k+1$ are likewise determined using the scheme of the semi-implicit Euler method:

$$\dot{\theta}(k+1) = \dot{\theta}(k) + \ddot{\theta}(k)t_\mathrm{S}, \quad (18)$$

$$\theta(k+1) = \theta(k) + \dot{\theta}(k+1)t_\mathrm{S}. \quad (19)$$

This procedure likewise allows the pitch behavior to be predicted as a function of the manipulated variables.

3.2. Optimization

Following the prediction of the vehicle dynamics as a function of the manipulated variables, the optimization is executed with regard to the control objectives. The optimization is performed using the entire prediction horizon n_p. The prediction horizon equals 0.15 s.

A major advantage of the non-linear model-based control algorithm is that constraints can be taken into account within the optimization. In this contribution, the manipulated variables are constrained. Thus, the physical limits of the actuators can be considered within the optimization. This results in the restriction of the counter roll torques u_1 and u_2 between a minimum counter roll torque T_min and a maximum counter roll torque T_max

$$T_\mathrm{min} \leq u_i \leq T_\mathrm{max}, \; i \in \{1,2\}, \tag{20}$$

as well as the restriction of the damping factors u_3, u_4, u_5 and u_6 in between a minimum damping factor d_min and a maximum damping factor d_max

$$d_\mathrm{min} \leq u_i \leq d_\mathrm{max}, \; i \in \{3,4,5,6\}. \tag{21}$$

For the description of the manipulated variables, temporal polynomials defined over the prediction horizon are used to take into account the temporal course within the optimization [5]. The definition of the polynomial degree is done with respect to the desired characteristics. The manipulated variables of the counter roll torques at the front and rear axle u_1 and u_2, respectively, are defined as cubic polynomials:

$$u_1(k) = a_{11} + a_{12}k + a_{13}k^2 + a_{14}k^3, \tag{22}$$

$$u_2(k) = a_{21} + a_{22}k + a_{23}k^2 + a_{24}k^3. \tag{23}$$

The manipulated variables representing the damping factors u_3, u_4, u_5 and u_6 are specified by quadratic polynomials:

$$u_3(k) = a_{31} + a_{32}k + a_{33}k^2, \tag{24}$$

$$u_4(k) = a_{41} + a_{42}k + a_{43}k^2, \tag{25}$$

$$u_5(k) = a_{51} + a_{52}k + a_{53}k^2, \tag{26}$$

$$u_6(k) = a_{61} + a_{62}k + a_{63}k^2. \tag{27}$$

For the further description, the following notation is used:

$$\boldsymbol{u}(k) = (u_1(k), \ldots, u_6(k))^\mathrm{T}, \tag{28}$$

$$\boldsymbol{a} = (a_{11}, \ldots, a_{63})^\mathrm{T}. \tag{29}$$

The manipulated variables are grouped in the vector $\boldsymbol{u}$ and the parameters of the polynomials are grouped in the vector $\boldsymbol{a}$. Furthermore, the predicted vehicle dynamic states of the roll angle φ, the pitch angle θ and the self-steering gradient SSG are summarized in the vector $\boldsymbol{x}$:

$$\boldsymbol{x}(k) = (\varphi(k), \theta(k), SSG(k))^\mathrm{T}. \tag{30}$$

The reference variables of the central predictive control are given in $\boldsymbol{x}_\mathrm{Ref}$. These result from the generation of the reference trajectories:

$$\boldsymbol{x}_\mathrm{Ref}(k) = (\varphi_\mathrm{Ref}(k), \theta_\mathrm{Ref}(k), SSG_\mathrm{Ref}(k))^\mathrm{T}. \tag{31}$$

A dynamic roll angle specification is used for the control objective of the active roll stabilization. A non-linear roll model with passive chassis elements is used for this purpose:

$$\begin{aligned}
&\tfrac{1}{J_{xx}}[h_{GR}ma_y\cos\varphi_p(k)+h_{GR}mg\sin\varphi_p(k)-2(s_{S,f}^2c_{S,f}+s_{S,r}^2c_{S,r})\sin\varphi_p(k)\\
&-\left(d_{p,f}s_{D,f}^2+d_{p,r}s_{D,r}^2\right)\dot{\varphi}_p(k)\cos\varphi_p(k)\\
&-2\tfrac{c_{St,f}s_{St,f}}{b_{St,f}}\arcsin\left(\tfrac{a_{St,f}}{2b_{St,f}}\sin\varphi_p(k)\right)\\
&-2\tfrac{c_{St,r}s_{St,r}}{b_{St,r}}\arcsin\left(\tfrac{a_{St,r}}{2b_{St,r}}\sin\varphi_p(k)\right)]=\ddot{\varphi}_p(k).
\end{aligned} \tag{32}$$

Here, $d_{\mathrm{p},i}$ represents the damping factors of the passive dampers. The passive stabilizers are characterized by the stiffnesses $c_{\mathrm{St},i}$, the effective lengths $a_{\mathrm{St},i}$ and the lever arms $b_{\mathrm{St},i}$. The double integration of the passive roll acceleration $\ddot{\varphi}_{\mathrm{p}}$ by the explicit Euler method yields the corresponding roll angle φ_{p}. This passive roll angle φ_{p} is then scaled by a scaling factor ζ in order to determine the dynamic roll angle specification φ_{Ref}:

$$\varphi_{\mathrm{Ref}}=\zeta\varphi_{\mathrm{p}}. \tag{33}$$

This dynamic reference roll angle specification improves comfort and safety overall, since the roll behavior is significantly reduced while still maintaining feedback of the lateral dynamics to the driver. For the pitch and self-steering behavior, static reference values are specified. The pitch angle specification θ_{Ref} corresponds to the stationary pitch angle of the vehicle. The specification of the self-steering gradient SSG_{Ref} is used to achieve an understeering vehicle behavior corresponding to the passive vehicle behavior.

In addition to maintaining the reference trajectories, the optimization also takes into account the energy requirements of the actuators, which should be set to a minimum. For this reason, the weighting factors $\boldsymbol{\lambda}$ are introduced within the optimization. Thus, the focus can be set on the control quality as well as on the energy demand:

$$\boldsymbol{\lambda}=(\lambda_{\mathrm{R}},\lambda_{\mathrm{P}},\lambda_{\mathrm{S}},\lambda_{\mathrm{u1}},\lambda_{\mathrm{u2}},\lambda_{\mathrm{u3}},\lambda_{\mathrm{u4}},\lambda_{\mathrm{u5}},\lambda_{\mathrm{u6}})^{\mathrm{T}}. \tag{34}$$

During the optimization, the cost function $f(\boldsymbol{u}(k),\boldsymbol{x}(k),\boldsymbol{x}_{\mathrm{Ref}}(k))$ is minimized for the entire prediction horizon n_{p} by adjusting the parameters of the polynomials $\boldsymbol{a}$. The optimization toolbox using the interior-point algorithm of MATLAB is used to solve the optimization [28,29]:

$$\begin{aligned}
&\min_{a} f\left(\boldsymbol{u}(k),\boldsymbol{x}(k),\boldsymbol{x}_{Ref}(k)\right)=\\
&\tfrac{1}{n_p}[\lambda_R\sum_{k=0}^{n_p}\left(\varphi_{Ref}(k)-\varphi(k)\right)^2\\
&+\lambda_S\sum_{k=0}^{n_p}\left(SSG_{Ref}(k)-SSG(k)\right)^2+\lambda_P\sum_{k=0}^{n_p}\left(\theta_{Ref}(k)-\theta(k)\right)^2\\
&+\lambda_{u1}\sum_{k=0}^{n_p}(u_1(k))^2+\lambda_{u2}\sum_{k=0}^{n_p}(u_2(k))^2+\lambda_{u3}\sum_{k=0}^{n_p}(u_3(k))^2\\
&+\lambda_{u4}\sum_{k=0}^{n_p}(u_4(k))^2+\lambda_{u5}\sum_{k=0}^{n_p}(u_5(k))^2+\lambda_{u6}\sum_{k=0}^{n_p}(u_6(k))^2].
\end{aligned} \tag{35}$$

The result of the optimization is a set of optimal polynomial parameters over the entire prediction horizon, from which optimal manipulated variable trajectories are obtained. Finally, the principle of the receding horizon is applied [22]. Only the manipulated variables for the next time step are taken from the optimal manipulated variable trajectories and passed on to the actuator models. In the next time step, the entire process of prediction and optimization is repeated. This allows the non-linear model-based predictive control to adapt to non-modeled disturbances in an optimal way. Furthermore, the warm-start method is used [30], in which the last determined optimal polynomial parameters are used as a starting point for the optimization in the following time step. As a result, the number of iterations within the optimization can be reduced.

4. Results

In a first step, the validation maneuvers are presented which are used to validate the central predictive vehicle dynamics control. Subsequently, the focus is on the evaluation of the control quality for the individual driving maneuvers. The central predictive control is evaluated against a vehicle using a conventional roll control as well as a skyhook damping according to [16] and a vehicle with passive chassis elements. The conventional roll control is based on a proportional integral derivative controller, which is parameterized by using the control system design toolbox of MATLAB. The section concludes with a summary of the results obtained.

4.1. Validation Maneuvers

In order to validate the central non-linear model-based predictive control algorithm, two driving maneuvers are utilized. First, the double lane change driving maneuver is used [31]. For this purpose, the vehicle is first accelerated from standstill to the target velocity before performing the double lane change at constant velocity. The first lane change is performed counterclockwise. The target velocity within this contribution equals 50 km/h. The track limits for the double lane change are defined in the ISO standard. By combining the acceleration phase and the dynamic lane changes, the central predictive control can be validated with regard to all control objectives. The second driving maneuver for validation is a sinusoidal steering. This maneuver also involves accelerating the vehicle to a velocity of 50 km/h and then performing sinusoidal steering, according to [32]. Three periods with a steering wheel angle amplitude of 68° and a frequency of 1 Hz are performed. In addition, a significantly reduced friction coefficient of 0.4 is used for this driving maneuver.

4.2. Double Lane Change

The evaluation of the control quality for the double lane change is carried out separately for each control objective. The evaluation is done qualitatively and quantitatively.

4.2.1. Roll Behavior

For the qualitative evaluation of the control quality of the non-linear model-based predictive control algorithm with respect to the main control objective, the roll angle is plotted over time for the validation maneuver. This is illustrated in Figure 3. The roll angle φ_{Mpc} resulting from the execution of the non-linear model-based predictive control algorithm is represented by a dotted black line. The dynamic reference variable φ_{Ref} is represented by a red solid line. Furthermore, a green dashed line illustrates the roll angle curve φ_{Pas} of the vehicle with passive chassis elements and an orange dotted line illustrates the roll angle curve φ_{PID} resulting from the conventional control approach.

The reference variable φ_{Ref} reduces the roll motion by about 75% compared to the passive vehicle φ_{Pas}. The roll angle resulting by the central predictive control follows the reference with an excellent accuracy. In contrast, the conventional control approach based on the proportional integral derivate controller and the skyhook damping results in increased control deviations. Thus, with the central predictive control, not only the safety but also the comfort is increased compared to the conventional control approach as well as to the passive vehicle behavior. The rolling movements that the passive vehicle exhibits during acceleration are not present for the vehicle utilizing the central predictive control and the vehicle utilizing the conventional control approach, respectively.

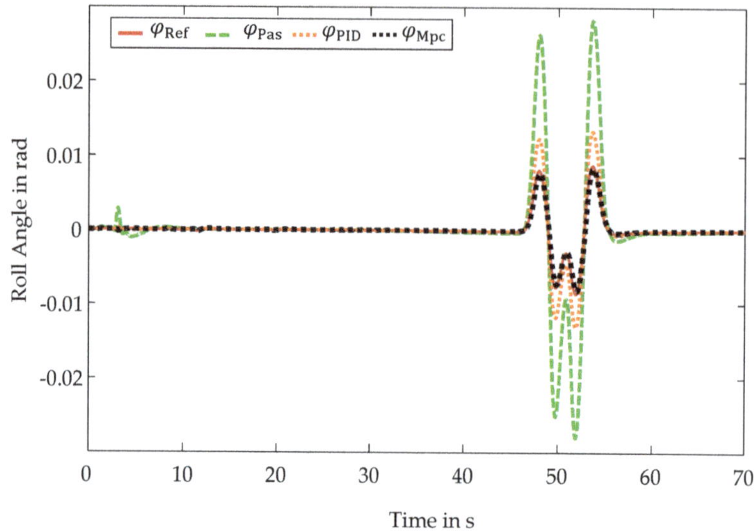

Figure 3. Control Quality Double Lane Change–Roll Behavior–Roll Angle Curves.

In addition, the impacts of the active roll stabilizations on the comfort are evaluated by examining the resulting frequency spectra. These are illustrated in Figure 4. Due to the use of the central predictive control, a stronger damping within the frequency spectrum is present compared to the passive vehicle. The frequency spectrum of the conventional approach also features a stronger damping than the passive vehicle. Compared to the central predictive control algorithm, however, a weaker damping is present.

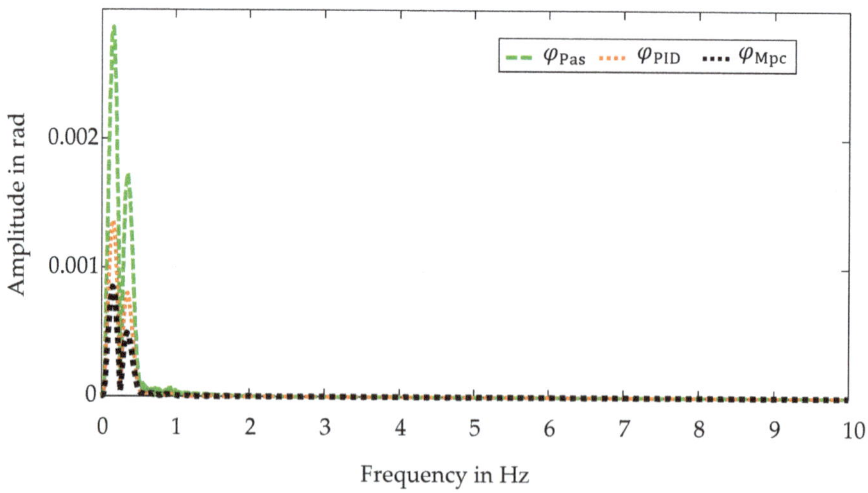

Figure 4. Control Quality Double Lance Change-Roll Behavior-Frequency Spectra.

Furthermore, the root mean squared errors are determined for the quantitative evaluation:

$$RMSE(x_{\mathrm{Ref}}, x_i) = \sqrt{\frac{\sum_{k=1}^{n}\left(x_{\mathrm{Ref}}(k) - x_{\mathrm{Mpc}}(k)\right)^2}{n}}, i \in \{\mathrm{PID}, \mathrm{Mpc}\}. \quad (36)$$

In relation to the active roll stabilization, the non-linear model-based predictive control exhibits a root mean squared error of 2.3906×10^{-4} rad. This corresponds to 0.0137°. In contrast, the conventional control approach results in a root mean squared error of 0.0010 rad, which is equivalent to 0.0573°. The quantitative evaluation, thus, confirms the qualitative analysis. The control quality of the central predictive control is excellent. In general, the central predictive control, thus, outperforms the conventional control approach.

4.2.2. Self-Steering Behavior

In comparison to controlling the roll angle, influencing the self-steering behavior is a subordinate control objective. Here, a constant understeering behavior SSG_{Ref} is pursued, which corresponds approximately to the one of passive vehicle. Because the vehicle is equipped with active stabilizers and semi-active dampers, this control objective can only be pursued to a limited extent. With regard to the representability for the evaluation, the pseudo quantity $\widetilde{\alpha}_i$ is introduced:

$$\widetilde{\alpha}_i = \alpha_{\mathrm{f},i} - \alpha_{\mathrm{r},i} = a_y SSG_i, i\epsilon\{\mathrm{Ref,\ Pas, PID, Mpc}\} \tag{37}$$

The pseudo quantity $\widetilde{\alpha}_i$ represents the difference between the slip angle of the front axle $\alpha_{\mathrm{f},i}$ and rear axle $\alpha_{\mathrm{r},i}$ and, thus, corresponds to the product of the self-steering gradient SSG_i and the lateral acceleration a_y. The evaluation is conducted in the following based on the pseudo quantity $\widetilde{\alpha}_i$. The qualitative evaluation is performed using the pseudo quantity courses for the validation maneuver. This is shown in Figure 5.

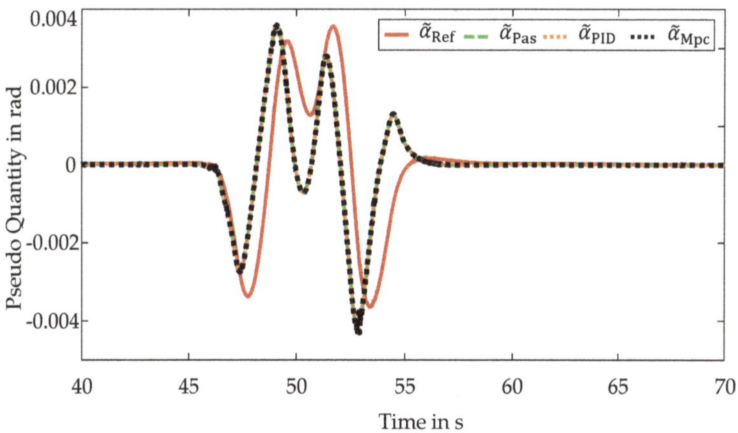

Figure 5. Control Quality Double Lane Change–Self-Steering Behavior–Pseudo Quantity Curves.

With regard to the absence of lateral dynamics in the acceleration phase, only the section of the double lane change is considered. The representation of the pseudo quantity suggests a dynamic reference variable $\widetilde{\alpha}_{\mathrm{Ref}}$. However, this results from the dynamics of the lateral acceleration a_y. With regard to safety, a constant self-steering gradient SSG_{Ref} is used. The representation remains consistent with the evaluation of the roll behavior. The vehicle with the central predictive control shows an almost identical course of the pseudo quantity to the passive vehicle. Due to the weighting of this control target and the limited possibility to manipulate, there is a deviation from the reference variable. Since the conventional control approach does not explicitly consider the influence on the self-steering behavior, only a limited evaluation can be performed. For the validation maneuver of the double lane change, the conventional control exhibits a self-steering behavior corresponding to the passive self-steering behavior, similar to the central predictive control.

In principle, the control objective is satisfied by the central predictive control, since the self-steering behavior corresponds to the behavior of the passive vehicle despite utilizing an active control system. This is confirmed by examining the frequency spectra, which are illustrated in Figure 6.

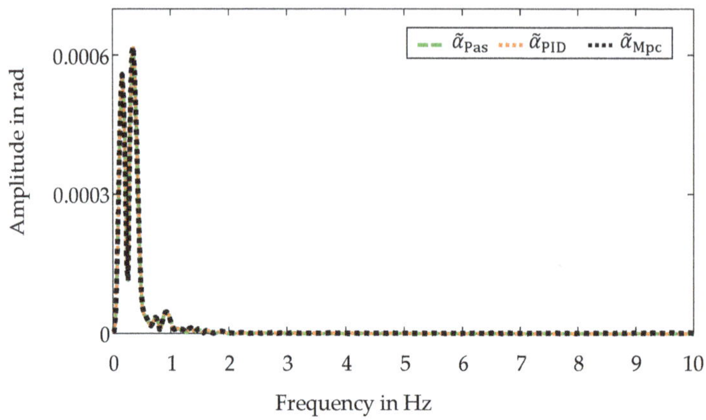

Figure 6. Control Quality Double Lane Change–Self-Steering Behavior–Frequency Spectra.

The frequency spectrum for the vehicle with the central predictive control is almost identical to that of the passive vehicle. The conventional control approach features a similar frequency spectrum.

The quantitative evaluation for the whole validation maneuver results in a root mean squared error for the central predictive control of 8.1009×10^{-4} rad, which is equivalent to $0.0464°$. The conventional control approach yields a root mean squared error of 8.1334×10^{-4} rad, which corresponds to $0.0466°$.

4.2.3. Pitch Behavior

Similar to the manipulation of the self-steering behavior, the reduction of pitching movements is a subordinate control objective within the non-linear model-based predictive control algorithm. Due to the vehicle's equipment with active stabilizers and semi-active dampers, only a limited influence on the pitch behavior is possible. In order to evaluate the central predictive control in a qualitative way with respect to the control objective of reducing the pitching movements, the pitch angle is plotted over time. The pitch angle curves are shown in Figure 7.

The focus is on the acceleration phase of the validation maneuver, since pitching movements are present here due to the gear changes and the present longitudinal accelerations. The pitching movements during the double lane change in contrast are negligibly small. The representation remains consistent. Analogous to the manipulation of the self-steering behavior, a constant reference variable θ_{Ref} is used. This reference variable represents the stationary pitch angle of the vehicle, which corresponds to the pitch angle that is present when the vehicle is at standstill. The reference variable cannot be adjusted fully by the present vehicle setup, but it is considered in order to implement the reduction of pitching movements. In comparison to the vehicle with passive chassis elements θ_{Pas}, pitching movements of the vehicle with the central predictive control θ_{Mpc} are, therefore, reduced. The implementation of the skyhook damping within the conventional control approach also reduces pitching movements compared to the passive vehicle. By exploiting synergies, however, the implementation of the central predictive control results in a greater reduction of pitching movements compared to the conventional control approach.

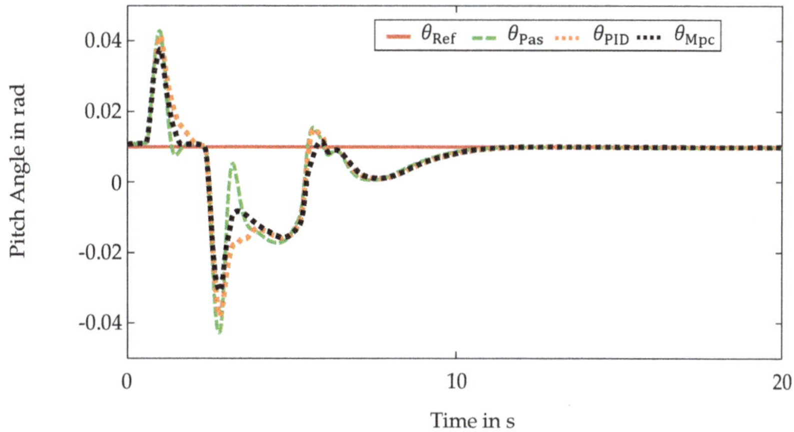

Figure 7. Control Quality Double Lane Change–Pitch Behavior–Pitch Angle Curves.

The corresponding frequency spectra are shown in Figure 8. Here, the positive influence of the central predictive control is also evident. The non-linear model-based predictive control results in greater damping within the frequency spectrum. The amplitudes of the conventional control approach present at very low frequencies are sometimes even greater than those of the passive vehicle. In terms of ride comfort, the central predictive control, thus, provides a significant improvement in comparison to the application of the conventional control approach.

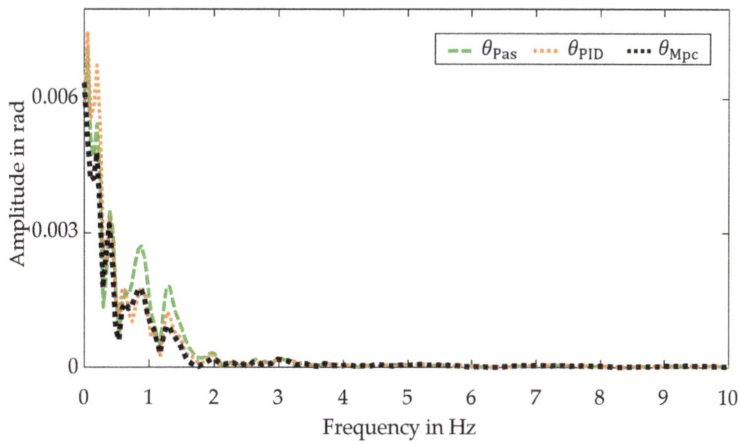

Figure 8. Control Quality Double Lane Change–Pitch Behavior-Frequency Spectra.

The quantitative evaluation confirms this statement. The non-linear model-based predictive control yields a root mean squared error of 0.0106 rad, which corresponds to 0.6073°. The conventional control approach results in a root mean squared error of 0.0119 rad corresponding to 0.6818°. By using the central predictive control, pitching movements are, thus, reduced compared to the conventional control approach as well as compared to the passive vehicle. As a result, the ride comfort is, thus, increased.

4.3. Sinusoidal Steering

The evaluation of the control quality for the sinusoidal steering maneuver is likewise performed qualitatively and quantitatively. The acceleration phase, which is already examined in Section 4.2, is neglected and only the sinusoidal steering phase is considered.

4.3.1. Roll Behavior

To evaluate the control quality of the central predictive control with a focus on the active roll stabilization, the roll angle curves during the sinusoidal steering are plotted. The present roll angle curves are shown in Figure 9.

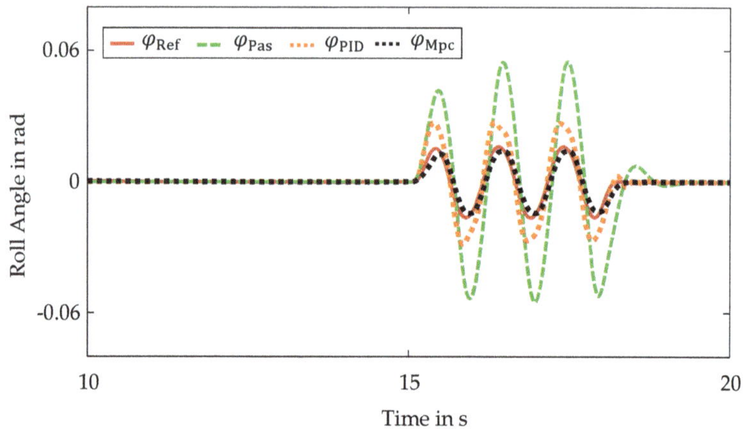

Figure 9. Control Quality Sinusoidal Steering–Roll Behavior–Roll Angle Curves.

As with the double lane change maneuver, the reference variable φ_{Ref} reduces the rolling motions by approximately 75% compared to rolling motions φ_{Pas} of the passive vehicle. The central predictive control φ_{Mpc} is able to follow the reference variable φ_{Ref} despite the minimal friction coefficient. Compared to the results for the validation maneuver of the double lane change, there are slightly larger control deviations due to this low friction coefficient. In comparison to the conventional control approach, the centralized predictive control nevertheless features an improved control quality.

This is confirmed by the evaluation of the root mean squared errors. The non-linear model-based predictive control yields a root mean squared error of 0.0013 rad for the sinusoidal steering, which is equivalent to 0.0745°. In contrast, the conventional control approach results in a root mean squared error of 0.0031 rad, which corresponds to 0.1776°. Despite the aggravated road conditions, the control quality of the central predictive control is excellent.

4.3.2. Self-Steering Behavior

For the evaluation of the control quality with respect to the self-steering behavior, the courses of the pseudo quantity $\widetilde{\alpha}_i$ are first considered again. The resulting curves are illustrated in Figure 10.

Comparable to the results for the double lane change, the self-steering behavior $\widetilde{\alpha}_{\text{Mpc}}$ implemented by the central predictive control system almost corresponds to the passive self-steering behavior $\widetilde{\alpha}_{\text{Mpc}}$. With regard to the reference variable $\widetilde{\alpha}_{\text{Ref}}$, there are larger deviations due to the very low friction coefficient. Through the limited adjustment possibilities caused by the vehicle equipment and the prioritization of the control objectives, only a limited improvement of the control quality is achieved by the central predictive control compared to the passive vehicle. A similar self-steering behavior is obtained by the conventional control approach, though slightly larger deviations from the reference variable are present.

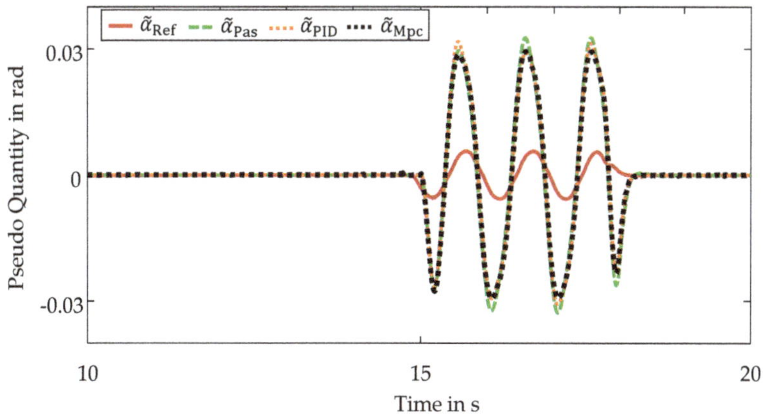

Figure 10. Control Quality Sinusoidal Steering–Self-Steering Behavior–Pseudo Quantity Curves.

The non-linear model-based predictive control algorithm results in a root mean squared error of 0.0077 rad corresponding to 0.4412°. The conventional control approach yields a root mean squared error of 0.0078 rad, which is equivalent to 0.4469°.

In general, the control objective of influencing the self-steering behavior is achieved as well. A self-steering behavior corresponding to the passive self-steering behavior is present despite the application of the central predictive control system.

4.3.3. Pitch Behavior

The qualitative evaluation for the control objective of reducing pitching movements is likewise realized by examining the pitch angle curves during the sinusoidal steering. The pitch angle curves are shown in Figure 11.

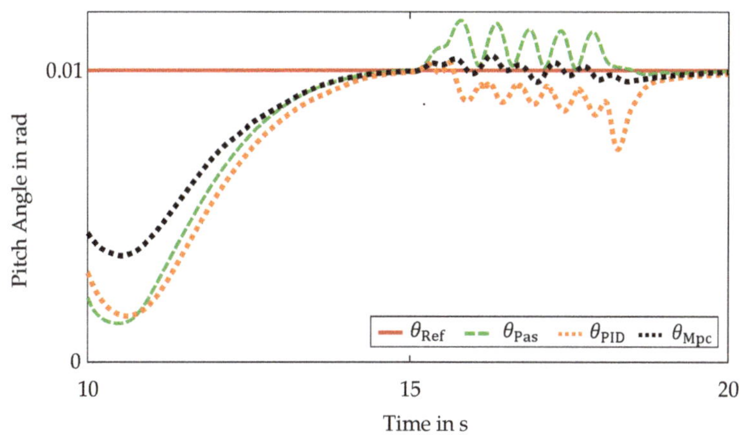

Figure 11. Control Quality Sinusoidal Steering–Pitch Behavior–Pitch Angle Curves.

The positive influence of central predictive control is likewise evident during the sinusoidal steering. Compared to the passive vehicle and the conventional control approach, the pitching movements are reduced.

The quantitative evaluation confirms these results. Whereas the vehicle utilizing the conventional control approach deviates from the stationary pitch angle θ_{Ref} by a root mean squared error of 0.0035 rad, which corresponds to 0.2005°, the non-linear model-based predictive control results in a root mean squared error of 0.0025 rad, which corresponds to 0.1432°.

In comparison to the passive vehicle and the application of the conventional control approach, the utilization of the central predictive control system, thus, increases the ride comfort.

4.4. Conclusions

The central predictive vehicle dynamics control fulfills the defined control objectives as desired. A summary of the control quality is given in Table 2. In addition to this, the results are compared with those of a vehicle with passive chassis elements and those obtained by a conventional control approach.

Table 2. Central Predictive Vehicle Dynamics Control—Root Mean Squared Errors.

Setup	Driving Maneuver	Vehicle Dynamics	RMSE	Unit
Passive Chassis	Double Lane Change	Roll Behavior	0.0043	rad
		Self-Steering Behavior	8.2703×10^{-4}	rad
		Pitch Behavior	0.0114	rad
	Sinusoidal Steering	Roll Behavior	0.0099	rad
		Self-Steering Behavior	0.0080	rad
		Pitch Behavior	0.0034	rad
Proportional Integral Derivative Control and Skyhook Damping	Double Lane Change	Roll Behavior	0.0010	rad
		Self-Steering Behavior	8.1334×10^{-4}	rad
		Pitch Behavior	0.0119	rad
	Sinusoidal Steering	Roll Behavior	0.0031	rad
		Self-Steering Behavior	0.0078	rad
		Pitch Behavior	0.0035	rad
Central Predictive Control	Double Lane Change	Roll Behavior	2.3906×10^{-4}	rad
		Self-Steering Behavior	8.1009×10^{-4}	rad
		Pitch Behavior	0.0106	rad
	Sinusoidal Steering	Roll Behavior	0.0013	rad
		Self-Steering Behavior	0.0077	rad
		Pitch Behavior	0.0025	rad

The main control objective of the central predictive vehicle dynamics control, the active roll stabilization, is met with an excellent control quality. In comparison to the conventional control approach, the control quality is improved by approximately 67.08%. Likewise, good results are achieved with regard to the further subordinate control objectives. As a result, pitching movements are reduced by approximately 19.75% compared to the conventional control approach. Despite the intervention of the control system, the self-steering behavior of the controlled vehicle corresponds to that of the passive vehicle.

The validation demonstrates the advantages of a centralized predictive control structure over a passive vehicle and a vehicle using a conventional control approach comprising a proportional integral derivative controller and a skyhook damping. The improvements in the control quality by of a non-linear model-based predictive control compared to a linear control approach utilizing a skyhook damping for a pure active roll stabilization are also presented in [33].

5. Conclusions and Outlook

This contribution presents a central predictive vehicle dynamics control system. The control objectives of an active roll stabilization, a manipulation of the self-steering behavior as well as a reduction of pitching movements are pursued. The control system is implemented by a non-linear model-based predictive control algorithm. The simulation framework comprising a co-simulation between IPG CarMaker and MATLAB & Simulink is used. The test vehicle utilized is a sport utility vehicle equipped with active stabilizers and semi-active dampers to realize the control. Due to the heightened center of gravity, the sport utility vehicle features an increased challenge in influencing rolling and pitching motions. Here, the implementation of active roll stabilization is the primary control objective. The weighting of the control objectives is done via corresponding weighting factors within the optimization of the non-linear model-based predictive control algorithm. The validation is performed for the driving maneuvers of the double lane change and the sinusoidal steering. In conclusion, the implementation of the central predictive control demonstrates an excellent control quality. The central predictive control increases safety and comfort significantly compared to a vehicle with passive chassis elements and a vehicle utilizing a conventional control approach.

The focus of this contribution is on the presentation of the central predictive vehicle dynamics control based on the non-linear model-based predictive control algorithm and the resulting excellent control quality. However, due to the numerical solution of the optimization problem, an increased computational effort is imposed at the same time. With regard to a future implementation of the central predictive control, a real-time implementation of this central predictive vehicle dynamics control with a reduced computational effort is required. Whereas classical approaches linearize the underlying prediction models and, thus, simplify them, making them analytically solvable, or limit the iteration steps of the optimization, future work will explore the use of artificial intelligence. Initial conceptual research has already shown a particular suitability of neuro-fuzzy inference systems to address and solve this issue [18].

Author Contributions: P.M.S.: conceptualization, methodology, software, validation, formal analysis, investigation, writing and editing, visualization. D.S.: supervision; project administration and critical revision. All authors have read and agreed to the published version of the manuscript.

Funding: This research received no external funding.

Institutional Review Board Statement: Not applicable.

Informed Consent Statement: Not applicable.

Acknowledgments: We acknowledge support by the Open Access Publication Fund of the University of Duisburg-Essen.

Conflicts of Interest: The authors declare no conflict of interest.

References

1. Johansson, R. Vision Zero—Implementing a policy for traffic safety. *Safety Sci.* **2009**, *47*, 826–831. [CrossRef]
2. Elbanhawi, M.; Simic, M.; Jazar, R. In the Passenger Seat: Investigating Ride Comfort Measures in Autonomous Cars. *IEEE Intell. Transp. Syst. Mag.* **2015**, *7*, 4–17. [CrossRef]
3. Chen, W.; Xiao, H.; Wang, Q.; Zhao, L.; Zhu, M. *Integrated Vehicle Dynamics and Control*; John Wiley & Sons Singapore Pte. Ltd.: Singapore, 2016.
4. Fan, Y.; Dao-Fei, L.; Crolla, D.A. Integrated Vehicle Dynamics Control—State-of-the art review. In Proceedings of the 2008 IEEE Vehicle Power and Propulsion Conference, Harbin, China, 3–5 September 2008; pp. 1–6.
5. Sieberg, P.M.; Blume, S.; Reicherts, S.; Schramm, D. Nichtlineare modellbasierte prädiktive Regelung der Fahrzeugdynamik in Bezug auf eine aktive Wankstabilisierung und eine Nickreduzierung. In *Forschung im Ingenieurwesen*; Springer: Berlin/Heidelberg, Germany, 2019; Volume 83, pp. 119–127.
6. Ikenaga, S.; Lewis, F.L.; Campos, J.; Davis, L. Active suspension control of ground vehicle based on a full-vehicle model. In Proceedings of the 2000 American Control Conference, ACC (IEEE Cat. No.00CH36334), Chicago, IL, USA, 28–30 June 2000; Volume 4016, pp. 4019–4024.
7. Chang, S.; Gordon, T.J. A flexible hierarchical model-based control methodology for vehicle active safety systems. *Veh. Syst. Dyn.* **2008**, *46*, 63–75. [CrossRef]
8. Bahr, M.; Reicherts, S.; Sieberg, P.; Morss, L.; Schramm, D. Application of Artificial Neural Networks for Active Roll Control Based on Actor-Critic Reinforcement Learning. In Proceedings of the Simulation and Modeling Methodologies, Technologies and Applications, Prague, Czech Republic, 29–31 July 2021; pp. 61–82.

9. Li, H.-m.; Wang, X.-b.; Song, S.-B.; Li, H. Vehicle Control Strategies Analysis Based on PID and Fuzzy Logic Control. *Procedia Eng.* **2016**, *137*, 234–243. [CrossRef]
10. Sands, T. Development of Deterministic Artificial Intelligence for Unmanned Underwater Vehicles (UUV). *J. Mar. Sci. Eng.* **2020**, *8*, 578. [CrossRef]
11. Camacho, E.F.; Bordons, C.A. *Model Predictive Control*; Springer: Berlin, Germany, 1998.
12. Grüne, L.; Pannek, J. Nonlinear Model Predictive Control. In *Nonlinear Model Predictive Control: Theory and Algorithms*; Grüne, L., Pannek, J., Eds.; Springer: Cham, Switzerland, 2017; pp. 45–69. [CrossRef]
13. Beal, C.E.; Gerdes, J.C. Predictive control of vehicle roll dynamics with rear wheel steering. In Proceedings of the 2010 American Control Conference, Baltimore, MD, USA, 30 June–2 July 2010; pp. 1489–1494.
14. Mehra, R.K.; Amin, J.N.; Hedrick, K.J.; Osorio, C.; Gopalasamy, S. Active suspension using preview information and model predictive control. In Proceedings of the 1997 IEEE International Conference on Control Applications, Hartford, CT, USA, 5–7 October 1997; pp. 860–865.
15. Canale, M.; Milanese, M.; Novara, C. Semi-Active Suspension Control Using "Fast" Model-Predictive Techniques. *IEEE Trans. Control Syst. Technol.* **2006**, *14*, 1034–1046. [CrossRef]
16. Hrovat, D. Survey of Advanced Suspension Developments and Related Optimal Control Applications. *Automatica* **1997**, *33*, 1781–1817. [CrossRef]
17. Giua, A.; Seatzu, C.; Usai, G. Semiactive Suspension Design with an Optimal Gain Switching Target. *Veh. Syst. Dyn.* **1999**, *31*, 213–232. [CrossRef]
18. Sieberg, P.M.; Hürten, C.; Schramm, D. Representation of an Integrated Non-Linear Model-Based Predictive Vehicle Dynamics Control System by a Co-Active Neuro-Fuzzy Inference System. In Proceedings of the 2020 IEEE Intelligent Vehicles Symposium (IV), Las Vegas, NV, USA, 19 October–13 November 2020; pp. 572–577.
19. Sieberg, P.M.; Blume, S.; Reicherts, S.; Maas, N.; Schramm, D. Hybrid State Estimation—A Contribution towards Reliability Enhancement of Artificial Neural Network Estimators. *IEEE Trans. Intell. Transp. Syst.* **2021**. [CrossRef]
20. Sieberg, P.M.; Blume, S.; Schramm, D. Side-Slip Angle Estimation by Artificial Neural Networks for Vehicle Dynamics Control Applications. In Proceedings of the AmE 2021—Automotive meets Electronics 12th GMM-Symposium, Online, 10–11 March 2021.
21. Blume, S.; Sieberg, P.M.; Maas, N.; Schramm, D. Neural Roll Angle Estimation in a Model Predictive Control System. In Proceedings of the IEEE Intelligent Transportation Systems Conference (ITSC)/ITSC 2019, Auckland, New Zealand, 27–30 October 2019; pp. 1625–1630.
22. Rossiter, J.A. *Model-Based Predictive Control—A Practical Approach*; CRC Press: Boca Raton, FL, USA, 2004; p. 344.
23. Cromer, A. Stable solutions using the Euler approximation. *Am. J. Phys.* **1981**, *49*, 455–459. [CrossRef]
24. Schramm, D.; Hesse, B.; Maas, N.; Unterreiner, M. *Vehicle Technology—Technical Foundations of Current and Future Motor Vehicles*; De Gruyter Oldenbourg: Berlin, Germay; Boston, MA, USA, 2020. [CrossRef]
25. Schramm, D.; Hiller, M.; Bardini, R. *Vehicle Dynamics: Modeling and Simulation*, 2nd ed.; Springer: Berlin/Heidelberg, Germany, 2018. [CrossRef]
26. Butcher, J.C. *Numerical Methods for Ordinary Differential Equations*, 3rd ed.; John Wiley & Sons, Ltd.: Hoboken, NJ, USA, 2016. [CrossRef]
27. Pacejka, H.B. *Tyre and Vehicle Dynamics*, 2nd ed.; Butterworth-Heinemann: Oxford, UK, 2006; p. 1.
28. Waltz, R.A.; Morales, J.L.; Nocedal, J.; Orban, D. An interior algorithm for nonlinear optimization that combines line search and trust region steps. *Math. Program.* **2006**, *107*, 391–408. [CrossRef]
29. Ypma, T.J. Historical Development of the Newton-Raphson Method. *SIAM Rev.* **1995**, *37*, 531–551. [CrossRef]
30. Yildirim, E.A.; Wright, S.J. Warm-Start Strategies in Interior-Point Methods for Linear Programming. *SIAM J. Optim.* **2002**, *12*, 782–810. [CrossRef]
31. ISO. *ISO 3888-1:2018 Passenger Cars—Test Track for a Severe Lane-Change Manoeuvre—Part 1: Double Lane-Change*; International Organisation for Standardization (ISO): Geneva, Switzerland, 2018.
32. ISO. *ISO 13674-1:2010 Road Vehicles—Test Method for the Quantification of on-Centre Handling—Part 1: Weave Test*; International Organisation for Standardization (ISO): Geneva, Switzerland, 2010.
33. Sieberg, P.M.; Reicherts, S.; Schramm, D. Nichtlineare modellbasierte prädiktive Regelung zur aktiven Wankstabilisierung von Personenkraftwagen. In Proceedings of the Vierte IFToMM D-A-CH Konferenz 2018, Lausanne, Switzerland, 15–16 February 2018; pp. 1–8.

applied sciences

MDPI

Article

Empirical Models for the Viscoelastic Complex Modulus with an Application to Rubber Friction

Marco Furlan Tassara †, Kyriakos Grigoriadis † and Georgios Mavros *

Department of Aeronautical and Automotive Engineering, Loughborough University, Loughborough LE11 3TU, UK; m.furlan-tassara@lboro.ac.uk (M.F.T.); kyriagri@gmail.com (K.G.)
* Correspondence: g.mavros@lboro.ac.uk
† These authors contributed equally to this work.

Abstract: Up-to-date predictive rubber friction models require viscoelastic modulus information; thus, the accurate representation of storage and loss modulus components is fundamental. This study presents two separate empirical formulations for the complex moduli of viscoelastic materials such as rubber. The majority of complex modulus models found in the literature are based on tabulated dynamic testing data. A wide range of experimentally obtained rubber moduli are used in this study, such as SBR (styrene-butadiene rubber), reinforced SBR with filler particles and typical passenger car tyre rubber. The proposed formulations offer significantly faster computation times compared to tabulated/interpolated data and an accurate reconstruction of the viscoelastic frequency response. They also link the model coefficients with critical sections of the data, such as the gradient of the slope in the storage modulus, or the peak values in loss tangent and loss modulus. One of the models is based on piecewise polynomial fitting and offers versatility by increasing the number of polynomial functions used to achieve better fitting, but with additional pre-processing time. The other model uses a pair of logistic-bell functions and provides a robust fitting capability and the fastest identification, as it requires a reduced number of parameters. Both models offer good correlations with measured data, and their computational efficiency was demonstrated via implementation in Persson's friction model.

Keywords: viscoelastic modulus; rubber; friction; empirical modeling

Citation: Tassara, M.F.; Grigoriadis, K.; Mavros, G. Empirical Models for the Viscoelastic Complex Modulus with an Application to Rubber Friction. *Appl. Sci.* **2021**, *11*, 4831. https://doi.org/10.3390/app11114831

Academic Editors: Aleksandr Sakhnevych, Andrea Genovese and Flavio Farroni

Received: 16 April 2021
Accepted: 20 May 2021
Published: 25 May 2021

1. Introduction

Viscoelastic materials exhibit both viscous and elastic characteristics when stressed so that their behaviour is in-between that of a purely viscous liquid and a perfectly elastic solid. When a viscoelastic material is deformed, part of the deformation energy dissipates, and the rest is stored as reversible elastic energy. The properties of viscoelastic materials such as rubber can vary widely according to temperature, frequency and chemical composition [1]. Rubber is a highly deformable material employed in a wide range of products, from everyday necessities, e.g., shoe soles, to industrial applications—notably, vehicle tyres. Such applications demand large deformations, vibration damping [2] and enhanced gripping characteristics (e.g., tyres and conveyor belts). With safety-critical components such as tyres, it is important to be able to predict friction for different types of rubber; however, the non-linear nature of rubber poses significant modelling challenges [3].

The response of a rubber material can be obtained from rheological models. Linear viscoelastic materials can be represented by combinations of springs and dashpots. The most common ones are the Maxwell and Voigt–Kelvin models [4] which use springs and dashpots connected in series or in parallel, respectively (see Figure 1). The standard linear solid (SLS) is an upgraded rheological model in which an extra spring is added in parallel with the Maxwell unit to allow for an accurate representation of a larger group of polymers [5]. Additionally, the Weichert model offers a relaxation spread over a longer period by using a single spring in parallel with several Maxwell units [6,7]. Xu et al. [8] proposed a non-linear rheological model to offer better accuracy in simulating the material's relaxation modulus and viscoelastic responses. A logistic-type function approximation was proposed by [8] to simulate the viscoelastic response of spring–dashpot networks. Another recent study employed a combination of a generalised Maxwell model and a relative fraction derivative model to reproduce viscoelastic material behaviour [9].

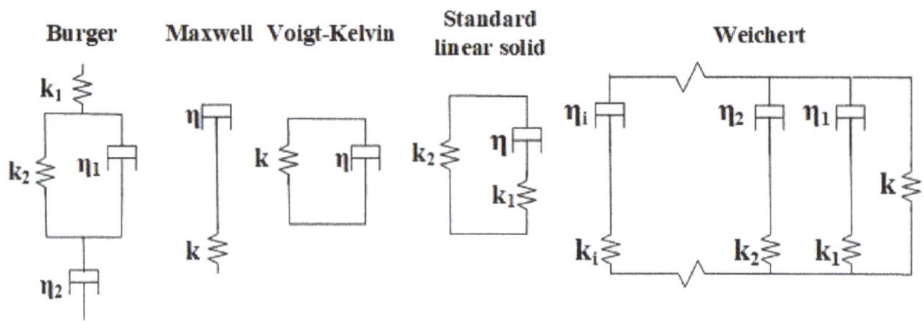

Figure 1. Basic rheological models. k is the stiffness of the spring and η the viscosity of the dash-pot, representing the linear and viscous parts of the material, respectively.

Both models proposed herein describe rubber's viscoelastic response in the frequency domain. As shown in [10], rubber friction strongly depends on the rubber's frequency response, and this is accounted for in Persson's friction model [11]. The first model uses a group of polynomials to capture the viscoelastic frequency response with great accuracy, whilst the other method uses a generic empirical formulation with few parameters to expedite the parameter identification process and to offer a compact set of equations. Thereafter, the proposed formulations are applied to Persson's friction model without thermal effects [11] to evaluate their computational efficiency and accuracy. The paper is structured as follows: In Section 2, the measured moduli used as the benchmark are presented. In Section 3, the polynomial equation method and the empirical model are described. In Section 4, the ability of both models to fit measured data and their performances as part of Persson's friction model are evaluated. Section 5 provides a summary of the conclusions.

2. Measured Viscoelastic Modulus

When a sinusoidal force is applied to a purely elastic material, the stress and deformation occur concurrently, so that both quantities are in phase. On the contrary, when the same sinusoidal force is applied to a purely viscous fluid, the deformation will lag the stress by one-quarter cycle ($\pi/2$ radians). This means that when the deformation is maximal, the force is minimal and vice versa. In the case of a viscoelastic material, the phase lag between the force and the deformation lies somewhere between zero and a quarter cycle. Mathematically, a sinusoidal strain can be written using Euler's formula, as shown in Equation (1), where ε_0 is the amplitude of the strain and ω is the frequency:

$$\varepsilon(t) = \varepsilon_0 e^{i\omega t} \tag{1}$$

Similarly, the stress can be written as:

$$\sigma(t) = \sigma_0 e^{i(\omega t+\delta)} \tag{2}$$

where σ_0 is the amplitude of the stress and δ is the phase angle. In Figure 2 the stress and strain phasors are shown as they rotate at a frequency ω, with the strain lagging behind the stress by an angle δ.

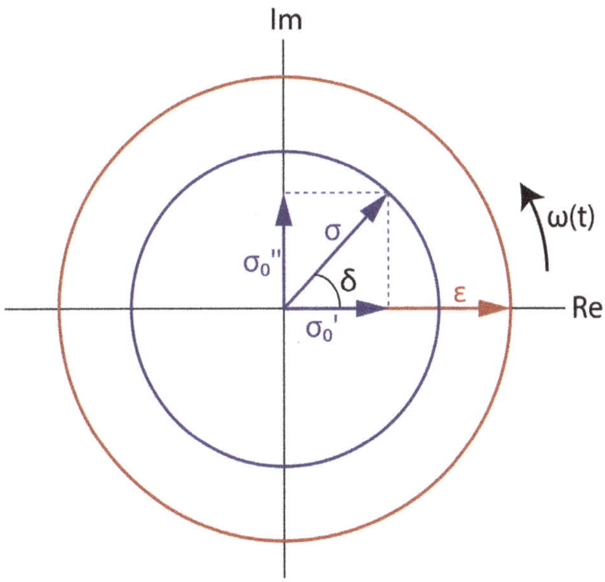

Figure 2. A representation of the harmonic stress and strain.

The following relationships can be deduced from Figure 2:

$$\sigma_0' = \sigma_0 \cos(\delta) \tag{3}$$

$$\sigma_0'' = \sigma_0 sin(\delta) \tag{4}$$

With the above definitions, the dynamic or complex modulus will have a real and an imaginary part. The real or storage modulus is defined as the ratio between the real part of the stress and the strain:

$$E' = \sigma_0' / \varepsilon_0 \tag{5}$$

The imaginary or loss modulus is defined as the ratio between the imaginary part of the stress and the strain:

$$E'' = \sigma_0'' / \varepsilon_0 \tag{6}$$

Equations (1)–(6) lead to the broadly used complex or dynamic modulus formula [12]:

$$E^* = E' + iE'' \tag{7}$$

Finally, the tangent modulus, also known as the loss tangent is defined as:

$$\tan \delta = E''/E' \tag{8}$$

By definition, the modulus of a material is considered as the overall resistance of the material to an applied deformation. Due to its viscoelastic nature, the rubber modulus is split into elastic (storage), E', and viscous (loss), E'', components, denoting the ability of the material to store and dissipate energy as heat, respectively. The real and imaginary components of a viscoelastic modulus are collectively referred to as the material's complex modulus. Equally important is the loss tangent or $\tan \delta$, which is defined as the ratio of energy loss to energy storage. The area near the peak value denotes high energy loss and is considered as the area where the tyre is designed to be operating most of the time. While spring–dashpot-based analytical models are useful for representing the physics of viscoelastic materials, the experimentally measured complex modulus as a function of excitation frequency is often used instead. In the case of rubber, the complex modulus is shown to be strongly linked to friction [13,14]. Figure 3 shows the typical complex modulus of a rubber used in tyre manufacturing.

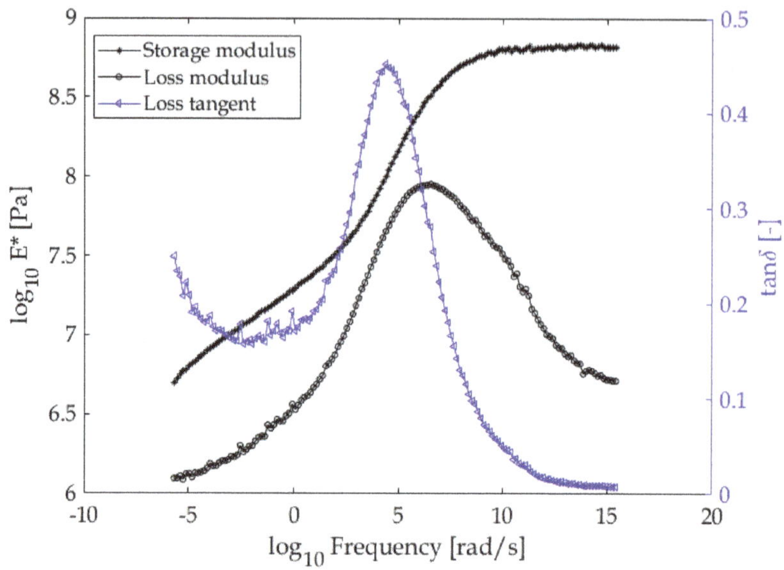

Figure 3. A typical complex modulus of a tyre. DMA material testing was commissioned by the authors for a tread rubber compound.

Such moduli of rubber materials are obtained using oscillatory strain in tension or shear form, while monitoring the resulting stress [15]. When rubber is dynamically stretched and released, the returned energy is less than the energy put into the rubber in the first place. This viscoelastic effect can only be tested dynamically. Dynamic mechanical analysis (DMA) instruments measure the viscoelastic modulus in response to applied oscillating strain. The stress response to the deformation is recorded as a function of time or frequency. The rubber sample is oscillated at different frequencies and the same process is repeated at various temperatures. Results are then shifted according to the time–temperature superposition principle [16], forming a master curve and covering a wide range of frequencies at a chosen reference temperature. The literature was scrutinised and DMA data for different viscoelastic materials were gathered [4,11,13,15,17–27]. While spring–dashpot networks can be used to reproduce DMA data, the required structure of the network is sometimes difficult to decide a priori, and the demand on parameter identification can be significant. In the following sections it is shown that the DMA responses of different materials can be approximated successfully using our proposed formulations with a limited set of parameters.

3. Proposed Empirical Models

Two empirical approaches for fitting the viscoelastic complex modulus using piece-wise polynomial fitting, or a pair of logistic-bell functions, are presented below.

3.1. Approximation Using Polynomials

The first model proposes a piece-wise fitting of both the storage and loss moduli using polynomials. As a first step, the two moduli curves (storage and loss) are divided piecewise into sections that will be described by different polynomials. Typically, both the storage and loss moduli curves include regions that can be described by linear functions, as shown in Figure 4a. These linear functions are estimated first using least squares fitting of all data points between the start and finish of each straight line. Linear regions are typically connected via higher-order polynomials—see Figure 4b. The requirement for any non-linear section is continuity up to—and including—the first derivative with the preceding and succeeding curves. In this context, greatest flexibility is offered by fourth-order polynomials whereby continuity at the start and finish points offers four equations for four parameters. A fifth equation is obtained by requesting that the function passes from a selected midpoint between the start and finish points. A pair of typical first and fourth-order equations used in the process is shown in Equation (9), below. The superscript i indicates the number of the piece-wise curve, with counting starting from the left of the frequency range. This method offers robustness against global optimisation methods and very easy tuning.

$$\begin{cases} y = \alpha^i x + b^i \\ y = c_4^i x^4 + c_3^i x^3 + c_2^i x^2 + c_1^i x + c_0^i \end{cases} \tag{9}$$

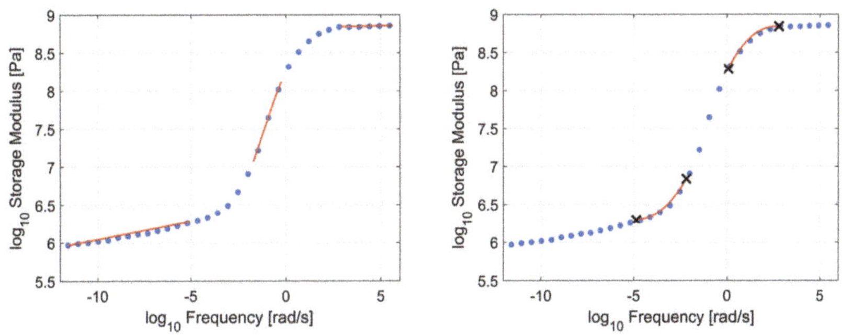

(**a**) Linear parts.

(**b**) Curved parts.

Figure 4. Demonstration of linear (**left**) and higher degree (**right**) polynomials with the critical boundary points that initiate curvature, for the storage modulus of SBR with no phr carbon black [20].

3.2. The Logistic-Bell Empirical Model

The logistic-bell empirical model (LBEM) is an alternative rheological model that has few parameters that can be easily linked to typifying quantities of the DMA data, but is still fast to compute and can describe the responses of real viscoelastic materials over a wide frequency range. From data gathered in the literature, it can be observed that in the logarithmic scale the storage modulus reassembles the shape of a logistic function and the loss tangent approximates a bell-shaped function. Therefore, a variation of the logistic function is proposed for modelling the storage modulus, as follows:

$$\log_{10}\left(E'(\omega)\right) = ry_o + \frac{L}{1 + e^{-k(\log_{10}(\omega) - r\omega_0)}} \tag{10}$$

where:

- E' is the real or storage modulus in (Pa) based on a logistic function.
- ry_o is a vertical shift.
- L is the curve's maximum value.
- k the logistic rate of increase or steepness of the curve.
- ω is the excitation frequency in rad/s.
- $r\omega_0$ is the logarithm in base 10 of the frequency value at the sigmoid midpoint.

The loss tangent is modelled using a piecewise bell-shaped function:

$$\tan\delta(\omega) = \begin{cases} sy_1 + (p - sy_1) sech^{\frac{\log_{10}(\omega) - t\omega_0}{r_1}} & if\ (\log_{10}(\omega) - t\omega_0) \leq 0 \\ sy_2 + (p - sy_2) sech^{\frac{\log_{10}(\omega) - t\omega_0}{r_2}} & if\ (\log_{10}(\omega) - t\omega_0) > 0 \end{cases} \tag{11}$$

where:

- $\tan\delta$ is the loss tangent.
- $t\omega_0$ is the logarithm in base 10 of the frequency at the bell's midpoint or peak.
- sy_1 is the bell function vertical shift at low frequencies before the peak.
- sy_2 is the bell function vertical shift at high frequencies after the peak.
- p is the bell function peak value.
- ω is the excitation frequency in rad/s.
- r_1 is the bell's rate of increase at low frequencies before the peak.
- r_2 is the bell's rate of increase at high frequencies after the peak.

Finally, the loss modulus can be derived from Equation (8).

3.3. Fitting Accuracy

Several published moduli from the literature were used to assess the quality of fitting of the polynomial method. The storage and loss moduli, and the loss tangent, were fitted separately, and the relevant coefficients are provided in Table 1. For all three materials, three linear parts were combined with two fourth-order polynomials, to obtain the overall curves. In Table 1, any subscript indicates the role of the parameter in Equation (9), and a superscript indicates the sequential number of the respective linear or fourth-order polynomial. Clearly, any two of the three complex modulus curves are sufficient for fully describing any tested material.

Table 1. Coefficients of polynomial fitting for the curves shown in Figure 5.

Coefficient	Loss Modulus			Storage Modulus		
	F14 [22]	L13 [26]	P01 [11]	F14 [22]	L13 [26]	P01 [11]
a^1	0.202	0.089	−0.167	0.003	0.058	0.072
b^1	5.059	6.340	4.028	6.362	7.222	6.910
a^2	0.708	0.356	0.504	0.647	0.361	0.423
b^2	3.208	6.050	8.696	3.284	6.181	8.602
a^3	−0.276	−0.044	−0.105	0.107	0.013	0.022
b^3	10.765	8.782	8.748	7.651	9.146	9.245
c_1^1	−0.020	−0.0010	−0.2660	0.0010	0	0.0020
c_2^1	0.2690	0.0020	−7.4390	0.0100	0.0060	0.0740
c_3^1	−1.1890	0.0390	−77.4850	−0.1350	−0.0020	0.8870
c_4^1	2.2830	0.1630	−356.3080	0.4370	0.0580	4.5900
c_5^1	3.8000	6.3010	−604.6300	5.9270	7.2220	15.2570
c_1^2	0.0060	−0.0010	0	0.0060	0	0
c_2^2	−0.1670	0.0250	0.0030	−0.1980	−0.0020	0.0020
c_3^2	1.5860	−0.4080	−0.0520	2.2840	−0.0390	−0.0440
c_4^2	−5.1930	2.9660	0.1770	−10.6870	0.9460	0.3270
c_5^2	9.9850	0.4520	8.2460	23.8790	4.3130	8.5520

The correlations between measured and fitted curves for three different rubbers using the polynomial method are shown in Figures 5 and 6. Figure 5 shows that the piecewise polynomial method is capable of following the curvature of the curves faithfully. A quantitative assessment of the quality of fit is provided via the relative percentage error in Figure 6, where it is shown that the maximum relative error for the storage modulus is 0.7%, and for the loss modulus the maximum relative error is 1.49%.

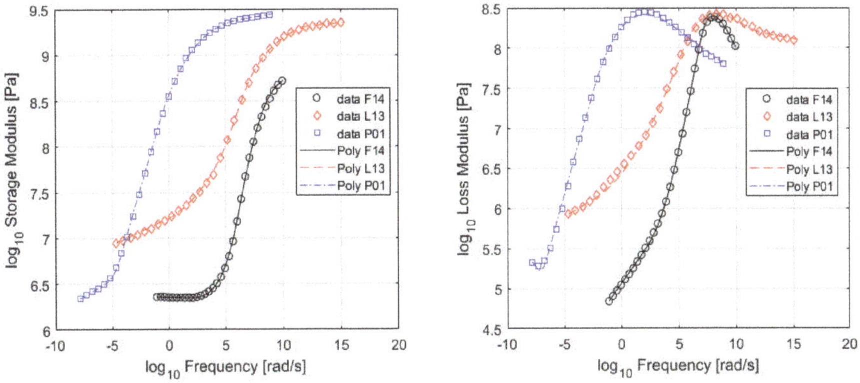

Figure 5. Storage (**left chart**) and loss (**right chart**) moduli for three rubber compounds. Experimental data (markers) vs. polynomial model (lines).

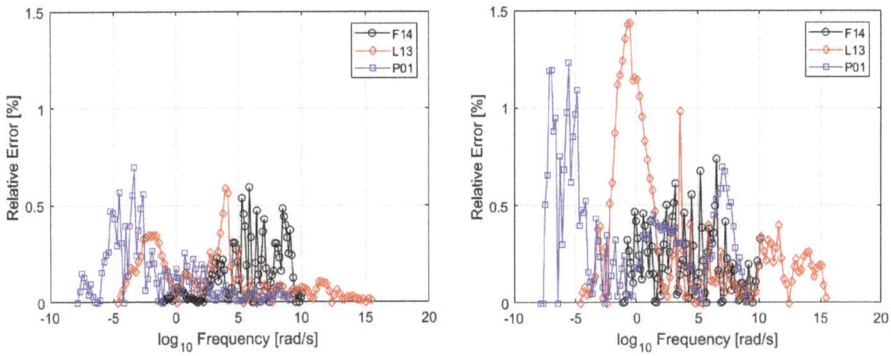

Figure 6. Storage (**left chart**) and loss (**right chart**) moduli relative error for three rubber compounds.

The same dataset was used for assessing the quality of the fit using the LBEM approach. The MATLAB function "lsqnonlin" with the default trust-region-reflective fitting algorithm [28] was used to identify the optimal parameter values of the proposed formulation by minimising the non-linear cost function that effectively consists of the sum of the squared differences between measured and estimated loss tangent and storage modulus. The resulting parameter values for three example datasets have been tabulated in Table 2.

Table 2. Parameter values for the LBEM curves shown in Figure 7.

Acronym	$r\omega_0$	ry_0	L	k	$t\omega_0$	sy_1	sy_2	r_1	r_2	p
F14 [22]	6.613	6.341	2.394	1.084	6.719	0.020	0.000	1.258	1.129	2.181
L13 [26]	5.490	7.039	2.338	0.524	5.216	0.105	0.051	1.834	1.838	0.686
P01 [11]	−1.785	6.204	3.209	0.555	−2.780	0.014	0.000	1.347	2.612	0.825

Figure 7 shows a comparison between the measurements and the fitted LBEM models. The results are very satisfactory, as the formula is able to describe all the measured characteristics and is continuous at all times. From Figure 8 it can be appreciated that the relative error for the LBEM is higher than that of the polynomial formulation, as expected. This is because the LBEM approach uses a generic formulation with fewer parameters that can, however, be easily linked to the typifying quantities of the data. For the storage modulus the maximum relative error increased to 1.4%. The loss tangent shows a higher error of up to 27%, but such discrepancies happen at a limited number of discrete points.

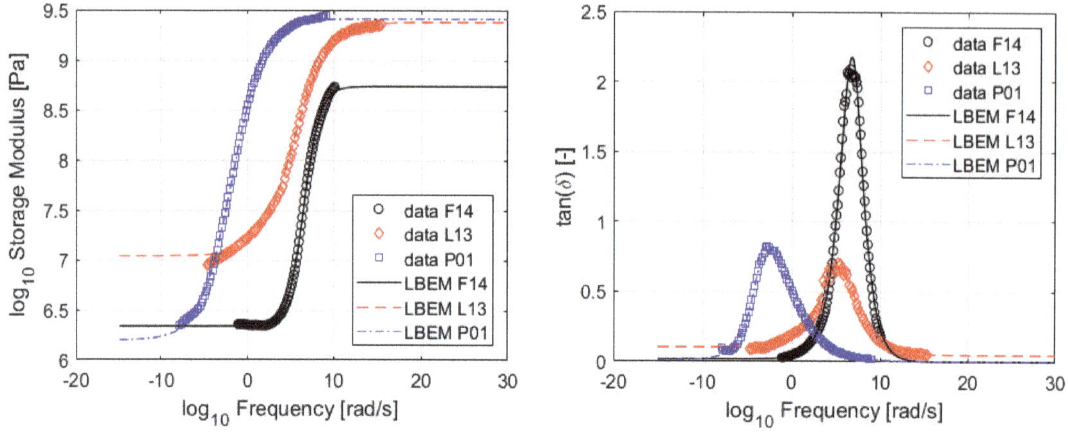

Figure 7. Storage moduli (**left chart**) and loss tangents (**right chart**) for three rubber compounds. Experimental data (markers) vs. LBEM model (lines).

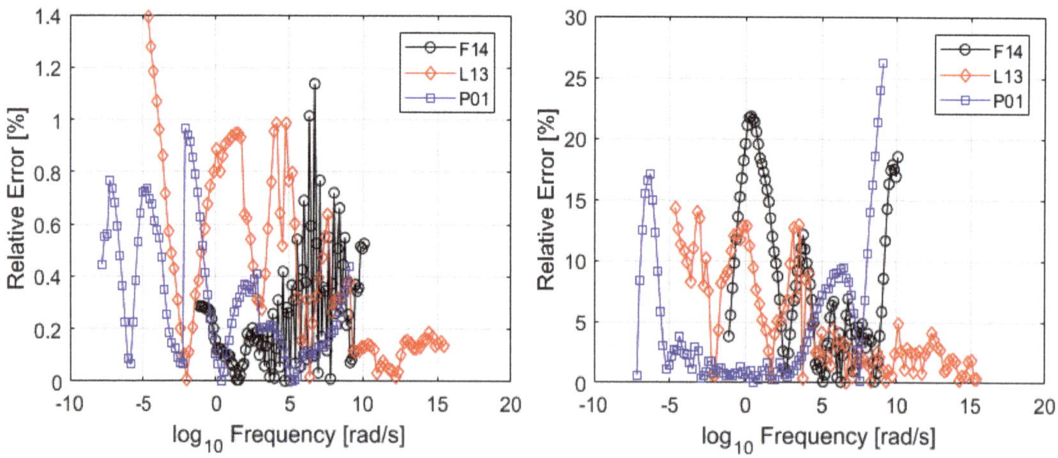

Figure 8. Storage moduli (**left chart**) and loss tangents (**right chart**) relative error for three rubber compounds.

The fitting quality of the polynomial and LBEM models was also evaluated by means of the normalised root mean squared error (NRMSE) and the coefficient of determination, denoted R^2 for both the storage modulus and the loss tangent. The results are summarised in Tables 3 and 4.

Table 3. Normalised root mean squared error for all the compounds using the polynomial and LBEM models.

Acronym	NRMSE E' (Poly)	NRMSE E'' (Poly)	NRMSE $\tan\delta$ (Poly)	NRMSE E' (LBEM)	NRMSE E'' (LBEM)	NRMSE $\tan\delta$ (LBEM)
F14 [22]	0.0064	0.0049	0.0104	0.0231	0.0314	0.0217
F15 [23]	0.0097	0.0129	0.0291	0.0255	0.0442	0.0373
HK8 [24]	0.0101	0.0131	0.0170	0.0199	0.0369	0.0205
L13 [26]	0.0053	0.0129	0.0232	0.0165	0.0306	0.0266
L15A [25]	0.0150	0.0075	0.0097	0.0434	0.0749	0.0216
L15B [25]	0.0128	0.0088	0.0100	0.0780	0.1046	0.0309
L15C [25]	0.0124	0.0090	0.0068	0.0298	0.1726	0.0265
P01 [11]	0.0047	0.0096	0.0095	0.0237	0.0363	0.0129
P06A [21]	0.0181	0.0108	0.0369	0.0227	0.1495	0.0847
P06B [21]	0.0078	0.0175	0.0216	0.0194	0.0517	0.0306
Se14 [27]	0.0069	0.0090	0.0187	0.0151	0.1070	0.0311
Su18A [29]	0.0079	0.0064	0.0488	0.0486	0.0755	0.0275
Su18B [29]	0.0081	0.0124	0.0445	0.0546	0.0928	0.0298
T19 [15]	0.0067	0.0044	0.0196	0.0295	0.0908	0.0232
I00 [20]	0.0097	0.0141	0.0212	0.0261	0.0775	0.0158
I20 [20]	0.0099	0.0115	0.0101	0.0350	0.0489	0.0098
I35 [20]	0.0085	0.0114	0.0090	0.0331	0.0654	0.0173
I40 [20]	0.0084	0.0132	0.0155	0.0312	0.0473	0.0121
I60 [20]	0.0077	0.0082	0.0102	0.0538	0.0517	0.0104

Table 4. Coefficient of determination for all the compounds using the polynomial and LBEM models.

Acronym	R^2 E' (Poly)	R^2 E'' (Poly)	R^2 $\tan\delta$ (Poly)	R^2 E' (LBEM)	R^2 E'' (LBEM)	R^2 $\tan\delta$ (LBEM)
F14 [22]	0.9997	0.9998	0.9984	0.9932	0.9919	0.9960
F15 [23]	0.9992	0.9988	0.9910	0.9926	0.9852	0.9837
HK8 [24]	0.9992	0.9986	0.9953	0.9972	0.9875	0.9957
L13 [26]	0.9998	0.9987	0.9943	0.9982	0.9926	0.9923
L15A [25]	0.9998	0.9996	0.9987	0.9826	0.9553	0.9936
L15B [25]	0.9986	0.9994	0.9986	0.9294	0.9129	0.9870
L15C [25]	0.9990	0.9994	0.9993	0.9936	0.7593	0.9866
P01 [11]	0.9998	0.9991	0.9991	0.9960	0.9891	0.9985
P06A [21]	0.9982	0.9989	0.9870	0.9969	0.7696	0.9528
P06B [21]	0.9996	0.9971	0.9946	0.9977	0.9694	0.9884
Se14 [27]	0.9997	0.9994	0.9947	0.9984	0.9177	0.9833
Su18A [29]	0.9996	0.9996	0.9712	0.9863	0.9288	0.9886
Su18B [29]	0.9996	0.9994	0.9757	0.9827	0.8916	0.9857
T19 [15]	0.9997	0.9999	0.9946	0.9929	0.9347	0.9920
I00 [20]	0.9994	0.9985	0.9931	0.9957	0.9400	0.9960
I20 [20]	0.9994	0.9991	0.9986	0.9909	0.9757	0.9988
I35 [20]	0.9995	0.9990	0.9987	0.9912	0.9562	0.9959
I40 [20]	0.9996	0.9986	0.9962	0.9935	0.9769	0.9982
I60 [20]	0.9996	0.9995	0.9982	0.9796	0.9739	0.9987

4. Results

Results show that the empirical formulations presented herein were able to model all the viscoelastic moduli that were examined with correlation levels similar to those presented in the previous section. More specifically, the two proposed approaches were compared against each other and against other methods from the literature, such as the zero/pole fitting presented in [9] and the fractal derivative generalised Maxwell Model [30]. Specific performance attributes of the models that are addressed include: (a) versatility, (b) number of parameters required, (c) computational efficiency and (d) integration with Persson's friction model [31]. Persson's model was selected as the "carrier" friction model, as it has been heavily relied upon in the literature for the prediction of rubber friction [32–38].

It was also used because it is a particularly computationally intensive model that will benefit from efficient implementations of the viscoelastic material model. Finally, the effects on model parameters of five SBR (styrene-butadiene rubber) compounds from the study presented in [20] are discussed and the results are linked to the CB (carbon black) content. In line with the findings in [20], the varying levels of CB in the rubber do not affect the peak values of the viscoelastic modulus, but do influence the gradient of the linear sections.

4.1. Versatility

Versatility of a model is interpreted as the ability to successfully fit the moduli of different materials. The performances of the models in this respect were demonstrated in the previous section, particularly in Figures 5–8 and Tables 3 and 4. Due to its piece-wise nature, the polynomial method is inherently more flexible in terms of fitting any of the three commonly occurring elements of the complex modulus (storage modulus, loss modulus and loss tangent). It is also more likely to fit more accurately a wider range of materials, albeit with increased pre-processing time and an increase in the number of fitting parameters. The LBEM method, on the other hand, requires less pre-processing, does not depend on manual selection of the relevant data segments and relies on fewer parameters than the polynomial method, as described below.

4.2. Number of Parameters

Both formulas make use of parameters that can be linked to the critical sections of the data, such as storage modulus transition regions or the peak values of the loss modulus and loss tangent. The LBEM method requires 10 parameters to be defined, whereas the polynomial method needs a minimum of 16 parameters for the storage and loss modulus to be evaluated.

An important quality of the LBEM approach is that the parameters are directly linked to typifying quantities of the data, such as the loss tangent frequency peak location, the curve's growth rate and the loss tangent peak value.

4.3. Computational Efficiency

In terms of solution efficiency, both formulas provide a greatly reduced computation time compared with the common data interpolation method, the proposed zero/pole formula of [39] and the FDGM (fractal derivative generalised Maxwell) model [9]. Table 5 compares solution times of Persson's isothermal friction model for a single friction point using a resolution of 200 frequency segments, and several complex modulus implementations. The commonly used data interpolation method was used as a benchmark and had a mean computation time of 0.318 s per friction point. The FDGM model (that utilises a zero/pole formulation) had a computational time of 1.371 s, which is an increase of 332% with respect to the interpolation method. The proposed LBEM model had a reduced computation time of 0.184 s, a reduction of 42% with respect to the interpolation method and 87% with respect to the zero/pole model. On the other hand, the polynomial method resulted in a 93% or 98% more efficient solution compared to interpolation and the zero/pole techniques, respectively, and is also 12% less computationally demanding than the LBEM method. The reduced efficiency of the LBEM approach in comparison to the polynomial method is related to the use of computationally expensive mathematical functions, such as exponentials and the hyperbolic secant.

Table 5. Computation times of the models.

	Interpolation	FDGM	LBEM	Poly
Computation time (sec/μ point)	0.318	1.371	0.184	0.022
% w.r.t interpolation	~	+332%	−42%	−93%
% w.r.t zero/pole	−77%	~	−87%	−98%

4.4. Application to Persson'S Friction Model

Figure 9 shows the friction coefficient estimated over a wide range of sliding speeds for the three different materials used previously. Persson's isothermal friction model has been used for this estimation [11] with the model parameters provided in Table 6.

Table 6. Commonly used parameters for Persson's isothermal friction model [11], employed in the present study.

Parameter	Description	Value	Units
ζ_{max}	Maximum magnification	10	[-]
H	Hurst exponent	0.85	[-]
ν	Poisson ratio	0.5	[-]
q_0	Lower distance cut-off	2000	[m^{-1}]
σ_0	Normal pressure	0.2×10^6	[Pa]

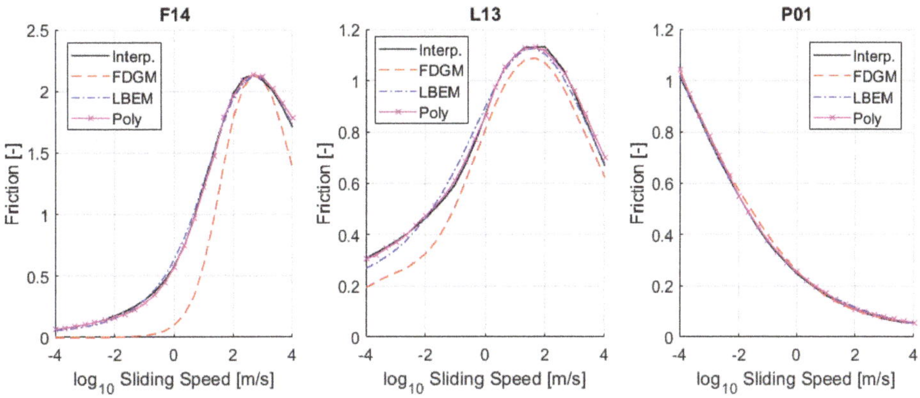

Figure 9. Estimated friction using 4 rheological models for 3 different DMA data sources.

All models gave similar friction results, with the LBEM model providing the best agreement with the interpolated data. A discrepancy was produced by the FDGM model, which is most likely a result of extrapolation at frequency extremes. However, the FDGM model may include an increased number of parameters or "zones"; therefore, there is room for improving its fitting accuracy [9].

4.5. PHR Content

We used the data published in [20] to study the effect of different concentrations of CB (carbon black) on the DMA data, and by extension, the effects on the identified parameters from both formulations. Five SBR compounds with different parts phr (per hundred rubber) of carbon black, ranging from 0 to 60 phr, are shown in Figure 10. Both models were able to fit the curves with a maximum normalised root mean squared error of 7.75% for the LBEM formulation and 2.12% for the polynomial formulation. From the raw data it can be seen that as the phr of CB increases, so does the storage modulus in the rubbery (low frequency) region. The storage modulus also increases with the phr of CB in the glassy (high frequency) region, albeit at a reduced rate. The peak of the loss tangent is broadly inversely proportional to the concentration of CB.

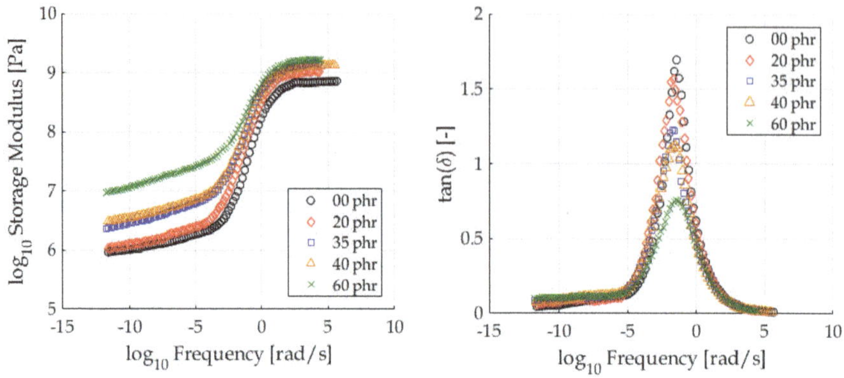

Figure 10. Raw data adapted from [20] for SBR material with different concentrations of CB.

With regard to the polynomial model, Figure 11 demonstrates the linear relationship between the vertical shift of the linear section (see Figure 4a) and the phr concentration of CB in rubber specimens. R-squared values are included in Table 7.

Table 7. Correlation between polynomial coefficients and CB concentration.

	Coefficient of Determination—R^2	
Coefficient	Storage Modulus	Loss Modulus
a^1	0.587	0.458
a^2	0.868	0.889
a^3	0.007	0.592
b^1	0.920	0.830
b^2	0.893	0.185
b^3	0.759	0.782

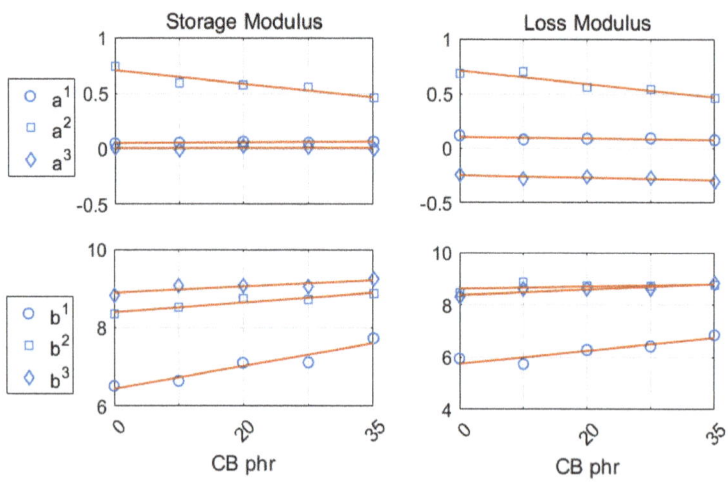

Figure 11. Linear regression between the coefficients of the linear polynomial equations ($y = ax + b$) and the CB concentration. Linear equations reproduce the linear sections of storage and loss modulus, respectively (see Figure 4). Coefficients (a^1, b^1), (a^2, b^2), (a^3, b^3) fit the first, second and third linear sections of the two modulus curves, from **left** to **right**.

From Figure 12 it can be observed that the parameters ry_0, k, sy_2, r_1, r_2 and p of the LBEM model have R-squared values equal to or higher than 90%, indicating a high level of correlation with the carbon black concentration. As expected from the raw data analysis, the parameter ry_0 that controls the vertical shift of the storage modulus shows a high correlation. However, the parameter k that controls the logistic rate of increase or steepness of the curve also shows a high correlation, and this was not easy to spot from a visual analysis of the data. In addition, as indicated from the visual analysis of raw data, the parameter p that controls the bell function peak value shows a clear correlation with CB concentration. The loss tangent parameter sy_2 that controls the bell function's vertical shift at high frequencies after the peak, together with r_1 and r_2 that determine the bell's rate of increase at low frequencies and high frequencies, respectively, also have R-squared values above 90%.

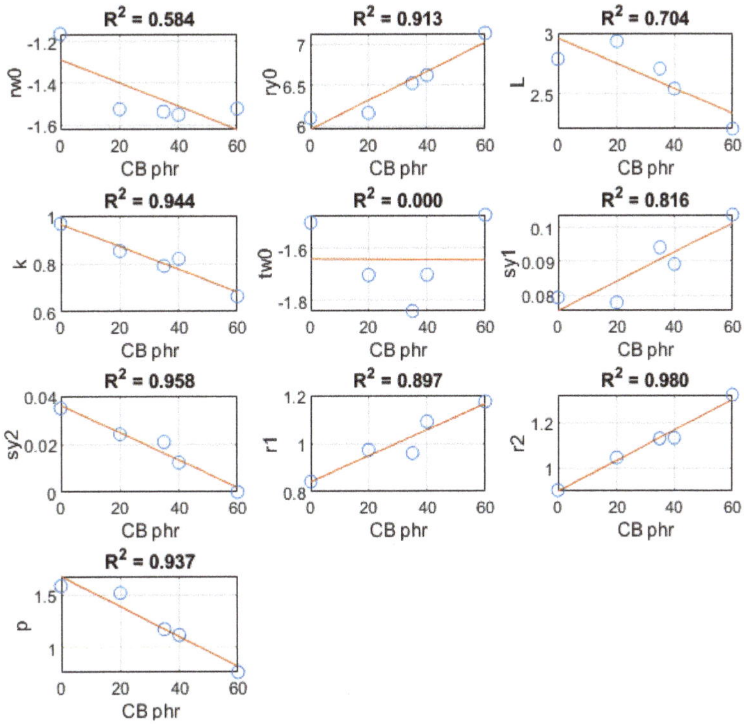

Figure 12. Correlations between the LBEM parameters and the CB concentration.

On the basis of the observed dependency of critical model parameters on CB content, the LBEM model and to a lesser extent the polynomial model, lend themselves to the creation of virtual hypothetical rubber materials to be simulated within friction models, so that the performance of a new tyre with new materials can be predicted.

5. Conclusions

The proposed empirical models are capable of not only accurately reproducing DMA data but also significantly reducing the computational time. This makes the models ideal for integration into advanced friction models, such as [11,13,14], where the viscoelastic complex modulus is evaluated multiple times and over a wide range of frequencies.

With the polynomial method, good predictions of the storage and loss moduli and the loss tangent can be obtained by selecting the piecewise linear/non-linear segments of the curves. Formulation is versatile in that in can fit a broad range of viscoelastic moduli because of the adjustability of the number of segments. The proposed LBEM formula uses parameters that can be easily linked to the typifying quantities of the data, such us the storage moduli in the rubbery and glassy regions or the frequency where the loss tangent is maximum. The fitting quality of both models was quantitatively evaluated against the data using the NRMSE and R^2 metrics. The R^2 for the polynomial method was above 0.97 for all cases, whereas the performance of the LBEM formula varied more widely, with R^2 varying from 0.99 to 0.75. Comparable results can be obtained with the FDGM model at the expense of computation time. The proposed models are primarily suited to calculations in the frequency domain, but they are not applicable the creep and relaxation tests because the relaxation time cannot be readily deduced from their formulation. With regard to the polynomial model, only a subset of model parameters can be linked to the CB content of the rubber, and by extension, to the chemical composition of the rubber. On the other hand, the LBEM model comes with a smaller parameter set that is linked to material composition; therefore, it can be used for the creation of hypothetical rubber materials to assess/optimise the grip of new tyres ahead of production.

Author Contributions: Conceptualization, M.F.T., K.G. and G.M.; investigation, M.F.T. and K.G.; methodology, M.F.T. and K.G.; software, M.F.T. and K.G.; validation, M.F.T., K.G. and G.M.; formal analysis, M.F.T. and K.G.; resources, M.F.T. and K.G.; writing—original draft preparation, M.F and K.G.; writing—review and editing, K.G. and G.M.; visualization, M.F.T., K.G. and G.M.; supervision, G.M. All authors have read and agreed to the published version of the manuscript.

Funding: This research received no external funding.

Conflicts of Interest: The authors declare no conflict of interest.

References

1. Ward, I.M.; Sweeney, J. *The Mechanical Properties of Solid Polymers*, 2nd ed.; John Wiley & Sons: Hoboken, NJ, USA, 2004
2. Samali, B.; Kwok, K. Use of viscoelastic dampers in reducing wind-and earthquake-induced motion of building structures. *Eng. Struct.* **1995**, *17*, 639–654. [CrossRef]
3. Gurvich, M.R. On challenges in structural analysis of rubber components: Frictional contact. *J. Elastomers Plast.* **2007**, *39*, 53–67. [CrossRef]
4. Selig, M. The Influence of Internal Tyre Pressure on Road Friction. Ph.D. Thesis, University of Huddersfield, Huddersfield, UK, 2014.
5. Roylance, D. *Engineering Viscoelasticity*; Department of Materials Science and Engineering, Massachusetts Institute of Technology: Cambridge, MA, USA, 2001; Volume 2139, pp. 1–37.
6. Askarbekov, R.; Herak, D.; Mizera, C. Mechanical behavior of rubber samples under relaxation. *Eng. Rural. Dev.* **2016**, *2016*, 745–749. [CrossRef]
7. Reinelt, J.; Wriggers, P. Multi-scale approach for frictional contact of elastomers on rough rigid surfaces. *Lect. Notes Appl. Comput. Mech.* **2010**, *51*, 53–94. [CrossRef]
8. Xu, Q.; Engquist, B. A mathematical and physical model improves accuracy in simulating solid material relaxation modulus and viscoelastic responses. *arXiv* **2014**, arXiv:1412.5225.
9. Genovese, A.; Carputo, F.; Maiorano, A.; Timpone, F.; Farroni, F.; Sakhnevych, A. Study on the Generalized Formulations with the Aim to Reproduce the Viscoelastic Dynamic Behavior of Polymers. *Appl. Sci.* **2020**, *10*, 2321. [CrossRef]
10. Grosch, K. The relation between the friction and visco-elastic properties of rubber. *Proc. R. Soc. Lond. A* **1963**, *274*, 21–39.
11. Persson, B.N.J. Theory of rubber friction and contact mechanics. *J. Chem. Phys.* **2001**, *115*, 3840–3861. [CrossRef]
12. Chawla, K.K.; Meyers, M. *Mechanical Behavior of Materials*; Prentice Hall: Upper Saddle River, NJ, USA, 1999.
13. Persson, B.N.J. Rubber friction: Role of the flash temperature. *J. Phys. Condens. Matter* **2006**, *18*, 7789–7823. [CrossRef]
14. Klüppel, M.; Heinrich, G. Rubber Friction on Self-Affine Road Tracks. *Rubber Chem. Technol.* **2000**, *73*, 578–606. [CrossRef]
15. Tolpekina, T.; Pyckhout-Hintzen, W.; Persson, B. Linear and nonlinear viscoelastic modulus of rubber. *Lubricants* **2019**, *7*, 22. [CrossRef]
16. Williams, M.L.; Landel, R.F.; Ferry, J.D. The temperature dependence of relaxation mechanisms in amorphous polymers and other glass-forming liquids. *J. Am. Chem. Soc.* **1955**, *77*, 3701–3707. [CrossRef]
17. Ido, T.; Yamaguchi, T.; Shibata, K.; Matsuki, K.; Yumii, K.; Hokkirigawa, K. Sliding friction characteristics of styrene butadiene rubbers with varied surface roughness under water lubrication. *Tribol. Int.* **2019**, *133*, 230–235. [CrossRef]
18. Emami, A.; Khaleghian, S. Investigation of tribological behavior of Styrene-Butadiene Rubber compound on asphalt-like surfaces. *Tribol. Int.* **2019**, *136*, 487–495. [CrossRef]
19. Rodriguez, N.; Dorogin, L.; Chew, K.; Persson, B. Adhesion, friction and viscoelastic properties for non-aged and aged Styrene Butadiene rubber. *Tribol. Int.* **2018**, *121*, 78–83. [CrossRef]

20. Isono, Y.; Aoyama, T. Filler effects on temperature shift factors in viscoelastic properties of carbon black filled rubbers. *Nihon Reoroji Gakkaishi* **2013**, *41*, 137–144. [CrossRef]
21. Persson, B.; Volokitin, A. Rubber friction on smooth surfaces. *Eur. Phys. J. E* **2006**, *21*, 69–80. [CrossRef]
22. Fina, E.; Gruber, P.; Sharp, R. Hysteretic rubber friction: Application of Persson's theories to Grosch's experimental results. *J. Appl. Mech.* **2014**, *81*, 121001. [CrossRef]
23. Fortunato, G.; Ciaravola, V.; Furno, A.; Lorenz, B.; Persson, B. General theory of frictional heating with application to rubber friction. *J. Phys. Condens. Matter* **2015**, *27*, 175008. [CrossRef]
24. Heinrich, G.; Klüppel, M. Rubber friction, tread deformation and tire traction. *Wear* **2008**, *265*, 1052–1060. [CrossRef]
25. Lorenz, B.; Oh, Y.; Nam, S.; Jeon, S.; Persson, B. Rubber friction on road surfaces: Experiment and theory for low sliding speeds. *J. Chem. Phys.* **2015**, *142*, 194701. [CrossRef] [PubMed]
26. Lorenz, B.; Persson, B.; Fortunato, G.; Giustiniano, M.; Baldoni, F. Rubber friction for tire tread compound on road surfaces. *J. Phys. Condens. Matter* **2013**, *25*, 095007. [CrossRef]
27. Selig, M.; Lorenz, B.; Henrichmöller, D.; Schmidt, K.; Ball, A.; Persson, B. Rubber friction and tire dynamics: A comparison of theory with experimental data. *Tire Sci. Technol.* **2014**, *42*, 216–262. [CrossRef]
28. Coleman, T.F.; Li, Y. An interior trust region approach for nonlinear minimization subject to bounds. *SIAM J. Optim.* **1996**, *6*, 418–445. [CrossRef]
29. Sunish, V.; Taheri, S. Multiscale Modeling of Rubber Friction under dry/wet condition. In Proceedings of the RPUG Conference 2018, Rapid City, SD, USA, 18–21 September 2018.
30. Arikoglu, A. A new fractional derivative model for linearly viscoelastic materials and parameter identification via genetic algorithms. *Rheol. Acta* **2014**, *53*, 219–233. [CrossRef]
31. Persson, B.N. On the theory of rubber friction. *Surf. Sci.* **1998**, *401*, 445–454. [CrossRef]
32. Dapp, W.B.; Prodanov, N.; Müser, M.H. Systematic analysis of Persson's contact mechanics theory of randomly rough elastic surfaces. *J. Phys. Condens. Matter* **2014**, *26*, 355002. [CrossRef] [PubMed]
33. Mavros, G. A thermo-frictional tyre model including the effect of flash temperature. *Veh. Syst. Dyn.* **2019**, *57*, 721–751. [CrossRef]
34. Genovese, A.; Farroni, F.; Papangelo, A.; Ciavarella, M. A Discussion on Present Theories of Rubber Friction, with Particular Reference to Different Possible Choices of Arbitrary Roughness Cutoff Parameters. *Lubricants* **2019**, *7*, 85. [CrossRef]
35. Motamedi, M.; Taheri, S.; Sandu, C. Rubber—Road contact: Comparison of physics-based theory and indoor experiments. *Tire Sci. Technol.* **2016**, *44*, 150–173. [CrossRef]
36. Tolpekina, T.; Persson, B. Adhesion and friction for three tire tread compounds. *Lubricants* **2019**, *7*, 20. [CrossRef]
37. Kanafi, M.M.; Tuononen, A.J.; Dorogin, L.; Persson, B.N. Rubber friction on 3D-printed randomly rough surfaces at low and high sliding speeds. *Wear* **2017**, *376*, 1200–1206. [CrossRef]
38. Ciavarella, M. A simplified version of Persson's multiscale theory for rubber friction due to viscoelastic losses. *J. Tribol.* **2018**, *140*, 011403. [CrossRef]
39. Renaud, F.; Dion, J.L.; Chevallier, G.; Tawfiq, I.; Lemaire, R. A new identification method of viscoelastic behavior: Application to the generalized Maxwell model. *Mech. Syst. Signal Process.* **2011**, *25*, 991–1010. [CrossRef]

applied sciences

MDPI

Article

Handling Enhancement of Autonomous Emergency Steering for Reduced Road Friction Using Steering and Differential Braking

Yu-Min Lin and Bo-Chiuan Chen *

Department of Vehicle Engineering, National Taipei University of Technology, Taipei 10608, Taiwan; t101669022@ntut.edu.tw
* Correspondence: bochen@ntut.edu.tw

Abstract: Steering has more potential than braking to prevent rear-end collisions at higher relative velocities. A path tracking controller based on multi-input multi-output (MIMO) model predictive control (MPC) is proposed to enhance the handling performance of autonomous emergency steering in this paper. A six-state MIMO bicycle model including actuator dynamics of steering and differential braking is used for model prediction. Two control inputs are front wheel steering angle and direct yaw moment. Two model outputs are lateral displacement and heading angle. According to the work load ratios at four wheels, control allocation is used to determine the optimal braking force distribution to prevent tire force saturation. The performance of a single-input single-output (SISO) MPC that uses only steering angle control to track the lateral displacement of the desired path is employed to benchmark the performance of the proposed algorithm. Simulation results show that both SISO MPC and MIMO MPC can track the path on nominal road surface with high road friction coefficient of 0.9. For a road surface with medium road friction coefficient of 0.7, the SISO MPC is unable to track the path and loses directional stability. However, the MIMO MPC can still track the path and demonstrate robust path tracking and handling enhancement against model uncertainty due to reduced road friction.

Keywords: autonomous emergency steering; multi-input multi-output model predictive control; actuator dynamics; control allocation; handling enhancement; road friction

Citation: Lin, Y.-M.; Chen, B.-C. Handling Enhancement of Autonomous Emergency Steering for Reduced Road Friction Using Steering and Differential Braking. *Appl. Sci.* **2021**, *11*, 4891. https://doi.org/10.3390/app11114891

Academic Editors: Flavio Farroni, Andrea Genovese and Aleksandr Sakhnevych

Received: 29 April 2021
Accepted: 24 May 2021
Published: 26 May 2021

1. Introduction

According to Traffic Safety Facts [1], rear-end collisions account for 18% of the fatal collisions between motor vehicles. Singh et al. [2] pointed out that driver error is the primary cause for most accidents. The main factor is the recognition error caused by driver inattention, internal and external distractions, and inadequate surveillance. Advanced driver assistance systems (ADAS) can assist the driver to prevent possible collisions with the obstacle in front of the vehicle via braking or steering. Bosch [3] reported that the autonomous emergency braking (AEB) system would reduce 30% of rear-end collisions. Alfred et al. [4] showed that the necessary minimum distance to avoid a collision by steering is smaller than that by braking for maneuvers with large relative velocities. Especially for a suddenly appearing obstacle, AEB is not able to avoid a collision due to insufficient braking distance. The European New Car Assessment Programme introduced safety ratings for AEB in 2014, and has announced that [5] emergency steering assist (ESA) and autonomous emergency steering system (AES) tests will start in 2020 and 2022, respectively. The goal of the tests is to deliver a further significant reduction in crashes and casualties.

1.1. Related Work

Both ESA and AES can make a sudden lane change or evasive maneuver via steering actions to avoid a collision. The main difference between ESA and AES is the trigger mechanism. When the ESA [6–9] detects the risk of collision and the steering torque from the driver, the system provides additional overlay torques to adjust the steering angle to track the evasive path without collision or directional instability. However, Shimizu et al. [10] showed that 37.9% of drivers did not take any avoidance maneuvers when encountering the approaching danger. Therefore, collision avoidance cannot be guaranteed for inattentive maneuvers with late steering actions from the driver. Unlike ESA, the AES [11–15] can steer automatically when the detected risk of collision is over the designed threshold.

Three types of planning approaches, search-based, optimization-based, and parameter-based, are commonly used for generating a desired path to avoid the collision. Search-based methods [16–18] are designed to generate an action set of multiple drivable trajectories, allowing an adjacent behavior planner to pick the most appropriate action for the global state in the scene. Optimization-based methods [19–21] plan the path from a mapping run and offline compute an optimal trajectory around the track using the trajectory optimization module to maximize the progress rate, and minimize slip angle cost and input rate. However, the first and second types of planning approaches are not easy to implement on production control units due to high computation loads. Parameter-based methods [7] including circular, sine/cosine, s-shape, polynomial, and trapezoidal-acceleration-profile (TAP) trajectories are designed by specifying the desired lateral displacement for a lane change maneuver that can also be employed for collision avoidance. Among the parameter-based methods, only TAP can consider transition time and ride comfort at the same time [22].

Two types of path tracking controllers, model-free and model-based, are commonly used for tracking an evasive path. If the dynamic model of the plant to be controlled is not needed for the controller design, it is classified as the model-free approach. The conventional model-free controller calculates the steering angle command based on the lateral displacement error and the heading error at the previewed point [23–25]. Machine learning can also be classified as the model-free approach. A neural network model can be trained to generate the steering action using the dataset of human drivers [26] or deep reinforcement learning [27]. However, the major challenge is the difficulty in proving the safety issue of these methods [28]. As for model-based approaches, the single-input single-output (SISO) model predictive control (MPC) is often used to obtain the vehicle steering angle based on a four-state bicycle model, which is used to predict the future lateral displacement of a vehicle [29–31].

Differential braking via electric hydraulic braking (EHB) can be used to generate the desired direct yaw moment (DYM) to improve path tracking performance. Falcone et al. [32] proposed a multi-input multi-output (MIMO) MPC controller that uses four-wheel longitudinal slip ratios and front wheel steering angle as the control input to minimize errors in heading angle, yaw rate and lateral displacement. Their simulation results showed that path error can be reduced by 50%. However, intensive computations are required for online linearization and optimization due to the complexities of the prediction model with nonlinear tire dynamics. Hajiloo et al. [33] proposed a MIMO MPC controller to determine the optimal front lateral force and the yaw moment to improve the yaw rate responsiveness for collision avoidance without decreasing the stability. Lin et al. [34] proposed a MIMO MPC controller for AES using front wheel steering angle and direct yaw-moment control (DYC) to enhance the performance of path tracking. The pseudo inverse control allocation (CA) is used to allocate the four-wheel brake forces for generating the desired DYM. Their simulation results showed that the path tracking performance of the MIMO MPC is better than that of the conventional SISO MPC. However, the actuator dynamics of steering and differential braking are not discussed in their proposed approach. The control performance might be diminished or control lost if the actuator dynamics are not fast enough to be negligible.

Electric power steering (EPS) can be used to generate the desired front wheel steering angle. However, the actuator dynamics might cause the actual steering angle to lag slightly behind the steering angle command. Chen et al. [35] proposed a MPC controller by considering EPS actuator dynamics as a first-order model to enhance the tracking performance and ride comfort. Kim et al. [36] proposed a MPC controller by considering the EPS as a second-order model. Simulation results showed that tracking performance is improved by incorporating steering actuator dynamics in the prediction model. Thornton et al. [37] proposed a MPC controller with steering input to evaluate the average and maximum yaw rates of different actuator dynamics. Their experiment results showed that the lumped first-order dynamics can meet design objectives with low computational complexity.

1.2. Main Contributions

The proposed path-tracking controller for AES is an extension of the previous work [34] in which the actuator dynamics were neglected in the vehicle model.

- The dynamic responses of the vehicle model in the previous work might deviate from those of the actual vehicle model with actuator dynamics for severe maneuvers. The EPS and EHB dynamics are integrated with the four-state bicycle model to enhance the accuracy of the prediction model.
- A diagonal matrix with components of tire workload ratios is used to replace the identity weighting matrix of CA to prevent saturating tire forces while generating the desired DYM.
- Different road friction coefficients are used for sensitivity analysis to evaluate the robustness and handling enhancement of the proposed algorithm at different vehicle velocities via simulation studies.

1.3. Organization

The remainder of this paper is organized as follows: system modeling for controller design is briefly introduced in Section 2, followed by the proposed algorithm in Section 3. Simulation results are presented in Section 4. Finally, conclusions are made in Section 5.

2. Modeling

2.1. Actuator Dynamics

First-order actuator dynamics of EPS and EHB as shown below are included in the plant model for MPC.

$$\dot{\delta}_f = -\frac{1}{\tau_{EPS}}\delta_f + \frac{1}{\tau_{EPS}}\delta_{f,r} \tag{1}$$

$$\dot{M}_z = -\frac{1}{\tau_{EHB}}M_z + \frac{1}{\tau_{EHB}}M_{z,r} \tag{2}$$

where $\delta_{f,r}$ and $M_{z,r}$ are the control commands of the front wheel steering angle and DYM, respectively; δ_f and M_z are the responses of front wheel steering angle and DYM, respectively; τ_{EPS} and τ_{EHB} are the time constants of EPS and EHB, respectively. τ_{EPS} is set to be 0.125 s [37] and τ_{EHB} is set to be 0.1 s [38] in this paper.

2.2. Bicycle Model

The six-state bicycle model is obtained by integrating the actuator dynamics with the four-state bicycle model that is linearized using small angle approximation and constant longitudinal velocity [39]. The state-space representation can be expressed as follows.

$$\begin{aligned} \dot{\mathbf{x}} &= \mathbf{A}_m\mathbf{x} + \mathbf{B}_m\mathbf{u} \\ \mathbf{y} &= \mathbf{C}_m\mathbf{x} \end{aligned} \tag{3}$$

where $\mathbf{x} = \begin{bmatrix} y & v_y & \psi & r & \delta_f & M_z \end{bmatrix}^T$ is the state vector, $\mathbf{u} = \begin{bmatrix} \delta_{f,r} & M_{z,r} \end{bmatrix}^T$ is the input vector, and $\mathbf{y} = \begin{bmatrix} y & \psi \end{bmatrix}^T$ is the output vector; v_y and r are the lateral velocity and yaw rate of the vehicle, respectively; y and ψ are the lateral displacement and heading angle of the vehicle, respectively; $\mathbf{A}_m$, $\mathbf{B}_m$ and $\mathbf{C}_m$ are system matrix, input matrix and output matrix as shown below.

$$\mathbf{A}_m = \begin{bmatrix} 0 & 1 & v_x & 0 & 0 & 0 \\ 0 & \frac{-(C_{\alpha f}+C_{\alpha r})}{m v_x} & 0 & \frac{l_r C_{\alpha r}-l_f C_{\alpha f}}{m v_x} - v_x & \frac{C_{\alpha f}}{m} & 0 \\ 0 & 0 & 0 & 1 & 0 & 0 \\ 0 & \frac{l_r C_{\alpha r}-l_f C_{\alpha f}}{I_z v_x} & 0 & \frac{-(l_f{}^2 C_{\alpha f}+l_r{}^2 C_{\alpha r})}{I_z v_x} & \frac{l_f C_{\alpha f}}{I_z} & \frac{1}{I_z} \\ 0 & 0 & 0 & 0 & -\frac{1}{\tau_{EPS}} & 0 \\ 0 & 0 & 0 & 0 & 0 & -\frac{1}{\tau_{EHB}} \end{bmatrix} \quad (4)$$

$$\mathbf{B}_m = \begin{bmatrix} 0 & 0 & 0 & 0 & \frac{1}{\tau_{EPS}} & 0 \\ 0 & 0 & 0 & 0 & 0 & \frac{1}{\tau_{EHB}} \end{bmatrix}^T \quad (5)$$

$$\mathbf{C}_m = \begin{bmatrix} 1 & 0 & 0 & 0 & 0 & 0 \\ 0 & 0 & 1 & 0 & 0 & 0 \end{bmatrix} \quad (6)$$

where m is the vehicle mass; v_x is the vehicle longitudinal velocity; $C_{\alpha f}$ and $C_{\alpha r}$ are the cornering stiffness of the front and rear axles, respectively; l_f and l_r are the distances from the C.G. to the front and rear axles, respectively; I_z is the yaw-plane rotational inertia of vehicle. The ranks of controllability matrix $\mathcal{C}$ and observability matrix $\mathcal{O}$ of the six-state bicycle model can be calculated as follows.

$$rank(\mathcal{C}) = rank\left(\begin{bmatrix} \mathbf{A}_m & \mathbf{A}_m\mathbf{B}_m & \mathbf{A}_m^2\mathbf{B}_m & \ldots & \mathbf{A}_m^5\mathbf{B}_m \end{bmatrix}\right) \quad (7)$$

$$rank(\mathcal{O}) = rank\left(\begin{bmatrix} \mathbf{C}_m & \mathbf{C}_m\mathbf{A}_m & \mathbf{C}_m\mathbf{A}_m^2 & \ldots & \mathbf{C}_m\mathbf{A}_m^5 \end{bmatrix}^T\right) \quad (8)$$

Since both $\mathcal{C}$ and $\mathcal{O}$ are full rank of 6, the proposed model is fully controllable and observable.

2.3. Model Discretization

A D-Class Sedan in CarSim is selected as the host vehicle to obtain the parameter values of m, a, b, and I_z. Nominal values of $C_{\alpha f}$ and $C_{\alpha r}$ are identified on a road surface with the nominal friction coefficient of 0.9 by minimizing the path errors between the trajectory responses of CarSim and the bicycle model. The vehicle parameters used in CarSim are shown in Table 1.

For the discrete controller design, the discretized plant model can be expressed as follows.

$$\begin{aligned} \mathbf{x}_{k+1} &= \mathbf{\Phi}_m \mathbf{x}_k + \mathbf{\Gamma}_m \mathbf{u}_k \\ \mathbf{y}_k &= \mathbf{H}_m \mathbf{x}_k \end{aligned} \quad (9)$$

where k denotes for the kth sample. The sample time T_s is selected to be 0.04 s in this paper; $\mathbf{u}_k$ is the discretized input vector; $\mathbf{\Phi}_m$, $\mathbf{\Gamma}_m$ and $\mathbf{H}_m$ can be expressed as follows.

$$\mathbf{\Phi}_m = \begin{bmatrix} 1 & T_s & v_x T_s & 0 & 0 & 0 \\ 0 & \frac{-(C_{\alpha f}+C_{\alpha r})T_s}{m v_x} + 1 & 0 & \frac{l_r C_{\alpha r}-l_f C_{\alpha f}}{m v_x} T_s - v_x T_s & \frac{C_{\alpha f} T_s}{m} & 0 \\ 0 & 0 & 1 & T_s & 0 & 0 \\ 0 & \frac{l_r C_{\alpha r}-l_f C_{\alpha f}}{I_z v_x} T_s & 0 & \frac{-l_f{}^2 C_{\alpha f}+l_f{}^2 C_{\alpha r} T_s}{I_z v_x} + 1 & \frac{l_f C_{\alpha f} T_s}{I_z} & \frac{T_s}{I_z} \\ 0 & 0 & 0 & 0 & -\frac{T_s}{\tau_{EPS}} + 1 & 0 \\ 0 & 0 & 0 & 0 & 0 & -\frac{T_s}{\tau_{EHB}} + 1 \end{bmatrix} \quad (10)$$

$$\mathbf{\Gamma}_m = \begin{bmatrix} 0 & 0 & 0 & 0 & \frac{T_s}{\tau_{EPS}} & 0 \\ 0 & 0 & 0 & 0 & 0 & \frac{T_s}{\tau_{EHB}} \end{bmatrix}^T \quad (11)$$

$$\mathbf{H}_m = \begin{bmatrix} 1 & 0 & 0 & 0 & 0 & 0 \\ 0 & 0 & 1 & 0 & 0 & 0 \end{bmatrix} \tag{12}$$

Table 1. Vehicle parameters.

Name	Value
Vehicle mass (m)	1530 kg
Yaw-plane rotational inertia of vehicle (I_z)	2315 kg–m^2
Distances from the C.G. to the front axles (l_f)	1.1 m
Wheel base (L)	2.78 m
Cornering stiffness of the front axles ($C_{\alpha f}$)	150.3 kN/rad
Cornering stiffness of the rear axles ($C_{\alpha r}$)	104.9 kN/rad

3. Controller Design

The block diagram of the proposed algorithm is shown in Figure 1. The path planning module generates the reference path $\mathbf{R}_s$ using the trapezoidal acceleration profile [7]. The six-state bicycle model is used to predict the future path. The PGC calculation module calculates the PGC index from $\mathbf{R}_s$. The optimizer module is used to compute the best set of future control commands. The first two elements of the set are used as the desired control commands of front wheel steering angle and DYM. The CA module distributes brake pressure at the four wheels to generate the desired DYM. After the AES is triggered, the driver is disengaged from the vehicle until the driver's intent to override the system is detected.

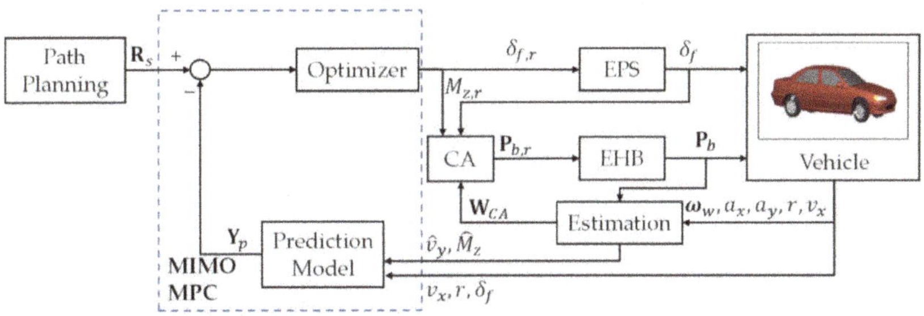

Figure 1. Block diagram of the proposed algorithm.

3.1. Model Predictive Controller

The predicted state vectors in the prediction horizon N_p with the future control commands in the control horizon N_c can be expressed as follows.

$$\begin{aligned} \mathbf{x}_{k,i+1} &= \mathbf{\Phi}_m \mathbf{x}_{k,i} + \mathbf{\Gamma}_m \mathbf{u}_{k,i} \\ \mathbf{x}_{k,i+2} &= \mathbf{\Phi}_m^2 \mathbf{x}_{k,i} + \mathbf{\Phi}_m \mathbf{\Gamma}_m \mathbf{u}_{k,i} + \mathbf{\Gamma}_m \mathbf{u}_{k,i+1} \\ &\vdots \\ \mathbf{x}_{k,i+N_p} &= \mathbf{\Phi}_m^{N_p} \mathbf{x}_{k,i} + \mathbf{\Phi}_m^{N_p-1} \mathbf{\Gamma}_m \mathbf{u}_{k,i} + \cdots + \mathbf{\Phi}_m^{N_p-N_c} \mathbf{\Gamma}_m \mathbf{u}_{k,i+N_c-1} \end{aligned} \tag{13}$$

where the initial condition of the state vector $\mathbf{x}_{k,i}$ can be expressed as follows.

$$\mathbf{x}_{k,i} = \begin{bmatrix} 0 & \hat{v}_y & 0 & r & \delta_f & \hat{M}_z \end{bmatrix}^T \tag{14}$$

where the initial lateral displacement and heading angle are zero in the body coordinates of the vehicle; r and δ_f can be measured using yaw rate and steering angle sensors, respectively; $\hat{v}_y$ is the estimated lateral velocity that can be estimated using localization-based [40] or inertial-measurement-unit-based [41] approaches; $\hat{M}_z$ can be estimated using estimated logitudinal tire forces as follows.

$$\hat{M}_z = \mathbf{B}_{CA}\hat{\mathbf{F}}_x \tag{15}$$

where $\hat{\mathbf{F}}_x$ is the estimated longitudinal tire force that can be estimated using a closed-loop disturbance estimator [42]; $\mathbf{B}_{CA}$ is the effective matrix as shown below.

$$\mathbf{B}_{CA} = \begin{bmatrix} -\frac{d}{2}cos\delta_f + l_f sin\delta_f & -\frac{d}{2} & \frac{d}{2}cos\delta_f + l_f sin\delta_f & \frac{d}{2} \end{bmatrix} \tag{16}$$

where d is the track width of the vehicle.

The predicted output vectors in the prediction horizon N_p can be expressed as follows.

$$\begin{aligned} \mathbf{y}_{p,k,i+1} &= \mathbf{H}_m\mathbf{\Phi}_m\mathbf{x}_{k,i} + \mathbf{H}_m\mathbf{\Gamma}_m\mathbf{u}_{k,i} \\ \mathbf{y}_{p,k,i+2} &= \mathbf{H}_m\mathbf{\Phi}_m^2\mathbf{x}_{k,i} + \mathbf{H}_m\mathbf{\Phi}_m\mathbf{\Gamma}_m\mathbf{u}_{k,i} + \mathbf{H}_m\mathbf{\Gamma}_m\mathbf{u}_{k,i+1} \\ &\vdots \\ \mathbf{y}_{p,k,i+N_p} &= \mathbf{H}_m\mathbf{\Phi}_m^{N_p}\mathbf{x}_{k,i} + \mathbf{H}_m\mathbf{\Phi}_m^{N_p-1}\mathbf{\Gamma}_m\mathbf{u}_{k,i} + \cdots + \mathbf{H}_m\mathbf{\Phi}_m^{N_p-N_c}\mathbf{\Gamma}_m\mathbf{u}_{k,i+N_c-1} \end{aligned} \tag{17}$$

From Equations (11) and (15), the augmented prediction model can be formulated as follows.

$$\mathbf{Y}_p = \mathbf{F}\mathbf{x}_{k,i} + \mathbf{G}\mathbf{U}_p \tag{18}$$

where $\mathbf{U}_p$ is the input vector; $\mathbf{Y}_p$ is the output vector; $\mathbf{F}$ and $\mathbf{G}$ are the system and input matrices, respectively. They are defined as follows.

$$\mathbf{U}_p = \begin{bmatrix} \mathbf{u}_{k,i}^T & \mathbf{u}_{k,i+1}^T & \cdots & \mathbf{u}_{k,i+N_c-1}^T \end{bmatrix}^T \tag{19}$$

$$\mathbf{Y}_p = \begin{bmatrix} \mathbf{y}_{p,k,i+1}^T & \mathbf{y}_{p,k,i+2}^T & \cdots & \mathbf{y}_{p,k,i+N_p}^T \end{bmatrix}^T \tag{20}$$

$$\mathbf{F} = \begin{bmatrix} \mathbf{H}_m\mathbf{\Phi}_m & \mathbf{H}_m\mathbf{\Phi}_m^2 & \cdots & \mathbf{H}_m\mathbf{\Phi}_m^{N_p} \end{bmatrix}^T \tag{21}$$

$$\mathbf{G} = \begin{bmatrix} \mathbf{H}_m\mathbf{\Gamma}_m & 0 & \cdots & 0 \\ \mathbf{H}_m\mathbf{\Phi}_m\mathbf{\Gamma}_m & \mathbf{H}_m\mathbf{\Gamma}_m & \cdots & 0 \\ \vdots & \vdots & \ddots & \vdots \\ \mathbf{H}_m\mathbf{\Phi}_m^{N_p-1}\mathbf{\Gamma}_m & \mathbf{H}_m\mathbf{\Phi}_m^{N_p-2}\mathbf{\Gamma}_m & \cdots & \mathbf{H}_m\mathbf{\Phi}_m^{N_p-N_c}\mathbf{\Gamma}_m \end{bmatrix} \tag{22}$$

The optimizer calculates the optimal $\mathbf{U}_p$ to minimize the cost function J_p as shown below.

$$J_p = (\mathbf{R}_s - \mathbf{Y}_p)^T\mathbf{Q}(\mathbf{R}_s - \mathbf{Y}_p) + \mathbf{U}_p^T\bar{\mathbf{R}}\mathbf{U}_p \tag{23}$$

where $\mathbf{R}_s$ is the reference input in the prediction horizon; $\mathbf{Q}$ and $\bar{\mathbf{R}}$ are weighting matrices. They are defined as follows.

$$\mathbf{R}_s = \begin{bmatrix} \mathbf{r}_{r,k,i+1}^T & \mathbf{r}_{r,k,i+2}^T & \cdots & \mathbf{r}_{r,k,i+N_p}^T \end{bmatrix}^T \tag{24}$$

$$\mathbf{r}_{r,k,i+1} = \begin{bmatrix} y_{r,k,i+1} & \psi_{r,k,i+1} \end{bmatrix}^T \tag{25}$$

$$\mathbf{Q} = \begin{bmatrix} Q_1 & 0 & \cdots & 0 & 0 \\ 0 & Q_2 & & 0 & 0 \\ & \vdots & \ddots & \vdots & \\ 0 & 0 & \cdots & Q_1 & 0 \\ 0 & 0 & & 0 & Q_2 \end{bmatrix}_{2N_p \times 2N_p} \tag{26}$$

$$\bar{\mathbf{R}} = \begin{bmatrix} R_1 & 0 & \cdots & 0 & 0 \\ 0 & R_2 & & 0 & 0 \\ \vdots & & \ddots & & \vdots \\ 0 & 0 & & R_1 & 0 \\ 0 & 0 & \cdots & 0 & R_2 \end{bmatrix}_{2N_c \times 2N_c} \quad (27)$$

where y_r is the desired lateral displacement; ψ_r is the desired heading angle; Q_1 is the weighting of lateral displacement error; Q_2 is the weighting of heading angle error; R_1 is the weighting of the steering angle; R_2 is the weighting of the DYM. These weightings are designed using Bryson's rule [43] as shown below.

$$Q_1 = \frac{1}{y_{e,max}^2};\ Q_2 = \frac{1}{\psi_{e,max}^2} \quad (28)$$

$$R_1 = \frac{1}{\delta_{f,max}^2};\ R_2 = \frac{1}{M_{z,max}^2} \quad (29)$$

where $y_{e,max}$ is the maximum allowable lateral displacement error; $\psi_{e,max}$ is the maximum allowable heading angle error; $\delta_{f,max}$ is the maximum allowable steering angle; $M_{z,max}$ is the maximum allowable DYM. The optimal $\mathbf{U}_p$ can be obtained as follows.

$$\mathbf{U}_p = \left(\mathbf{G}^T\mathbf{Q}\mathbf{G} + \bar{\mathbf{R}}\right)^{-1}\mathbf{G}^T\mathbf{Q}(\mathbf{R}_s - \mathbf{F}\mathbf{x}_{k,i}) \quad (30)$$

The current optimal control input $\mathbf{u}_{k,i}$ is equal to the vector that consists of the first and second components of $\mathbf{U}_p$ as follows.

$$\mathbf{u}_{k,i} = \begin{bmatrix}\mathbf{U}_p(1) & \mathbf{U}_p(2)\end{bmatrix}^T \quad (31)$$

3.2. Evasive Path Generation

The desired evasive path is shown in Figure 2. A Trapezoidal Acceleration Profile (TAP) is used to design the desired global lateral displacement of the evasive path [7]. The lateral acceleration can be obtained by specifying the maximum allowable $a_{y,max}$, the maximum allowable jerk $J_{y,max}$, and the evasive lateral displacement y_{eva}. t_1 and t_2 of the TAP can be expressed as follows [7].

$$t_1 = \frac{a_{y,max}}{J_{y,max}},\ t_2 = \frac{-t_1^2 + \sqrt{t_1^4 + 4t_1\frac{y_{eva}}{J_y}}}{2t_1} \quad (32)$$

where $a_{y,max} = \hat{\mu}g$; g is the gravity; $\hat{\mu}$ is the estimated road friction coefficient that is assumed to be obtained from other approaches [44–46]. The desired global lateral position Y_d of the evasive path can then be obtained as follows by integrating the global lateral acceleration $\ddot{Y}_d$ twice [7].

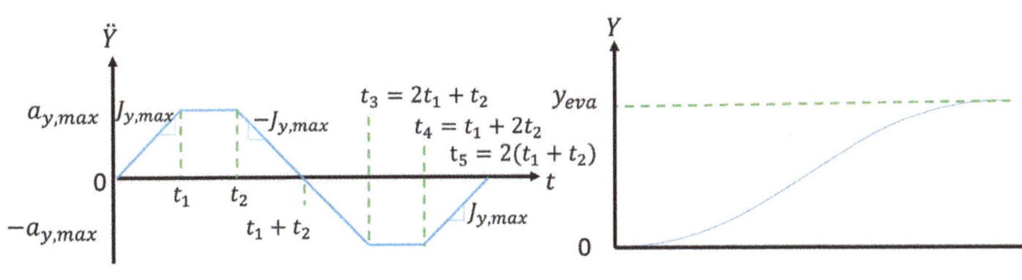

Figure 2. Evasive path planning.

$$Y_d(t) = \begin{cases} \frac{1}{6}J_{y,max}t^3 & for\ 0 \le t \le t_1 \\ Y_{d,1} + v_{y,1}(t-t_1) + \frac{1}{2}a_{y,max}(t-t_1)^2 & for\ t_1 < t \le t_2 \\ Y_{d,2} + v_{y,2}(t-t_2) + \frac{1}{2}a_{y,max}(t-t_2)^2 - \frac{1}{6}J_{y,max}(t-t_2)^3 & for\ t_2 < t \le t_3 \\ Y_{d,3} + v_{y,3}(t-t_3) - \frac{1}{2}a_{y,max}(t-t_3)^2 & for\ t_3 < t \le t_4 \\ Y_{d,4} + v_{y,4}(t-t_4) - \frac{1}{6}J_{y,max}(t-t_4)^3 & for\ t_4 < t \le t_5 \\ y_{eva} & for\ t > t_5 \end{cases} \quad (33)$$

where $v_{y,i} = v_y(t_i), Y_{d,i} = Y_d(t_i),\ i \in [0,5]$. By assuming the vehicle longitudinal velocity v_x is constant along the evasive path, the desired global longitudinal velocity $\dot{X}_d$ of the evasive path can then be obtained as follows.

$$\dot{X}_d(t) = \sqrt{v_x^2 - \dot{Y}_d(t)} \quad (34)$$

where $v_{x,0}$ is the initial longitudinal velocity of vehicle; $\dot{Y}_d$ is the global lateral velocity that can be obtained by integrating $\ddot{Y}_d$ in Equation (31). We can then obtain the desired global X_d by integrating $\dot{X}_d$.

The station number of the evasive path S_d that is the distance along the designed evasive path can be defined as follows.

$$S_{d,j+1} = S_{d,j} + \sqrt{\left(X_{d,j} - X_{d,j-1}\right)^2 + \left(Y_{d,j} - Y_{d,j-1}\right)^2} \quad (35)$$

where j denotes the jth sample point of the evasive path. The initial values of X_d, Y_d, $\dot{Y}_d$, and S_d are set to be zero. The desired evasive path can then be expressed as functions of the station number as follows.

$$X_d = f_{SX}(S_d), Y_d = f_{SY}(S_d) \quad (36)$$

The station number in the prediction horizon can be defined as follows with a constant v_x.

$$S_{p,k+i+1} = S_{p,k+i} + v_x i T_s, i \in [0,\ 1, \cdots,\ N_p - 1] \quad (37)$$

where the initial station number $S_{p,k}$ is the current station number S_c. The desired evasive path in the preview horizon can then be expressed as follows.

$$X_{p,k} = f_{SX}\left(S_{p,k}\right), Y_{p,k} = f_{SY}\left(S_{p,k}\right) \quad (38)$$

The desired evasive path in the body coordinates of the vehicle can be calculated using coordinate transformation as shown below.

$$\begin{bmatrix} x_{r,k} \\ y_{r,k} \end{bmatrix} = \begin{bmatrix} cos(\Psi_c) & sin(\Psi_c) \\ -sin(\Psi_c) & cos(\Psi_c) \end{bmatrix} \begin{bmatrix} X_{p,k} - X_c \\ Y_{p,k} - Y_c \end{bmatrix} \quad (39)$$

where Ψ_c is the current global heading angle of vehicle; X_c and Y_c are the current global position of the vehicle, respectively; x_r and y_r are the local coordinates of the evasive path. X_c, Y_c, Ψ_c and S_c are assumed to be available from the localization system. The desired lateral displacement of the evasive path can be expressed as $y_r = f_r(x_r)$ in the body coordinates of the vehicle. The heading angle reference, which is the first derivative of the desired path, can then be approximated as follows.

$$\psi_{r,k+i} = f_r'(x_{r,k+i}) = \frac{f_r(x_{r,k+i+1}) - f_r(x_{r,k+i})}{x_{r,k+i+1} - x_{r,k+i}} \quad (40)$$

3.3. Control Allocation

Weighted pseudo inverse with constraints [47] is used to obtain the distributed brake forces at four wheels. The DYM command $M_{z,r}$ can be expressed as follows.

$$M_{z,r} = \mathbf{B}_{CA}\mathbf{F}_x \tag{41}$$

where $\mathbf{F}_x$ is the tire force vector that consists of four tire forces.

$$\mathbf{F}_x = \begin{bmatrix} F_{x,fl} & F_{x,rl} & F_{x,fr} & F_{x,rr} \end{bmatrix}^T \tag{42}$$

where the subscripts *fl*, *rl*, *fr*, and *rr* denote quantities at front-left, rear-left, front-right, and rear-right tires, respectively.

The pseudo inverse solution is a two-norm solution to the *CA* problem that can be formulated as follows.

$$\underset{\mathbf{F}_x}{min} J_{CA} = \underset{\mathbf{F}_x}{min}\frac{1}{2}(\mathbf{F}_x + \mathbf{C}_{CA})^T \mathbf{W}_{CA}(\mathbf{F}_x + \mathbf{C}_{CA}) \tag{43}$$

subject to:

$$M_{z,r} = \mathbf{B}_{CA}\mathbf{F}_x \tag{44}$$

$$\mathbf{F}_{x,min} \le \mathbf{F}_x \le 0 \tag{45}$$

where $\mathbf{C}_{CA}$ is the offset vector used to represent off-nominal condition with one or more tire forces; $\mathbf{F}_{x,min}$ is the vector used to represent the minimum allowable tire forces. The maximum allowable tire forces are set to be 0 such that only braking forces are used. The optimal distribution can then be obtained as follows.

$$\mathbf{F}_x = -\mathbf{C}_{CA} + \mathbf{B}_{CA}^{\#}(M_{z,r} + \mathbf{B}_{CA}\mathbf{C}_{CA}) \tag{46}$$

$$\mathbf{B}_{CA}^{\#} = \mathbf{W}_{CA}^{-1}\mathbf{B}_{CA}^{T}\left(\mathbf{B}_{CA}\mathbf{W}_{CA}^{-1}\mathbf{B}_{CA}^{T}\right)^{-1} \tag{47}$$

where $\mathbf{B}_{CA}^{\#}$ is the pseudo inverse of $\mathbf{B}_{CA}$. The $\mathbf{C}_{CA}$ is zero in the normal condition, so (44) can be simplified as follows.

$$\mathbf{F}_x = \mathbf{B}_{CA}^{\#} M_{z,r} \tag{48}$$

where the diagonal weighting matrix $\mathbf{W}_{CA}$ is designed with elements equal to the estimated tire workload ratios as follows [42] to prevent saturating tire forces. The larger the ratio is, the closer to saturation the corresponding tire force is. Thus, tire forces that are needed for generating the desired DYM are distributed to tires with lower tire work load ratios.

$$\mathbf{W}_{CA} = \begin{bmatrix} \hat{\rho}_{fl} & 0 & 0 & 0 \\ 0 & \hat{\rho}_{rl} & 0 & 0 \\ 0 & 0 & \hat{\rho}_{fr} & 0 \\ 0 & 0 & 0 & \hat{\rho}_{rr} \end{bmatrix} \tag{49}$$

where $\hat{\rho}$ is the estimated tire workload ratio of each tire that can be obtained as follows [42].

$$\hat{\rho} = \sqrt{\frac{\hat{F}_x^2 + \hat{F}_y^2}{\hat{F}_z^2}} \tag{50}$$

where $\hat{F}_x$, $\hat{F}_y$ and $\hat{F}_z$ are the estimated tire forces at longitudinal, lateral and normal directions, respectively. If the ith element of $\mathbf{F}_x$ exceeds the boundary values, the corresponding ith element in $\mathbf{C}_{CA}$ is set to be the opposite number of the boundary value and the corresponding ith element in $\mathbf{B}_{CA}$ is set to be zero. The adjusted vectors $\mathbf{B}_{CA,a}$ and $\mathbf{C}_{CA,a}$ can be obtained using the Algorithm 1 as shown below.

Algorithm 1 Saturation Prevention

$\mathbf{B}_{CA,a} = \mathbf{B}_{CA}$; $\mathbf{C}_{CA,a} = \mathbf{C}_{CA}$;
if $\mathbf{F}_x(i) > 0$
  $\mathbf{B}_{CA,a}(i) = 0$
elseif $\mathbf{F}_x(i) < \mathbf{F}_{x,min}(i)$
  $\mathbf{B}_{CA,a}(i) = 0$
  $\mathbf{C}_{CA,a}(i) = -\mathbf{F}_{x,min}(i)$
end

After obtaining the adjusted vectors, we can replace $\mathbf{B}_{CA}$ and $\mathbf{C}_{CA}$ with $\mathbf{B}_{CA,a}$ and $\mathbf{C}_{CA,a}$, respectively, and obtain the adjusted tire force vector using Equations (44) and (45).

The four-wheel brake pressure command $\mathbf{P}_{b,r}$ can be expressed as follows.

$$\mathbf{P}_{b,r} = \mathbf{F}_x diag\left(\begin{bmatrix} \frac{r_w}{G_{b,f}} & \frac{r_w}{G_{b,r}} & \frac{r_w}{G_{b,f}} & \frac{r_w}{G_{b,r}} \end{bmatrix}\right) \quad (51)$$

where $G_{b,f}$ and $G_{b,r}$ are the brake gains of front and rear wheels, respectively; r_w is the wheel radius; diag is a function that creates a square diagonal matrix with the input elements on the main diagonal. The EHB module is used to generate four-wheel brake pressure $\mathbf{P}_b$.

4. Simulation Results

The proposed algorithm was evaluated using CarSim in MATLAB/Simulink. Figure 3 shows the test scenario used in this paper. There was one moving vehicle and one stationary vehicle in front of the host vehicle. After the front moving vehicle changed lanes at the last moment [48], the stationary vehicle was detected as a suddenly appearing obstacle by the CarSim Sensor module of the host vehicle at 5 m ahead the last point to steer (LPS) [4], which was defined as the last location to avoid a collision by steering. The longitudinal vehicle velocity was set at 80 km/h which is the highest longitudinal velocity for testing ESA. Since the stationary vehicle suddenly appeared in front of the host vehicle, the last point to brake (LPB), which was defined as the last location to avoid the collision by braking, was already behind the host vehicle. A single lane-change path generated using TAP was used as the evasive path to avoid the collision. When the host vehicle arrived at the LPS, the AES was trigged automatically.

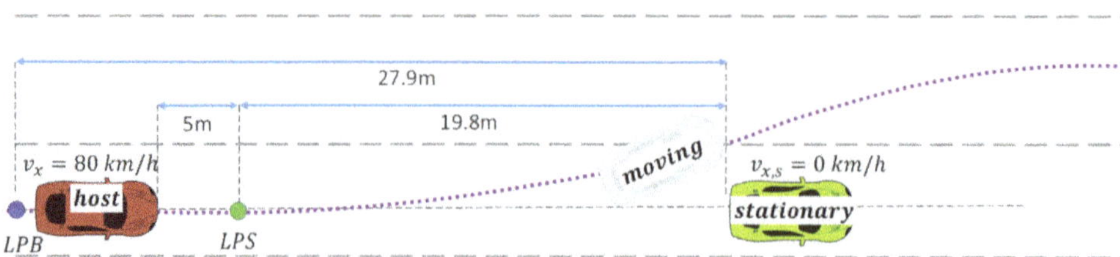

Figure 3. Test scenario at 80 km/h.

The maximum angular rate of the front wheel steering angle was limited to 42°/s due to the EPS according to the requirement of the fishhook test [49]. The front wheel steering angle was limited between −35° and 35° due to the pinion-rack configuration. Simulation results of the proposed MIMO MPC were compared with those of the SISO MPC [35]. For the first test scenario, the nominal road friction coefficient was 0.9 which was used to obtain the nominal cornering stiffness of the prediction model in MPC. The second test scenario was used to test the robustness of the proposed algorithm with a medium road friction coefficient of 0.7 which is 22% less than the nominal value; i.e., the actual cornering stiffness should be smaller than the nominal values used in the prediction model.

4.1. Comparison of Prediction Models

An open-loop test was used to compare the responses of three bicycle models. The structures of these models are listed in Table 2. The four-state bicycle model [34] has two inputs and two outputs that are the same as those of the proposed six-state bicycle model in (3). However, the actuator dynamics of EPS and EHB were not considered for the four-state bicycle. The five-state bicycle model that includes the EPS actuator dynamics has one steering angle control input and the lateral displacement output. The DYM input was not considered in the five-state bicycle model. Dynamic responses of these models are shown in Figure 4. The step steering angle input was applied at the beginning of simulation, and the DYM step input was applied at 1 s. CarSim denotes the dynamic responses of CarSim with two inputs of steering angle and DYM. Before 1 s, the main difference between the four-state model and other models was the transient response due the lack of actuator dynamics. The steady-state responses of three bicycle models were almost the same. The responses of the five-state and the six-state bicycle models were similar to those of CarSim. After the DYM was applied at 1 s, only the responses of the six-state bicycle model were close to those of CarSim. Since the six-state model with actuator dynamics can achieve both transient and steady-state responses closest to those of CarSim, it was used as the prediction model in the proposed algorithm.

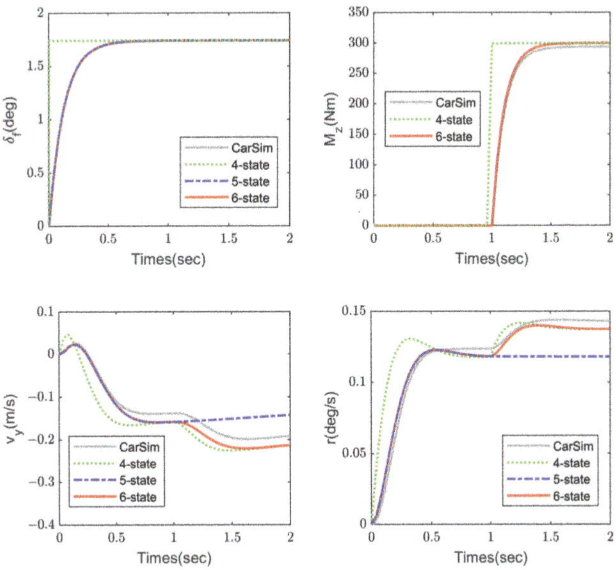

Figure 4. Open-loop test of three bicycle models.

Table 2. Structures of three bicycle models.

Model	State	Input	Output
Four-states	$y,\ v_y, \psi,\ r$	$\delta_f,\ M_z$	$y,\ \psi$
Five-states	$y,\ v_y, \psi,\ r,\ \delta_f$	$\delta_{f,r}$	y
Six-states	$y,\ v_y, \psi,\ r,\ \delta_f,\ M_z$	$\delta_{f,r},\ M_{z,r}$	$y,\ \psi$

4.2. Nominal Road Friction Coefficient of 0.9

The simulation results with accurately estimated road-friction coefficient of 0.9, which is used to obtain the maximum allowable lateral acceleration for designing the desired evasive path, are shown in Figures 5–7. The AES is assumed to be triggered at

3 s. SISO MPC and MIMO MPC can track the desired evasive path and complete the lane change maneuver.

Path tracking responses are shown in Figure 5. SISO MPC and MIMO MPC have similar responses of global lateral displacement. However, their global heading angle responses are less similar to each other. As can be seen from the vehicle dynamic responses in Figure 6, the values of y_e of MIMO MPC are more negative than those of SISO MPC between 3.1 and 3.7 s. Thus, the lane change response of MIMO MPC is faster than that of SISO MPC at the beginning of lane change due to the additional DYM input. Meanwhile, the heading angle errors of MIMO MPC are smaller than those of SISO MPC during the lane change maneuver.

Figure 7 shows control commands and actuator responses. At the beginning of AES, MIMO MPC and SISO MPC have the same steering angle responses due to the angular rate limitation and the actuator dynamics of EPS. However, MIMO MPC has an additional DYM input, which can accelerate the response of r and reduce the maximum steering angle to be smaller than that of SISO MPC. Since the heading angle is the integration of yaw rate, faster response of r can lead to smaller ψ_e. The comparison of the $\beta - r$ phase plane plots is shown in Figure 8. MIMO MPC has better convergences due to a smaller circling area around the origin. This means that MIMO MPC can reduce β in order to enhance handling performance and directional stability. Although DYM is generated using braking forces, the vehicle longitudinal velocity of MIMO MPC is only reduced slightly compared to that of SISO MPC.

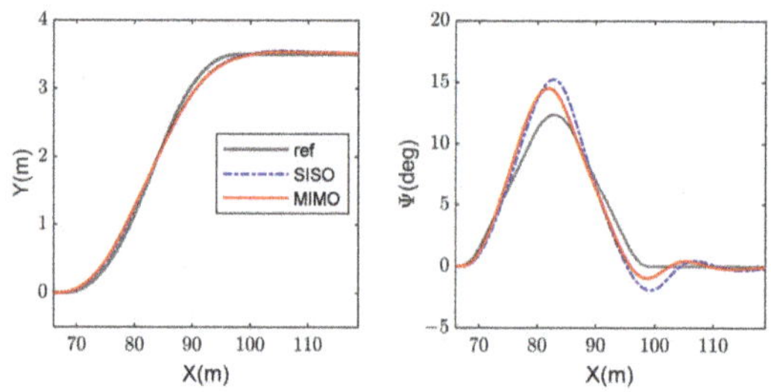

Figure 5. Path tracking responses for $\mu = 0.9$.

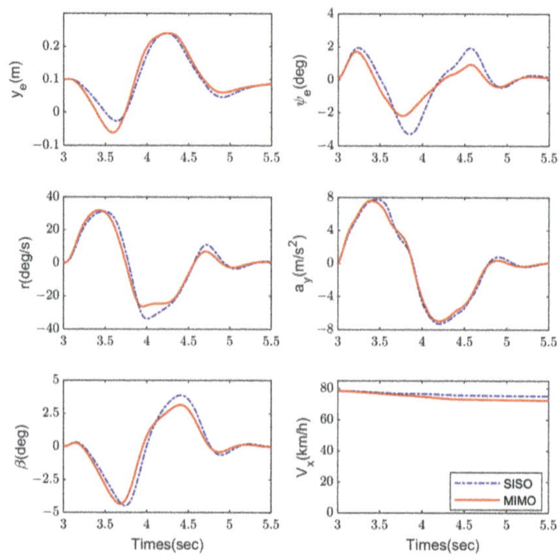

Figure 6. Path tracking responses for $\mu = 0.9$.

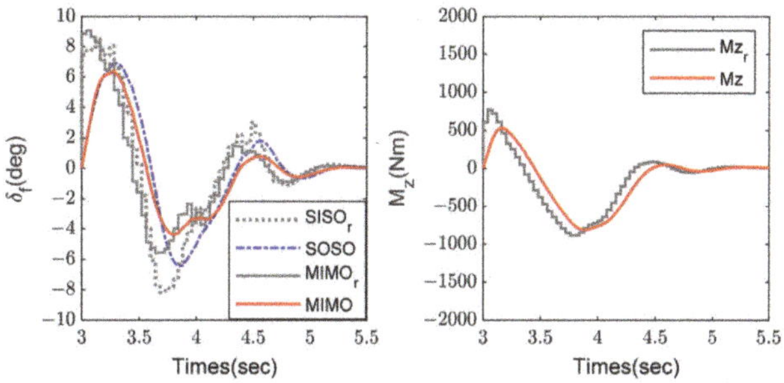

Figure 7. Actuator responses for $\mu = 0.9$.

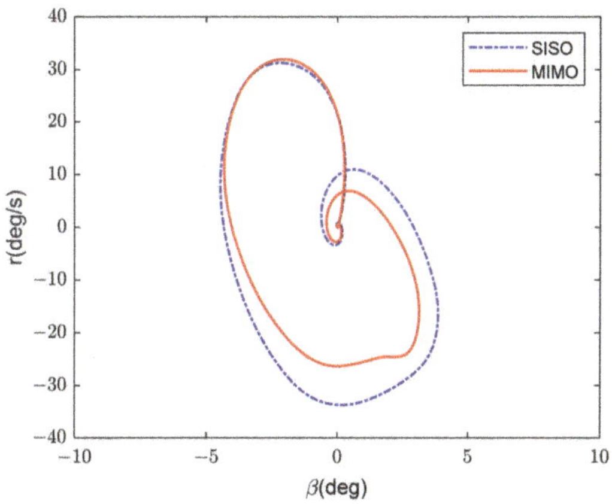

Figure 8. $\beta - r$ phase plane plots for $\mu = 0.9$.

4.3. Nominal Road Friction Coefficient of 0.7

The simulation results with a medium road friction coefficient of 0.7, which is 22% less than the nominal value, are shown in Figures 9–13. Assuming the AES is not aware of the overestimation of the road friction coefficient, the nominal value of 0.9 is still used to design the evasive path and the nominal cornering stiffnesses are still used in the prediction model.

Path tracking responses are shown in Figure 9. The AES is assumed to be triggered at 3 s. SISO MPC cannot complete the lane change maneuver and loses the directional stability. Conversely, MIMO MPC can track the desired evasive path and complete the lane change maneuver successfully. Similar to the analysis of Figure 8 in Section 4.1, we found that MIMO MPC has a lane change response faster than that of SISO MPC at the beginning of lane change, as shown in Figure 10. Meanwhile, the peak heading angle error of MIMO MPC is smaller than that of SISO MPC at about 4 s due to minimizing y_e and ψ_e simutaneously via both steering and DYM. From the yaw rate response, we can observe that the vehicle with SISO MPC starts to spin continuously after 4 s.

As can be seen from Figure 11, SISO MPC requires steering angles much larger than those of MIMO MPC after 4 s. Although SISO MPC tries to correct the steering angle as fast as possible, the steering angle response is not fast enough due to the angular rate limitation and actuator dynamics. Figure 12 shows that the side-slip angle of SISO MPC increases rapidly with large yaw rate after passing $\beta = 0$. It also confirms the directional instability of SISO MPC. On the other hand, the $\beta - r$ phase plane plot of MIMO MPC converges to the origin after completing the lane change maneuver.

Figures 13 and 14 show the responses of tire slip angle and lateral tire forces at four wheels. The dashed lines indicate the thresholds of the tire slip angles to saturate lateral tire forces. The front lateral tire forces of SISO MPC are found to be saturated at 4.1 s. Even though the steering angle of SISO MPC in Figure 11 is still increasing after 4 s, the corresponding lateral tire forces are not increased due to saturation as shown in Figure 14. Since the lateral tire force is no longer increased with the increment of the tire slip angle, the prediction model used in MPC is not accurate anymore. Thus, SISO MPC cannot effectively control the vehicle to track the desired evasive path and prevent directional instability. Although the rear tire slip angles of MIMO MPC are saturated around 4.5 s, the front tire slip angles are not saturated. Thus, the front tires can still provide more lateral tire forces via increasing the steering angle input if necessary. Meanwhile, MIMO MPC has additional DYM which can improve the responses of r and prevent saturating the reference command of the steering angle. Figure 15 shows the responses of tire workload ratios. The CA distributes the longitudinal tire forces as shown in Figure 16 to the wheels with lower tire workload ratios in order to generate the required DYM without saturating the tire forces; thus, demonstrating better, more robust control performance and handling enhancement against model uncertainty due to reduced road friction coefficient.

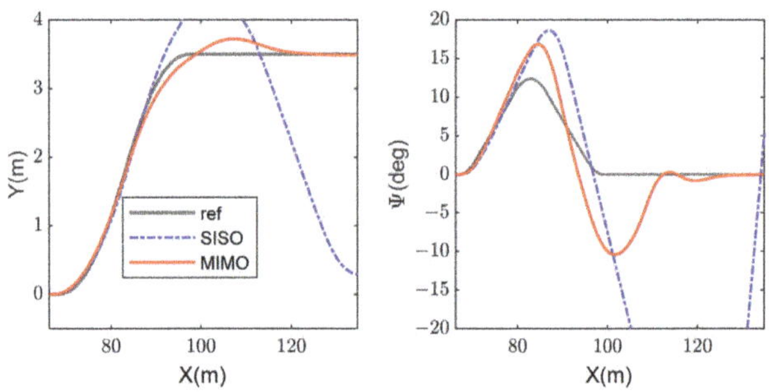

Figure 9. Path tracking responses for $\mu = 0.7$.

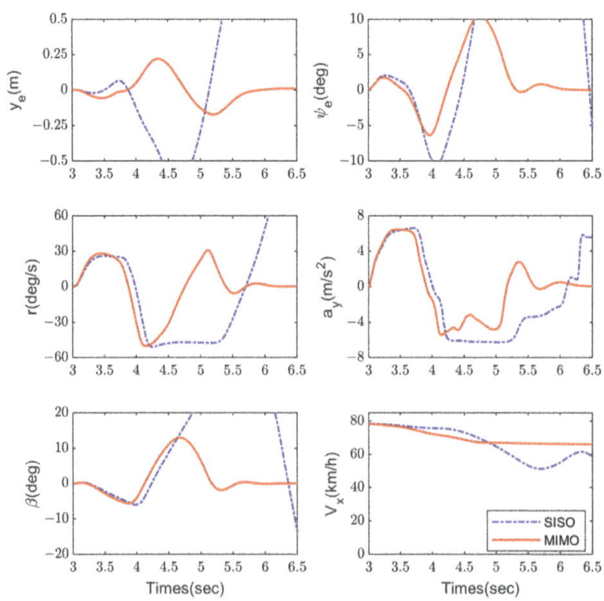

Figure 10. Responses of vehicle dynamics for $\mu = 0.7$.

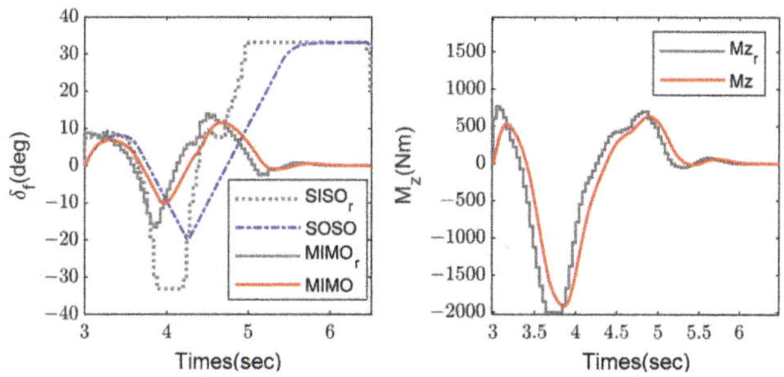

Figure 11. Actuator responses for $\mu = 0.7$.

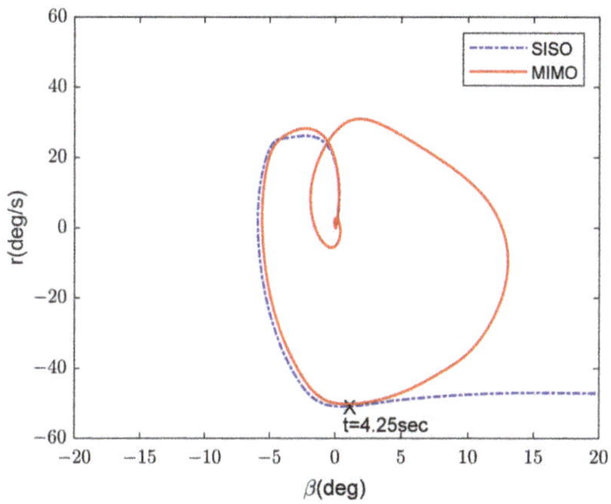

Figure 12. $\beta - r$ phase plane comparisons for $\mu = 0.7$.

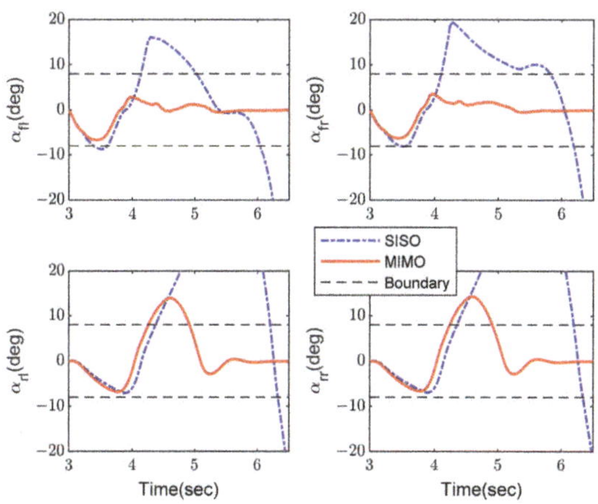

Figure 13. Responses of tire slip angles for $\mu = 0.7$.

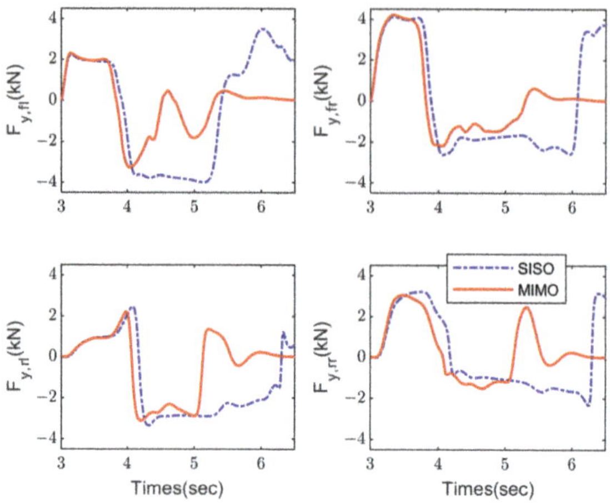

Figure 14. Responses of lateral tire forces for $\mu = 0.7$.

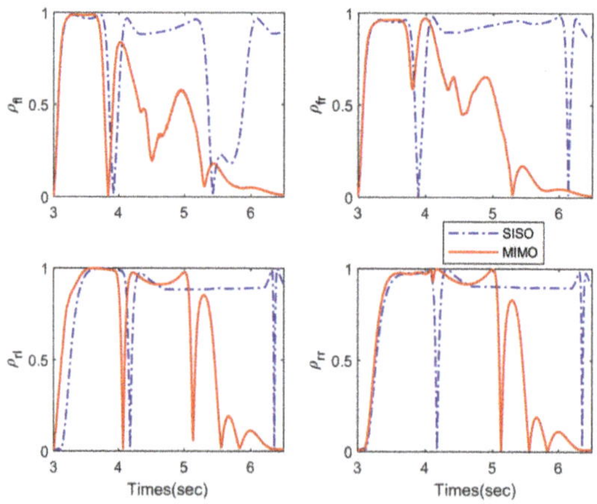

Figure 15. Responses of tire workload ratios for $\mu = 0.7$.

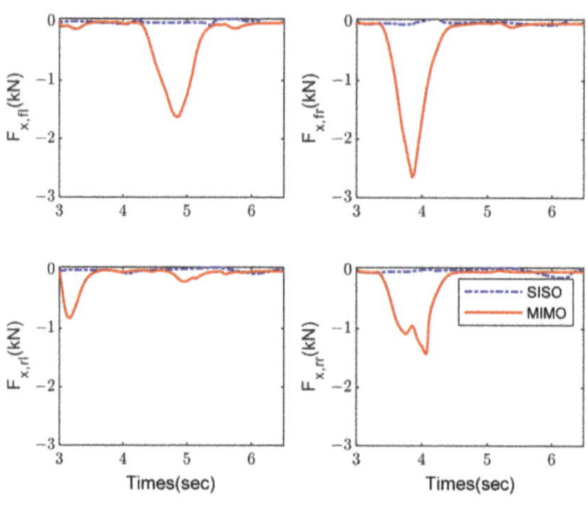

Figure 16. Responses of longitudinal tire forces for $\mu = 0.7$.

4.4. Sensitivity Analysis

The estimated road friction coefficient be used not only for evasive path generation, but also for obtaining the equivalent front and rear cornering stiffnesses of the bicycle model used in MPC via a look-up table. However, the estimation might not be accurate due to the lack of enough excitation. Different road friction coefficients deviated from the estimated nominal value of 0.9 are used for sensitivity analysis to evaluate the robustness of the proposed algorithm. The minimum distance D_L [50] between the host vehicle and the stationary vehicle to be avoided as shown in Figure 17 can be defined as follows.

$$D_L = \sqrt{(x_1 - x_2)^2 + (y_1 - y_2)^2} - \frac{W_h}{2} - \frac{W_s}{2} \tag{52}$$

where $P_h(x_1, y_1)$ is the front center point of host vehicle; $P_s(x_2, y_2)$ is the front center point of stationary vehicle; W_h is the width of host vehicle; W_s is the width of stationary vehicle. D_L at different road friction coefficients are shown in Table 3. The collision can be avoided for all simulated scenarios.

The vehicle sideslip angle can be used to evaluate the yaw stability of the vehicle. The root-mean-square (rms) and maximum (max) values of sideslip angle responses at different road friction coefficients are shown in Table 4. Those values are obtained using the responses from the beginning of control at 3–10 s. The more the road friction coefficient deviates from the nominal value of 0.9, the larger the rms and max values of sideslip angles are. MIMO can stabilize the vehicle in the new target lane for all scenarios. However, SISO exhibits large lateral responses as shown in Figure 10 and cannot stabilize the vehicle in the new target lane for $\mu \leq 0.75$. The rms and max values of sideslip angle suddenly increase for $\mu \leq 0.75$ which are indications of yaw instabilities. Thus, MIMO demonstrates better robustness of yaw stability and handling performance for reduced road friction coefficients.

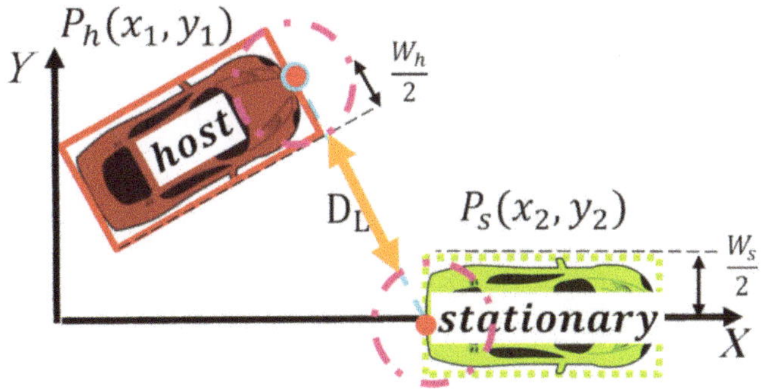

Figure 17. The minimal distance between host vehicle and stationary vehicle.

Table 3. The minimal distance between host vehicle and stationary vehicle at 80 km/h.

	Road Friction Coefficient, μ				
	0.9	0.85	0.8	0.75	0.7
SISO	0.51 m	0.53 m	0.52 m	0.47 m	0.36 m
MIMO	0.48 m	0.49 m	0.51 m	0.51 m	0.45 m

Table 4. Sideslip angle at different road friction coefficients at 80 km/h.

		Road Friction Coefficient, μ				
		0.9	0.85	0.8	0.75	0.7
SISO	rms	1.1°	1.5°	3.2°	62.3°	77.2°
	max	4.4°	6°	14.2°	180°	180°
MIMO	rms	1°	1.2°	1.6°	2.2°	3.1°
	max	4.3°	4.7°	5.9°	9°	13°

In order to evaluate the robustness of the proposed algorithm at different velocities, the longitudinal velocity of host vehicle is increased to 120 km/h in Figure 18. The suddenly appearing obstacle is assumed to be detected at 7.5 m ahead of the LPS. D_L at different road friction coefficients is shown in Table 5. Similar to the results in Table 3, the collision can be avoided for all simulated scenarios. The rms and max values of sideslip angle responses at different road friction coefficients are shown in Table 6. Similar to the results in Table 4, the more the road friction coefficient deviates from the nominal value of 0.9, the larger the rms and max values of sideslip angles are. MIMO can stabilize the vehicle in the new target lane for all scenarios. However, the rms and max values of sideslip angle suddenly increase for $\mu \leq 0.65$ which are indications of yaw instabilities. Thus, MIMO demonstrates better robustness of yaw stability and handling performance for reduced road friction coefficients even at a much higher longitudinal vehicle velocity of 120 km/h.

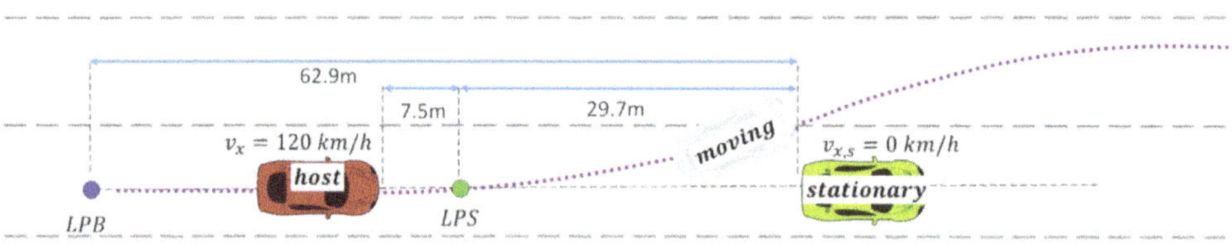

Figure 18. Test scenario at 120 km/h.

Table 5. The minimal distance between host vehicle and stationary vehicle at 120 km/h.

	Road Friction Coefficient, μ						
	0.9	**0.85**	0.8	0.75	0.7	0.65	0.6
SISO	0.44 m	0.45 m	0.47 m	0.47 m	0.45 m	0.39 m	0.27 m
MIMO	0.40 m	0.42 m	0.43 m	0.43 m	0.42 m	0.36 m	0.24 m

Table 6. Sideslip angle at different road friction coefficients at 120 km/h.

		Road Friction Coefficient, μ						
		0.9	**0.85**	0.8	0.75	0.7	0.65	0.6
SISO	rms	1.7°	2.2°	3.1°	4.3°	6.2°	66.7°	73.2°
	max	5.1°	6.9°	9.9°	13.8°	19.1°	180°	180°
MIMO	rms	1.5°	1.7°	2.2°	3.1°	4.2°	5.6°	6.7°
	max	4.7°	5.4°	6.8°	9.8°	13.4°	17°	19.8°

5. Conclusions

MIMO MPC was employed to design the path tracking controller for autonomous emergency steering using steering and differential braking. The prediction model was based on a six-state road following bicycle model including the actuator dynamics of EPS and EHB. For nominal road friction, the path tracking responses of SISO MPC (using steering only) and MIMO MPC (using both steering and differential braking) are very similar, but the directivity of MIMO MPC can be improved by DYM. The usage of steering angle can be reduced, and the handling performance of MIMO MPC is better than SISO MPC. For medium road friction, SISO MPC is unable to track the path and loses directional stability. However, MIMO MPC can still reduce the maximum required steering angle and accelerate the response of r via DYM. In order to prevent tire force saturation, the CA can distribute the longitudinal tire forces to wheels with lower tire work load ratios for generating the desired DYM. Thus, MIMO MPC can achieve robust path tracking and handling enhancement against model uncertainty due to road friction reduced by 22% and 33% at 80 and 120 km/h, respectively. In this paper, the handling performance was mainly discussed for avoiding the collision with a stationary vehicle via lateral dynamics control. For a slow moving vehicle in front of the host vehicle, collision avoidance via AES should be further investigated to demonstrate the effectiveness the proposed algorithm. Since the lateral tire forces at front steering wheels are affected by load transfer that can be created via acceleration or braking, coordinated longitudinal and lateral dynamics control might further enhance the handling performance for collision avoidance.

Author Contributions: Conceptualization, B.-C.C. and Y.-M.L.; methodology, B.-C.C. and Y.-M.L.; software, Y.-M.L.; validation, Y.-M.L.; formal analysis, B.-C.C. and Y.-M.L.; writing—original draft preparation, Y.-M.L.; writing—review and editing, B.-C.C.; project administration, B.-C.C. All authors have read and agreed to the published version of the manuscript.

Funding: This research was funded by Ministry of Science and Technology (MOST, Taipei, Taiwan), grant number MOST-109-2218-E-027-004.

Institutional Review Board Statement: Not applicable.

Informed Consent Statement: Not applicable.

Data Availability Statement: Not applicable.

Acknowledgments: The authors would like to thank the Ministry of Science and Technology (MOST, Taipei, Taiwan) for supporting this research.

Conflicts of Interest: The authors declare no conflict of interest. The funding sponsors had no role in the design of the study; in the collection, analysis, or interpretation of data; in the writing of the manuscript, and in the decision to publish the results.

References

1. NHTSA. *Traffic Safety Facts 2017 a Compilation of Motor Vehicle Crash Data*; NHTSA: Washington, DC, USA, 2019.
2. Singh, S. Critical Reasons for Crashes Investigated in the National Motor Vehicle Crash Causation Survey. Available online: https://crashstats.nhtsa.dot.gov/Api/Public/ViewPublication/812506 (accessed on 24 May 2021).
3. Lich, T.; Mettler, A.; Georgi, A.; Danz, C.; Doopyo, L.; Sooncheol, L.; Sul, J. Benefit Analysis of Predictive Rear End Collision Avoidance and Mitigation Systems for South Korea Using Video Documented Accident Data. In Proceedings of the 11th International Symposium on Advanced Vehicle Control, Seoul, Korea, 9–12 September 2012.
4. Eckert, A.; Hartmann, B.; Sevenich, M.; Rieth, P. Emergency Steer & Brake Assist—A Systematic Approach for System Integration of Two Complementary Driver Assistance Systems. In Proceedings of the 22nd ESV Enhanced Safety of Vehicles Conference, Washington, WA, USA, 13–16 June 2011.
5. EuroNCAP. *Euro NCAP 2025 Roadmap*; EuroNCAP: Leuven, Belgium, 2018.
6. Choi, J.; Yi, K. Design and Evaluation of Emergency Driving Support Using Motor Driven Power Steering and Differential Braking on a Virtual Test Track. *SAE Int. J. Passeng. Cars–Mech. Syst.* **2013**, *6*, 691–704. [CrossRef]
7. Choi, J.; Yi, K.; Suh, J.; Ko, B. Coordinated Control of Motor-Driven Power Steering Torque Overlay and Differential Braking for Emergency Driving Support. *IEEE Trans. Veh. Technol.* **2014**, *63*, 566–579. [CrossRef]
8. Keller, M.; Hass, C.; Seewald, A.; Bertram, T.; Keller, M. Driving simulator study on an emergency steering assist. In Proceedings of the 2014 IEEE International Conference on Systems, Man, and Cybernetics (SMC), San Diego, CA, USA, 5–8 October 2014; pp. 3008–3013.
9. Zhao, Z.; Zhou, L.; Luo, Y.; Li, K. Emergency Steering Evasion Assistance Control Based on Driving Behavior Analysis. *IEEE Trans. Intell. Transp. Syst.* **2018**, *20*, 457–475. [CrossRef]
10. Shimizu, M.; Usami, M.; Fujinami, H. Development of Collision-Avoidance Assist System Operating at Driver Steering. *Trans. Soc. Automot. Eng. Jpn.* **2008**, *39*, 441–446. [CrossRef]
11. Choi, C.; Kang, Y. Simultaneous braking and steering control method based on nonlinear model predictive control for emergency driving support. *Int. J. Control. Autom. Syst.* **2017**, *15*, 345–353. [CrossRef]

12. Ferdinand, J.; Yi, B. Trajectory planning for collision avoidance in urban area. In Proceedings of the 2016 IEEE Intelligent Vehicles Symposium (IV), Gothenburg, Sweden, 19–22 June 2016; pp. 202–207.
13. Hayashi, R.; Isogai, J.; Raksincharoensak, P.; Nagai, M. Autonomous collision avoidance system by combined control of steering and braking using geometrically optimised vehicular trajectory. *Veh. Syst. Dyn.* **2012**, *50*, 151–168. [CrossRef]
14. Zhou, L.; Zhao, Z.; Shen, P.; Li, M.; Li, K. Emergency Steering Evasion Control by Combining the Yaw Moment with Steering Assistance. In Proceedings of the SAE 2018 World Congress and Exhibition, Detroit, MI, USA, 10–12 April 2018. [CrossRef]
15. Cui, Q.; Ding, R.; Wu, X.; Zhou, B. A new strategy for rear-end collision avoidance via autonomous steering and differential braking in highway driving. *Veh. Syst. Dyn.* **2019**, *58*, 955–986. [CrossRef]
16. Zhang, Y.; Chen, H.; Waslander, S.L.; Gong, J.; Xiong, G.; Yang, T.; Liu, K. Hybrid Trajectory Planning for Autonomous Driving in Highly Constrained Environments. *IEEE Access* **2018**, *6*, 32800–32819. [CrossRef]
17. Stahl, T.; Wischnewski, A.; Betz, J.; Lienkamp, M. Multilayer Graph-Based Trajectory Planning for Race Vehicles in Dynamic Scenarios. In Proceedings of the 2019 IEEE Intelligent Transportation Systems Conference (ITSC), Paris, France, 9–12 June 2019.
18. McNaughton, M.; Urmson, C.; Dolan, J.M.; Lee, J.-W. Motion planning for autonomous driving with a conformal spatiotemporal lattice. In Proceedings of the 2011 IEEE International Conference on Robotics and Automation, Shanghai, China, 9–13 May 2011; pp. 4889–4895.
19. Srinivasan, S.; Giles, S.N.; Liniger, A. A Holistic Motion Planning and Control Solution to Challenge a Professional Racecar Driver. *arXiv* **2021**, arXiv:2103.00358. Available online: https://arxiv.org/abs/2103.00358 (accessed on 24 May 2021).
20. Qian, X.; Altché, F.; Bender, P.; Stiller, C.; Fortelle, A. La Optimal Trajectory Planning for Autonomous Driving Integrating Logical Constraints: An MIQP Perspective. In Proceedings of the 2016 IEEE 19th International Conference on Intelligent Transportation Systems, Rio de Janeiro, Brazil, 1–4 November 2016; pp. 205–210.
21. Dolgov, D.; Thrun, S.; Montemerlo, M.; Diebel, J. Path Planning for Autonomous Vehicles in Unknown Semi-structured Environments. *Int. J. Robot. Res.* **2010**, *29*, 485–501. [CrossRef]
22. Nelson, W. Continuous-curvature paths for autonomous vehicles. In Proceedings of the 1989 International Conference on Robotics and Automation, Scottsdale, AZ, USA, 14–19 May 1989; Volume 3, pp. 1260–1264.
23. Sotelo, M.A. Lateral control strategy for autonomous steering of Ackerman-like vehicles. *Robot. Auton. Syst.* **2003**, *45*, 223–233. [CrossRef]
24. Cremean, L.B.; Foote, T.B.; Gillula, J.H.; Hines, G.H.; Kogan, D.; Kriechbaum, K.L.; Lamb, J.C.; Leibs, J.; Lindzey, L.; Rasmussen, C.E.; et al. Alice: An information-rich autonomous vehicle for high-speed desert navigation. *J. Field Robot.* **2006**, *23*, 777–810. [CrossRef]
25. Chatzikomis, C.I.; Spentzas, K.N. A path-following driver model with longitudinal and lateral control of vehicle's motion. *Forsch. Ing.* **2009**, *73*, 257–266. [CrossRef]
26. Chen, B.; Li, L.; Zhang, W.; Dong, H.; Wang, Y.; Xiao, L. A Data-Based Approach to Path Following Controller Design for Autonomous Vehicles. In Proceedings of the CICTP 2019, Nanjing, China, 6–8 July 2019; pp. 367–378.
27. Chen, I.-M.; Chan, C.-Y. Deep reinforcement learning based path tracking controller for autonomous vehicle. *Proc. Inst. Mech. Eng. Part D J. Automob. Eng.* **2020**, 1–11. [CrossRef]
28. Grigorescu, S.; Trasnea, B.; Cocias, T.; Macesanu, G. A survey of deep learning techniques for autonomous driving. *J. Field Robot.* **2020**, *37*, 362–386. [CrossRef]
29. Chen, B.-C.; Luan, B.-C.; Lee, K. Design of lane keeping system using adaptive model predictive control. In Proceedings of the 2014 IEEE International Conference on Automation Science and Engineering (CASE), Taipei, Taiwan, 18–22 August 2014; pp. 922–926.
30. Skarpetis, M.G.; Koumboulis, F.N.; Papanikolaou, P. Vehicle lateral control using a robust tracking controller based on vision look ahead system. In Proceedings of the 2015 International Conference on Electrical, Electronics, Signals, Communication and Optimization (EESCO), Visakhapatnam, India, 24–25 January 2017; pp. 000213–000218.
31. Song, L.; Guo, H.; Wang, F.; Liu, J.; Chen, H. Model predictive control oriented shared steering control for intelligent vehicles. In Proceedings of the 2017 29th Chinese Control and Decision Conference (CCDC), Chongqing, China, 28–30 May 2017; pp. 7568–7573.
32. Falcone, P.; Borrelli, F.; Asgari, J.; Tseng, H.E.; Hrovat, D. A model predictive control approach for combined braking and steering in autonomous vehicles. In Proceedings of the 2007 Mediterranean Conference on Control & Automation, Athens, Greece, 27–29 June 2007; pp. 1–6.
33. Hajiloo, R.; Abroshan, M.; Khajepour, A.; Kasaiezadeh, A.; Chen, S.-K. Integrated Steering and Differential Braking for Emergency Collision Avoidance in Autonomous Vehicles. *IEEE Trans. Intell. Transp. Syst.* **2021**, *22*, 3167–3178. [CrossRef]
34. Lin, Y.; Chen, B.; Chan, Y. Design of Autonomous Emergency Steering Using Multi-Input Multi-Output Model Predictive Control. In Proceedings of the Advanced Vehicle Control Proceedings of the 14th International Symposium on Advanced Vehicle Control (AVEC'18), Beijing, China, 14–18 July 2018.
35. Chen, B.-C.; Tsai, C.-T.; Lin, Y.-M.; Lee, K. Design of an automated steering controller with steering actuator dynamics and adaptive preview time. In Proceedings of the Advanced Vehicle Control AVEC'16, Munich, Germany, 13–16 September 2016; pp. 163–168.
36. Kim, E.; Kim, J.; Sunwoo, M. Model predictive control strategy for smooth path tracking of autonomous vehicles with steering actuator dynamics. *Int. J. Automot. Technol.* **2014**, *15*, 1155–1164. [CrossRef]

37. Thornton, S.; Zhang, V.; Varnhagen, S.; Gerdes, J. Comparative Analysis of Steering System Models in Model Predictive Control of Automated Vehicles. In Proceedings of the 14th International Symposium on Advanced Vehicle Control (AVEC 2018), Beijing, China, 16–20 July 2018.
38. Soltani, A.; Assadian, F.; Soltani, A. New Slip Control System Considering Actuator Dynamics. *SAE Int. J. Passeng. Cars–Mech. Syst.* **2015**, *8*, 512–520. [CrossRef]
39. Ulsoy, A.G.; Peng, H.; Cakmakci, M. *Automotive Control Systems*; Cambridge University Press: Cambridge, UK, 2012.
40. Wischnewski, A.; Stahl, T.; Betz, J.; Lohmann, B. Vehicle Dynamics State Estimation and Localization for High Performance Race Cars. *IFAC-PapersOnLine* **2019**, *52*, 154–161. [CrossRef]
41. Chen, B.-C.; Hsieh, F.-C. Sideslip angle estimation using extended Kalman filter. *Veh. Syst. Dyn.* **2008**, *46*, 353–364. [CrossRef]
42. Chen, B.-C.; Kuo, C.-C. Electronic stability control for electric vehicle with four in-wheel motors. *Int. J. Automot. Technol.* **2014**, *15*, 573–580. [CrossRef]
43. Bryson, A.E. *Control of Spacecraft and Aircraft*; De Gruyter: Berlin, Germany, 1994.
44. Ahn, C.S. *Robust Estimation of Road Friction Coefficient for Vehicle Active Safety Systems*; University of Michigan: Ann Arbor, MI, USA, 2011.
45. Wielitzka, M.; Dagen, M.; Ortmaier, T. State and maximum friction coefficient estimation in vehicle dynamics using UKF. In Proceedings of the 2017 American Control Conference (ACC), Seattle, WA, USA, 24–26 May 2017; pp. 4322–4327. [CrossRef]
46. Hermansdorfer, L.; Betz, J.; Lienkamp, M. A Concept for Estimation and Prediction of the Tire-Road Friction Potential for an Autonomous Racecar. In Proceedings of the 2019 IEEE Intelligent Transportation Systems Conference (ITSC), Auckland, New Zealand, 27–30 October 2019; pp. 1490–1495.
47. Oppenheimer, M.W.; Doman, D.B.; Bolender, M.A. Control Allocation for Over-actuated Systems. In Proceedings of the 2006 14th Mediterranean Conference on Control and Automation, Ancona, Italy, 28–30 June 2006; pp. 1–6.
48. Car Insurers Warn on "Autonomous" Vehicles—BBC News. Available online: https://www.bbc.com/news/technology-44439523 (accessed on 21 May 2021).
49. NHTSA. *Laboratory Test Procedure for Dynamic Rollover the Fishhook Maneuver Test Procedure*; NHTSA: Washington, DC, USA, 2013.
50. He, X.; Liu, Y.; Lv, C.; Ji, X.; Liu, Y. Emergency steering control of autonomous vehicle for collision avoidance and stabilisation. *Veh. Syst. Dyn.* **2018**, *57*, 1163–1187. [CrossRef]

applied sciences

MDPI

Article

Multi-Objective Optimisation of Tyre and Suspension Parameters during Cornering for Different Road Roughness Profiles

Georgios Papaioannou [1,2,*], Jenny Jerrelind [1,2] and Lars Drugge [1,2]

[1] Department of Engineering Mechanics, KTH Royal Institute of Technology, Teknikringen 8, SE-100 44 Stockholm, Sweden; jennyj@kth.se (J.J.); larsd@kth.se (L.D.)
[2] The Centre for ECO2 Vehicle Design, KTH Royal Institute of Technology, Teknikringen 8, SE-100 44 Stockholm, Sweden
* Correspondence: papaioa@kth.se

Abstract: Effective emission control technologies and novel propulsion systems have been developed for road vehicles, decreasing exhaust particle emissions. However, work has to be done on non-exhaust traffic related sources such as tyre–road interaction and tyre wear. Given that both are inevitable in road vehicles, efforts for assessing and minimising tyre wear should be considered. The amount of tyre wear is because of internal (tyre structure, manufacturing, etc.) and external (suspension configuration, speed, road surface, etc.) factors. In this work, the emphasis is on the optimisation of such parameters for minimising tyre wear, but also enhancing occupant's comfort and improving vehicle handling. In addition to the search for the optimum parameters, the optimisation is also used as a tool to identify and highlight potential trade-offs between the objectives and the various design parameters. Hence, initially, the tyre design (based on some chosen tyre parameters) is optimised with regards to the above-mentioned objectives, for a vehicle while cornering over both Class A and B road roughness profiles. Afterwards, an optimal solution is sought between the Pareto alternatives provided by the two road cases, in order for the tyre wear levels to be less affected under different road profiles. Therefore, it is required that the tyre parameters are as close possible and that they provide similar tyre wear in both road cases. Then, the identified tyre design is adopted and the optimum suspension design is sought for the two road cases for both passive and semi-active suspension types. From the results, significant conclusions regarding how tyre wear behaves with regards to passenger comfort and vehicle handling are extracted, while the results illustrate where the optimum suspension and tyre parameters have converged trying to compromise among the above objectives under different road types and how suspension types, passive and semi-active, could compromise among all of them more optimally.

Keywords: wear; tyre; suspension; semi-active; handling; comfort; optimisation

Citation: Papaioannou, G.; Jerrelind, J.; Drugge, L. Multi-Objective Optimisation of Tyre and Suspension Parameters during Cornering for Different Road Roughness Profiles. *Appl. Sci.* **2021**, *11*, 5934. https://doi.org/10.3390/app11135934

Academic Editor: Flavio Farroni

Received: 28 May 2021
Accepted: 23 June 2021
Published: 25 June 2021

1. Introduction

Due to increased environmental issues, the automotive industry has increased its focus on energy efficient driving, without neglecting the aspects of comfort and vehicle stability. In this direction, effective emission control technologies and novel propulsion systems have been developed, decreasing exhaust particle emissions. However, work also has to be done on decreasing non-exhaust traffic related sources (i.e., the tyre–road interaction and the tyre wear, which both are inevitable in road vehicles).

The non-exhaust traffic-related sources have a great impact on the pavement condition (i.e., degradation and permeability) [1] and environmental pollution. Regarding the latter, in 2013, it was estimated that the tyre wear, in a couple of European countries (Germany, Netherlands, Sweden, Italy, United Kingdom, Denmark and Norway) was around 300,000 tonnes in total with about 40% coming from passenger vehicles. At the same time, a similar amount of wear is disposed per year in the environment from vehicles in India, where the population size is 5.5 times larger [2]. According to Grigoratos et al. [3], a significant percentage of these fall in the PM_{10} fraction, which means that they have a diameter larger than 10 μm and they finally end up in air, water, soils, etc. [4]. Hence, the need to develop more environmentally friendly vehicle systems that can decrease tyre wear has risen. This is especially the case now that the more environmentally friendly, but much heavier, electric vehicles are expected to have increased particle pollution from tyre wear compared to conventional vehicles, a fact that could potentially cancel the benefits of removing the exhaust emissions [5,6].

Tyre wear occurs because of friction during the sliding between the tyre tread surface and the road [7]. With the tread being the component responsible for the vehicle–road interaction, the aspect of minimising wear is a crucial criterion during the tyre design. According to Huang et al. [8], one of the existing categorisations of wear is between normal and abnormal. The first leads to uniform wear along the tyre circumference and over its width, while the latter is defined by uneven and irregular wear. Uneven wear mostly describes the non-uniform wear distribution over the tyre width, whereas the irregular wear mainly considers the circumferential wear. The total amount of wear is related to internal (tyre design, manufacturing, etc.) and external (vehicle, road, driving condition, environmental circumstances, etc.) factors (Figure 1) [9]. According to Maitre et al. [10], the driving condition is the most dominant in terms of its impact to wear while the tyre design, the environmental circumstances, the vehicle and the road follow. This work covers the majority of these factors, as highlighted in Figure 1. Specifically, the optimisation of tyre and suspension design of a passenger vehicle is investigated to minimise wear under multiple road roughness surfaces.

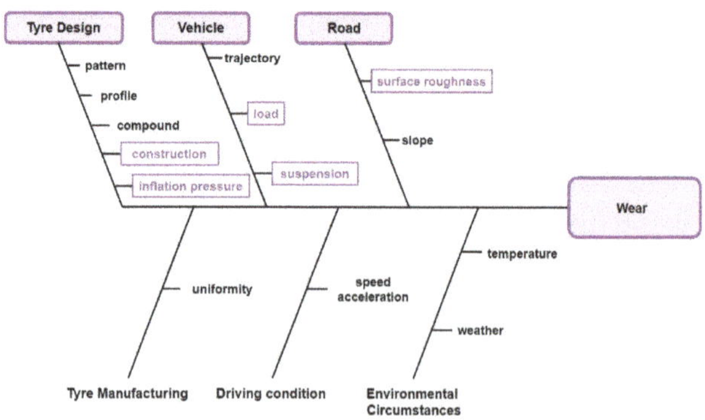

Figure 1. Internal and external factors related to tyre wear. The highlighted ones are where the emphasis of this work is.

Various wear models have been presented in the literature, to investigate how different factors influence wear and to study the wear behaviour in detail. Bin Ma et al. [11] simplified the kinetic sliding friction coefficient taking into consideration the road roughness. Afterwards, they coupled the contact model with a 9DOF vehicle model to study tyre marks using wear quantity during the vehicle's pro and post-crash phases. Da Silva et al. [12] developed a qualitative formula for tyre wear evaluation and conducted a sensitivity analysis of a few tyre and vehicle parameters during cornering manoeuvres using a simplified single-track model. Huang et al. [8] proposed a theoretical tyre model for predicting the 3D tyre wear with regards to the roughness of the road and vehicle dynamic characteristics. While most models focus on the lateral and longitudinal direction, Sueoka et al. [13] presented a computationally efficient analytical model to study tyre wear due to vertical excitations and considered the suspension systems as well. Later, Li et al. [14] incorporated in this model a formula of tyre wear considering the temperature effect and the dynamic characteristics of the vehicle, which allowed them to analyse the effects of speed, ambient temperatures, tyre pressure and sprung mass on tyre wear. Thereby, this model is able to evaluate more accurately the wear performance while also considering the suspension effect, which is not widely discussed in the literature as far as the wear performance is concerned. This model is chosen in this work.

The estimation of tyre wear has been extensively experimentally studied in order to identify effective methods and to capture the effects of different factors on it. For example, Stalnaker et al. [15] developed a methodology for estimating tyre wear indoors, in order to establish consistent test results. Similarly, Knuth et al. [16] described a simulation method of indoor testing which could accurately capture the tyre–vehicle–driving interaction. In the same direction, Lupker et al. [17] provided a tool which could estimate numerically global tyre wear as well as qualitatively determine the wear distribution. This wear model was later further validated and investigated in a sensitivity analysis [18]. Only recently, Farroni et al. [19] developed a physical model of tyre wear to analyse the impact of thermal and frictional effects on vehicle performance. Similarly, Emami et al. [20] designed and developed a new portable test setup to study friction and wear, while Lepine et al. [21] presented a novel empirical tyre wear model for heavy vehicles that can be used to predict the wear for multi-axle vehicles based on route data and a vehicle model. In addition to the wear models and testing methods, Yamazaki et al. [22] investigated experimentally the impact of alignments such as camber angle and toe angle to the wear performance. Case studies were considered for various alignment configurations simulated real-life configurations observed in road tests. Tandy et al. [23] studied how increased shoulder wear is finally affecting the driving behavior, illustrating that tyre lateral force and overturning moment capacities increase significantly with the usage of the vehicle and hence its wear. However, in order to reduce the time-consuming experimental procedures, Tamada [24] considered the prediction of uneven tyre wear conducting progress simulation in a wearing out finite element (FE) tyre model.

Even if the modelling and estimation of tyre wear has been extensively investigated, very few works have considered the optimisation towards tyre wear minimisation. Up to now, most of these works used FEA models to assess the wear performance. For example, Koishi et al. [25] investigated the trade-off between uneven wear and wear life using multi-objective optimisation, where the objectives were evaluated using response surface methodology in order to save computational time as wear simulation was significant high, even for a super computer. Similarly, Serafinska et al. [26] suggested a multi-objective optimisation approach for uniform wear, by minimising the ratio of the contact pressure in the tyre shoulder and the contact pressure in the tyre footprint central part. However, the FEA models focus mainly on the detailed tyre modelling and they require high computational power, but they do not consider the rest of the vehicle subsystems and their interaction with the tyre. This has led to an unclear understanding regarding the trade-off between important vehicle performance aspects and wear. For instance, the conflict with regards to comfort and vehicle stability is widely studied [27,28]; however, only Anderson et al. [29] investigated the trade-off between wear performance and handling. To the authors knowledge, there is not extensive literature on the optimisation of the vehicle and tyre parameters with regards to comfort, vehicle stability and wear performance where this work focuses. This is a considerably critical subject, as, during the conceptual design, the effort is placed upon efficient simulation and optimisation of both tyre and suspension parameters in order to decrease the development costs for physical testing.

Considering the above, in this work, the trade-off among comfort, vehicle stability and wear performance is investigated using a vehicle model combined with a wear model, which considers the vehicle dynamic characteristics, temperature effects, tyre dimensions, vehicle velocity and tyre slip angles. The emphasis is on the optimisation of both tyre and suspension parameters for minimising wear on a passenger vehicle, which is equipped with either passive or semi-active suspensions. The aim is to seek for a tyre design that is not significantly affected by different road profiles and to investigate how different control algorithms in the semi-active suspension influence the tyre wear. Hence, initially, tyre parameters (e.g., pressure, tyre width, outside radius, crown thickness and chordwise radius) are optimised (Scenario 1) for a vehicle being equipped with a passive suspension and driven over an S-Path, on which road roughness of Class A and B are assigned. Afterwards, a common optimal solution is sought among the alternatives provided by the two optimisation cases, requiring to have close design variable values and provide similar wear in both cases. Then, the identified tyre design is adopted and the optimum suspension design is sought (Scenario 2) for the two cases, but also for different suspension types. For each suspension type, a common solution with regards to their design variables is identified among the optimal alternatives provided by the two optimisation cases, requiring to have close values in their design variables. Significant conclusions regarding how tyre wear behaves with regards to passenger comfort and vehicle stability are extracted, while the results illustrate where the optimum suspension and tyre parameters have converged trying to compromise among the above objectives under different road types, and which suspension could compromise among all of them more optimally.

This paper is organised as follows. Firstly, all the models (vehicle, suspensions, tread and wear) are described and the road path and profiles used as excitations are displayed. Secondly, the validation of the models with IPG/CarMaker is illustrated. Thirdly, the formulation of the multi-objective optimisation is displayed. Then, the results are outlined. Finally, conclusions are extracted.

2. Methods and Materials

2.1. Vehicle Model

The vehicle model is a one-dimensional model with three degrees of freedom which only includes vertical motions (Figure 2). The simulation model considers four basic subsystems of the vehicle: the chassis, the suspension systems, the unsprung mass and the tyres. The chassis is considered as a rigid body of mass (m_s) and is connected with the unsprung mass (m_u) through the suspension system, which is modelled as a spring and a damper (K_u and C_u). In this work, the damper is considered both passive and semi-active. Regarding the tyres, a more advanced modelling is applied compared to the common quarter car model. Firstly, the tyre sidewall is modelled as a spring and a damper (K_T and C_T) and connects the unsprung mass with the tyre tread (m_t), whose mass is varying dynamically according to analytical equations presented later in details. Finally, the tread element consists of linear springs and dampers (K_t and C_t), which receive the unevenness of the road profile (z_R) as an excitation.

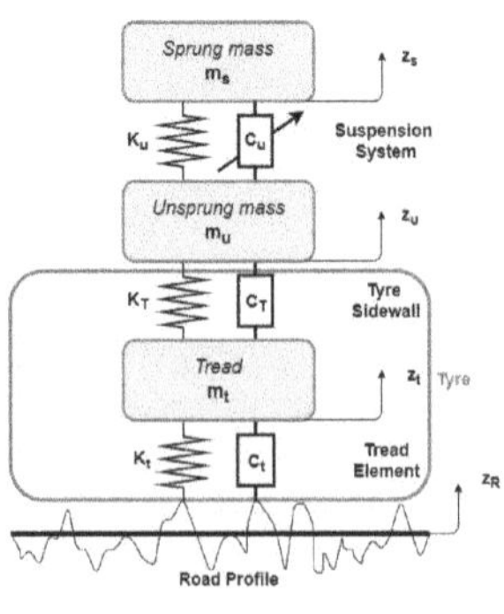

Figure 2. Quarter car model.

The governing equations of motion of the vehicle model are the following (Equations (1)–(3)):

$$Sprung\ mass : m_s\ddot{z}_s + C_u(\dot{z}_s - \dot{z}_u) + K_u(z_s - z_u) = 0 \tag{1}$$

$$\begin{aligned} Unsprung\ mass: m_u\ddot{z}_u - C_u(\dot{z}_s - \dot{z}_u) - K_u(z_s - z_u) \\ +C_T(\dot{z}_u - \dot{z}_t) + K_T(z_u - z_t) = 0 \end{aligned} \quad (2)$$

$$\begin{aligned} Tread\ mass: m_t\ddot{z}_t - C_T(\dot{z}_u - \dot{z}_t) - K_T(z_u - z_t) \\ +C_t(\dot{z}_t - \dot{z}_R) + K_t(z_t - z_R) = 0 \end{aligned} \quad (3)$$

while the parameters used are selected according to IPG/CarMaker 8.0 to generate a digital twin of the software demo vehicle. The parameter values are presented in Table 1.

Table 1. Vehicle model parameters.

Masses [kg]			Springs [N/m]			Dampers [Nm/s]		
m_s	=	1301/4	K_u	=	25000	C_u	=	2500
m_u	=	43	K_T	=	0.80*K_t	C_T	=	2508
m_t	=	Equation (6)	K_t	=	367240	C_t	=	508

2.2. Suspension Designs

Semi-active suspensions do not consume significant amount of energy and can provide a more energy efficient solution compared to fully active ones. At the same time, the reliability of passive suspensions can be maintained. Therefore, different control strategies have been developed exploiting the benefits of semi-active suspensions. In this work, the suspension system of the vehicle model operates as passive (PS) and semi-active. Several different control algorithms are used in the semi-active suspension case: SH-2 [30], ADD [31], SH-ADD-2 [32], PDD [33], SH-PDD [34] and GH-2 [35]. The operational conditions of the semi-active control algorithms are illustrated in Table 2.

Table 2. Semi-active control algorithms applied in the vehicle suspension systems (Figure 2).

Case	Description	Control Law [a]
1	SH-2	$C_u = \begin{cases} C_{min} & ,if\ \dot{d}_{ST}\ \dot{z}_s < 0 \\ C_{max} & ,if\ \dot{d}_{ST}\ \dot{z}_s > 0 \end{cases}$
3	ADD	$C_u = \begin{cases} C_{min} & ,if\ \dot{d}_{ST}\ddot{z}_s \le 0 \\ C_{max} & ,if\ \dot{d}_{ST}\ddot{z}_s > 0 \end{cases}$
4	SH-ADD-2 [b]	$C_u = \begin{cases} C_{min} & ,if\ \ddot{z}_s^2 - a_f^2\dot{z}_s^2 \ge 0 \\ C_{max} & ,if\ \ddot{z}_s^2 - a_f^2\dot{z}_s^2 < 0 \end{cases}$
5	GH-2	$C_u = \begin{cases} C_{min} & ,if\ \dot{d}_{ST}\ \dot{z}_u > 0 \\ C_{max} & ,if\ \dot{d}_{ST}\ \dot{z}_u < 0 \end{cases}$
6	PDD	$C_u = \begin{cases} C_{min} & ,if\ K_u d_{ST} + C_{min}\dot{d}_{ST}^2 \ge 0 \\ C_{max} & ,if\ K_u d_{ST} + C_{min}\dot{d}_{ST}^2 < 0 \\ \frac{C_{min}+C_{max}}{2} & ,if\ d_{ST} \ne 0 \text{ and } \dot{d}_{ST} = 0 \\ \frac{-K_u d_{ST}}{\dot{d}_{ST}} & ,otherwise \end{cases}$
7	SH-PDD	$C_u = \begin{cases} C_{SH-2} & ,if\ \ddot{z}_s^2 - \dot{z}_s^2 \ge 0 \\ C_{PDD} & ,otherwise \end{cases}$

[a] The parameter d_{ST} refers to the vehicle suspension travel, i.e., $z_s - z_u$, while $\dot{d}_{ST}$ is the vehicle suspension velocity, i.e., $\dot{z}_s - \dot{z}_u$. [b] The parameter α_f is the crossover frequency between SH-2 and ADD $\in$ [10–60] $\frac{rad}{sec}$.

2.3. Tread Model

In order to evaluate the wear quantity, the energy loss has to be calculated, hence the contact between the tyre and the road should be modelled with regards to that and not only with spring and dampers as usual. Therefore, this model considers that the tread. which is in contact with the road surface (z_R), is a cuboid element with a, b and h being its length, width and height, respectively (Figure 3a). Regarding the first two, they are evaluated according to empirical equations (Equation (4)) [9,36], which consider tyre dimensions, i.e., the outside radius (d) and the crown thickness (b_0), and its deformation (δ):

$$a = 4d\left(\frac{\delta}{2d}\right)^s, b = b_0(1 - e^{-\psi}) \tag{4}$$

where ψ (=115) and s (=0.67) are empirical coefficients. The tyre deformation (δ) is evaluated according to Equation (5):

$$\delta = \frac{\alpha_1 F_z}{2P} + \sqrt{\left(\frac{\alpha_1 F_z}{2P}\right)^2 + \alpha_2 F_z} \tag{5}$$

where α_1 is $1/(2\pi d r_w)$, with r_w being the tyre chordwise curvature radius; α_2 (=3.70 $*$ 10^{-8}) is a constant for car tyre; and F_z is the wheel load. According to δ (Equation (5)), the tread deformation is related with the pressure (P) (Figure 3b), the tyre radius (d) and the chordwise radius (r_w). The force–deflection curves that occur through Equation (5) are plotted in Figure 3b for different tyre pressures (P). Finally, the mass of the tread element can be derived by Equation (6), where ρ (=1156 kg/m^3) is the rubber density of the tread.

$$m_t = abh\rho \tag{6}$$

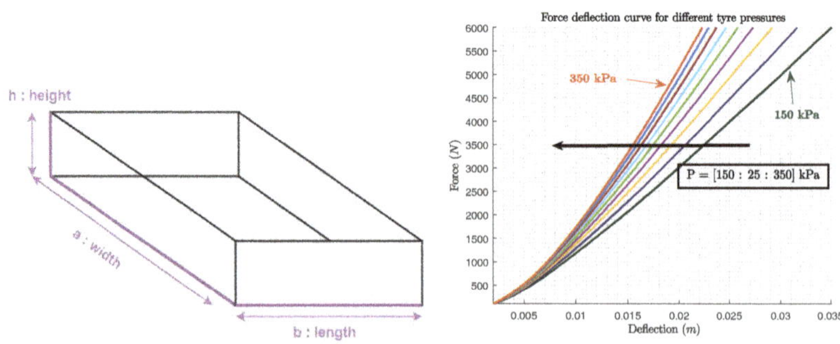

(a) (b)

Figure 3. (**a**) The modelling of the tread element as a cuboid and (**b**) the force deflection curves according to Equation (5) for pressures ranging [150:25:350] kPa.

2.4. Vehicle Performance Assessment

2.4.1. Tyre Wear Quantity

Tyre wear is defined according to the frictional or wear energy, E_w[Nm/m^2], which is the energy due to slippage per unit area and can be evaluated as the product of the shear stress ($\vec{\tau}$) and slip ($\vec{S}$):

$$E_w = \int_0^b \vec{\tau} d\vec{S} \tag{7}$$

where b is the contact length; τ_x and τ_y are, respectively, shear stresses in x- and y-directions; and S_x and S_y are, respectively, slip distances in x- and y-directions.

Wear energy during cornering: In order to calculate the wear energy (E_w) during cornering, a solid tyre is considered using a brush model (Figure 4a), while, for the wear evaluation during cornering, a wear model is used (Figure 4b). In the considered wear model, parabolic pressure distribution (q_z) and small slip angles are assumed. The contact pressure is defined as:

$$q_z = 4p_m \frac{x}{b}\left(1 - \frac{x}{b}\right) \quad (8)$$

where p_m is the maximum pressure that occurs from the vertical load F_z and is equal to $3F_z/2ab$ if the contact patch is assumed as cuboid (Figure 3a). According to the wear model (Figure 4b), the lateral slip angle is α , while the lateral shear stress at a distance x measured from the the contact patch (0–b) is τ_y and is expressed by Equation (9):

$$\tau_y = C_y x tan\alpha \quad (9)$$

where C_y [N/m] is the lateral shear elastic tread coefficient per unit area. The shear force τ_z (=$\mu_s q_z b_h$) applied to the contact patch per unit area is equal to the lateral shear stress applied to the sliding point b_h (Equation (9)), i.e., the maximum frictional force per unit area ($\tau_{y_{max}}$). Therefore, combining Equations (8) and (9):

$$\tau_{y_{max}} = C_y b_h tan\alpha = \mu_s q_z(b_h) = 4\mu_s p_m \frac{x}{b_h}\left(1 - \frac{x}{b_h}\right) \quad (10)$$

where μ_s is the static friction coefficient, while $\zeta \left(= \frac{C_y b tan\alpha}{4\mu_s p_m} \right)$ is introduced as a new coefficient, to help in the simplification of the equations. Based on ζ, the sliding point (b_h) is defined as:

$$\begin{aligned} b_h =& b(1 - \zeta_h) \quad ,0 \le \zeta_h \le 1 \\ b_h =& 0 \quad\quad\quad\quad ,1 > \zeta_h \end{aligned} \quad (11)$$

where the first part is when the tyre is in the adhesion region, while the second part considers that sliding occurs across the whole contact area when ζ_h is greater than or equal to unity. Then, considering the force equilibrium in the sliding region ($b_h \le x \le b$), the sliding distance is defined according to Equation (12), where μ_d is the kinematic friction coefficient:

$$S_y = x tan\alpha - \frac{\mu_d q_z(x)}{C_y} \quad (12)$$

Considering Equation (7), the wear energy per unit area is defined as:

$$E_y{}^w = \int_{b_h}^{b} \mu_d q_z(x) dS_y = \int_{b_h}^{b} \mu_d q_z(x) \frac{dS_y}{dx} dx = \mu_d tan\alpha \int_{b_h}^{b} q_z(x)dx + \frac{1}{2}\frac{\mu_d^2}{C_y} q_z^2(b_h) \quad (13)$$

where Equations (8) and (11) can be substituted, and the equation of wear energy can be derived at the two conditions explained in Equation (11). In addition, in this work, μ_d and μ_s are assumed to be equal to μ, while F_z is considered from the vehicle model:

$$F_z = F_{Z_0} + C_t(\dot{z}_t - \dot{z}_R) + K_t(z_t - z_R) \quad (14)$$

where F_{Z_0} is the static load applied.

Mass Loss: The wear energy per contact area described above can be related with the mass loss according to Equation (15), which was derived by Lupker et al. [17] from experiments:

$$\Delta m = f_1 E^{f_2} \quad (15)$$

where f_1 and f_2 are constants at a given temperature. The mass loss (Δm) evaluated at each time step is used to calculate the m_t dynamically. As an extension of this expression, Li et al. [14] redefined f_1 by incorporating in it the effect of temperature:

$$\Delta m = f_0 1.02^{T_t - T_0} E^{f_2} \tag{16}$$

where f_2 is 1.5106 and f_0 (=2.010^{-10}) is the constant f_1 at T_0 (=60°). T_t is defined as the steady state temperature of the tyre tread:

$$T_t = \frac{0.0447\gamma(3.6 * V)^{0.16} F_z d^{-0.5} \delta^{0.5}}{\pi\beta[2dc + 0.4(d^2 - r^2) + 0.4(d^2 + dr + r^2) - 0.6r(d+r)]} + T_\infty \tag{17}$$

where V[m/s] is the vehicle velocity; r is rim radius; c is the tyre width; γ (=0.12) is the tyre hysteresis coefficient; β (=1.40) is the correction coefficient; and T_∞ is the ambient temperature.

Wear quantity: The total wear quantity that is caused by the mass loss (Equation (16)) is defined as the height change (Δh):

$$\Delta h = \frac{\Delta m}{\rho a b} \tag{18}$$

Finally, the total tread wear is defined:

$$\begin{aligned} H(t) &= \Delta h && , t \leq t_0 \\ H(t) &= H(t - t_0) + \Delta h && , t > t_0 \end{aligned} \tag{19}$$

where $H(t - t_0)$ is the total tread wear occurred during the first tyre revolution taking place as long as $t \leq t_0$, with t_0 (=$2\pi r_r / V$) being the rotating period and r_r the rolling radius.

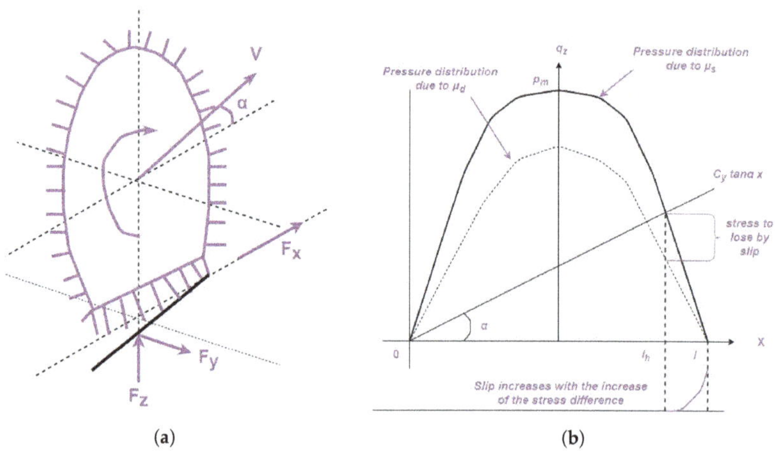

Figure 4. (**a**) Tyre brush model and (**b**) tyre wear model in cornering.

2.4.2. Ride Comfort

Ride comfort (RC) is assessed via the root mean square of the sprung mass' vertical acceleration ($\ddot{z}_s$), when measurements are not applicable to the seat or the occupant's body [37]. More specifically, RC is calculated as follows:

$$RC = \left[\frac{1}{T}\left(\int_0^T \ddot{z}_{w_z}(t)^2 dt\right)\right]^{\frac{1}{2}} = RMS(\ddot{z}_{w_z}) \tag{20}$$

where T ([s]) is the duration and $\ddot{z}_{w_z}$ (m/s^2) is the weighted acceleration as a function of time. In this work, no weighting is applied in the accelerations measured at the sprung mass. Therefore, $\ddot{z}_{w_z}$ is considered equal to $\ddot{z}_s$.

2.4.3. Vehicle Handling

The vehicle handling is normally assessed through cornering, braking and traction abilities, which all are enhanced if the tyre loads/deflections are maintained at a low level. Therefore, the root mean square of the tyre deflection ($z_u - z_{Road}$) is considered as an index (Equation (21)) for vehicle stability (Figure 2), taking advantage of the available outputs from the vehicle model:

$$TD = \left[\frac{1}{T}\left(\int_0^T (z_u - z_{Road})^2 dt\right)\right]^{\frac{1}{2}} \tag{21}$$

At the same time, the suspension travel depicts the ability of the suspension system to support and hold the vehicle load. This is achieved when rattle space requirements of the vehicle are kept small. Therefore, the root mean square of the suspension travel ($z_s - z_u$) is considered as an index (Equation (22)) for vehicle stability as well:

$$ST = \left[\frac{1}{T}\left(\int_0^T (z_s - z_u)^2 dt\right)\right]^{\frac{1}{2}} \tag{22}$$

2.5. Road Profiles and Path

The combination of the vehicle model with the above analysis on the wear quantity allows the assessment of the wear under different conditions (i.e., road surfaces, paths and vehicle velocities), tyres and vehicle types. More specifically, different road surfaces can be used for exciting the vehicle model and different paths can be considered by applying in the wear model the vehicle's corresponding slip angles and velocities while driving on any path. The vehicle's corresponding slip angles and velocities have to be obtained from offline simulations prior to the simulation of the vehicle model.

In this work, road profiles of Class A and B (Figure 5a) are used to excite the vehicle model and are designed according to ISO-8608 [38]. Then, the S-path illustrated in Figure 5b, which consists of two turns of different radius (70 and 60 m), is used to extract the slip angles and the vehicle velocity while driving on it. In this work, both are obtained from simulating the demo vehicle of IPG/CarMaker 8.0, IPG AUTOMOTIVE, Braunschweig, Germany (Figure 6) while driving over the S-path, having assigned on it the two road profiles.

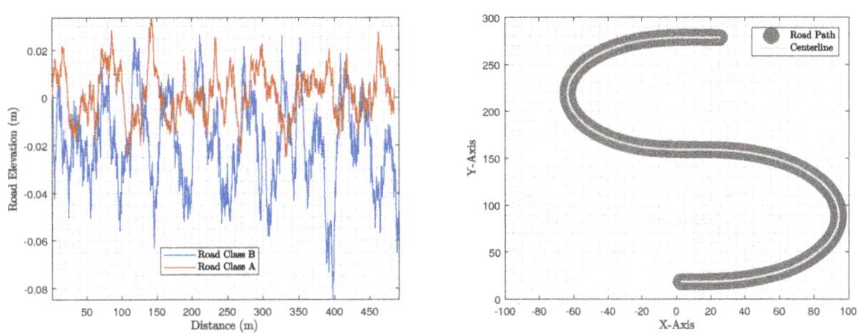

(**a**) (**b**)

Figure 5. (**a**) Class A and B road profiles assigned on the (**b**) road path.

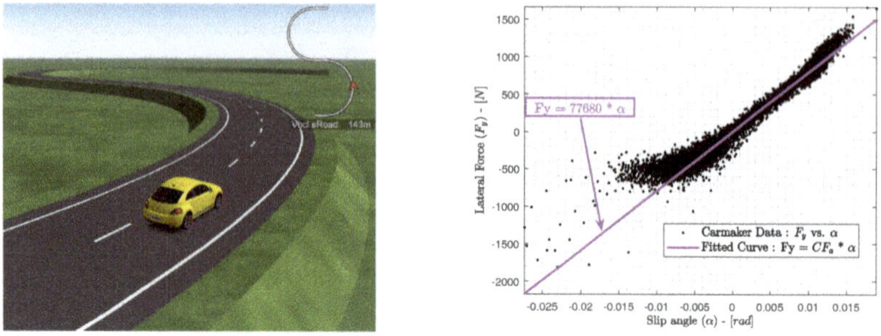

(**a**) (**b**)

Figure 6. (**a**) The IPG/CarMaker demo vehicle used in this work and (**b**) the evaluation of its tyre cornering stiffness.

3. Validation of the Model with IPG CarMaker 8.0

In order to secure that the optimisation results in this work are realistic, the model is validated using IPG/CarMaker 8.0 for the path illustrated in Figure 5b and the road profile of Class B assigned to it (Figure 5a).

As mentioned above, the vehicle parameters selected for the model are extracted from the software demo vehicle to build its digital twin. The suspension and tyre parameters of the models are adjusted according to Tables 1 and 3. More specifically, the suspension parameters (Table 1) of the front wheels are adjusted into the section Car/VehicleDataSet/Suspensions of the software. Regarding the suspensions, for both the stiffness and the damping coefficient, a characteristic value option is selected, while the values of the buffer are set zero in the software, as the buffer is not included in the model. Then, regarding the tyres, the parameters related with them and considered in the vehicle and wear model (P, d, c, r, K_T and C_T), are adjusted according to section Tires/TireDataSet/GeneralandModelParameters of the software. The only parameters that are not provided are the crown thickness (b_0), the chordwise curvature (r_w) and the lateral shear elastic tread coefficient (C_y). Regarding the first two, its consideration is not possible, but C_y is extracted after evaluating it using the cornering stiffness obtained from the software simulation results. More specifically, the side force applied in the wheel for a rectangular contact area, as in this case, can be evaluated for small slip angles:

$$F_y = \int_0^b f_y dx = C_y a \int_0^b V dx = C_y a \int_0^b x\alpha dx = \frac{C_y a b^2}{2}\alpha \tag{23}$$

Thus, according to Equation (23), the cornering stiffness C_{Fa} is equal to $\frac{C_y a b^2}{2}\alpha$. Therefore, after fitting the simulation results (lateral slip angles and side forces) to a linear equation, as shown in Figure 6b, the value of C_y is extracted.

Table 3. Wear and tread model parameters.

Wear and Tread Model Parameters				
P[KPa]	c[m]	d[m]	b_0[m]	r_w[m]
250	0.1950	0.3175	0.1660	0.1550

Having created the digital twin of the demo vehicle (Figure 6), the wear energy (E_y[Nm/m^2]) is selected for comparison in order to validate the simulation results. Therefore, the obtained measurements are used as follows for calculating the wear energy ($E^w_{y_{carmaker}}$):

- The sliding distance (s_d) is calculated according to Equation (12).

- The parameter (ϕ) is calculated as follows:

$$\phi = \frac{C_{Fa}|tan\alpha|}{\mu F_z}. \quad (24)$$

- The sliding force is obtained by integrating the maximum possible frictional force distribution (μq_z) over the sliding region $b_h \leq x < b$:

$$F_{slid} = \int_b^{b_h} \mu q_z dx = \mu F_z \left(\frac{1}{3}\phi^2 - \frac{2}{27}\phi^3\right) sign(a). \quad (25)$$

- Finally, the wear energy per contact area according to Equation (26) is evaluated by multiplying the sliding force with the slipping distance (S_y) and dividing it by the contact area:

$$E^w_{y_{carmaker}} = \frac{F_{slid} S_y}{a * b}. \quad (26)$$

The comparison of $E^w_{y_{carmaker}}$ with E^w_y is illustrated in Figure 7, which proves a good convergence in the two simulation results. More specifically, Figure 7 illustrates high similarity in terms of the trend of the two curves. Slight differences occur in the magnitude with the model used in this work underestimating a few of the peaks. However, this underestimation is not an issue as the trend is accurately followed, and, given the simplicity of the model in comparison with IPG/CarMaker 8.0, the results should be considered valid.

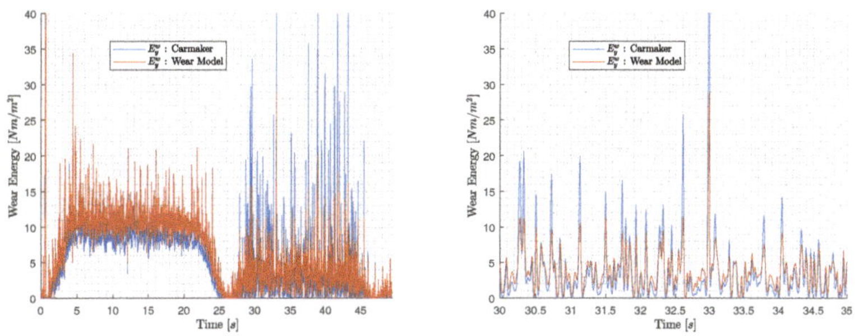

(a) (b)

Figure 7. The comparison of $E^w_{y_{carmaker}}$ with E^w_y for: (**a**) the complete simulation time; and (**b**) the time interval 30–35 s.

4. Optimisation Configuration

In this work, the emphasis is on the optimisation of tyre and suspension parameters for minimising tyre wear. However, as comfort and vehicle stability are still important criteria in the design process, they are also included as objectives by configuring a multi-objective optimisation problem. In order to decrease the complexity of the optimisation, the tyre and suspension parameters are optimised separately considering two optimisation scenarios. Initially, the optimal tyre design solutions (Scenario 1) are sought for the vehicle equipped with a passive suspension, while driving over a road class A (Case 1) and B (Case 2) S-path (Figure 5). Then, after selecting one of the optimal tyre design solutions, which can provide similar tyre wear for both road profiles, the optimal suspension designs (Scenario 2) are sought for different suspension types, both passive and semi-active (PS, SH-2, ADD, SH-ADD-2, PDD, SH-PDD and GH-2), for the two road cases.

4.1. Objectives

The multi-objective optimisation problem is formulated in order to minimise tyre wear, enhance ride comfort and increase vehicle handling at the same time. In both scenarios, wear and comfort are represented from the same objective function (F_1 and F_2), whereas vehicle handling (F_3) is represented with the most suitable function according to the scenario. In Scenario 1, where the tyre is optimised, TD, defined in Equation (21), is selected as the objective function, while, in Scenario 2, where the focus is on the suspension, ST is selected, as defined in Equation (22).

$$\text{Scenario 1: } F_1 = max(H(t)); F_2 = RC_z; F_3 = TD; \tag{27}$$

$$\text{Scenario 2: } F_2 = max(H(t)); F_2 = RC_z; F_3 = ST; \tag{28}$$

In both scenarios, the optimisation is conducted using the GAMULTIOBJ toolbox from MATLAB 2017b, where the *PopoulationSize* is selected as 250 and the *EliteCount* as 5. The rest of the options are selected according to the default suggestions from the software.

4.2. Design Variables: Scenario 1

In this scenario, the following design variables are selected:

$$P, c, d, b_0 \text{and} r_w \tag{29}$$

while the upper and lower bounds set are illustrated in Table 4. In addition, K_t and K_T might be considered indirect design variables as they are evaluated according to the problem design variables. More specifically, during the optimisation, the force–deflection curve (Figure 3b) is evaluated according to Equation (5), which is a relation of P, d and r_w, for F_z varying from 500 to 7000 N for each potential optimal solution. Then, having obtained the force–deflection curve, the K_t is evaluated as the curve slope at the linear region, while the sidewall stiffness (K_T) is evaluated by $0.8 * K_t$ as the sidewall is always less stiff than the tread.

Table 4. Design variables and their bounds for Cases 1 and 2.

Scenario 1			Scenario 2		
Design Variable	Bounds		Design Variable	Bounds	
	LB	*UB*		*LB*	*UB*
P[KPa]	150	350	K_u [N/m]	15,000	60,000
c[m]	0.14	0.22	C_u [Ns/m]	500	5000
d[m]	0.27	0.35	C_{min}[Ns/m]	500	2500
b_0[m]	0.12	0.19	C_{max} [Ns/m]	2500	5000
r_w [m]	0.12	0.20	a_f [rad/s] [a]	10	60

[a] The parameter α_f is included as a design variable for Case 2 only when the SH-ADD2 is optimised.

4.3. Design Variables: Scenario 2

Regarding the second scenario, where the optimum designs are sought for the different suspension types, the design variables shown in Equations (30)–(32) are selected:

For PS:

$$K_u, C_u \tag{30}$$

For SH-2, ADD, GH-2, PDD and SH-PDD:

$$K_u, C_{min} \text{and} C_{max} \tag{31}$$

For SH-ADD2:

$$K_u, C_{min}, C_{max} \text{and} a_f \tag{32}$$

while the upper and lower bounds set in the optimisation are illustrated in Table 4.

5. Results

As described above, in this work, two optimisation cases, with regards to different excitations, are studied in two optimisation scenarios where the focus is on tyre and suspension design, respectively. In this section, the results of the two multi-objective optimisation scenarios are displayed. More specifically:

- Scenario 1 investigates the tyre optimisation in a vehicle being employed with a passive suspension and driven over an S-Path, in which road surfaces of Class A (Case 1) and B (Case 2) are assigned. The optimisation results are illustrated in Figures 8 and 9, which display the Pareto fronts with regards to the objectives (Figure 8a,b) and the design variables (Figure 9a–e). Having obtained the optimal solution alternatives for the two cases, common solutions are sought between them, requiring to provide similar wear in the two cases (<8% difference) and have close design variable values (<9% difference). The common solutions identified with the above characteristics are illustrated in Figures 8 and 9 alongside the alternatives, while their values are listed in Table 5. In addition, in Figure 10, the final circumferential tread profiles for the optimum designs are compared with the initial tyre design, after driving over the Class B S-Path twenty times back and forth.
- Scenario 2 explores the suspension optimisation for two road cases and various suspension types (PS, SH-2, ADD, SH-ADD-2, GH-2, PDD and SH-PDD). The optimisation results are illustrated in Figure 11, which presents the Pareto fronts of the optimum solutions for the two cases. For each suspension type, a common solution is identified among the optimal alternatives provided by the two optimisation cases, requiring them to have close design variables values (<3–15% difference according to the cases). Finally, the values of the identified common solutions are presented in Table 6 alongside with the threshold of difference that was allowed.

The goal of these scenarios is for conclusions to be extracted regarding: (a) how tyre wear behaves with regards to passenger comfort and vehicle handling; (b) where optimum tyre and suspension parameters have converged trying to compromise among the above objectives under different road types; and (c) which suspension could compromise among all of them optimally.

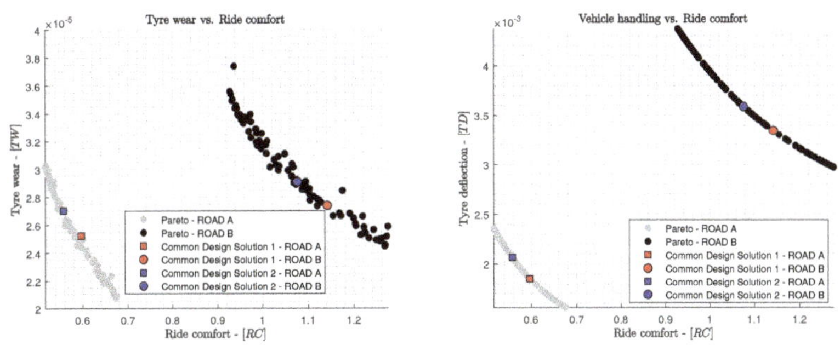

(**a**) (**b**)

Figure 8. Comparison of the optimum solutions obtained for the two optimisation cases (class A and B road profiles) in Scenario 1: (**a**) TW vs. RC; and (**b**) TD vs. RC.

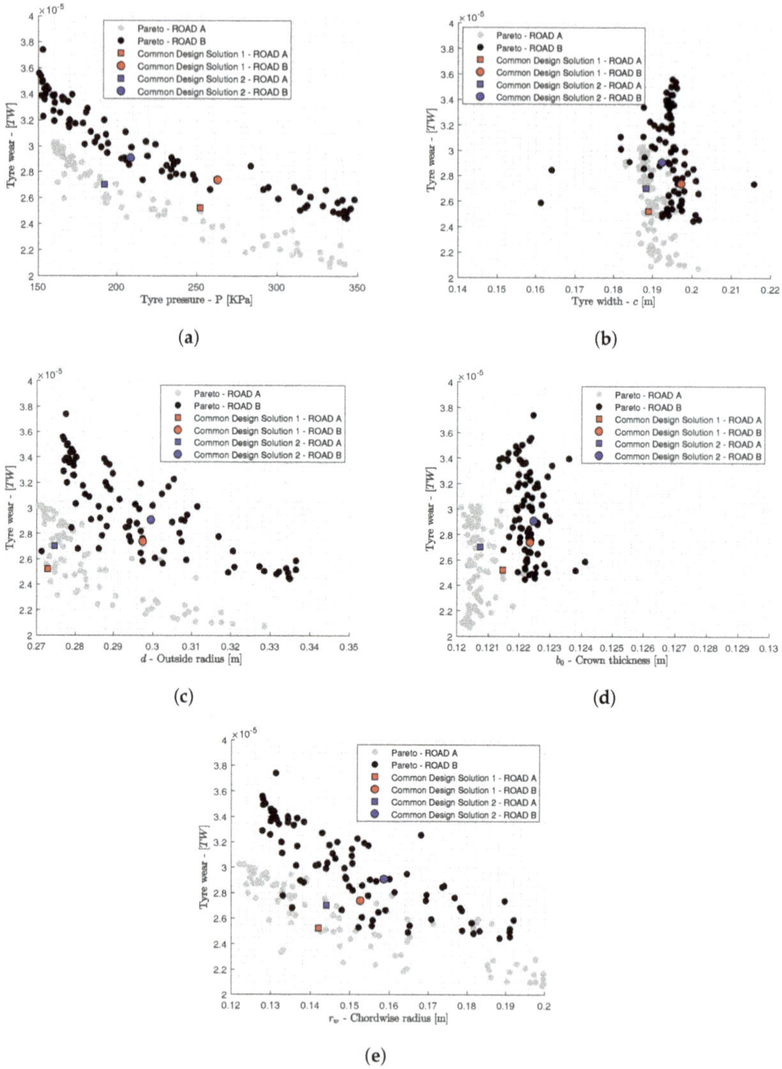

Figure 9. Comparison of the optimum solutions obtained for the two optimisation cases (Class A and B road profiles) in Scenario 1: (**a**) TW vs. P; (**b**) TW vs. c; (**c**) TW vs. d; (**d**) TW vs. b_0; and (**e**) TW vs. r_w.

5.1. Optimum Tyre Design Solutions

Optimisation Results

Regarding Scenario 1, Figure 8a,b illustrates the relation between the optimisation objectives and reveals the trade-offs among them. More specifically, Figure 8a reveals the conflicting relation between wear and comfort levels, where the increase of the RC index, which yields the increase of discomfort, leads to the decrease of tyre wear. At the same time, the conflicting relation of comfort and handling, captured by the optimisation in Figure 8b, serves as an indication of the validity of the results. In addition, the conflicting relation of the two objectives, handling and wear, with comfort reveal the linear relation between them, implying that increased levels of tyre deflections lead to increasing wear.

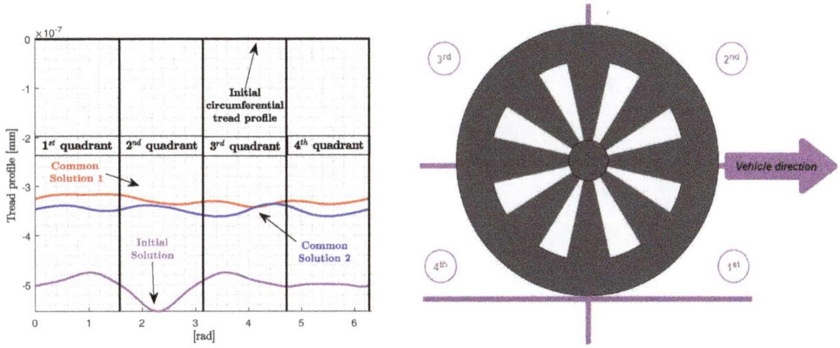

Figure 10. The tread profile induced to the tyre for the two common optimal tyre designs, after the vehicle is driven over the Class B S-path 20 times back and forth.

Table 5. Optimal tyre design variables of the identified common solutions among the two cases of different road profiles.

Solution	Optimum Design Variables				
	P[KPa]	c[m]	d[m]	b_0[m]	r_w[m]
1	252	0.1889	0.2728	0.1215	0.1420
2	192	0.1883	0.2746	0.1206	0.1441

Table 6. Common optimal suspension parameters among the two road profile cases for each suspension type.

Solution	Optimum Design Variables				
	K_3[N/m]	C_{min}[Nm/s]	C_{max}[Nm/s]	a_f [rad/sec]	Threshold (ε)
PS	16,042	3675	-	-	6.0%
SH-2	43,181	2275	4394	-	9.5%
GH-2	16,160	2415	3327	-	3.0%
ADD	41,406	2452	4530	-	5.0%
PDD	15,534	1442	4045	-	5.0%
SH-PDD	42,281	2416	4749	-	5.0%
SH-ADD2	17,246	684	3515	23.5	15.0%

As far as the design variables are concerned, according to Figure 9a–e, the tyre pressure (P), the outside radius (d) and the chordwise curve (r_w) illustrate a conflicting relation with wear. For these, the algorithm has provided optimal alternatives ranging from the lower bound to the upper bound, which allowed capturing the conflicting relation between these variables and wear. On the other hand, the Pareto alternatives of tyre width (c) and crown thickness (b_0), as shown in Figure 9b,d, respectively, have not captured their relation with wear, as the algorithm converged close to one value. In the case of the tyre width, this value is different from the upper or lower bound, whereas the crown thickness alternatives have converged close to the lower bound trying to decrease the length of the tread (Figure 3a). However, both values are realistic according to the literature.

Regarding the different road cases, Case 2 alternatives are providing larger values in all the objectives, as the more intense road roughness profile has increased discomfort, vehicle instability and tyre wear. Regarding the latter, according to Figure 8a, Case 2 Pareto alternatives have increased wear around ~25% compared to Case 1, which illustrates how road profiles increase wear significantly if different tyre designs are adopted. In terms of convergence, Case 2 has provided more scattered fronts in both the objective and the design variables fronts, illustrating that the less intense excitation (Case 1 with Class A road roughness) allowed the algorithm to converge to a more solid Pareto front. However, in both cases, the alternatives mostly have captured the same conflicting relations, while the design variables are within the same range of values.

Finally, as mentioned above, common solutions are sought requiring for them to provide similar wear in the two cases (<8% difference) and have close design variables values (<9% difference). This search led to two common solutions between the optimal alternatives of the two cases, as pointed out in Figures 8 and 9 and Table 5. Based on these, in Case 1 where road Class A is considered, the common solutions are closer to the upper edge of the Pareto front, being more comfort oriented in this case, while in Case 2 they are closer to the lower edge, being more stability oriented. Regarding their design variable values (Table 4), they are all the same with only the pressure (P_t) being significantly different. This outcome of the optimisation implies that the optimum tyre design has two pressures (P) with which it can drive optimally over both a road Class A and B S-path with the same levels of wear. The similarity of these two solutions in terms of the overall wear is illustrated in Figure 10, where the circumferential tread profile that occurs with these solutions is compared along with the one that would occur with the initial tyre design, as shown in Table 3. In addition, Figure 10 displays how the optimisation has improved the tread circumferential profile by not only minimising the overall wear, but also smoothening the variations compared to the initial solution.

5.2. Optimum Suspension Design

Regarding Scenario 2 and the suspension optimisation, the Pareto fronts of the different suspension types with regards to the three optimisation objectives are illustrated in Figure 11, while the common solutions identified are pointed out in Figure 11a.

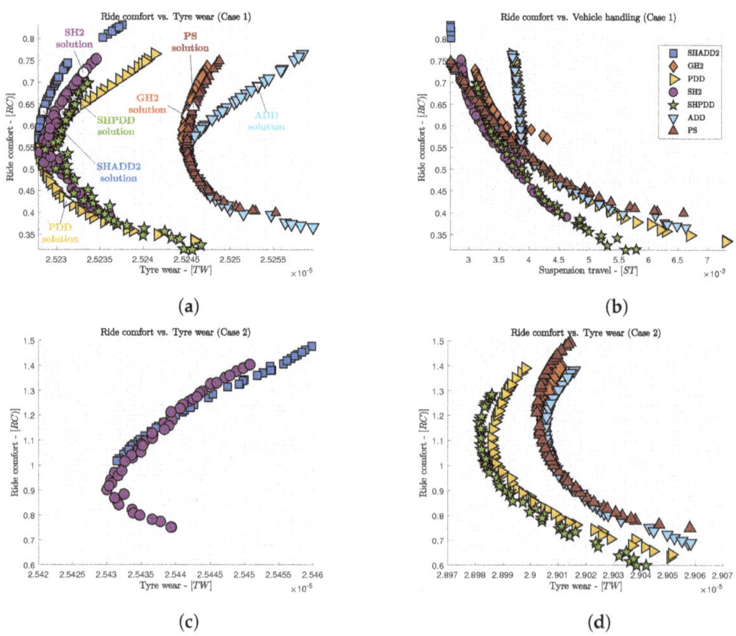

Figure 11. *Cont.*

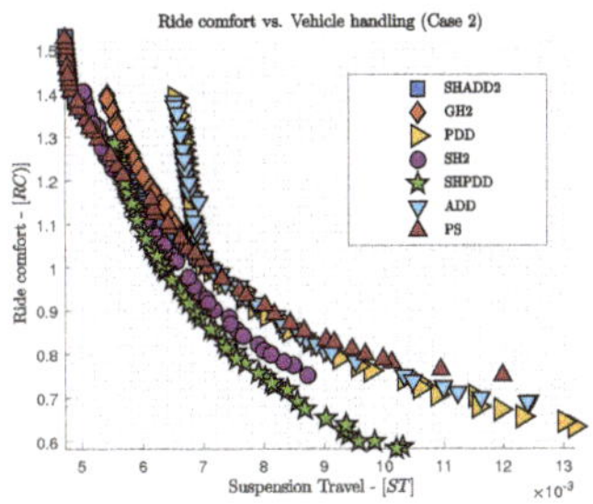

(e)

Figure 11. Comparison of the optimum suspension designs obtained for the two optimisation cases (Class A and B road profiles) for the various suspension types (PS, SH-2, ADD, SH-ADD-2, GH-2, PDD and SH-PDD) in Scenario 2: (**a**) RC vs. TW for Case 1; (**b**) RC vs. ST for Case 1; (**c**) RC vs. TW for Case 2 (first part); (**d**) RC vs. TW for Case 2 (second part); and (**e**) RC vs. ST for Case 2.

Based on the results in Figure 11b,e, the relation between comfort and handling is consistent with what was expected, illustrating the conflict between the two objectives. At the same time, the optimisation algorithm has converged to solutions with similar wear characteristics for most of the suspension types as the wear objective has small differences (Figure 11a,c,d), especially for Case 1. On the one hand, the above outcomes imply that for a suspension system, after having decided the appropriate suspension type and the optimal tyre design with regards to wear, the engineers can focus on the rest of the objectives to tune the suspension optimally. On the other hand, different control algorithms could make a significant difference in terms of the tyre wear performance. According to the Pareto fronts, the increase of the wear from Case 1 (Figure 11a, Class A) to Case 2 (Figure 11d, Class B) is $\sim$13% for most of the semi-active suspensions, illustrating their lack of adaptability in different road profiles. However, SH-2 and SH-ADD-2 (Figure 11c) provide the minimum levels of wear, with insignificant differences between them, and maintained the same tyre wear levels for the two cases.

Regarding the compromise between comfort and vehicle handling, the Pareto fronts of all the suspension types in both cases have similar trends, and they validate the theoretical basis of each control algorithm. More specifically, SH-ADD2 is trying to balance the characteristics of SH-2 and ADD, with its Pareto alternatives being between the SH-2 and ADD Pareto fronts. In this direction, the SH-ADD2 Pareto front connects the upper part of the SH-2 front (the stability oriented) and the lower part of the ADD front (the comfort oriented). Similarly, SH-PDD is trying to balance the characteristics of SH-2 and PDD with its Pareto front alternatives being between those of SH-2 and PDD, while it also connects the upper part of the PDD front (the comfort oriented) and the lower part of the SH-2 front (the stability oriented). At the same time, GH-2 is providing solutions with more vehicle stability but less comfort. Finally, the PS Pareto alternatives in both cases include solutions that could replace few semi-active algorithms (SH-ADD2, PDD and ADD) but with slightly either less comfort or less vehicle handling.

As far as the common solutions are concerned, in most of the suspension types, the common identified solution between the two road cases is one that enhances comfort more, being in the top part of the Pareto font (Figure 11a). Only in PDD, the common solution is located at the bottom part of the Pareto front, as shown in Figure 11a, which is the one that emphasises vehicle handling. The design variables are are illustrated in Table 6. Based on Table 6, for a few of the suspension types (PS, GH-2, PDD and SH-ADD2), the design variables have converged to a soft spring, while the rest (SH-2, ADD and SH-PDD) to a much stiffer one. Regarding, SH-2 and SH-ADD-2, which are the ones that provide similar tyre wear levels for both road profiles, they have converged to different suspension tuning and also the common solution was identified with a high allowed threshold, which could be the reason these two algorithms provide minimum tyre wear for the two road cases.

The first has converged to a stiff spring and damper coefficient for both C_{min} and C_{max}, however the comfort levels are maintained to good levels for both road cases. On the other hand, SH-ADD-2 has converged to a soft suspension design, with spring (K_3) and damping coefficients (C_{min}) close to the lower bounds allowed by the optimisation. However, the vehicle handling is maintained to similar levels with the SH-2 solution.

6. Conclusions

To sum up, this work considers a vehicle model combined with a tyre wear model in order to optimise tyre and suspension parameters to minimise tyre wear, enhance comfort and improve stability. In addition to the search for the optimum tyre and suspension design, the optimisation is also used as the tool to identify any conflicting or not relations between the design variables and the optimisation objectives.

In summary, the following conclusions are extracted.

- Comfort illustrates the same conflicting relation with wear as with vehicle stability. This means that the increase of the suspension travel, hence degradation of handling, leads to an increase in wear but to an improvement in comfort.
- Wear could increase up to 21% in different road profiles, while the appropriate tyre design could provide only 2% increase if the road roughness changes from Class A to B. The appropriate tyre design could be extracted using the optimisation method proposed in the current work.
- For the same tyre design, two pressures could optimally combine comfort, wear and stability in two road roughness cases but also maintain wear at the same levels in these two cases. The recursive feasibility of this outcome should be tested, but it is an interesting remark for the current case study.
- Regarding suspension types, according to the results, the type and their control algorithm should be selected with regards to tyre wear damage. In the current case study, SH-2 and SH-ADD-2 seemed to be able to provide the best wear performance. After the selection of the control algorithm, the tuning of the suspension parameters, i.e., stiffness and damping coefficient, should take place mainly with regards to comfort and vehicle handling. This is because according to the results it seems that different configurations do not affect wear to a great extent.

Further work is in progress to develop active suspension systems using novel control algorithms to decrease wear. At the same time, tyre and suspension design should be investigated in depth with regards to the tyre wear using more detailed wear and vehicle models. Finally, their experimental validation and investigation is an aspect that should be considered further.

Author Contributions: Conceptualisation, G.P., J.J. and L.D.; methodology, G.P.; software, G.P.; validation, G.P., J.J. and L.D.; formal analysis, G.P., J.J. and L.D.; investigation, G.P.; resources, G.P., J.J. and L.D.; data curation, G.P., J.J. and L.D.; writing—original draft preparation, G.P.; writing—review and editing, J.J. and L.D.; visualisation, G.P.; supervision, J.J. and L.D.; project administration, J.J.; and funding acquisition, J.J. and L.D. All authors have read and agreed to the published version of the manuscript.

Funding: This research was funded by TRENoP (Transport Research Environment with Novel Perspectives) at KTH Royal Institute of Technology and the Centre for ECO2 Vehicle Design, which is funded by Vinnova (the Swedish innovation agency) through grant number 2016-05195.

Institutional Review Board Statement: Not applicable.

Informed Consent Statement: Not applicable.

Conflicts of Interest: The authors declare no conflict of interest.

References

1. Sambito, M.; Severino, A.; Freni, G.; Neduzha, L. A systematic review of the hydrological, environmental and durability performance of permeable pavement systems. *Sustainability* **2021**, *13*, 4509. [CrossRef]

2. Jan Kole, P.; Löhr, A.J.; Van Belleghem, F.G.; Ragas, A.M. Wear and tear of tyres: A stealthy source of microplastics in the environment. *Int. J. Environ. Res. Public Health* **2017**, *14*, 1265. [CrossRef]
3. Grigoratos, T.; Gustafsson, M.; Eriksson, O.; Martini, G. Experimental investigation of tread wear and particle emission from tyres with different treadwear marking. *Atmos. Environ.* **2018**, *182*, 200–212. [CrossRef]
4. Wik, A.; Dave, G. Occurrence and effects of tire wear particles in the environment—A critical review and an initial risk assessment. *Environ. Pollut.* **2009**, *157*, 1–11. [CrossRef]
5. Webster, B. Tyres of electric cars add to air pollution, experts warn. *The Times*, 7 December 2020. Avaliable online: https://www.thetimes.co.uk/article/electric-car-tyres-are-growing-source-of-air-pollution-28276fmdp (access on 25 June 2021).
6. Fuller, G. Pollutionwatch: How smart braking could help cut electric car emissions. *The Guardian*, 29 January 2021. Avaliable online: https://www.theguardian.com/environment/2021/jan/29/pollutionwatch-how-smart-braking-could-help-cut-electric-car-emissions (access on 25 June 2021).
7. Knisley, S. A correlation between rolling tire contact friction energy and indoor tread wear. *Tire Sci. Technol. TSTCA* **2002**, *30*, 83–99. [CrossRef]
8. Huang, H.; Chiu, Y.; Wang, C.; Jin, X. Three-dimensional global pattern prediction for tyre tread wear. *Proc. Inst. Mech. Eng. Part D J. Automob. Eng.* **2015**, *229*, 197–213. [CrossRef]
9. Nakajima, Y. *Advanced Tire Mechanics*; Springer: Singapore, 2019; pp. 1–1265. [CrossRef]
10. Maître, O.L.; Süssner, M.; Zarak, C. *Evaluation of Tire Wear Performance*; SAE Technical Papers; SAE International: Warrendale, PA, USA, 1998. [CrossRef]
11. Ma, B.; Xu, H.g.; Chen, Y.; Lin, M.y. Evaluating the tire wear quantity and differences based on vehicle and road coupling method. *Adv. Mech. Eng.* **2017**, *9*, 168781401770006. [CrossRef]
12. da Silva, M.M.; Cunha, R.H.; Neto, A.C. A simplified model for evaluating tire wear during conceptual design. *Int. J. Automot. Technol.* **2012**, *13*, 915–922. [CrossRef]
13. Sueoka, A.; Ryu, T.; Kondou, T.; Togashi, M.; Fujimoto, T. Polygonal Wear of Automobile Tire. *JSME Int. J. Ser. C* **1997**, *40*, 209–217. [CrossRef]
14. Li, Y.; Zuo, S.; Lei, L.; Yang, X.; Wu, X. Analysis of impact factors of tire wear. *J. Vib. Control* **2011**, *18*, 833–840. [CrossRef]
15. Stalnaker, D.; Turner, J.; Parekh, D.; Whittle, B.; Norton, R. Indoor Simulation of Tire Wear: Some Case Studies. *Tire Sci. Technol. TSTCA* **1996**, *24*, 94–118. [CrossRef]
16. Knuth, E.F.; Stalnaker, D.O.; Turner, J.L. *Advances in Indoor Tire Tread Wear Simulation*; SAE Technical Papers; SAE International: Warrendale, PA, USA, 2006. [CrossRef]
17. Lupker, H.; Cheli, F.; Braghin, F.; Gelosa, E.; Keckman, A. Numerical prediction of car tire wear. *Tire Sci. Technol. TSTCA* **2004**, *32*, 164–186. [CrossRef]
18. Braghin, F.; Cheli, F.; Melzi, S.; Resta, F. Tyre wear model: Validation and sensitivity analysis. *Meccanica* **2006**, *41*, 143–156. [CrossRef]
19. Farroni, F.; Sakhnevych, A.; Timpone, F. Physical modelling of tire wear for the analysis of the influence of thermal and frictional effects on vehicle performance. *Proc. Inst. Mech. Eng. Part J. Mater. Des. Appl.* **2017**, *231*, 151–161. [CrossRef]
20. Emami, A.; Khaleghian, S.; Bezek, T.; Taheri, S. Design and development of a new portable test setup to study friction and wear. *Proc. Inst. Mech. Eng. Part J J. Eng. Tribol.* **2020**, *234*, 730–742. [CrossRef]
21. Lepine, J.; Na, X.; Cebon, D. An Empirical Tire-Wear Model for Heavy-Goods Vehicles. *Tire Sci. Technol.* **2021**, *49*. [CrossRef]
22. Yamazaki, S.; Fujikawa, T.; Hasegawa, A.; Ogasawara, S. Indoor Test Procedures for Evaluation of Tire Treadwear and Influence of Suspension Alignment. *Tire Sci. Technol.* **1989**, *17*, 236–273. [CrossRef]
23. Tandy, D.F.; Pascarella, R.J.; Neal, J.W.; Baldwin, J.M.; Rehkopf, J.D. Effect of Tire Wear on Tire Force and Moment Characteristics. *Tire Sci. Technol.* **2010**, *38*, 47–79. [CrossRef]
24. Tamada, R.; Shiraishi, M. Prediction of Uneven Tire Wear Using Wear Progress Simulation. *Tire Sci. Technol.* **2017**, *45*, 87–100. [CrossRef]
25. Koishi, M.; Shida, Z. Multi-Objective Design Problem of Tire Wear and Visualization of Its Pareto Solutions. *Tire Sci. Technol.* **2006**, *34*, 170–194. [CrossRef]
26. Serafinska, A.; Kaliske, M.; Zopf, C.; Graf, W. A multi-objective optimization approach with consideration of fuzzy variables applied to structural tire design. *Comput. Struct.* **2013**, *116*, 7–19. [CrossRef]
27. Papaioannou, G.; Koulocheris, D. An approach for minimizing the number of objective functions in the optimization of vehicle suspension systems. *J. Sound Vib.* **2018**, *435*, 149–169. [CrossRef]
28. Papaioannou, G.; Koulocheris, D. Multi-objective optimization of semi-active suspensions using KEMOGA algorithm. *Eng. Sci. Technol. Int. J.* **2019**, *22*, 1035–1046. [CrossRef]
29. Anderson, J.R.; McPillan, E. Simulation of the Wear and Handling Performance Trade-off by Using Multi-objective Optimization and TameTire. *Tire Sci. Technol.* **2016**, *44*, 280–290. [CrossRef]
30. Karnopp, D.; Crosby, M.J.; Harwood, R.A. Vibration control using semi-active force generators. *J. Manuf. Sci. Eng. Trans.* **1974**, *96*, 619–626. [CrossRef]
31. Savaresi, S.M.; Silani, E.; Bittanti, S. Acceleration-Driven-Damper (ADD): An Optimal Control Algorithm For Comfort-Oriented Semiactive Suspensions. *J. Dyn. Syst. Meas. Control* **2005**, *127*, 218. [CrossRef]

32. Savaresi, S.M.; Spelta, C. A Single-Sensor Control Strategy for Semi-Active Suspensions. *IEEE Trans. Control Syst. Technol.* **2009**, *17*, 143–152. [CrossRef]
33. Morselli, R.; Zanasi, R. Control of port Hamiltonian systems by dissipative devices and its application to improve the semi-active suspension behaviour. *Mechatronics* **2008**, *18*. [CrossRef]
34. Liu, Y.; Zuo, L. Mixed Skyhook and Power-Driven-Damper: A New Low-Jerk Semi-Active Suspension Control Based on Power Flow Analysis. *J. Dyn. Syst. Meas. Control* **2016**, *138*, 081009. [CrossRef]
35. Valasek, M.; Kortum, W.; Sika, Z.; Magdolen, L.; Vaculín, O. Development of semi-active road-friendly truck suspensions. *Control Eng. Pract.* **1998**, *6*, 735–744. [CrossRef]
36. Li, Y.; Zuo, S.; Duan, X.; Guo, X.; Jiang, C. Theory analysis of the steady-state surface temperature on rolling tire. *Proc. Inst. Mech. Eng. Part C J. Mech. Eng. Sci.* **2012**, *226*, 1278–1289. [CrossRef]
37. ISO2631. *Mechanical Vibration and Shock-Evaluation of Human Exposure to Whole-Body Vibration—Part 1: General Requirements*; ISO 2631-1:1997; Technical Report; ISO: Geneva, Switzerland, 1997.
38. ISO8608. *Mechanical Vibration-Road Surface Profiles-Reporting of Measured Data*; Technical Report; ISO: Geneva, Switzerland, 1995.

applied sciences

MDPI

Article

Assessment of Tire Features for Modeling Vehicle Stability in Case of Vertical Road Excitation

Vaidas Lukoševičius *, Rolandas Makaras and Andrius Dargužis

Faculty of Mechanical Engineering and Design, Kaunas University of Technology, Studentų Str. 56, 44249 Kaunas, Lithuania; rolandas.makaras@ktu.lt (R.M.); andrius.darguzis@ktu.lt (A.D.)
* Correspondence: vaidas.lukosevicius@ktu.lt

Abstract: Two trends could be observed in the evolution of road transport. First, with the traffic becoming increasingly intensive, the motor road infrastructure is developed; more advanced, greater quality, and more durable materials are used; and pavement laying and repair techniques are improved continuously. The continued growth in the number of vehicles on the road is accompanied by the ongoing improvement of the vehicle design with the view towards greater vehicle controllability as the key traffic safety factor. The change has covered a series of vehicle systems. The tire structure and materials used are subject to continuous improvements in order to provide the maximum possible grip with the road pavement. New solutions in the improvement of the suspension and driving systems are explored. Nonetheless, inevitable controversies have been encountered, primarily, in the efforts to combine riding comfort and vehicle controllability. Practice shows that these systems perform to a satisfactory degree only on good quality roads, as they have been designed specifically for the latter. This could be the cause of the more complicated car control and accidents on the lower-quality roads. Road ruts and local unevenness that impair car stability and traffic safety are not avoided even on the trunk roads. In this work, we investigated the conditions for directional stability, the influence of road and vehicle parameters on the directional stability of the vehicle, and developed recommendations for the road and vehicle control systems to combine to ensure traffic safety. We have developed a refined dynamic model of vehicle stability that evaluates the influence of tire tread and suspensions. The obtained results allow a more accurate assessment of the impact of the road roughness and vehicle suspension and body movements on vehicle stability and the development of recommendations for the safe movement down the road of known characteristics.

Keywords: directional stability; road profile; road unevenness; vehicle-road interaction; vertical vehicle excitation; tire models; tire tread

Citation: Lukoševičius, V.; Makaras, R.; Dargužis, A. Assessment of Tire Features for Modeling Vehicle Stability in Case of Vertical Road Excitation. *Appl. Sci.* **2021**, *11*, 6608. https://doi.org/10.3390/app11146608

Academic Editors: Flavio Farroni, Andrea Genovese and Aleksandr Sakhnevych

Received: 29 June 2021
Accepted: 16 July 2021
Published: 18 July 2021

1. Introduction

The vehicle-road subsystem includes various aspects of the vehicle-road interaction. This subsystem is emphasized where the technical part of transport operations is investigated. Considerable emphasis has already been placed on the description of the vehicle-road interaction, vehicle movement under various conditions, and substantiation of the requirements for vehicles both in the theoretical studies and in the experimental investigations where the theory has not provided any viable solutions [1–3]. Basic solutions could mistakenly be considered to have been found, with further inquiry allegedly being limited to observations that are necessary to correct the solutions. Nonetheless, certain limitations have been introduced increasingly by a lot of countries in response to the aggressive development of this type of transport. These limitations have an effect both on individual components of the subsystem and on the subsystem itself and promote the improvement of both technical and organizational aspects thereof. The road and vehicle as the elements of the vehicle-road subsystem may be described individually as they are viewed and may be explored as individual elements of the system in an investigation of transport systems [4].

Pavement unevenness values are not completely random. An attempt was made to systematize these values. A method enabling assessment of the unevenness height dispersion and correlation function that has been used for a fairly long time is already more appropriate for the modeling [5–8]. The specifics of unevenness lie in that certain patterns between the unevenness height and unevenness wavelength become established during the pavement laying process and use. Hence, the random process model cannot be applied to the road description, as waviness and ruts are registered in the road pavement, in particular, after the use [9–12].

In the stability tasks, two types of unevenness have the greatest effect: road microprofile that causes excitation of the suspension operation [13,14] and road roughness due to its relation to the grip ratio [15–17]. The vehicle-road interaction is determined by two intermediate links: the tire and the suspension [18,19].

Vehicle stability modeling tasks involve multiple parameters and need to be split into several individual tasks during the analysis. Nonetheless, a more detailed inquiry has shown that the tasks are interrelated, and integrated analysis is needed when dealing with more complex tasks. Analysis of the evolution of trends in traffic accidents and vehicle design has shown that controllability is one of the key operational properties of a vehicle that influences safety. Vehicle controllability is the ability to move following the trajectory designed by the driver [20,21]. The driver sets the desired direction using the steering wheel mechanism, and the controllability tasks are focused on the steering wheel mechanism [22,23].

Deeper inquiry into the potential excitation caused by road unevenness has shown that it is not enough to know the slip characteristics, as the wheel direction vector is influenced by the wheel spatial position, i.e., camber, toe alignment, etc. The dynamic equations cannot be used directly due to the specifics of the tasks and to the fact that the characteristics of individual elements are clearly nonlinear. Semi-empirical formulae are used to describe the effect of the change of the forces acting on the wheel and the spatial position thereof on the direction vector. Dependencies of the link between the slip angle and lateral force, lateral force and relative wheel slippage, slip angle, and stabilizing moment are required in order to be able to perform the modeling. During the modeling, the dependencies are expressed by either linear or higher-degree dependencies [24–30].

The majority of the currently available methodologies employ simplified models. These models individually assess the kinematics of the steering wheel mechanism, suspension kinematics, stiffness of the suspension elements, and forces acting on the vehicle while making a turn [31–35]. This kind of analysis does not enable the identification of the key factors. Hence, the models consisting of multiple elements, the characteristics of which can be known only after specific vehicle design details are available, need to be used for the integrated analysis. This complicates the integrated analysis and generalization in the investigation of directional stability viewed as the criterion for assessment of the technical condition of the vehicle and the condition of the pavement. Hence, the focus was first turned to another driver-vehicle-road system element, i.e., road, prior to designing the integrated directional stability model for the vehicle and performing the analysis. In this context, the road is associated with the effects which may cause vehicle instability, while the purpose of improvement of stability is related to the solution of the vehicle-road interaction.

A moving vehicle is subject to external effects and reacts to the driver's actions (Figure 1). In the investigation of vehicle directional stability, all these factors need to be considered. However, the vehicle model becomes very complex, making the analysis of the effect of individual factors more difficult. The present paper does not account for the driver's actions, environmental impact, aerodynamic effects, and vehicles design features.

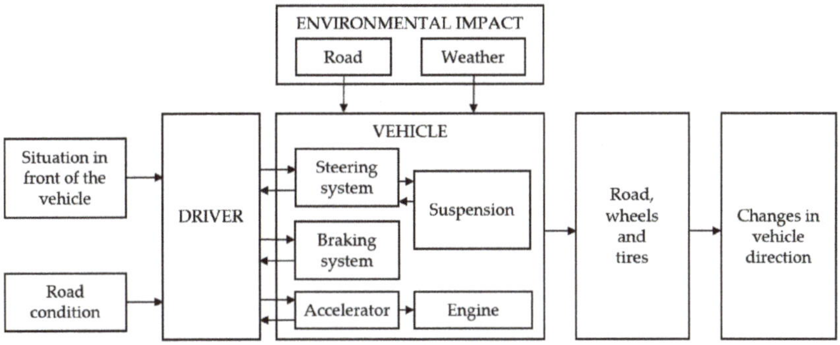

Figure 1. Influence of external factors on vehicle directional stability.

This paper provided a comprehensive review of the key parameters of road-vehicle interaction that have an effect on vehicle stability and design the movement model, consisting of a finite number of parameters, for investigation of stability. The required tire parameters were determined and revised using the theoretical and experimental methods. Following the numerical investigation using the dynamic models, experimental studies were performed to verify the model. The effect of the road parameters on the driving characteristics of the vehicle was investigated. A revised dynamic model of the vehicle directional stability was developed assessing the effect of vehicle suspensions on vehicle stability and traffic safety. The generated results have enabled a more accurate assessment of the influence of road unevenness and vehicle tire and car body vibrations on vehicle stability, and the development of the recommendations on safe movement on the roads using the available characteristics.

Based on the topics discussed above, the main contributions of this paper are as follows: (1) we examine the vehicle stability dynamic models and develop a refined methodology for estimating the key parameters of these models; (2) we produce a revised tire smoothing function model that takes into account tread deflections; (3) we investigate vertical dynamics models and develop recommendations for their use in-vehicle stability studies; (4) we verify the validity of the model of the vertical dynamics using experimental studies; and (5) we establish the vehicle stability model, allowing evaluation of vehicles' technical performance and safe driving conditions of vehicles on roads with different characteristics.

2. Investigation the Tire-Road Interaction

2.1. Investigation of the Tire Properties

When dealing with the vehicle stability tasks, the issue of tire-road interaction is one of the most complex tasks. Analysis of the vehicle wheel-to-road contact requires taking into account the deformation properties of the tire. The complexity of the task stems from the uncertainty of the operating conditions of the tire and the complex structure of the tire. The complexity of the problem is defined by the mere fact that no consensus has been reached on the shape of the tire.

The performed studies of the tire-road interaction [36] confirmed that it would be necessary to consider the deformation properties of the band in the investigation of the compensating function of the band. For this purpose, an additional element is added to the quarter-car model (Figure 2). Additional studies are needed in order to identify the deformation properties of the element.

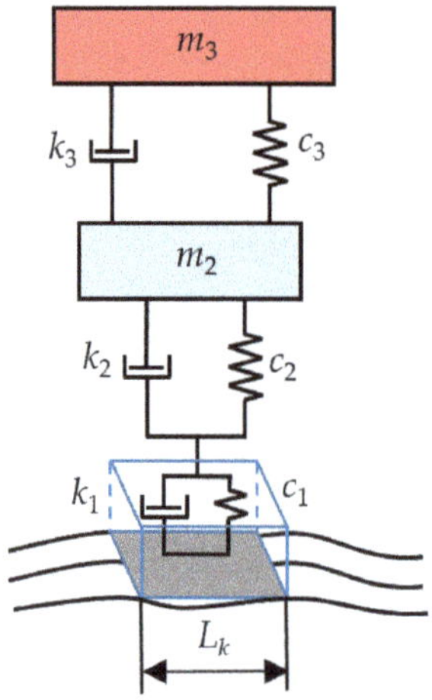

Figure 2. Quarter-car model: L_k—tire-road contact length; m_1, c_1 and k_1—mass, stiffness, and damping ratio of the tire tread part; c_2, k_2—tire stiffness and damping ratio; c_3, k_3—suspension stiffness and damping ratio; m_2—unsprung mass; m_3—sprung mass.

The methodology may be used where deformation of an additional element, i.e., the tread, is needed in the quarter-car model (Figure 2). Whereas both tread layers (with and without the pattern) are not made of composite material, the equations applicable to rubber parts are used here as they consider that ν = 0.5 [37]. The tread model is comprised of two consistently joined layers: the one with a pattern and the one without a pattern. The deformation of the part with a pattern was assessed using three models.

This paper analyzes radial tires of three types that differ considerably in terms of structure: passenger car (175/70R13; 195/50R15), light-duty truck (185/75R14C), and heavy-duty truck (12.00R20) tires. In the assessment of the vehicle tire-road interaction, it is first necessary to describe the deformation properties of the tire under the vertical load. The static load consists of the vehicle and cargo mass. The loaded tire is subject to deformation when forming the contact area with the road pavement. The tire-road contact is depicted in Figure 3, and tire element characteristics are presented in Table 1. The following assumptions were used when determining the area of the contact with the road:

- the contact with the road has an elliptical shape and is limited by the tire width and length of the contact with the road;
- the layers under the tread are sufficiently stiff, and their length does not change during deformation of the tire ($l_d \approx l_0$) (Figure 3a);
- there is the linear dependency between the radial load of the tire and radial deformation Δh;
- the pressure is distributed evenly across the contact surface.

The second and third assumptions enable determining the length of the tire-road contact:

$$L_k/2 = r \cdot \arccos((1 - C_R R_z)/r) \tag{1}$$

where C_R—radial stiffness of the tire and r—free radius of the tire.

More accurate information on the tire element loads is provided by the model 'elastic ring with the outer elastic layer' (Figure 4a,b) [38]. The ring is considered to be the elastic band; the tensile force N_0 caused within it by the inner pressure does not change during deformation; and the ring points are displaced only in the radial direction by the measure w.

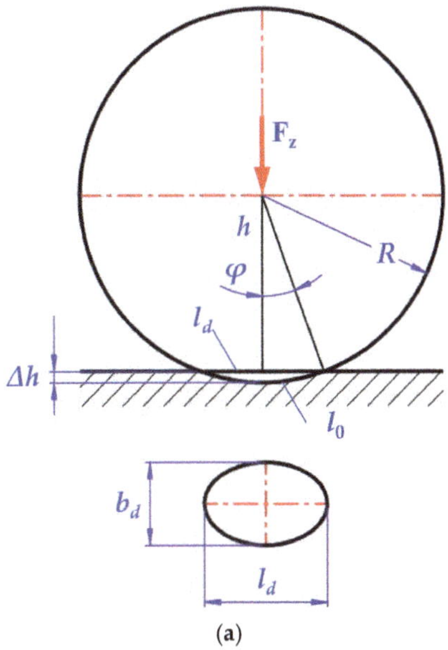

Figure 3. Tire load scheme: (**a**) tire deformation not assessed; (**b**) tire deformation assessed.

Table 1. Data of the tire components.

	Tire	Tire Element	
		Thickness, mm	Fiber Angle
A	175/70R13 Breaker, 3 layers Cord, 1 layer	$\delta_{1a} = 8.0$ $\delta_{1b} = 7.0$ $\delta_2 = 4.1$ $\delta_3 = 1.2$ $\delta_4 = 1.0$	– – $\theta_2 = 20°$ $\theta_3 = 90°$ –
B	195/50R15 Breaker, 2 layers Cord, 2 layers	$\delta_{1a} = 8.1$ $\delta_{1b} = 7.8$ $\delta_2 = 3.9$ $\delta_3 = 1.7$ $\delta_4 = 1.3$	– – $\theta_2 = 20°$ $\theta_3 = 90°$ –
C	185/75R14C Breaker, 2 layers Cord, 2 layers	$\delta_{1a} = 12.0$ $\delta_{1b} = 11.0$ $\delta_2 = 3.2$ $\delta_3 = 1.8$ $\delta_4 = 1.0$	– – $\theta_2 = 20°$ $\theta_3 = 90°$ –
D	12.00R20 Breaker, 4 layers Cord, 6 layers	$\delta_{1a} = 17.0$ $\delta_{1b} = 15.0$ $\delta_2 = 6.4$ $\delta_3 = 7.2$ $\delta_4 = 1.0$	$\theta_2 = 20°$ $\theta_3 = 90°$ –

Note: θ—cord angle in relation to the rolling direction. The indexes of the measures have the following meaning: 1*a*—tread part with a pattern, 1*b*—continuous tread part, 2—breaker, 3—cord, 4—inner sealing liner.

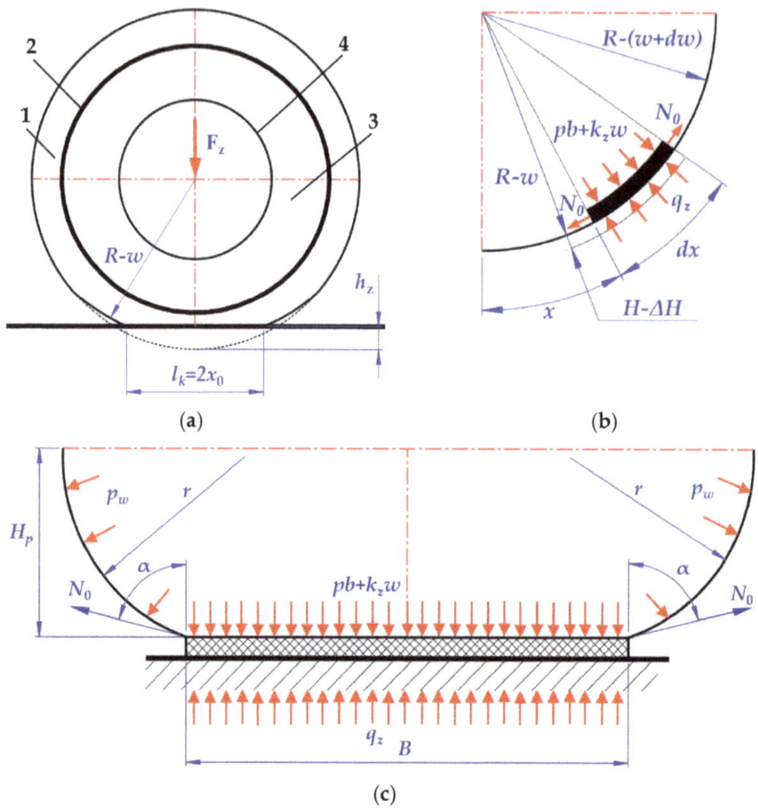

Figure 4. Model of the deformable ring of the tire: (**a**) deformable ring; (**b**) deformable ring element; (**c**) cross-section scheme for assessing the impact of the cord; 1—elastic layer (tread); 2—breaker layers; 3—elastic base (side walls); 4—disk.

Where the impact of the cord threads is assessed, the radial deformation model is used, and the tire reaction is assessed using single pressure $q_z = C_z w$. The sidewalls react only to the displacements in the radial direction, and the reaction per length unit equals $q_s = C_{zs} w$. An elastic outer layer with the thickness equal to H before deformation deforms by ΔH in the contact area. Inelastic forces are not subject to assessment. Tread reaction:

$$q_z = C_{zs} \Delta H0 \quad (2)$$

Arc length x (Figure 4b) is measured from the center of contact. The main measures sought are [38]:

- model deflection h_z assuming that the band does not deform in the radial direction:

$$h_z = (w + \Delta H)_{\backslash x=0} \quad (3)$$

- contact zone length—$2x_0$;
- contact pressure distribution—$q_z(x)$;
- load F_z:

$$F_z = 2\int_0^{x_0} q_z(x)dx \quad (4)$$

As a result of integration and rearrangement of Equation (4), the following is obtained:

$$F_z = 2k_{zs}x_0\left[a_0\left(1-\frac{th\alpha x_0}{\alpha x_0}\right)-a_1x_0^2\left(\frac{1}{3}-\frac{th\alpha x_0}{\alpha x_0}\right)\right] \quad (5)$$

The unknown measures h_z and x_0 are determined using the following dependencies [38]:

$$h_z=\frac{x_0^2}{2R}\left(1+m\frac{2}{\beta x_0}\right),\ \beta=\sqrt{k_z/N_0},\ m=\left(1-\frac{\beta^2}{\alpha^2}\right)\frac{1-(th\alpha x_0/\alpha x_0)}{1+(\beta/a)th\alpha x_0} \quad (6)$$

The described methodology validates the specifics of the tire ring load, namely, that the tire pressure in the contact area is not equal to the internal pressure of the tire due to the breaker and the cord layers. In order to refine the model, the deformation characteristics of the ring consisting of the cord and breaker layers and the tread as a separate deformable element are determined.

2.2. Deformation Properties of the Ring

In order to determine the deformation characteristics of the tire, the tire is divided into elements in the band consisting of the tread, breaker, cord, and protective layers attached by a flexible joint to the sides by the cord-reinforced rubber. The deformation characteristics of the individual tire elements were determined by breaking them down into monolayers made of rubber or rubber reinforced with unidirectional fiber. The elastic characteristics of the monolayer were determined according to the composite mechanic's equations [39–42]:

$$E_{C1}=E_f\phi_f+E_m(1-\phi_f),\ E_{C2}=E_m\left(\left(1-\phi_f\right)+\frac{E_m}{E_{f2}}\phi_f\right)^{-1} \quad (7)$$

$$G_{C12}=G_m\left(\left(1-\phi_f\right)+\frac{G_m}{G_{f12}}\phi_f\right)^{-1},\ \nu_{C12}=\varphi_f\nu_{f12}+\varphi_m\nu_m \quad (8)$$

where E_c and G_c—modulus of elasticity and shear of the composite; E_m and G_m—modulus of elasticity and shear of the matrix; E_f and G_f—the modulus of elasticity and shear of the filler; φ_f and φ_m—volumetric parts of the filler and matrix; ν_C, ν_f, and ν_m—Poisson's coefficients of the composite, filler and matrix; indices 1, 2—the direction along the fiber and the direction transverse to the fiber, respectively.

The properties of the cord and breaker layers in the longitudinal and transverse directions were determined using expressions of the plane stress state for the anisotropic monolayer [39–42]:

$$\begin{Bmatrix}\varepsilon_x\\ \varepsilon_y\\ \gamma_{xy}\end{Bmatrix}=\begin{Bmatrix}\frac{1}{E_x} & -\frac{\nu_{yx}}{E_y} & \frac{\eta_{xy}}{G_{xy}}\\ -\frac{v}{E_x} & \frac{1}{E_y} & \frac{\mu_{xy}}{G_{xy}}\\ \frac{\eta_x}{E_x} & \frac{\mu_y}{E_y} & \frac{1}{G_{xy}}\end{Bmatrix}\begin{Bmatrix}\sigma_x\\ \sigma_y\\ \tau_{xy}\end{Bmatrix} \quad (9)$$

The components of the elasticity matrix were determined according to [39–42] using the axes provided in Figure 5:

$$
\begin{aligned}
E_x(\theta) &= (\tfrac{c^4}{E_1} + \tfrac{s^4}{E_2} + c^2 s^2(\tfrac{1}{G_{12}} - 2\tfrac{\nu_{21}}{E_2}))^{-1}, \\
E_y(\theta) &= (\tfrac{s^4}{E_1} + \tfrac{c^4}{E_2} + c^2 s^2(\tfrac{1}{G_{12}} - 2\tfrac{\nu_{21}}{E_2}))^{-1}, \\
G_{xy}(\theta) &= (4c^2 s^2(\tfrac{1}{E_1} + \tfrac{1}{E_2} + 2\tfrac{\nu_{21}}{E_2}) + \tfrac{(c^2-s^2)^2}{G_{12}})^{-1}, \\
\tfrac{\nu_{yx}}{E_y}(\theta) &= \tfrac{\nu_{21}}{E_2}(c^4 - s^4) - c^2 s^2(\tfrac{1}{E_1} + \tfrac{1}{E_2} - \tfrac{1}{G_{12}}), \\
\tfrac{\eta_{xy}}{G_{xy}}(\theta) &= 2cs\{\tfrac{s^2}{E_2} + \tfrac{c^2}{E_1} - (c^2 - s^2)(\tfrac{\nu_{21}}{E_2} - \tfrac{1}{2G_{12}})\}, \\
\tfrac{\mu_{xy}}{G_{xy}}(\theta) &= 2cs\{\tfrac{c^2}{E_2} + \tfrac{s^2}{E_1} + (c^2 - s^2)(\tfrac{\nu_{21}}{E_2} - \tfrac{1}{2G_{12}})\}, \\
\eta_x &= \tfrac{\eta_{xy} E_x}{G_{xy}},\ \mu_y = \tfrac{\mu_{xy} E_y}{G_{xy}}
\end{aligned}
\quad (10)
$$

where $c = \cos\theta$ and $s = \sin\theta$; x-direction forming an angle θ with the reinforcement direction (Figure 5).

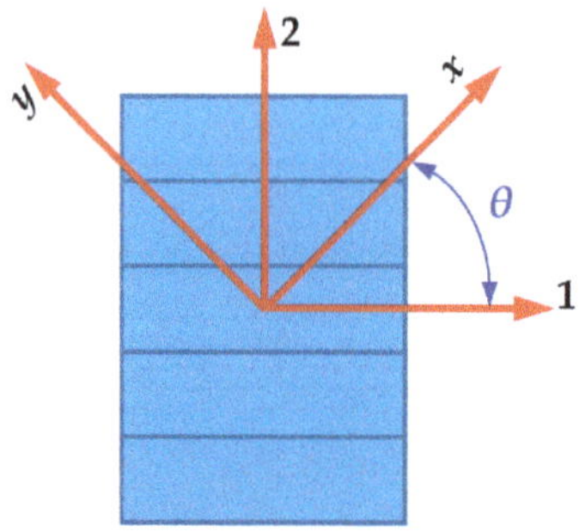

Figure 5. Layered composite element: 1-the direction along the fiber; 2-the direction transverse to the fiber.

The modulus of elasticity of the components and the Poisson's coefficients were taken from the works [36,43]. The methodology used for this purpose is presented in [39–42]. Direct modulus of elasticity calculations does not provide an answer in relation to the characteristics of the tread and unreinforced ring section that are required for the model provided in Figure 4. To determine the stiffness of the tread C_z, tread deformations are analyzed additionally. The obtained data are presented in Table 2.

2.3. *Investigation of Tread Deformation*

Tread deformations were investigated using several models: porous composite and rubber prism models. The research studies analyzing the pneumatic tires specify fairly wide limits of the modulus of elasticity of rubber E_r = 2 … 20 MPa. In the study [36], a simplified model of a passenger car tire was used, E_r = 18 MPa, in order to obtain adequate deformation characteristics of the whole tire. In order to specify the initial characteristics, an experiment was performed to determine the 313 mm wide section of the tread band of the truck tire corresponding to 12.00R20 tire treads. The height of the tread part with a pattern was δ_{1a} = 17.8 mm, without pattern δ_{1b} = 3.9 mm. At compression, the tread was under the action of the load corresponding to the nominal radial load R_{znom} of the 12.00R20 type tires in terms of the pressure. An experimental setup for monotonous tension compression tests consisted of a 50 kN testing machine Instron (Norwood, MA, USA) 8801 series and an electronic device that was designed to record the stress-strain curves. The mechanical characteristics were measured with an error not exceeding ±1% of the deformation scale. The data collection was performed by using the Bluehill Universal Software. Compression tests were performed in accordance with ASTM D575-91: Standard Test Methods for Rubber Properties in Compression.

The obtained load-deformation dependence (Figure 6) is very close to the linear (correlation coefficient r_{xy} = 0.947), proving the applicability of the description of the linear elastic material to the study of tread deformations. The obtained modulus of tread elasticity determined on the basis of the experimental results was 7.84 MPa, taking into consideration the tread part with a pattern according to the filling factor, E_r = 6.16 MPa, which was much lower than in the study [36].

Table 2. Modulus of elasticity of the tire elements, MPa.

Characteristics		Tire			
		A	B	C	D
Breaker (with nylon)	E_l	$1667 \cdot 10^3$	$1667 \cdot 10^3$	—	$1367 \cdot 10^3$
	E_t	26,736	26.736	—	24,568
	E_r	26,736	26.736	—	24,568
Breaker (with steel)	E_l	73,119	73,119	73,119	73,119
	E_t	21,737	21,737	21,737	21,737
	E_r	23,684	23,684	23,684	23,684
Cord (with viscose)	E_l	24,625	24,625	26,819	—
	E_t	$3553 \cdot 10^3$	$3553 \cdot 10^3$	$4339 \cdot 10^3$	—
	E_r	24,625	24,625	26,819	—
Inner sealing layer E		18	18	18	18
Part of the ring up to the cord E		16,601	16,630	17,118	16,511
Whole ring E		16,890	16,906	17,311	16,541
Sidewall elasticity modulus during radial tension E,		507.5	531.15	788.1	551.7
Equivalent sidewall elasticity modulus during radial bending E flat band,		23.32	24.37	29.77	67.89
ring.		42.69	43.04	61.51	113.4
Equivalent modulus of elasticity of the band bending E_c flat band in the circular direction,		15.89	17.74	19.63	15.92
perpendicular to the circular direction ring,		15.35	17.37	16.34	15.90
in the circular direction.		23.73	24.21	22.50	18.50
Relative position of the neutral layer $e/\sum\delta$ at bending flat band in the circular direction,		0.288	0.281	0.383	0.377
perpendicular to the circular direction ring,		0.108	0.11	0.088	0.124
in the circular direction.		0.286	0.277	0.376	0.372

Note: E—in all directions, E_l—the rolling direction of the tire, E_t—perpendicular to the rolling direction of the tire, and E_r—in the radial direction (in relation to the tire).

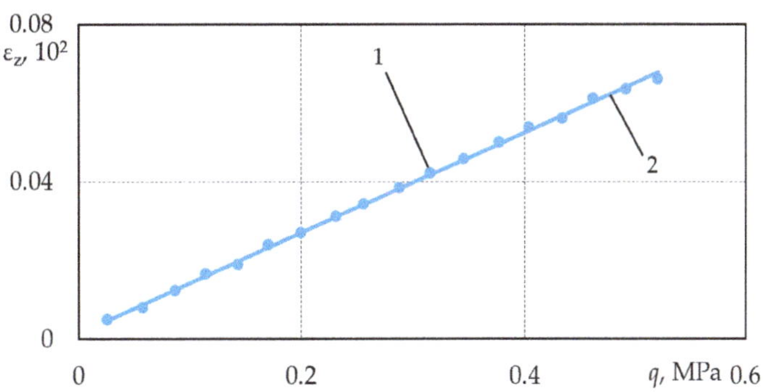

Figure 6. Summary experimental data: 1—experimental data; 2—generalized line.

Whereas the tread consists of the parts with and without a pattern and the deformation properties of the tread are evaluated in the model as a single generalized layer, the possibilities of assessment of the tread part with a pattern were examined. The simplest model assesses the filling of the part with a pattern (Figure 7), which can be executed using the expression of porous composites [44]:

$$E = k_u E_g \tag{11}$$

where k_u—tread filling factor, where $k_u = A_i/A_p$, A_i—the area of the continuous pattern in the part of the tread part selected according to the repetition of the pattern, A_p—area of the selected part of the tread; and E_g—tread rubber modulus of elasticity.

(**a**) (**b**) (**c**)

Figure 7. Tire tread patterns: (**a**) for a passenger car; (**b**) for a light-duty truck; and (**c**) for a truck.

Since the tread pattern is different across differed parts of the tread (Figure 7), the filling factor for each tread band was calculated separately. In this case, the equivalent modulus of elasticity of the tread was equal to:

$$E_1 = E_g \frac{\Sigma k_{ui} k_{bi}}{1 - k_a(1 - \Sigma k_{ui} k_{bi})} \tag{12}$$

where E_g—tread rubber modulus of elasticity; $k_a = \delta_{1a}/\delta_1$; $\delta_1 = \delta_{1a} + \delta_{1b}$; k_{bi}—share of the width of the i-th band in the width of the tread, δ_1—tire tread thickness, δ_{1a}—tread part with a pattern, and δ_{1b}—continuous part of the tread.

The results obtained for different tires are presented in Table 3. The modulus of elasticity of the whole tread was calculated using the equation in [44]:

$$\frac{1}{E_1} = \left(\frac{\delta_{1a}}{E_{1a}} + \frac{\delta_{1b}}{E_{1b}}\right) \cdot \frac{1}{\delta_1} \tag{13}$$

Table 3. Tread modulus of elasticity obtained by assessment using the first model, MPa.

Tire Type	Part with a Pattern E_{1a},	Continuous Part E_{1b},	Tread E_1,
A	5.72	7.84	6.16
B	5.88	7.84	6.22
C	7.08	7.84	6.51
D	5.33	7.84	5.37

This model does not accurately reflect the operating conditions of the tread. Deformation of the tread part with a pattern may be restricted by the contact with the road or a continuous breaker. Moreover, the deformation of the tread part with a pattern does not fully correspond to the ideal compression conditions, as the deformation of the ends at compression leads to a change in the element shape. The course of deformation would be more accurately estimated using the model of a prism subject for compression by applying a rubber-shock-absorber calculation methodology.

The deformations of the tread part with a pattern can be assessed more precisely using the calculation methods applicable to the rubber shock absorbers. Under this approach, the tread element is depicted as a rubber prism, deformed with more or less constrained end deformations depending on the conditions. If the deformations of the both ends of the prism-shaped rubber shock absorber are constrained, the relationship between the load and the deformation is expressed by the equation:

$$R_z = \frac{1}{3}\beta EA\left(\frac{1}{\lambda^2} - \lambda\right),\ \lambda = 1 - \frac{\Delta\delta}{\delta},\ \beta = \frac{4}{3}\left(1 + \eta^2 - \eta\right)\left(1 - \frac{2}{\alpha\delta}th\frac{\alpha\delta}{2}\right)^{-1} \quad (14)$$

where A—prism base area; δ—prism height; $\Delta\delta$—prism deformation; and α and η—shape change parameters. Whereas α and η are interrelated, a system of equations derived from the energy minimum condition is proposed in the study [45]:

$$\begin{gathered} \alpha^2 = \frac{1}{b^2}\frac{48(1+\eta^2-\eta)}{\eta^2(\frac{a}{b})+(1-\eta)^2}, \\ \frac{\alpha\delta-2th\frac{\alpha\delta}{2}}{\alpha\delta(th^2\frac{\alpha\delta}{2}-1)+2th\frac{\alpha\delta}{2}} = \frac{1}{2} + \frac{1+\eta^2-\eta}{1-2\eta}\frac{\eta(\frac{a}{b})^2+\eta-1}{\eta^2(\frac{a}{b})^2+(1-\eta)^2} \end{gathered} \quad (15)$$

where a, b—dimensions of the prism base.

Equations (14) and (15) were solved by approximation.

This model of the tread part with a pattern was applied to two cases: with the end deformations not constrained and with the end deformations completely constrained. If the deformations of the shock absorber ends were not constrained, β = 1. In case the deformations were constrained, the stiffness of the individual bands of the tread part with a pattern would increase by up to 4.5 times (Table 4).

Table 4. The values of the stiffening coefficient of the tread elements for individual tread bands.

Tire Type	Band Number	Length a, mm	Width b, mm	β Band
A	1	39	29	3.83
	2	32	29.6	3.67
B	1	37	29	3.75
	2	31	29	3.67
	3	239	24	4.58
C	1	33	20	2.62
	2	36	18	2.08
D	1	116	42	3.33
	2	90	41	3.17

Under actual conditions, the constraints of the end deformations of the tread part with a pattern were not absolute. In relation to the contact with the road, they are determined by the tire-road friction, while the other end is in contact with the tread part without a pattern. Deformations of the tread part without a pattern are restricted by the breaker layers that have the anisotropic properties (Table 2), and the modulus of elasticity of the layer is only 1.2–1.5 times higher than the modulus of elasticity of rubber in the direction transverse to the reinforcement. Under these conditions, the deformation conditions of the tread part with a pattern are closer to those of a shock absorber with freely deforming ends. This option was used for further calculations.

Both models offer different characteristics: the modulus of elasticity E_1 or deformation $\Delta\delta$ corresponding to the radial load R_z value. For comparison of the results, the generalized characteristic describing the relationship between load and deformation—stiffness C_1 was used:

$$\Delta\delta = C_1 \cdot R_z \tag{16}$$

where C_1—tread stiffness and R_z—radial load of the tire.

When determining the measure C_1 according to the first or second models, it should be taken into the account that the road contact area and the volume of the deformable tread part depend on the load. The following assumptions were used when determining the area of the contact with the road:

- the contact with the road has an elliptical shape and is limited by the tire width and length of the contact with the road;
- the layers under the tread are sufficiently stiff, and their length does not change during deformation of the tire;
- there is the linear dependency between the radial deformation and radial load;
- the pressure is distributed evenly across the contact surface.

The provided assumptions enable determination of the pressure in the contact area. Nonlinear $q(R_z)$ dependence was obtained for all the tested tires (Figure 8). The tire load is calculated using the nominal load of the tire R_{znom}, which is specified in the catalogs, by setting the limits R_{zmin} = 0.2R_{znom} and R_{zmax} = 1.4R_{znom}. The both models were compared to the first model by calculating the following:

$$\Delta\delta = q \cdot \delta / E_1 \tag{17}$$

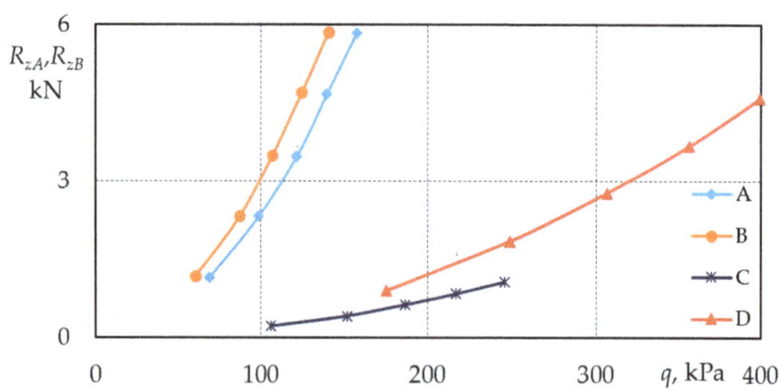

Figure 8. Radial loads of the tires (marking according to Table 2).

Nonlinear dependence was observed $C_{1a}(R_z)$ and was related to nonlinear dependence between R_z and the length of the contact with the road. This generated the nonlinear $q(R_z)$ dependence that reduced tread deformation in the area of higher loads. The specifics of the tire-road contact are explained by the fact that the type of tire $C_{1a}(R_z)$ has no significant effect on the nature of the dependence (Figure 9).

The influence of tread wear on its stiffness was determined using the second model that provides more accurate reflection of the deformation conditions of short rubber prisms [36]. The effect of prism end constraint was not assessed when calculating tread stiffness. For the calculation of the tread stiffness of type A tire, the tread height was 8.1 mm for a new tire and 1.6 mm for a worn tire, and an intermediate wear level (4.9 mm) was provided. The tread wear changed the C-R_z dependence slightly (Figure 10) but had a significant effect on the stiffness value. The effect was not monotonic as the stiffness would be subject to particular change at the beginning of the wear. For the tested tires, tread wear could increase the stiffness of the tread in relation to the nominal load by up to 1.3 times.

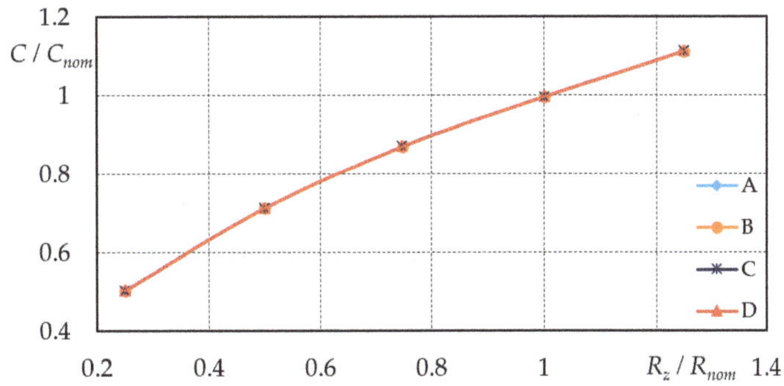

Figure 9. Dependence of relative stiffness of the tread part with a pattern on the relative radial load for different tires (compressed prism model; marking according to Table 2).

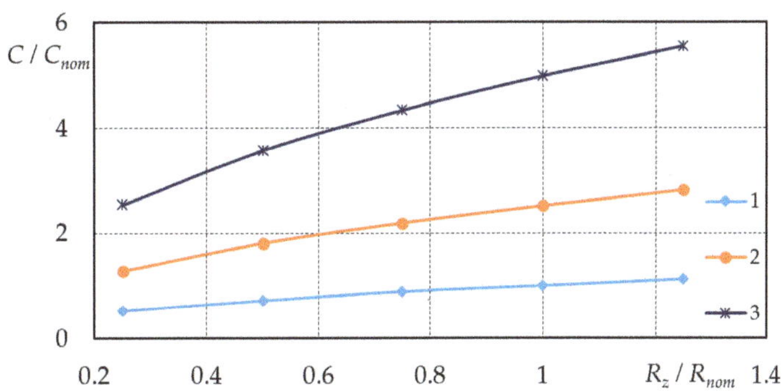

Figure 10. Dependence of stiffness of *B* tire tread part with a pattern on the tread height: 1—new tire; 2—worn by 50%; and 3—worn out.

Depending on the constraint conditions, the second model provided very different results. For more accurate assessment, the effect of the latter constraint, given that the properties of the ring consisting of the cord and breaker are anisotropic, and the influence of deformation characteristics of the tire ring consisting of the tread, breaker, cord and protective layer, on the tread properties were determined, and the road-to-tread contact model was adjusted.

The tape for Nokian passenger car tires 175/70R13 82T was modeled. Its more detailed structure is provided in Table 5.

Table 5. Layer distribution of 175/70R13 82T tire bands.

Layer No.	Layer Designation	Thickness, mm
1	Tread with a pattern	2.9
2	Tread without a pattern	2.1
3	Nylon-rubber	0.8
4	Steel-rubber	0.9
5	Rubber	0.9
6	Steel-rubber	0.9
7	Viscose-rubber	0.9
8	Sealing layer	0.9

The properties of the cord and breaker layers in the longitudinal and transverse directions were determined using expressions of the plane stress state for the anisotropic monolayer [39–42] by applying them to the axes presented in Figure 5:

$$\left\{\begin{array}{c}\varepsilon_x\\ \varepsilon_y\\ \gamma_{xy}\end{array}\right\} = [T']\left\{\begin{array}{ccc}\frac{1}{E_1} & -\frac{\nu_{21}}{E_2} & 0\\ -\frac{\nu_{21}}{E_2} & \frac{1}{E_2} & 0\\ 0 & 0 & \frac{1}{G_{12}}\end{array}\right\}[T]\left\{\begin{array}{c}\sigma_x\\ \sigma_y\\ \tau_{xy}\end{array}\right\}$$
$$[T'] = \left[\begin{array}{ccc}c^2 & s^2 & cs\\ s^2 & c^2 & -cs\\ -2cs & 2cs & (c^2-s^2)\end{array}\right], [T] = \left[\begin{array}{ccc}c^2 & s^2 & -2sc\\ s^2 & c^2 & -2sc\\ sc & -sc & (c^2-s^2)\end{array}\right] \quad (18)$$

where $c = \cos\theta$ and $s = \sin\theta$.

Deformation properties of the layers were determined according to the methodology set out in Section 2.2. An assumption was made that nylon and viscose fibers were anisotropic, with a different modulus of longitudinal elasticity and transverse to the fiber.

The issue of identification of the layers emerged in the investigation of the structure of the modeled tire band. There is a rubber layer between the individual bands of the breaker reinforced with steel wire. As a result, the band was investigated additionally by identifying the rubber layer between the reinforced layers (Figure 11a; Table 5) and using simplified separation into two layers (Figure 11b).

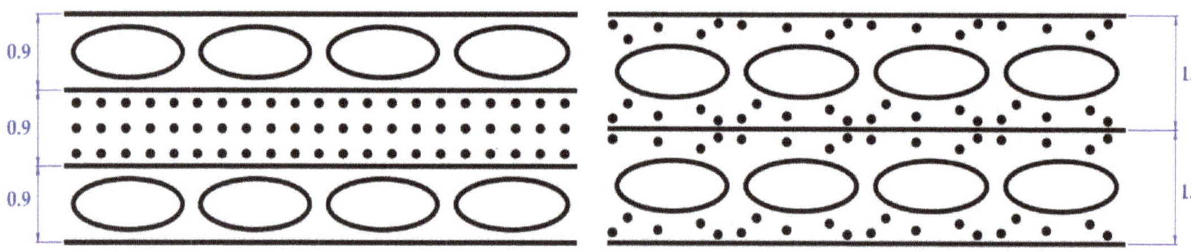

(a) (b)

Figure 11. Breaker layer calculation schemes: (**a**) by identifying the intermediate rubber layer and (**b**) without identifying the intermediate rubber layer.

In the investigation of the deformation properties of the whole band, the model used in layered composites was applied initially. The calculated modulus of elasticity of the band as a composite was calculated by taking into consideration the modulus of elasticity of the layers joined in parallel (X and Y directions) or consistently (Z direction).

Since the reinforcement angles of the adjacent breaker layers differed by the sign only (+20° and −20°), it was assumed that the resulting shear forces in the layers were counterbalanced and were not transmitted to other layers.

Considering the large difference between the modulus of elasticity of the reinforcing fiber and the rubber, it was determined that the modulus of elasticity of the layer in the direction of reinforcement was not significantly affected by the modulus of elasticity of the rubber. The method of identification of the layers (a and b) also did not have any effect in this case. At the same time, it was found that the model of elasticity of the layer would be more dependent on the modulus of elasticity of the rubber in the breaker fiber direction X and Z. The modulus of elasticity in X direction was less dependent on the total modulus of elasticity E. By changing the modulus of elasticity of the rubber in the range of 2 … 20 MPa, E_x differed by 3.7% for model (a) and 3.5% for model (b). E_z dependence was particularly evident: rubber stiffness E_r would change 10 times, while E_z would change 9.97 times (Figure 12). Considering that the deformation properties of the band in Z direction differed considerably from the data provided by other authors, the assumption was that stiffer cord and breaker layers may restrict tread deformations, and in the state of constrained deformation, tread deformation could be significantly influenced by the Poisson's ratio close to 0.5. Therefore, tread deformation under compression was investigated additionally under the assumptions that reflected the interaction between the individual layers in the identified tire band more accurately.

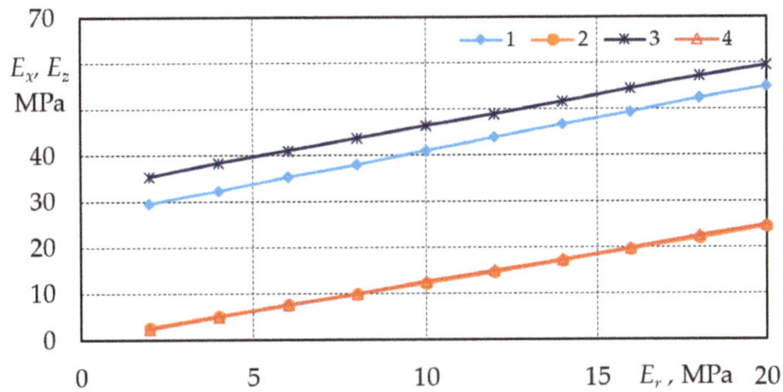

Figure 12. Tire band elastic modulus in X and Z directions: 1—model a (8 layers) in X direction, 2—model a in Z direction, 3—model b (7 layers) in X direction, and 4—model b in Z direction.

Two cases were examined for this purpose. In the first case, it was assumed that by deforming the band in Z direction, deformation of all layers was the same in X and Y directions:

$$\varepsilon_{x1} = \varepsilon_{x2} = \ldots = \varepsilon_{xn}, \ \varepsilon_{y1} = \varepsilon_{y2} = \ldots = \varepsilon_{yn} \quad (19)$$

The sum of the resulting single axial forces in the individual layers in the X and Y axis directions equaled zero:

$$\begin{gathered} \delta_1\sigma_{x1} + \delta_{x2}\sigma_{x2} + \ldots + \delta_n\sigma_{xn} = 0, \\ \delta_1\sigma_{y1} + \delta_{y2}\sigma_{y2} + \ldots + \delta_n\sigma_{yn} = 0, \\ \sigma_{z1} = \sigma_{z2} = \ldots = \sigma_{zn} = \sigma_z = q. \end{gathered} \quad (20)$$

The numerical experiment showed that the modulus of elasticity of rubber in Z direction would increase greatly and result in an increase in the modulus of elasticity of the entire band in Z direction (Figure 13).

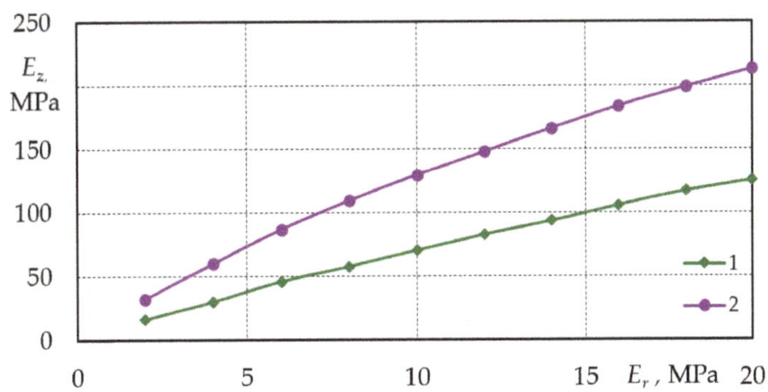

Figure 13. Dependence of the tire band modulus of elasticity on the modulus of elasticity of rubber (σ_z = 0.2 MPa): 1–7 layers and 2–8 layers.

The numerical experiment confirmed that the models predicting uniform layer deformation were particularly sensitive to components with the Poisson's ratio close to 0.5. The conditional modulus of elasticity of the rubber with the layer structure corresponding to the 175/70R1382T tire, depending on the layer, was 86.3 MPa. For this reason, the model with the identified intermediate layer of rubber predicted an unrealistically high stiffness.

In the second case, the tread-to-road friction was assessed, and an assumption was made that the tread deformation would be constrained as long as the shear force developing in directions *X* and *Y* did not exceed the frictional force. Whereas the control experiment was anticipated, the numerical model assumed that the frictional forces would also act on the other side of the band, i.e., in contact with the sealing liner. The constraint provided that the stresses in *X* and *Y* directions could not exceed the limit values:

$$\sigma_{xr} = \sigma_{yr} = \mu\sigma_z \quad (21)$$

where μ—coefficient of friction between rubber and steel.

The numerical experiment confirmed that friction had an impact on the E_z measure (Figure 14). As the coefficient of friction increased, the stiffness of the band in the transverse direction would increase rapidly. The identified additional rubber layers had a smaller influence in this case (Figure 14).

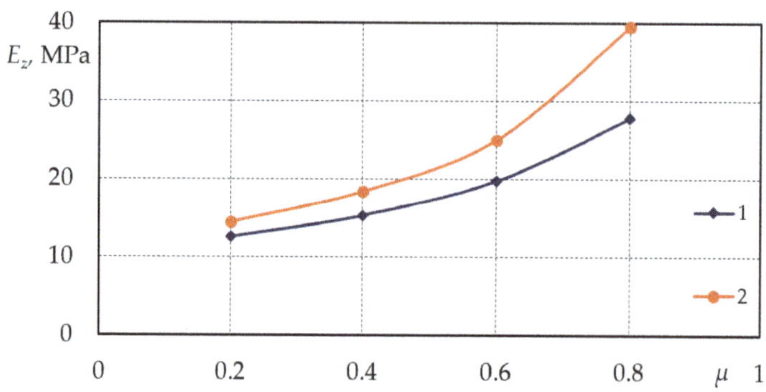

Figure 14. Dependence of tire band deformation on the coefficient of friction (σ_z = 0.2 MPa, E_r = 6 MPa) using different breaker models: 1—7 layers and 2—8 layers.

To revise the results obtained, it was attempted to deform the band cut from the tire in the *Z* direction. The experiment was performed using the 175/70R1382T Nokian tire band used for modeling. The summarized experimental data are presented in Figure 15.

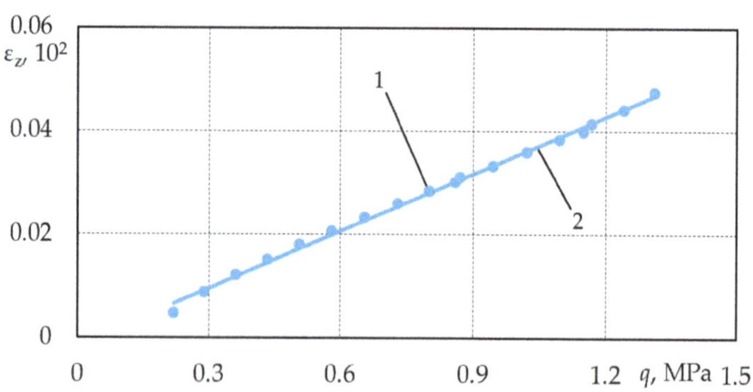

Figure 15. Summary experimental data: 1—experimental data and 2—generalized line.

The experiment confirmed that the linear relationship between pressure and deformation remained up to the pressures above the operating pressure range (tested up to q = 1.5 MPa). The calculated modulus of elasticity of the whole band was 27 MPa, which corresponded to the numerical model of the simplified model structure (7 layers) assessing the friction between the plates and the band.

In the simplified tire models, the cord was assessed during the calculation of radial loads. Therefore, the breaker layers could also be considered as conditionally involved in the radial deformation of the tread. Thereby, the calculation expressions of the layered composites with sequentially arranged layers would result in the following:

$$\frac{1}{C^*_{prz}} = \frac{\delta_1}{C^*_{1z}} + \frac{\delta_2}{C^*_{2z}} + \frac{\delta_3}{C^*_{3z}} \tag{22}$$

where C_i^*—member of the stiffness matrix accounting for the deformation constraint.

If the stiffness of the band used in the experiment was modeled with a layered composite with sequentially arranged layers, its stiffness could be estimated by Equation (22). In this case, the member of the stiffness matrix of the tested tire tread E_z^* would be equal to 27 MPa. This value was used in the research for determination of the C_z measure in Equation (22).

For real tread modeling, tread loads and operating conditions need to be adjusted by bringing the numerical model closer to real conditions. The third tread deformation model was used to evaluate the specifics of the tire-road texture interaction. Road texture means unevenness of the road pavement. For asphalt pavement, the height of the unevenness was 1–5 mm. The irregularities were assumed to be hemispherical protrusions of regular spacing (Figure 16). This assumption enabled modeling of the interaction of an individual unevenness with the tread, assuming that the unevenness acted on the continuous part of the tread. The coefficient of filling with spheres necessary for the conversion of the reference pressure was equal to $\pi/4$ = 0.785. Based on the geometrical conditions, it was determined that the gaps between the spheres would be filled at the sphere penetration into the tread equal to: Δ = 0.455r, where r—radius of the sphere. The known spheres used for the investigation of interaction were the solution for the elastic deformations of the plane contact. The interaction scheme is shown in Figure 17.

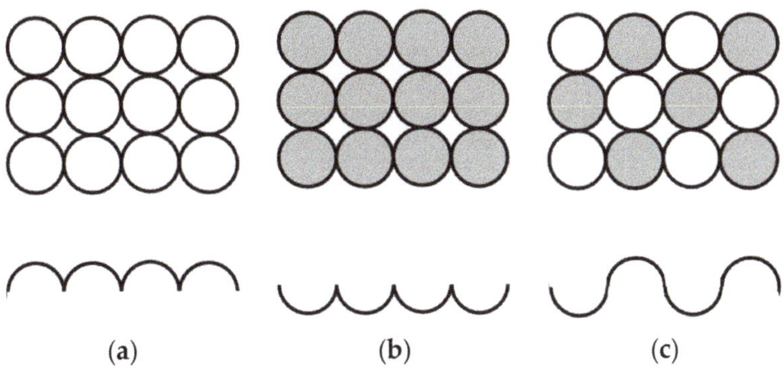

Figure 16. Road unevenness schemes: (**a**)—convexities; (**b**)—concavities; (**c**)—combination of convexities and concavities.

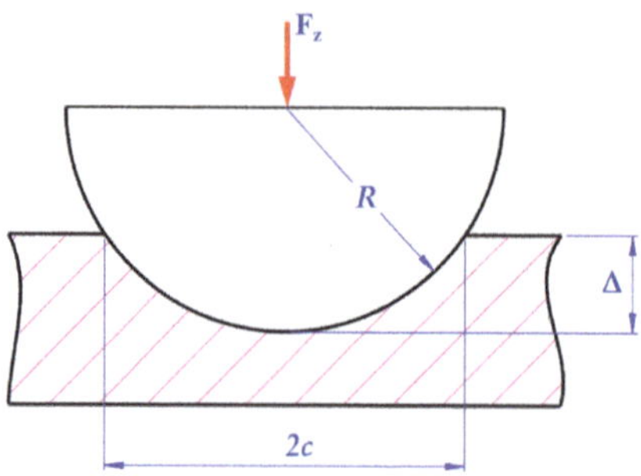

Figure 17. Unevenness-to-tread interaction scheme.

The hemisphere-shaped unevenness with radius r pressed against the tread located on a rigid base with force F. As a result, the unevenness penetrated into the tread to depth Δ. The radius of the depression was equal to c. The elastic solution for the task was known and taken from [45]. The depression radius c was equal to:

$$c = 0.721 \cdot \sqrt[3]{2 \cdot F \cdot r \cdot (k_1 + k_2)} \tag{23}$$

where $k_i = 1 - v_i^2/E_i$; E_i,ν_i—the modulus of elasticity and Poisson's ratio of the spherical and planar material (E_1 = 18 MPa, v_1 = 0.5, E_2 = 6 MPa, v_1 = 0.3) [36].

Average contact pressure:

$$q_0 = 0.918 \cdot \sqrt[3]{F/(2 \cdot r \cdot (k_1 + k_2))}^{-2} \tag{24}$$

F measure was to make sure that pressure q acting at the tire-road when the tire was subjected to force R_z would be equal to the mean pressure at the sphere-to-tread contact q_0, taking into account the coefficient of filling with spheres at the contact. Considering that, according to the elastic solution:

$$\Delta = 0.8255 \cdot \sqrt[3]{\frac{F^2}{r}(k_1 + k_2)^2},\ q_0 = \frac{F}{\pi \cdot c^2} \tag{25}$$

and after introduction of the condition $q_0 = q$, the result would be:

$$\Delta = 5.552 q^2 r (k_1 + k_2)^2 \tag{26}$$

The tread stiffness values obtained for the individual models are presented in Figure 18.

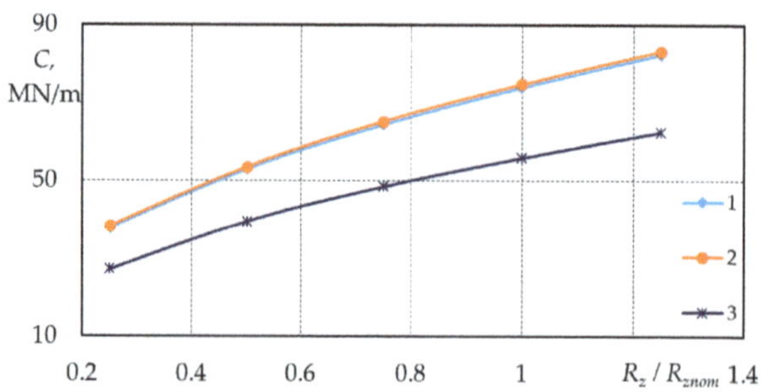

Figure 18. Dependence of the stiffness of the patterned part of tire 175/70R13 on the relative radial load using different models: 1—porous composite model, 2—compressible prism model, and 3—road-tire contact model.

The calculations confirmed that in the case of 5 mm high unevenness, condition $\Delta < 0.455r$ (the tread does not fill the entire volume between the unevennesses) would be met for the pressures at the tire-road contact. This result demonstrates that the gaps between the road unevenness elements of this shape and size are not completely filled by the tread and enable better adhesion to the road and removal of moisture from the tire-road contact area. Road texture interaction model (r = 5 mm) indicates lower stiffness values; however, dependence CR_z remains similar and is related to nonlinear qR_z dependence. In addition, it was confirmed that a simplified model estimating the modulus of elasticity of the tread band could be used for the assessment of the tread deformation. The modulus of elasticity of the tread material should be determined by assessing the deformation constraint.

3. Vehicle Excitation by Road Unevenness

Vehicle stability was assessed using several criteria. One of them was the stability of the linear motion of the steered wheels measured at a critical speed, above which the wheels started to vibrate about the vertical axis. The cause of the vibrations was the suspension vibrations. Therefore, the issue of car stability would be analyzed starting with the formation of the differential equations for a system consisting of a steering wheel mechanism, suspension, and wheels. At the initial moment, the number of elements comprising the system was reduced, i.e., the body vibrations were not taken into account. Body vibrations were not evaluated because its free vibration frequency was significantly lower than the suspension and wheel vibration frequency.

3.1. Suspension Models

When considering road excitation, it was necessary to specify which of the car's vertical motion models best met the requirements for stability models. Three vertical dynamics models are known: (1) quarter-car, (2) flat (2D), and (3) spatial (3D), examined in the studies [46–57]. The 2D model best describes the longitudinal vehicle dynamics, i.e., braking, acceleration, vertical obstacle clearance, etc., and will not be used in future investigations.

The simplest model used for road-vehicle interaction analysis is the quarter-car model. It consists of three masses: tire element m_1, unsprung masses (suspensions) m_2, and sprung masses (body) m_3 (Figure 2). This model enables the calculation of the maximum and characteristic values of the vertical displacements of the wheel and to specify the frequency characteristic of the suspension. More accurate models were found to require a more accurate assessment of tire performance, such as damping. The dimensions of the tire tread element were determined according to Figure 3, while the length of the contact was according to Equation (1) expression. The tread stiffness coefficient was determined using Equation (2). Quarter-car model behavior under kinematic excitation by road unevenness q was described using the following system of equations:

$$\begin{aligned} m_1\ddot{z}_1 + k_1(\dot{z}_1 - \dot{q}) + c_1(z_1 - q) = 0, \\ m_2\ddot{z}_2 + k_2(\dot{z}_2 - \dot{z}_1) + c_2(z_2 - z_1) = 0, \\ m_3\ddot{z}_3 + k_3(\dot{z}_3 - \dot{z}_2) + c_3(z_3 - z_2) = 0. \end{aligned} \tag{27}$$

where m_1, m_2, m_3—tread part of the tire, unsprung, and sprung mass, respectively; z_1, z_2, z_3—tire, suspension and body displacements, respectively; k_1, k_2, k_3—tread, tire, and suspension damping factors, respectively; c_1, c_2, c_3—stiffness of the tread, tire, and suspension, respectively; and q—height of road profile unevenness.

When developing the 3D model of the car, the following assumptions were made: the car body would be symmetrical with respect to the longitudinal and transverse axes, and its center of gravity would be on the car symmetry plane. For the first approach, it was assumed that the masses of the wheels would move only in the vertical direction, with the track of the front and rear axes being the same. The vertical oscillations of the 3D car model on the longitudinal plane were calculated using the system of Equation (27).

The transverse tilt of the car was expressed by the system of equations:

$$[M_\beta] \cdot \{\ddot{\beta}\} + [K_\beta] \cdot \{\dot{\beta}\} + [C_\beta] \cdot \{\beta\} = [M_q] \tag{28}$$

where {β}—transverse displacement vector consisting of $\beta_{1g} = \frac{1}{t_2}(z_{12k} - z_{11d})$, $\beta_{1p} = \frac{1}{t_2}(z_{11k} - z_{11d})$; indexes 1*i*—front, 2*i*—rear, *k*—left, *d*—right, t_1—front track of the car, and t_2—rear track of the car.

Mass matrix:

$$[M_\beta] = \begin{bmatrix} I_x & 0 & 0 \\ 0 & 2m_{11}\left(\frac{t_1}{2}\right)^2 & 0 \\ 0 & 0 & 2m_{12}\left(\frac{t_2}{2}\right)^2 \end{bmatrix} \tag{29}$$

where I_x—moment of inertia of the sprung mass of the vehicle around the longitudinal axis of the tilt and *t*—wheel track of the respective axis.

Damping matrix:

$$[K_\beta] = \begin{bmatrix} \frac{1}{2}\left(k_{21}t_1^2 + k_{22}\right)\left(\frac{t}{2}\right)^2 & -\frac{k_{21}t^2}{2} & -\frac{k_{22}t^2}{2} \\ -\frac{k_{21}t^2}{2} & \frac{(k_{11}+k_{12})t_1^2}{2} & 0 \\ -\frac{k_{22}t^2}{2} & k_{11} + k_{21} & \frac{(k_{12}+k_{22})t_2^2}{2} \end{bmatrix} \tag{30}$$

Stiffness matrix:

$$[C_\beta] = \begin{bmatrix} \frac{1}{2}\left(c_{21}t_1^2 + c_{22}\right)\left(\frac{t}{2}\right)^2 & -\frac{c_{21}t^2}{2} & -\frac{c_{22}t^2}{2} \\ -\frac{c_{21}t^2}{2} & \frac{(c_{11}+c_{12})t_1^2}{2} & 0 \\ -\frac{c_{22}t^2}{2} & c_{11} + c_{21} & \frac{(c_{12}+c_{22})t_2^2}{2} \end{bmatrix} \tag{31}$$

The pavement influence matrix must take into account that the excitation on the left and right sides may be unequal and was therefore expressed as:

$$[M_q] = \begin{bmatrix} 0 & 0 \\ 2c_{12}\left(\frac{t_1}{2}\right)^2 & 0 \\ 0 & 2c_{12}\left(\frac{t_2}{2}\right)^2 \end{bmatrix} \cdot \begin{Bmatrix} q_{\Delta 1} \\ q_{\Delta 2} \end{Bmatrix} \tag{32}$$

where $q_{\Delta 1} = \frac{1}{t_1}(q_{1k} - q_{2d})$, $q_{\Delta 2} = \frac{1}{t_2}(q_{2k} - q_{2d})$, with $q_{\Delta 1}$- and $q_{\Delta 2}$—angular differences in road unevenness at the front and rear axes of the vehicle, respectively.

The quarter-car and 3D car models (the characteristics of the model elements are presented in Table 6, and modeling results are presented in Figure 19) were compared numerically by modeling the compact car movement on a city-type road at the speed of 40 km/h and 80 km/h, taking into account the nonlinearity of the characteristics of the tire and the elastic members.

Figure 19 shows that in the case of the 3D model, larger vertical displacements of the sprung masses (up to 20%) were obtained, but the overall curve characteristic remained almost without any changes.

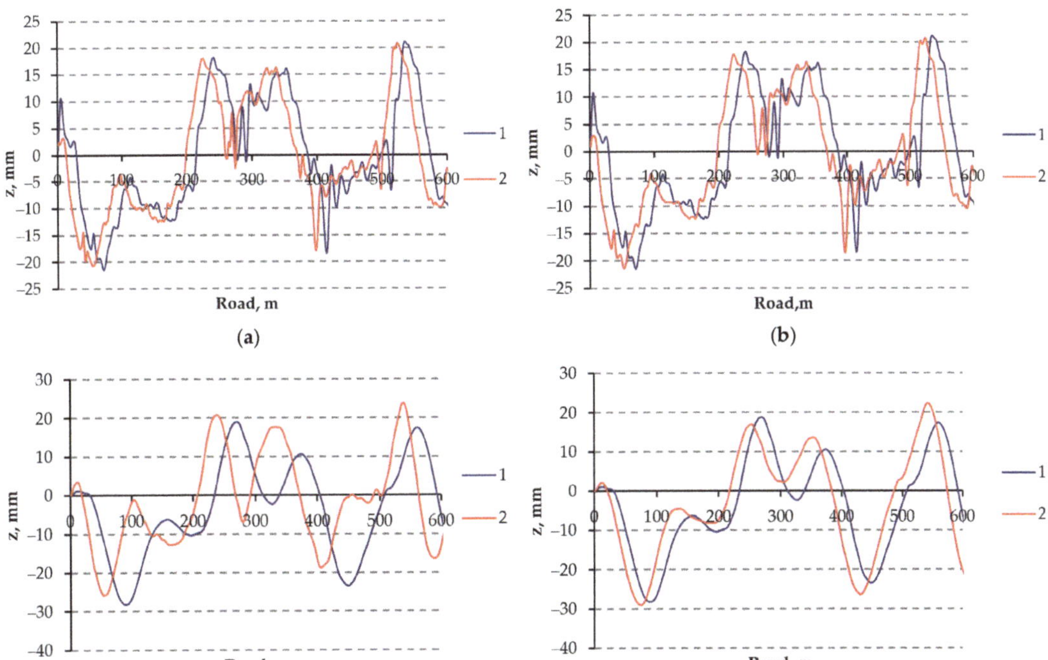

(a)

(b)

(c)

(d)

Figure 19. Displacements of unsprung (**a**,**b**) and sprung (**c**,**d**) masses (front) of the compact car, calculated using the quarter-car and 3D models: 1—quarter-car model; 2—3D model; (**a**,**c**)—v = 40 km/h; and (**b**,**d**)—v = 80 km/h.

Table 6. Quarter-car and 3D model element characteristics.

Parameter	Quarter-Car Model	3D Model	
		Front	Rear
Sprung mass m_3, kg	246	246	205
Unsprung mass m_2, kg	27.5	27.5	32
Mass of the tire element m_1, kg	0.3	0.3	0.3
Suspension spring stiffness c_3, kN/m	18	18	24
Tire stiffness c_2, kN/m	157	157	157
Tread stiffness c_1, kN/m	15,000	15,000	15,000
Shock absorber damping k_3, Ns/m	1835	1835	1835
Tire damping k_2, Ns/m	475	475	475
Tread damping k_1, Ns/m	0	0	0

3.2. Excitation during Movement on the Asphalt Pavement

The influence of nonlinear characteristics of the elastic elements of suspension on the vertical displacements of the body and wheels caused by the excitation frequency was investigated during modeling of the stability of the compact car by combining different quality road profile and car speed, as well as assessing the influence of the tire tread. The results were compared by assessing the dependence of the coefficient of dynamism (ratio of sprung and unsprung mass displacements to the height of unevenness) on the road excitation. Figure 20a,b shows the influence of nonlinear damping of shock absorbers and nonlinear stiffness of suspension on the sprung and unsprung mass displacements.

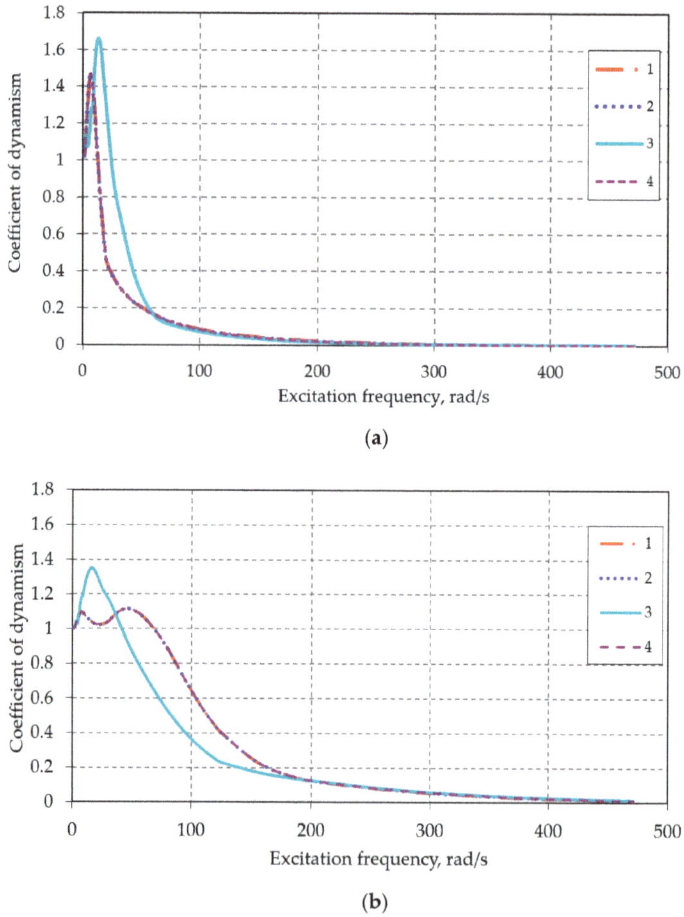

Figure 20. Influence of nonlinear damping of shock absorbers and nonlinear stiffness of suspension on the sprung (**a**) and unsprung (**b**) mass displacements: 1—linear shock absorber and linear suspension, tread not assessed; 2—tread assessed; 3—nonlinear shock absorber; and 4—nonlinear suspension.

With nonlinear elements, the frequency dependences shifted towards higher frequency because the damping of the shock absorbers due to nonlinearity increased suddenly, and the displacement (coefficient of dynamism) increased as well. The damping of the vibrations of the unsprung masses was fairly effective, but at a certain excitation frequency, the second peak would occur, causing certain damping deficiencies.

Figure 21a,b shows the influence of road unevenness on the tread displacements of the sprung and unsprung masses and the tire tread. As the unevenness increased, the coefficient of dynamism could be observed to increase as well. As long as the amplitude was small, the suspension and shock absorber would operate in a linear section. As the amplitude increased, the influence of the nonlinearity of the elements would be observed (Figure 21b).

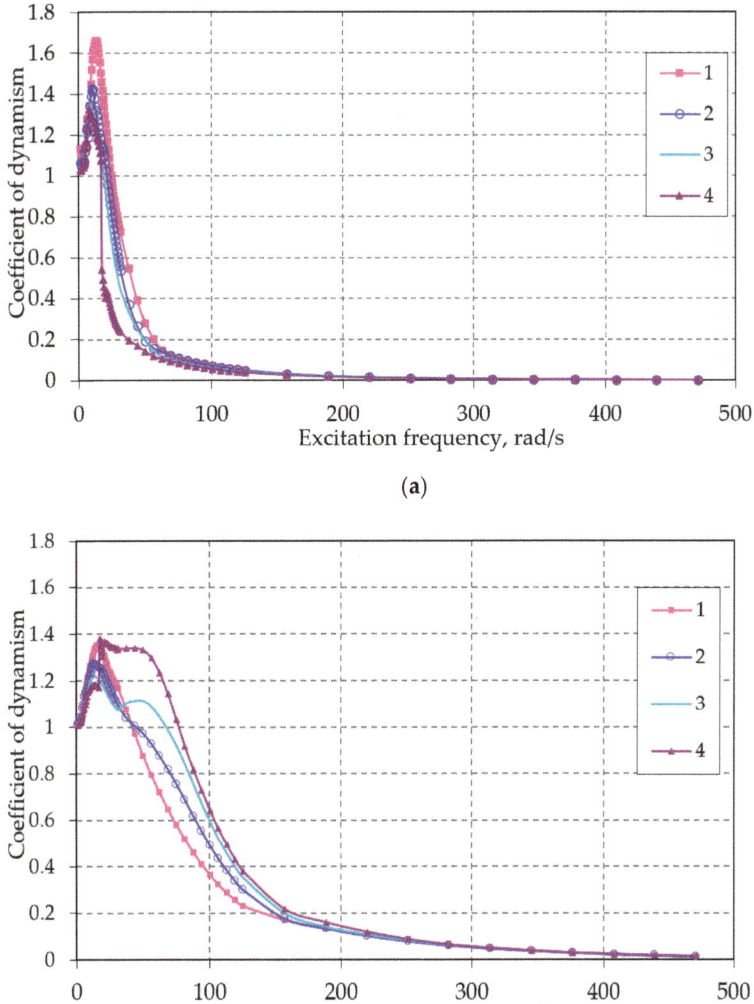

(a)

(b)

Figure 21. The influence of road unevenness on the sprung (**a**) and unsprung (**b**) mass displacements: 1—amplitude of irregularities 10.5 mm; 2—21 mm; 3—31.5 mm; and 4—42 mm.

Figure 22a–c shows the influence of the car speed on the displacements of the sprung (body), unsprung (tire) masses, and tread displacements.

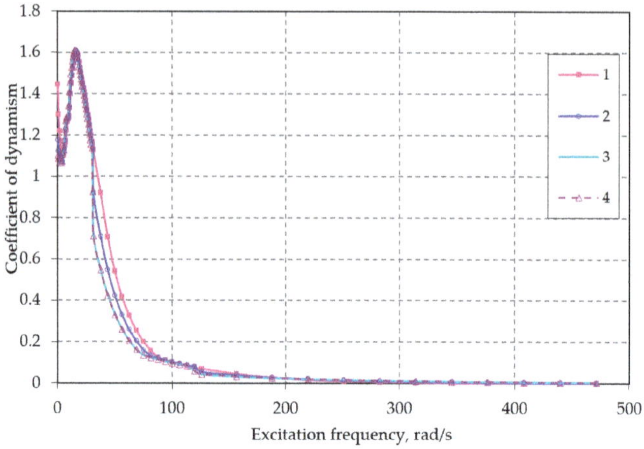

(a)

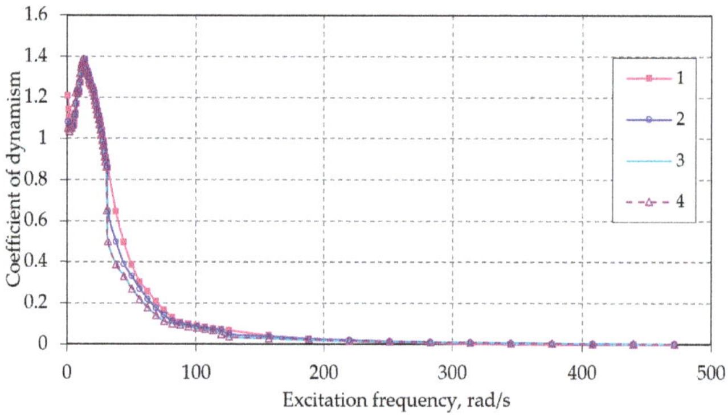

(b)

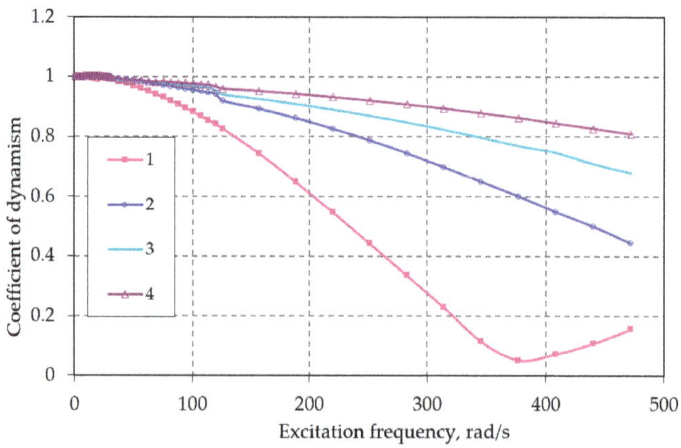

(c)

Figure 22. Influence of vehicle driving speed on the displacements of sprung mass (**a**), unsprung mass (**b**), and tire tread (**c**): 1—vehicle speed 10 m/s; 2—20 m/s; 3—30 m/s; and 4—40 m/s.

Damping of the sprung mass was effective at different speeds. The displacement reached its peak at low frequencies and afterward would decrease considerably. Tire deformations were the most sensitive to the effect of speed.

3.3. Experimental Investigation

In an investigation of a car stability, the characteristics of the road microprofile must be available. To measure the road microprofile, a measuring device consisting of the Volkswagen Transporter with DYNATEST road profile measuring equipment and laser sensors mounted on the front of the car were used (Figure 23).

Figure 23. Laser profilograph DYNATEST 5051 RSP mounted on the VW Transporter. Reprinted from ref. [58].

Measurement equipment: VAS-21 vibration analysis and processing system [59], PicoScope 3424 signal converter (Table 7), two low-frequency (WILCOXON Model 793L SN7688) and two medium-frequency (WILCOXON Model 784A SN12040) sensors (Table 8). The data were registered real-time on a PC, recorded, and processed using PICOSCOPE 5.13.7 software.

Table 7. Characteristics of the PicoScope 3424 signal converter. Reprinted from ref. [60].

Incoming signal values	±20 mV to ±20 V
Measured frequencies, max	10 MHz
Error	1%
Overload protection	±100 V
Incoming signal values	1 MΩ in parallel to 20 pF
Input	BNC

Table 8. Sensor characteristics. Reprinted from ref. [61].

Characteristics	Sensor	
	WILCOXON Model 793L	**WILCOXON Model 784A**
Sensitivity, ±20%, 25 °C	500 mV/g	100 mV/g
Measurement limits	±10 g	±50 g
Measured frequencies	0.6–700 Hz	4–7000 Hz
Supply voltage	18–30 V	18–30 V
Sensor mass	142 g	45 g

Measurements were performed in the city and on the highway. Road unevenness was measured using the Lithuanian Transport Safety Administration and Transport Competence Agency a VW Transporter car equipped with the measuring devices. The driving speeds in the city were and on the highway: 50, 60, 70, and 80 km/h. Measurement duration: 50 s.

To determine the frequency characteristics of the sprung and unsprung mass, compact car was selected for the study (Table 9).

Table 9. Compact car data.

Overall Dimensions, mm			Base, mm	Mass, kg		Tire Dimensions	Static Radius, mm	Tire Diameter, mm
Length	Width	Height		Tare	Payload			
3985	1665	1415	2475	870	1400	175/70R13	261	293

The measurement sensors for compact city car unsprung mass vibrations were mounted on the suspension arms (Figure 24a,b), while the sensors for measuring sprung mass vibrations were mounted on the front fender and car roof (Figure 24c,d).

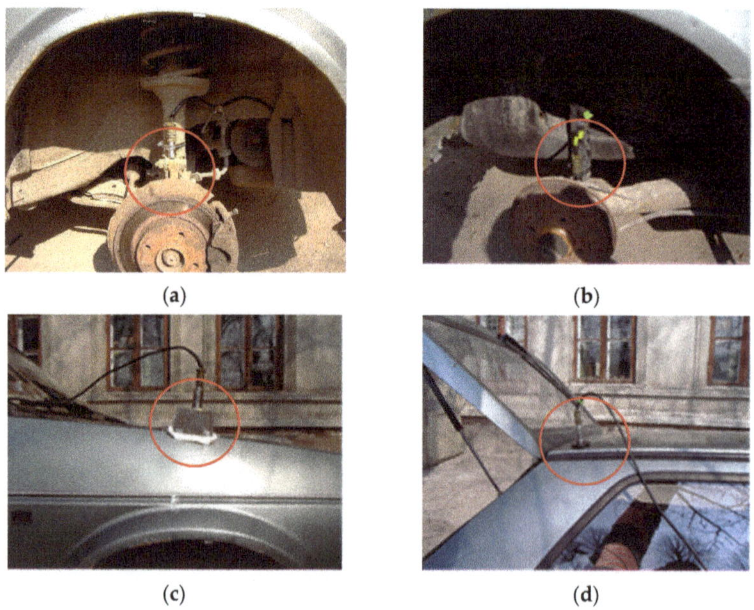

(a) (b) (c) (d)

Figure 24. Mounting of sensors for measuring vibrations of unsprung mass: (**a**)—front suspension, (**b**)—rear suspension; and sprung mass: (**c**)—on the front fender, and (**d**)—on the roof.

During the experimental studies, the driving speed down the road of the investigated profile was different during the 50 s drive and the accelerations of the sprung (body) and unsprung (suspension) masses of the cars. The test results for the compact car are shown in Figures 25 and 26. High-intensity noise and the nonlinear reaction of the car suspension to excitation made the result analysis more complex. While the movements of the unsprung mass show relatively little variation at the speeds of 50 and 70 km/h, the reaction of the sprung mass was very different at the same speeds, as individual unevenness may excite considerable body displacements without causing any significant suspension deformations.

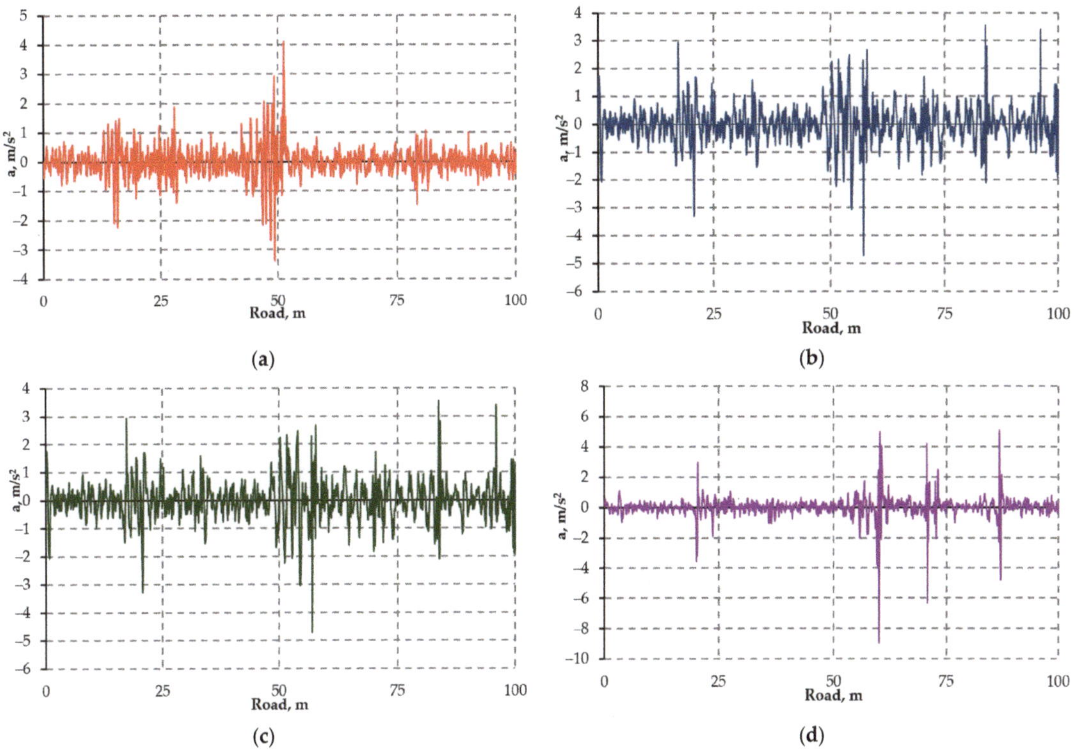

Figure 25. Accelerations of the sprung masses of compact car measured on a city-type road at different speeds: (**a**,**b**)—front; (**c**,**d**)—rear; (**a**,**c**)—50 km/h; and (**b**,**d**)—70 km/h.

Frequencies close to the frequency of own oscillations of the suspension (for passenger cars 0.8–1.3 Hz) and own oscillation frequency of unsprung masses (for passenger cars 8–13 Hz) were of particular interest in terms of the stability characteristic. It was the travel of the suspension that was defined as the difference between the displacements of the sprung and unsprung masses that had the greatest influence on the changes in the spatial position of the wheel. The equipment registered the peaks of the lowest frequencies in the range of 5–7 Hz (Figure 27), i.e., the frequencies characteristic of own oscillation frequency of the suspensions. In addition, the vibration sensors registered additional vibrations caused by the vibrations of the suspension elements (Figure 27a). Their frequencies exceeded the characteristic vibration frequencies of the tire (80–100 Hz). The oscillations were transmitted to the body (Figure 27b); however, comparing the frequency spectra of sprung and unsprung masses, the oscillations in the body were observed to be influenced by the structure of elements selected as sensor mounting points, namely, the wing at the suspension fixing point was less sensitive to high-frequency vibrations than the roof panel.

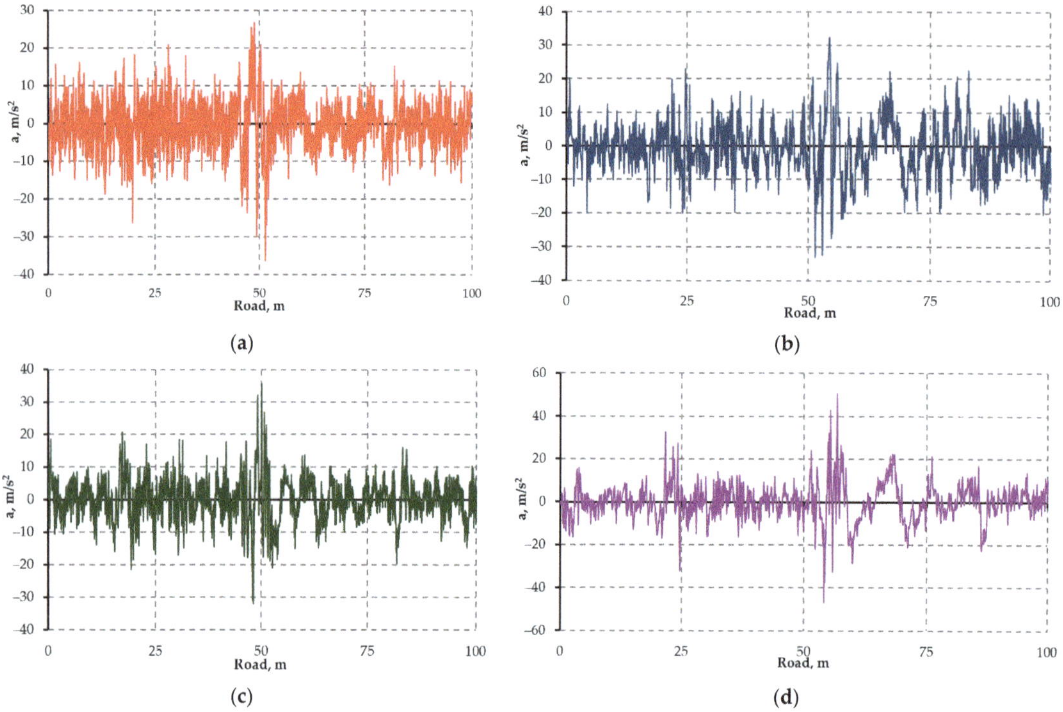

Figure 26. Accelerations of the unsprung masses of compact car measured on a city-type road at different speeds: (**a**,**b**)—front; (**c**,**d**)—rear; (**a**,**c**)—50 km/h; and (**b**,**d**)—70 km/h.

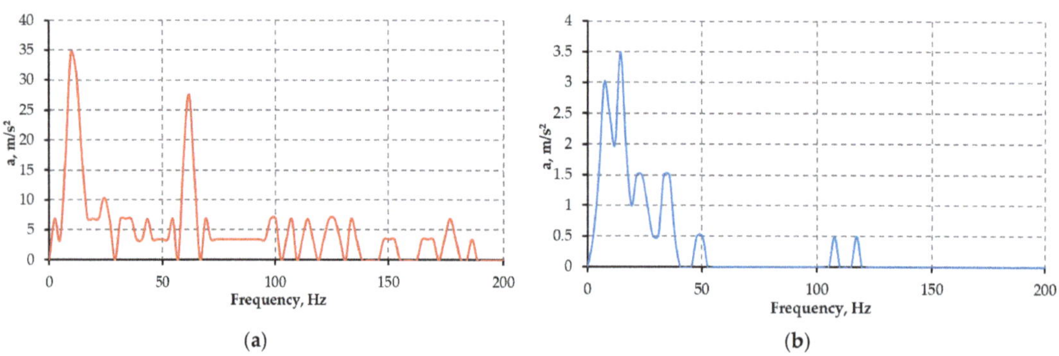

Figure 27. Frequency spectra of compact car on city type road at the speed of 70 km/h: (**a**)—front, unsprung mass and (**b**)—front, sprung mass.

The aim of the experiment was to verify the modeling results and investigate their applicability to stability analysis. A Fourier series was used for frequency analysis of the results of direct measurements used for further studies (Figure 27). The Fourier analysis provides a wider spectrum but also does not highlight the characteristic body—and suspension-specific frequencies. This may be related to the specifics of the road. Further research identified how the results of simulation of the suspension operation could be used to define stability.

3.4. Comparison of Excitation from Road Unevenness and Experimental Study

During the modeling, it was assumed that the compact car was moving on a road of a known profile in the city and on the highway at the speed of 50, 60, 70, and 80 km/h. To compare the results of modeling and experiment, a 100 m long road section was used. Road unevenness of the examined section was measured using profilograph DYNATEST 5051 RSP. The vehicle and suspension data used for the modeling are presented in Table 7. Matlab program (3D model with nonlinear suspension elements) was used for the modeling. The modeling results are presented in Figures 28–30.

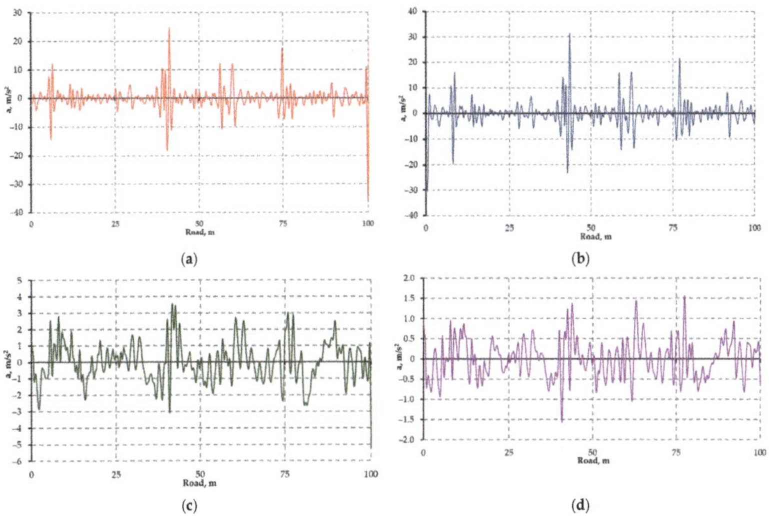

Figure 28. Accelerations of the unsprung (**a**,**b**) and sprung (**c**,**d**) masses of compact city measured by modeling on a city-type road at 50 km/h speed: (**a**,**c**)—front and (**b**,**d**)—rear.

Comparison of the curves of experimental and numerical modeling results suggested that in both cases very similar acceleration values were obtained (acceleration peak values differed by 5–7% for sprung masses and 20–25% for unnsprung masses). Comparison of the dependences of accelerations by means of Fourier analysis (Figure 29) with the experimental results suggested that a share of the vibrations caused by the resonance and stability of the suspension and body parts may be omitted from the assessment in the research findings.

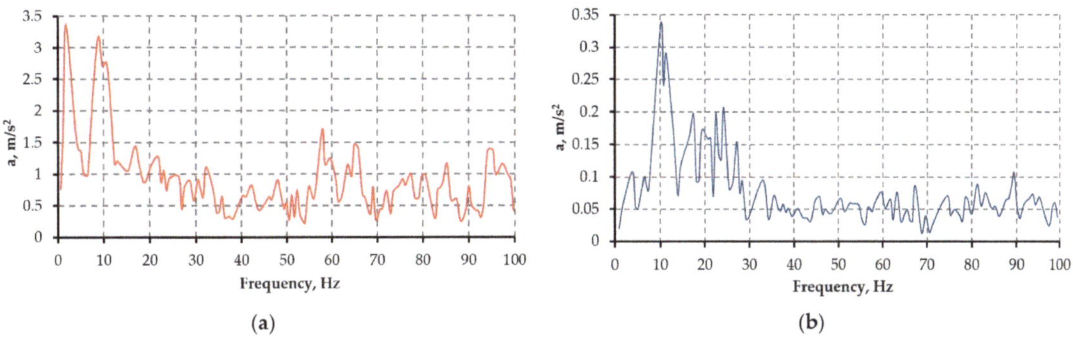

Figure 29. Frequency dependence of amplitudes of compact city, obtained by Fourier analysis: (**a**)—front damped unsprung mass and (**b**)—front damped sprung mass (city-type road, v = 70 km/h).

The obtained results show that the recording and analysis of signals required for active suspensions is an area of separate studies, as the additional noise and signal sources were recorded during the experiment, which also interfered with the analysis of directional stability. Therefore, the differences between the displacements of the sprung and unsprung masses were examined for further analysis (Figure 30). The maximum amplitude of the difference between the suspension and body displacements was 47 mm for the experiment and 40 mm for the modeling.

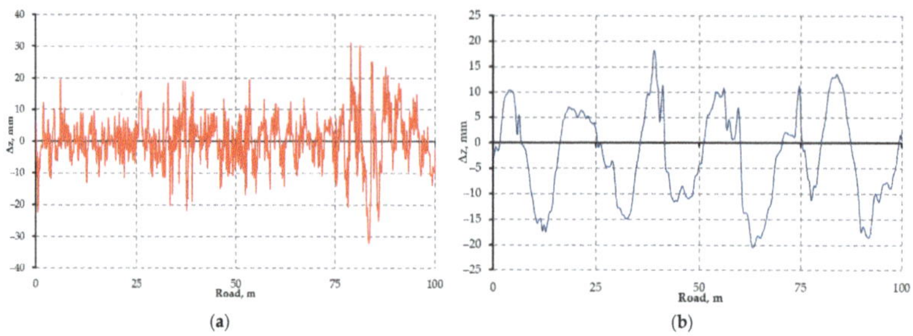

Figure 30. Difference between the sprung and unsprung mass displacements Δz simulating driving at the speed of 70 km/h: (**a**)—experiment and (**b**)—model.

4. Discussion

The analysis of the tire tread as the elastic band model using the numerical models has shown that due to the specificity of rubber deformability, the cord and breaker layers have effect on its stiffness in the radial direction by restricting the tread deformations.

The experiment confirmed the theoretical assumptions and showed that the reduced modulus of elasticity of the tread was 4.5 times higher than the modulus of elasticity of the tread material.

It was found that when assessing the tread deformation for different types of tires, summative expressions could be used for the relative coordinates as they use the relative load determined on the basis of the nominal tire load, thus greatly simplifying the assessment of tire deformation.

The analysis of the interaction between the tread and the road texture showed that the tread would even out the unevenness in the texture level, but the depressions would not be completely filled, thereby ensuring better adhesion between the tire and the road.

The specifics of the tire in the road-to-car interaction were determined. The deformation characteristics of the tread in the radial direction did not depend much on the tire or, consequently, on the type of vehicle.

Comparison of the quarter-car and 3D models showed that more detailed information necessary for stability assessment was provided by the 3D model, which predicted the sprung mass displacements higher by up to 20%.

The results of the experiment on the asphalt pavement were compared with the model. Certain issues were observed in the analysis of experimental results due to additional oscillations and vibrations occurring in the body and suspension elements, which could be avoided by modeling the movement after recording the road microprofile with a profilograph. The parameters that have the greatest impact on the stability of the direction—the difference between the displacement of the sprung and unsprung masses in the model versus the field experiment—did not exceed 15%.

Further Investigations

Based on the tire deformation properties obtained in this paper, numerically and experimentally investigated vertical excitation results, the changes in wheel position due to road roughness during suspension excitation can be determined and an analysis of the vehicle's yaw process can be performed. The modeling of the vehicle's directional stability needs to consider the specifics of the suspensions, such as the redistribution of radial loads under lateral force and the changes in wheel geometry due to the kinematic properties of the suspension. The obtained results justify further refinements of the directional stability numerical models:

- development of a dynamic model of a comprehensive integrated assessment of the suspension characteristics, suspension kinematics, steering mechanism, and pavement condition. The model would provide the development of a methodology for determination of the vehicles handling characteristics on roads of known quality using the numerical models.
- investigation of the sensitivity of the individual suspension types to vertical excitation in terms of directional stability criteria and to evaluate the technical condition of the vehicle if the tendency of changes in suspension parameters during known operation. To evaluate the technical condition of the vehicle by modifying the stiffness of the suspension components.
- analysis of vehicle stability in cornering, taking into account the changing position of the wheels as the vehicle is steered, which may have an effect on the vehicle stability. The stability of a vehicle in a turning situation requires the consideration of two criteria: the loss of stability of the vehicle when skidding, and the loss of stability of the vehicle when overturning. In addition, it is necessary to specify the effect of changes in the vehicle's trajectory in a corner due to the deformation of the suspension and the changes in the position of the wheels as the vehicle is steered.
- integrating driver reactions into the vehicle stability model. Without a driver model, such a simulation is not effective, as the vehicle model changes direction due to random effects and requires frequent correction to simulate steering course corrections.

5. Conclusions

1. The revised model of the tire equalizing function was developed. The coordinates of relative measures were found to cause no essential difference in assessment of tire deformations of different vehicle tires. This means that the number of tire groups may be reduced for assessment of their impact on stability.
2. The tread deformation characteristics of the tire were investigated by numerical modeling and experiment and showed that the tread deformations could be assessed by simplified models instead of the modulus of elasticity of the tread material, $E_r = 6$ MPa using the calculated modulus of elasticity $E_{1z} = 27$ MPa.
3. Studies on different types of cars (passenger cars, light-duty trucks and SUVs) have shown that the 3D suspension models more accurately provide reflection of the real situation in terms of stability assessment, while simpler quarter-car models can be used for comfort assessment. Comparison of the quarter-car model with the 3D model has shown that, in case of the latter, the resulting wheel travel is higher by up to 20%.
4. The results of the experiment on the asphalt pavement were compared with the model. Certain issues were observed in the analysis of experimental results due to additional oscillations and vibrations occurring in the body and suspension elements, which could be avoided by modeling the movement after recording the road microprofile with a profilograph. The parameters that have the greatest impact on the stability of the direction—the difference between the displacement of the sprung and unsprung masses in the model versus the field experiment—did not exceed 15%.
5. Changes in the car behavior at changing speed are related not only to the slip effect but also to changes in the wheel spatial position due to the suspension kinematics and suspension incompatibility with the steering mechanism. This effect can be particularly pronounced when road defects (pits, ruts) lead to an increase in the suspension travel.

Author Contributions: Conceptualization, V.L., R.M. and A.D.; methodology, V.L., R.M. and A.D.; software, V.L., R.M. and A.D.; validation, V.L., R.M. and A.D.; formal analysis, V.L., R.M. and A.D.; investigation, V.L., R.M. and A.D.; resources, V.L., R.M. and A.D.; data curation, V.L., R.M. and A.D.; writing—original draft preparation, V.L., R.M. and A.D.; writing—review and editing, V.L., R.M. and A.D.; visualization, V.L., R.M. and A.D.; supervision, V.L., R.M. and A.D.; project administration, V.L., R.M. and A.D.; funding acquisition, V.L., R.M. and A.D. All authors have read and agreed to the published version of the manuscript.

Funding: This research received no external funding.

Institutional Review Board Statement: Not applicable.

Informed Consent Statement: Not applicable.

Data Availability Statement: Not applicable.

Conflicts of Interest: The authors declare no conflict of interest.

References

1. Misaghi, S.; Tirado, C.; Nazarian, S.; Carrasco, C. Impact of pavement roughness and suspension systems on vehicle dynamic loads on flexible pavements. *Transp. Eng.* **2021**, *3*, 100045. [CrossRef]
2. Mahajan, G.R.; Radhika, B.; Biligiri, K.P. A critical review of vehicle-pavement interaction mechanism in evaluating flexible pavement performance characteristics. *Road Mater. Pavement Des.* **2020**. [CrossRef]
3. Zoccali, P.; Loprencipe, G.; Lupascu, R.C. Acceleration measurements inside vehicles: Passengers' comfort mapping on railways. *Measurement* **2018**, *129*, 489–498. [CrossRef]
4. Rill, G.; Castro, A. *Road Vehicle Dynamics: Fundamentals and Modeling with MATLAB*, 2nd ed.; CRC Press: Boca Raton, FL, USA, 2020. [CrossRef]
5. Yechen, Q.; Hong, W.; Huang, Y.; Xiaolin, T. *Real-Time Road Profile Identification and Monitoring: Theory and Application*; Morgan & cLaypool Publishers: San Rafael, CA, USA, 2019. [CrossRef]
6. Yang, S.; Chen, L.; Li, S. *Dynamics of Vehicle-Road Coupled System*; Springer: Beijing, China, 2015.
7. Leitner, B.; Decký, M.; Kováč, M. Road pavement longitudinal evenness quantification as stationary stochastic process. *Transport* **2019**, *34*, 195–203. [CrossRef]
8. Loprencipe, G.; Zoccali, P.; Cantisani, G. Effects of Vehicular Speed on the Assessment of Pavement Road Roughness. *Appl. Sci.* **2019**, *9*, 1783. [CrossRef]
9. Yan, J.; Zhang, H.; Hui, B. Driving Safety Analysis Using Grid-Based Water-Filled Rut Depth Distribution. *Adv. Mater. Sci. Eng.* **2021**, 5568949. [CrossRef]
10. Yan, J.; Zhang, H.; Hui, B. Analysis of the lateral slope's impact on the calculation of water-filled rut depth. *PLoS ONE* **2020**, *15*, e0243952. [CrossRef]
11. Ružinskas, A.; Giessler, M.; Gauterin, F.; Wiese, K.; Bogdevičius, M. Experimental investigation of tire performance on slush. *Eksploat. I Niezawodn. Maint. Reliab.* **2021**, *23*, 103–109. [CrossRef]
12. Gogoi, R.; Das, A.; Chakroborty, P. Rut depth measurement of an asphalt pavement from its original profile. *Aust. J. Civ. Eng.* **2020**, *18*, 119–125. [CrossRef]
13. Zhang, Z.; Sun, C.; Sun, M.; Bridgelall, R. Application of a Machine Learning Method to Evaluate Road Roughness from Connected Vehicles. *J. Transp. Eng. Part B Pavements* **2018**, *144*, 04018043. [CrossRef]
14. Zhang, Z.; Sun, C.; Bridgelall, R.; Sun, M. Road profile reconstruction using connected vehicle responses and wavelet analysis. *J. Terramech.* **2018**, *80*, 21–30. [CrossRef]
15. Hussan, S.; Kamal, M.A.; Hafeez, I.; Ahmad, N.; Khanzada, S.; Ahmed, S. Modelling asphalt pavement analyzer rut depth using different statistical techniques. *Road Mater. Pavement Des.* **2020**, *21*, 117–142. [CrossRef]
16. Prażnowski, K.; Mamala, J.; Śmieja, M.; Kupina, M. Assessment of the Road Surface Condition with Longitudinal Acceleration Signal of the Car Body. *Sensors* **2020**, *20*, 5987. [CrossRef] [PubMed]
17. Lenkutis, T.; Čerškus, A.; Šešok, N.; Dzedzickis, A.; Bučinskas, V. Road Surface Profile Synthesis: Assessment of Suitability for Simulation. *Symmetry* **2021**, *13*, 68. [CrossRef]
18. Pacejka, H.B.; Besselink, I. *Tire and Vehicle Dynamics*, 3rd ed.; Butterworth-Heinemann: Oxford, UK, 2012. Available online: http://scholar.google.com/scholar_lookup?hl=en&publication_year=2012&author=HB+Pacejka&author=I.+Besselink&title=Tire+and+vehicle+dynamics (accessed on 5 January 2021).
19. Shaohua, L. *Dynamics of Vehicle-Road Coupled System*; Science Press: Beijing, China; Springer: Heidelberg/Berlin, Germany, 2015. Available online: https://scholar.google.lt/scholar?hl=lt&as_sdt=0%2C5&q=Dynamics+of+Vehicle%E2%80%93Road+Coupled+System&btnG (accessed on 10 March 2021).
20. Mavromatis, S.; Laiou, A.; Yannis, G. Safety assessment of control design parameters through vehicle dynamics model. *Accid. Anal. Prev.* **2019**, *125*, 330–335. [CrossRef] [PubMed]

21. Yin, Y.; Wen, H.; Sun, L.; Hou, W. Study on the Influence of Road Geometry on Vehicle Lateral Instability. *J. Adv. Transp.* **2020**, 7943739. [CrossRef]
22. Saeedi, M.A. Simultaneous improvement of handling and lateral stability via a new robust control system. *Mech. Based Des. Struct. Mach.* **2021**. [CrossRef]
23. Vijaykumar, V.; Anand, P. Design Optimization of Suspension and Steering Systems for Commercial Vehicles. In Proceedings of the ICDMC 2019, Chenai, India, 28–29 March 2019; Lecture Notes in Mechanical Engineering. Yang, L.J., Haq, A., Nagarajan, L., Eds.; Springer: Singapore, 2020. [CrossRef]
24. Cheng, S.; Li, L.; Yan, B.; Liu, C.; Wang, X.; Fang, J. Simultaneous estimation of tire side-slip angle and lateral tire force for vehicle lateral stability control. *Mech. Syst. Signal Process.* **2019**, *132*, 168–182. [CrossRef]
25. Joa, E.; Yi, K.; Hyun, Y. Estimation of the tire slip angle under various road conditions without tire–road information for vehicle stability control. *Control. Eng. Pract.* **2019**, *86*, 129–143. [CrossRef]
26. Huang, Y.; Liang, W.; Chen, Y. Estimation and analysis of vehicle lateral stability region. In Proceedings of the American Control Conference (ACC), Seattle, WA, USA, 24–26 May 2017; pp. 4303–4308. [CrossRef]
27. Zhang, X.; Wang, P.; Li, Z.; Wang, F.; Chen, H. Estimation and Analysis of Vehicle Stability Region under Complex Road Conditions. In Proceedings of the Chinese Automation Congress (CAC), Shanghai, China, 6–8 November 2020; pp. 3036–3041. [CrossRef]
28. Xu, N.; Huang, Y.; Askari, H.; Tang, Z. Tire Slip Angle Estimation Based on the Intelligent Tire Technology. *IEEE Trans. Veh. Technol.* **2021**, *70*, 2239–2249. [CrossRef]
29. Cheli, F.; Sabbioni, E.; Pesce, M.; Melzi, S. A methodology for vehicle side slip angle identification: Comparison with experimental data. *Veh. Syst. Dyn.* **2007**, *45*, 549–563. [CrossRef]
30. Shekh, M.; Umrao, O.P.; Singh, D. Kinematic Analysis of Steering Mechanism: A Review. In Proceedings of the International Conference in Mechanical and Energy Technology. Smart Innovation, Systems and Technologies, Greater Noida, India, 7–8 November 2019; Yadav, S., Singh, D., Arora, P., Kumar, H., Eds.; Springer: Singapore, 2019; Volume 174. [CrossRef]
31. Pradhan, D.; Ganguly, K.; Swain, B.; Roy, H. Optimal kinematic synthesis of 6 bar rack and pinion Ackerman steering linkage. *Proc. Inst. Mech. Eng. Part D J. Automob. Eng.* **2021**, *235*, 1660–1669. [CrossRef]
32. Lv, T.; Zhang, Y.; Duan, Y.; Yang, J. Kinematics & compliance analysis of double wishbone air suspension with frictions and joint clearances. *Mech. Mach. Theory* **2021**, *156*, 104127. [CrossRef]
33. von Wysocki, T.; Chahkar, J.; Gauterin, F. Small Changes in Vehicle Suspension Layouts Could Reduce Interior Road Noise. *Vehicles* **2020**, *2*, 18–34. [CrossRef]
34. Ciampaglia, A.; Santini, A.; Belingardi, G. Design and analysis of automotive lightweight materials suspension based on finite element analysis. *Proc. Inst. Mech. Eng. Part C J. Mech. Eng. Sci.* **2021**, *235*, 1501–1511. [CrossRef]
35. Böhm, F. Dynamic rolling process of tires as layered structures. *Mech. Compos. Mater.* **1996**, *32*, 568–576. [CrossRef]
36. Sapragonas, J.; Dargužis, A. Model of radial deformations of protector of vehicle tire. *Mechanics* **2011**, *17*, 21–29. [CrossRef]
37. Sapragonas, J.; Keršys, A.; Makaras, R.; Lukoševičius, V.; Juodvalkis, D. Research of the influence of tire hydroplaning on directional stability of vehicle. *Transport* **2013**, *28*, 374–380. [CrossRef]
38. Levin, M.A.; Fufaev, N.A. *Theory of Deformable Wheel Rolling*; Nauka: Moscow, Russia, 1989. (In Russian)
39. Berthelot, J.M. *Matériaux Composites*, 5th ed.; Masson: France, Paris, 2012; (In French). Available online: https://scholar.google.com/scholar_lookup?author=M.+Berthelot&journal=Mat%C3%A9riaux+Composites.+Comportement+m%C3%A9canique+et+analyse+des+structures&publication_year=1999 (accessed on 15 July 2020).
40. Sharma, S. *Composite Materials*, 1st ed.; CRC Press: New York, NY, USA, 2021.
41. Cristensen, R.M. *Mechanics of Composite Materials*; Dover Publications, Inc.: New York, NY, USA, 2005.
42. Gay, D. *Matériaux Composites*; Hermes: Paris, France, 1997; Available online: https://scholar.google.com/scholar_lookup?author=D.+Gay&journal=Mat%C3%A9riaux+Composites&publication_year=1997 (accessed on 15 July 2020).
43. Pakalnis, A. Investigation of elasticity characteristics in a tyre ring model. *Mechanics* **2000**, *3*, 59–62.
44. Hilyard, N.C. *Mechanics of Cellular Plastics*; Macmillan: New York, NY, USA, 1982.
45. Matlin, M.; Kazankina, E.; Kazankin, V. Mechanics of initial dot contact. *Mechanics* **2009**, *76*, 20–23. Available online: https://web.a.ebscohost.com/ehost/pdfviewer/pdfviewer?vid=0&sid=412aad69-e974-47ff-ab94-820fa51cb935%40sdc-v-sessmgr01 (accessed on 15 March 2021).
46. Haddar, M.; Chaari, R.; Baslamisli, S.C.; Chaari, F.; Haddar, M. Intelligent optimal controller design applied to quarter car model based on non-asymptotic observer for improved vehicle dynamics. *Proc. Inst. Mech. Eng. Part I J. Syst. Control Eng.* **2021**, *235*, 929–942. [CrossRef]
47. Negash, B.A.; You, W.; Lee, J.; Lee, C.; Lee, K. Semi-active control of a nonlinear quarter-car model of hyperloop capsule vehicle with Skyhook and Mixed Skyhook-Acceleration Driven Damper controller. *Adv. Mech. Eng.* **2021**. [CrossRef]
48. Jain, S.; Saboo, S.; Pruncu, C.I.; Unune, D.R. Performance Investigation of Integrated Model of Quarter Car Semi-Active Seat Suspension with Human Model. *Appl. Sci.* **2020**, *10*, 3185. [CrossRef]
49. Filipozzi, L.; Assadian, F.; Kuang, M.; Johri, R.; Velazquez Alcantar, J. Estimation of Tire Normal Forces including Suspension Dynamics. *Energies* **2021**, *14*, 2378. [CrossRef]
50. Aljarbouh, A.; Fayaz, M. Hybrid Modelling and Sliding Mode Control of Semi-Active Suspension Systems for Both Ride Comfort and Road-Holding. *Symmetry* **2020**, *12*, 1286. [CrossRef]

51. Shahid, Y.; Wei, M. Comparative Analysis of Different Model-Based Controllers Using Active Vehicle Suspension System. *Algorithms* **2020**, *13*, 10. [CrossRef]
52. Khan, M.A.; Abid, M.; Ahmed, N.; Wadood, A.; Park, H. Nonlinear Control Design of a Half-Car Model Using Feedback Linearization and an LQR Controller. *Appl. Sci.* **2020**, *10*, 3075. [CrossRef]
53. Ahmad, E.; Iqbal, J.; Arshad Khan, M.; Liang, W.; Youn, I. Predictive Control Using Active Aerodynamic Surfaces to Improve Ride Quality of a Vehicle. *Electronics* **2020**, *9*, 1463. [CrossRef]
54. Zhang, Q.; Hou, J.; Duan, Z.; Jankowski, Ł.; Hu, X. Road Roughness Estimation Based on the Vehicle Frequency Response Function. *Actuators* **2021**, *10*, 89. [CrossRef]
55. Perrelli, M.; Farroni, F.; Timpone, F.; Mundo, D. Analysis of Tire Temperature Influence on Vehicle Dynamic Behaviour Using a 15 DOF Lumped-Parameter Full-Car Model. In *Advances in Service and Industrial Robotics; Mechanisms and Machine Science*; Zeghloul, S., Laribi, M., Sandoval Arevalo, J., Eds.; Springer: Cham, Switzerland, 2020; Volume 84. [CrossRef]
56. Mei, T.; Vanliem, N. Control performance of suspension system of cars with PID control based on 3D dynamic model. *J. Mech. Eng. Autom. Control. Syst.* **2020**, *1*, 1–10. [CrossRef]
57. Blekhman, I.; Kremer, E. Vertical-longitudinal dynamics of vehicle on road with unevenness. *Procedia Eng.* **2019**, *199*, 3278–3283. [CrossRef]
58. Dimaitis, M. *Modern Technologies for Evaluating the Condition of Road Surfaces*; Ex Arte: Vilnius, Lithuania, 2008; pp. 32–35. (In Lithuanian)
59. *Vibration Processing Collection and Analysis System VAS–21. User Guide*; UAB Elintos Prietaisai: Kaunas, Lithuania, 2003. (In Lithuanian)
60. PicoScope 3000 Series High–Performance Oscilloscopes. Available online: http://www.picotech.com/picoscope3000.html (accessed on 15 July 2020).
61. Wilcoxon Low Frequency Accelerometers. Available online: http://www.wilcoxon.com (accessed on 15 July 2020).

applied sciences

MDPI

Article

A Screw-Axis Approach to the Stability of Two-Wheeled Vehicles

Matteo Bova and Matteo Massaro *

Department of Industrial Engineering, University of Padova, Via Venezia 1, 35131 Padova, Italy; matteo.bova@phd.unipd.it
* Correspondence: matteo.massaro@unipd.it

Abstract: The stability of two-wheeled vehicles is predominantly characterized by the well-known weave and wobble vibration modes, which have been extensively investigated in the literature, mainly in terms of their frequencies and damping ratios. In this work the focus is towards their mode shapes, which are investigated using the screw-axis (also called Mozzi-axis), instead of the classic compass diagrams, for a better understanding of their three-dimensional patterns. The analysis is then carried out using the velocity centres for a characterization from the top, rear and side view of the vehicle. The multibody vehicle model employed for the numerical analysis is built in Adams. The dataset resembles that of a 250cc sport motorcycle, and has been derived from laboratory tests. The stability analysis is carried out in the frequency domain. It is found that, depending on the selected plane for the projection of the three-dimensional vibration motion, the trajectories of the velocity centres of the weave and wobble can cross either aft or fore the centre of mass, which has been associated to the under- and over-steering behaviour in the literature.

Keywords: motorcycle; rider; screw axis; weave; wobble; multibody

Citation: Bova, M.; Massaro, M. A Screw-Axis Approach to the Stability of Two-Wheeled Vehicles. *Appl. Sci.* **2021**, *11*, 7393. https://doi.org/10.3390/app11167393

Academic Editors: Flavio Farroni, Andrea Genovese and Aleksandr Sakhnevych

Received: 30 June 2021
Accepted: 10 August 2021
Published: 11 August 2021

1. Introduction

The lateral stability of two-wheeled vehicles has been investigated extensively in the last 50 years, both numerically and experimentally. It is well known that there are two main vibration modes, namely weave and wobble [1–6]. Weave is characterized by steering oscillations combined with yaw, roll and lateral motion of the vehicle, with frequencies usually below 4–5 Hz, while wobble is dominated by steering oscillation, with frequency usually in the range 6–10 Hz. Weave is usually lightly damped at high speed, say above 100 km/h, while wobble tends to be lightly damped at low to medium speed, although such range is affected by the structural stiffness of the vehicle [7,8].

The effect of a rider on the vehicle stability has also been investigated. Historically, the rider has been assumed to be rigidly attached to the chassis, with the inertial properties estimated using prediction-regression formulas from biomechanical databases. A number of different sources are available in the literature, with a review of the most used given in [9].

Starting from the mid 1980s, the rider's passive motion on the saddle has been modelled using spring-damper elements tuned to give the typical modal properties obtained from experiments [10–13]. More recently, the effect related to 'hands-on' and 'hands-off' the handlebar has been investigated. Experiments aimed at the identification of the rider–steering interaction has been carried out [14,15]. The numerical simulations and the experiments suggest that with hands on the handlebar wobble becomes more stable, while weave destabilizes [16–18].

Most of the analyses mentioned above focus on the effect that the main vehicle, tyre and rider parameters have on the frequency and damping of the vibration modes. Such modes are usually presented in terms of the real and imaginary part of the eigenvalues—which are related to the damping and frequency of the vibration modes—as a function of the vehicle speed. In contrast, the focus of this work is on the shape of the vibration modes.

The underlying idea is that, given the same combination of frequency and damping ratio, the vehicle may be perceived to be more or less stable by the rider depending on the shape of the vibration modes. In order to characterize the mode shape, the screw-axes associated to the weave and wobble are computed, in addition to the compass diagrams, which are the standard to describe the mode shape. It is recalled that each motion in the space, e.g., that of the chassis of a two-wheeled vehicle, can be represented (instantaneously) as a rotation around and a translation along a screw-axis, also called Mozzi axis [19,20]. The direction of the screw axis coincides with the (instantaneous) direction of the angular velocity vector. The intersection between the screw axis and a reference plane gives the velocity centres of the motion projected on that reference plane. Two planes are considered in this work: transverse plane (motion observed from the top of the vehicle) and frontal plane (motion observed from the front of the vehicle). The side view is not interesting in the case of straight-motion analysis. Indeed, it is well know that in such a condition weave and wobble modes do not involve either suspension oscillations or tyre radial oscillations, as long as the vehicle is symmetric (which is the case on most models, including the one considered in this work).

The screw axis has been previously employed to analyse the motion of motorcycles during slalom [21], lane-change and turning [22], and lap simulation [23]. More recently, velocity centres have been used to investigate the mode shape of the weave mode [24]. In this work, such analysis is generalized through the screw-axis and extended to include also the front view and wobble.

The work is organized as follows. Section 2 describes the vehicle model employed and discusses its vibration modes, in order to show its consistency with other results reported in the literature. Section 3 recalls the formulas related to the computation of the screw axis and velocity centres. Section 4 is devoted to the analysis of the shape of the weave mode, while in Section 5 the wobble mode is considered.

2. Multibody Model

The multibody model is built in Adams, which is one of the most used environments for vehicle dynamic analysis. The motorcycle model is standard, including the eleven degrees of freedom associated to the gross motion (chassis position and orientation, steering rotation, suspension travels and wheels rotation). The modelling of the structural compliance is limited to the lateral bending and torsion of the chassis (one additional degree of freedom), which is well known to be essential for a proper modelling of the vibration modes [5,7,8]. The tyres are modelled using the Magic Formula approach [5], which is included in Adams by default. The inertial properties of the bike were measured in the laboratory of the University of Padova, while the rider inertial properties were estimated using the Zatsiorsky–DeLeva dataset, which is one the most used [9]. The hands are assumed to be off the handlebar [16].

The vibration properties are usually extracted in two alternative ways: Using a time domain simulation or a frequency domain simulation. The former case resembles the approach employed also during experimental tests, where the system is excited using an impulse (e.g., on the handlebar) and the resulting oscillation is analysed in order to extract its modal component. The latter approach—frequency domain—is the preferred approach in this work, because it allows to compute the vibration properties more efficiently. In practice, all of the vibration modes are extracted simultaneously without applying any impulse on the system. The following procedure is carried out at all the speeds of interest.

A time domain simulation is started at the speed of interest using a proportional-derivative-integral (PID) engine torque controller to keep the vehicle at constant speed (the observed variable is the speed error): Indeed, at least the aerodynamic drag and the rolling resistance need to be balanced. The simulation is stopped when a steady-state condition is reached, i.e., when the transient oscillations have disappeared (which usually happen after

3–5 s). At this stage the PID is disengaged and the plant linearised in order to obtain the standard state space formulation

$$\dot{x} = Ax + Bu \tag{1}$$

$$y = Cx + Du \tag{2}$$

where A is the state matrix, B is the input matrix, C is the output matrix and D is the feed-through matrix; y denotes the observed variables, u are the inputs (steering torque and engine torque) and x are the model states:

$$x = [\dot{X}, X, \dot{Y}, Y, \dot{Z}, Z, \dot{\phi}, \phi, \dot{\theta}, \theta, \dot{\psi}, \psi, \dot{\delta}, \delta, \dot{\xi}_r, \xi_r, \dot{\xi}_f, \xi_f, \dot{\beta}_r, \beta_r, \dot{\beta}_f, \beta_f, \dot{\gamma}, \gamma, \sigma_1, \cdots, \sigma_9]^T. \tag{3}$$

The model has 31 states: The first 24 states in (3) consists of the 12 user selected position states (which are manually defined through the PSTATE command in Adams) and their derivatives, while the latter seven states are automatically generated by Adams and are likely associated to the (transient) Magic Formula tyre model (e.g., longitudinal and lateral relaxation equations of the front and rear tyres) and other internal variables not directly controlled by the user. In more detail, in the user-selected states X, Y and Z are the longitudinal, lateral and vertical displacements of the saddle position (however any other point of the chassis works equally fine), ϕ, θ and ψ are the roll, pitch and yaw angles of the saddle reference marker (i.e., of the chassis, since it is assumed rigid), δ is the steering angle, ξ_r and ξ_f are the rear and front suspension deformations, β_r and β_f are the rear and front wheel spin angles and γ is the bending deflection of the front frame [7,8], while $\sigma_1, \cdots, \sigma_9$ are the states automatically generated by Adams. It is noted that the controller is disengaged before linearising the system, in order to obtain the open-loop dynamics, i.e., the dynamics of the bike and not those of the bike-controller combination (in addition, one would have more states in (3) in the case the PID is still engaged at the time of the linearization). The procedure is automated using an Adams *macro*, which is then called from a Matlab script. The vibration modes consist of the (complex) eigenvalues of the state matrix A, with the mode shape being the related (complex) eigenvectors.

Figure 1 shows the root-locus (imaginary vs. real part of the eigenvalues) of the vehicle under investigation, for speeds between 5 (circles) and 50 m/s (squares). It goes without saying that there are as many eigenvalues as the number of states in (3)—as usual only the upper half of the root-locus diagram is shown because it is symmetric with respect to the x-axis. Two 'high' frequency modes can be seen above 13.5 Hz (85 rad/s), corresponding to the front and rear wheel hop, respectively—the former mainly involves the bouncing of the front wheel on the road, while the latter mainly involves the bouncing of the rear wheel on the road. They are marginally affected by the speed variations, the rear hop mode being mainly affected in terms of frequency (which slightly decreases at high speed), the front hop mode being mainly affected in terms of damping (which slightly decreases with speed). The rear wobble mode is visible below the hop modes, with a damping increasing with the speeds, while its frequencies decrease with speed from 9.5 Hz (60 rad/s) to 0 Hz (non-vibrating). This mode is usually not very interesting because it is always stable. Wobble is a mode mainly involving steering oscillations, with frequencies decreasing from 8.5 Hz (53 rad/s) to 5.5 Hz (35 rad/s) in this vehicle. Weave is a mode involving the lateral degrees-of-freedom of the vehicle, and has frequencies rising with speeds from 0.5 Hz (3 rad/s) up to 5 Hz (31 rad/s). On the horizontal axis there is the capsize mode: It does not vibrate, is lightly unstable at low speeds (in this vehicle), and mainly involves the roll and lateral degrees of freedom. Finally, the modes associated with the suspension dynamics are visible at lower frequencies: Bounce (or heave) has frequencies around 2.4 Hz (15 rad/s), while pitch has frequencies around 3.2 Hz (20 rad/s). As a recap, the modes of the vehicle in straight running can be divided into two categories: In-plane modes (namely bounce, pitch and hops) and out-of-plane modes (namely capsize, rear wobble, wobble and weave). The latter modes are the most interesting for the current analysis, because they have implications on the running stability of two-wheeled vehicles. Weave is also one of the most investigated modes, because it is lightly damped in many vehicles when the speed is high enough.

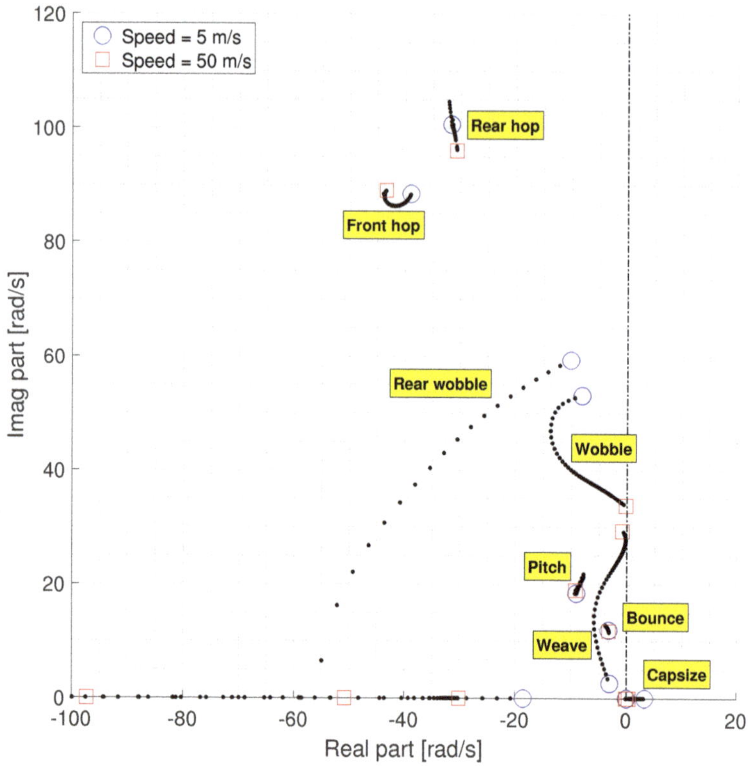

Figure 1. Vibration modes for speeds between 5 m/s (circle) and 50 m/s (square).

Figure 2 shows the compass plots of the weave mode shape: The length of each arrow represents the magnitude of the associated degree of freedom, with the related phase being the angle with respect to the horizontal axis. It is recalled that, as is common in vibration analysis, only the ratio between the magnitudes and the relative angles matter. The shape of the weave mode at high speed is similar to other results reported in the literature, with yaw almost in phase opposition with respect to the steering, roll in between the steer and yaw, while the steer lags with respect to the lateral motion [8,12,24]—an SAE conventions is employed, i.e., the z axis (both yaw and steer) points downward and the x axis points forward, while the y (lateral) axis completes the orthogonal triad. At low speeds all the four main degrees of freedom involved have similar magnitudes, with the roll (ϕ) prevailing on the others, while at high speeds the steer angle (δ) becomes dominant (however, there are still visible components of the other degrees-of-freedom).

It is recalled that in [24] two main types of weave shapes are defined: When the yaw (ψ) is in phase with the lateral displacement (y) the weave is called "under-steering" (the centrode intersection falls behind the centre of mass, as will be discussed in Section 3), otherwise the weave is called "over-steering" (the centrode intersection falls ahead of the centre of mass) and represents the case discussed in this work and shown in Figure 2. It is also said that "under-steering" scenarios are associated with an increased perception of safety.

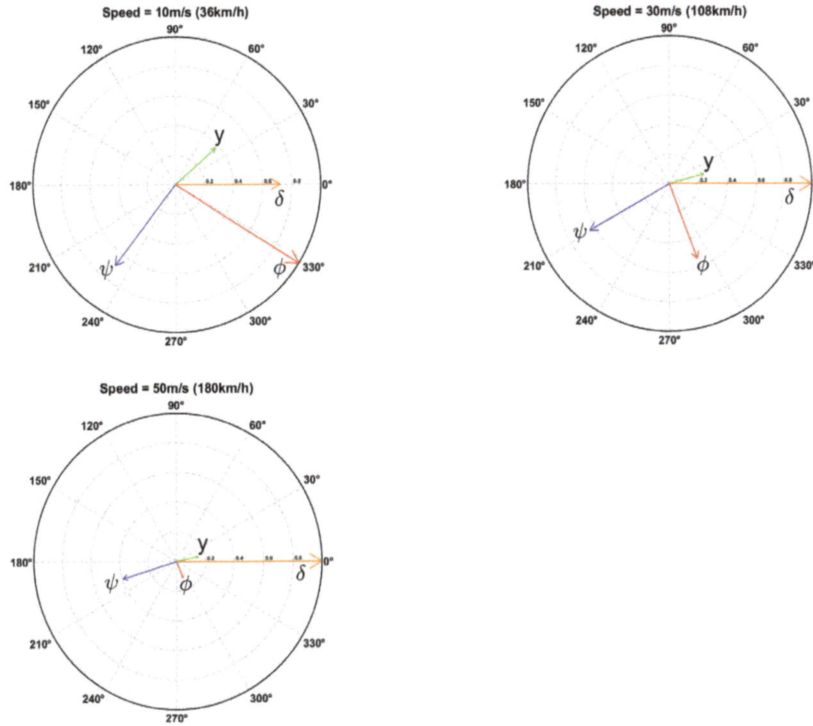

Figure 2. Weave compass diagrams at different speeds: y is the saddle point lateral displacement, δ is the steer angle, ϕ is the roll angle and ψ is the yaw angle.

3. Screw-Axis and Velocity Centres

The screw axis of the chassis is computed from the eigenvectors of the state matrix A in (1), which can be 'animated' in the time domain through

$$x(t) = x_0 e^{-\omega_n \zeta t} \sin\left[\left(\sqrt{1-\zeta^2}\right)\omega_n t + \phi\right], \tag{4}$$

where x_0 is the magnitude of the eigenvector state (i.e., the length of the arrow of the compass plot), ω_n is the (undamped) frequency of the corresponding eigenvalue, ζ is the damping ratio of the corresponding eigenvalue, ϕ is the phase (i.e., the angle of the arrow of the compass plot), and t is the time. Each point P_0 of the chassis has a velocity v_0 and angular velocity ω during the oscillation—the latter is the same for each point of the chassis, which is here treated as a rigid body. The velocity v_0 can be divided in two different components: $v_\parallel$ parallel to ω and $v_\perp$ normal to ω, which are given by

$$v_\parallel = \left(v_0 \cdot \frac{\omega}{\|\omega\|}\right)\frac{\omega}{\|\omega\|} \tag{5}$$

$$v_\perp = v_0 - v_\parallel \tag{6}$$

The normal component can also be written as

$$v_\perp = \omega \times d, \tag{7}$$

where d is the distance between the screw axis and P_0. When multiplying both sides by ω one obtains

$$v_\perp \times \omega = (\omega \times d) \times \omega, \tag{8}$$

which can be rewritten using triple-product identity as

$$\begin{aligned} v_\perp \times \omega &= (\omega \cdot \omega)d - (d \cdot \omega)\omega \qquad (9) \\ &= \|\omega\|^2 d. \qquad (10) \end{aligned}$$

Therefore, the distance d can be computed at each instant from

$$d = \frac{v_\perp \times \omega}{\|\omega\|^2}, \tag{11}$$

while the equation of the screw axis is given by

$$\begin{aligned} S_a &= P_0 - d + \mu\omega, \qquad (12) \\ \begin{pmatrix} x_a \\ y_a \\ z_a \end{pmatrix} &= \begin{pmatrix} x_0 \\ y_0 \\ z_0 \end{pmatrix} - \frac{1}{\omega_x^2 + \omega_y^2 + \omega_z^2} \begin{pmatrix} -v_z\omega_y + v_y\omega_z \\ -v_x\omega_z + v_z\omega_x \\ -v_y\omega_x + v_x\omega_y \end{pmatrix} + \mu \begin{pmatrix} \omega_x \\ \omega_y \\ \omega_z \end{pmatrix} \qquad (13) \end{aligned}$$

where μ is a scalar associated with the infinite points of the screw axis line.

The three-dimensional motion of the chassis can be projected on an $x - y$ transverse plane (parallel to the road plane, i.e., top view), a $y - z$ frontal plane (i.e., front view) or an $x - z$ sagittal plane (i.e., side view). In this case the screw axis is replaced by the velocity centre, whose trajectory on the selected plane can be obtained by intersection between the screw line and the selected plane. The scalar μ corresponding to the transverse plane is obtained when solving the third equation in (13) for $z_a = z_S$, with z_S being the height of the selected plane:

$$\mu = \frac{1}{\omega_z}\left(z_S - z_0 - \frac{v_y\omega_x - v_x\omega_y}{\omega_x^2 + \omega_y^2 + \omega_z^2}\right). \tag{14}$$

Similarly, the scalar μ corresponding to the projection on the frontal plane is given by

$$\mu = \frac{1}{\omega_x}\left(x_M - x_0 - \frac{v_z\omega_y - v_y\omega_z}{\omega_x^2 + \omega_y^2 + \omega_z^2}\right), \tag{15}$$

which is obtained from the second of (13) in the case $x_a = x_S$, with x_S being the coordinate of the selected plane. Finally, the intersection with the sagittal plane is not considered, since in straight motion the lateral modes are decoupled from the longitudinal modes: As a consequence $\omega_y = 0$ (pitch rate) during weave and wobble, and the screw axis remains parallel to the vehicle sagittal plane.

In principle, the approach presented can also be applied to experimental data, i.e., logged from the IMU of the vehicle, in addition to numerical data. The approach has shown to be able to work in two-dimensional scenarios (velocity centres), as demonstrated in [24]. From the practical point of view, the mode shape need be extracted first. When the problem is attacked numerically, it is convenient to extract the mode shape from the linearisation of the system (frequency domain approach). When the problem is attacked experimentally, a time domain approach is usually the preferred choice, i.e., the time signals are fitted to build the mode shape. Once the mode shape is obtained, the method makes no distinction on their numerical or experimental origin. In general, the approach presented is expected to work well if the mode shape is computed well.

4. Weave Analysis

In this section the weave mode shape is analysed in three transverse (top view) and three frontal (front view) planes, at different speeds. For the graphical representation of the velocity centre trajectories, the eigenvector has been normalized to have a saddle point lateral displacement equal to 0.5 m. However, the scaling does not affect the position of the cross points either on the transverse or frontal planes.

4.1. Top-View

Figure 3 shows the trajectory of the velocity centre (centrode) during the weave oscillation, when the transverse plane considered is through the saddle (which in this case is at 0.8 m from the road on static condition). The pattern changes with the speed: At 10 m/s the trajectories cross between the handlebar and the centre of mass at 0.795 m from the rear contact point, while at 30 m/s and 50 m/s the cross point shifts towards the rear contact point at 0.633 m and 0.623 m, respectively. Following the nomenclature introduced in [24], such a condition is called over-steering weave (centrode intersection ahead of the centre of mass, which is at 0.6 m from the rear contact point on static condition).

However, the centrode intersection changes when considering different transverse planes. Another option is to choose a plane through the centre of mass (which in this case is at 0.58 m from the road on static condition), see Figure 4. In this case the intersections exhibit the same behaviour mentioned above, moving from 0.709 m at 10 m/s to 0.615–0.613 m at 30 and 50 m/s, respectively, getting closer to the centre of mass while remaining still ahead of it (over-steering weave).

Yet another option is to consider the projection on the road plane. In this case the results are those in Figure 5: The intersection appears behind the centre of mass, moving closer to it from 0.479 m to 0.568–0.587 m at 10, 30 and 50 m/s, respectively—following the nomenclature introduced in [24] this weave is called under-steering.

The saddle plane may be a good choice if one is willing to associate the shape of the mode to the rider's feeling and perceived stability. On the other hand, the centre of mass plane has been apparently employed in [24], while the road plane is the choice in [21].

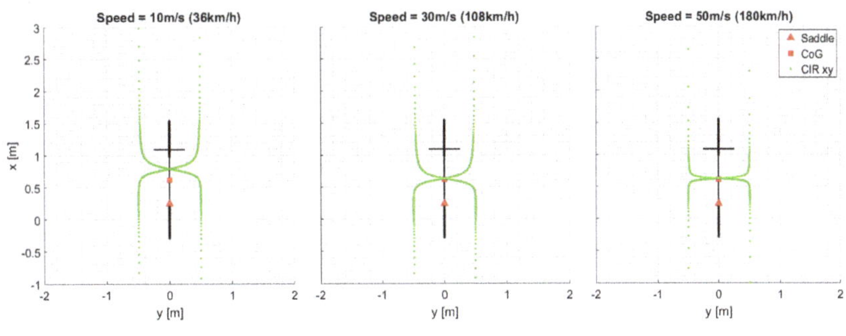

Figure 3. Velocity centres of weave in a transverse plane through the saddle point.

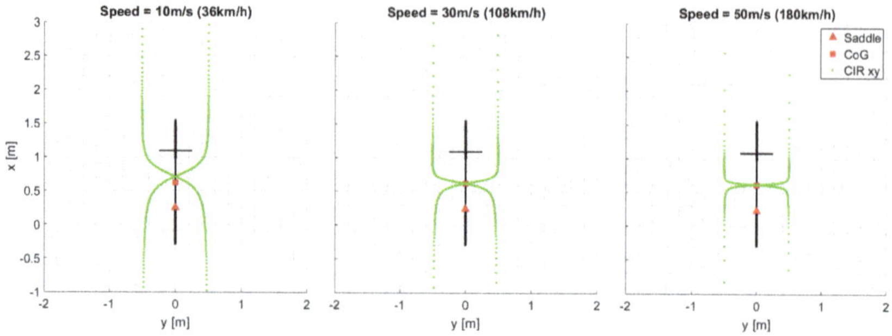

Figure 4. Velocity centres of weave in a transverse plane through the centre of mass.

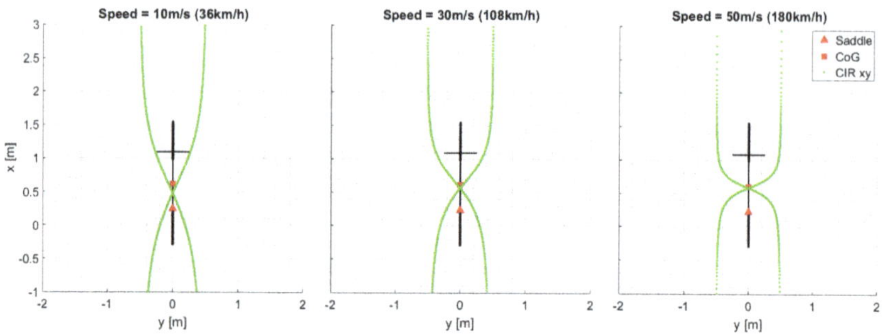

Figure 5. Velocity centres of weave in the road plane.

4.2. Front-View

Similarly to the analysis carried out in the top view, different choices are possible for the analysis from the frontal view as well.

When the plane through the saddle point (0.25 m ahead of the rear contact point in static condition) is considered, see Figure 6, the centrode intersection occurs below the road plane ($z = 0$), with the distance from the road plane increasing with the speed from 0.586 m up to 3.949–7.926 m at 10, 30 and 50 m/s, respectively—indeed the roll component reduces with respect to the lateral component.

When the reference is a plane through the centre of mass (0.62 m ahead of the rear contact point in static condition), see Figure 7, then the centrode intersection moves above the road plane, shifting from −0.349 to −0.586 m as the speed increases from 10 to 30 m/s, and then moving back towards the road plane until −0.555 m at 50 m/s.

Finally, in the case the reference plane being through the rear contact point, see Figure 8, the diagrams get closer to those with a reference on the saddle; this is not surprising since the two point are close in the longitudinal direction. The cross point moves from 1.209 m at 10 m/s up to 6.972–13.200 m at 30 and 50 m/s, respectively.

The same considerations related to the selection of the reference plane and reported at the end of the previous section still hold, although there are no references in the literature to compare the different options available.

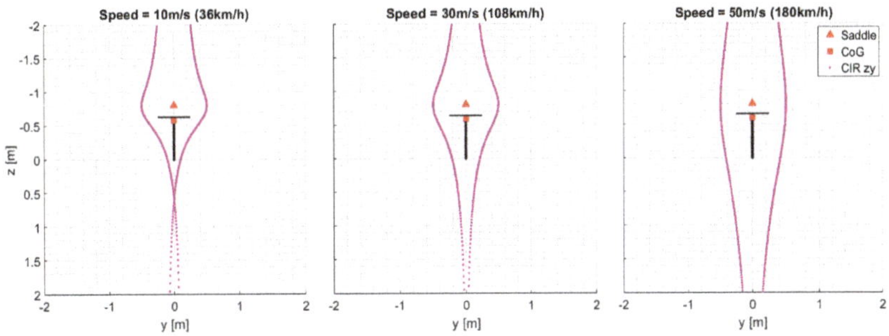

Figure 6. Velocity centres of weave in a frontal plane through the saddle.

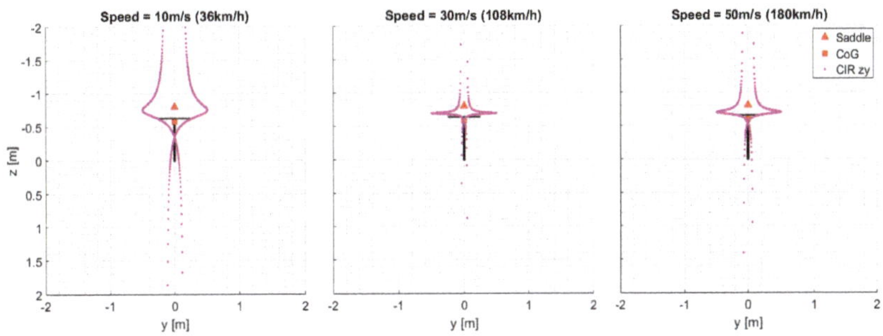

Figure 7. Velocity centres of weave in a frontal plane through the centre of mass.

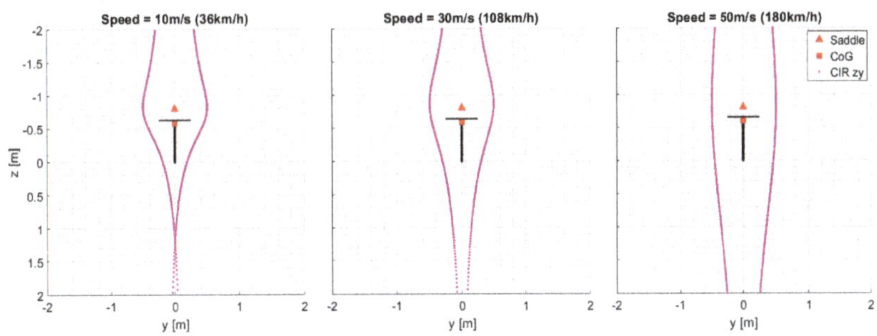

Figure 8. Velocity centres of weave in a frontal plane through the rear contact point.

5. Wobble

In this section the wobble mode shape is analysed in three transverse (top view) and three frontal (front view) planes, at different speeds. These are exactly the same planes considered for the weave in Section 4. The same scaling has also been applied to the eigenvectors, with the maximum saddle lateral displacement component equal to 0.5 m.

5.1. Top-View

Figure 9 shows the trajectory of the velocity centre (centrode) during the wobble oscillation, when the plane considered is through the saddle. The pattern undergoes minor variations when varying the speed: The trajectory cross point moves forward from 0.626 m up to 0.644–0.650 m at 10, 30 and 50 m/s, respectively—a slightly over-steer scenario, if we keep using the same convention employed in Section 4 (cross point ahead of the centre of mass).

Figure 10 shows the trajectory of the velocity centre projected on a transverse plane through the centre of mass. The results are similar to those described above, with the trajectory cross point again almost speed independent and moving from 0.587 m at 10 m/s to 0.571–0.585 m at 30 and 50 m/s, respectively—a slightly under-steering scenario (cross point behind the centre of mass).

Finally, the road plane is considered. In this case the results are those in Figure 11: The intersection appears behind the centre of mass, moving with a non-linear trend from 0.485 m at 10 m/s to 0.373–0.405 m at 30 and 50 m/s, respectively—following the nomenclature in Section 4 this wobble is called under-steering (cross point behind the centre of mass).

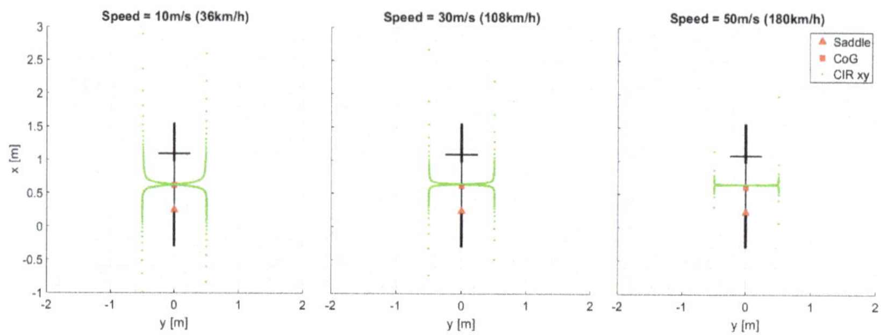

Figure 9. Velocity centres of wobble in a transverse plane through the saddle point.

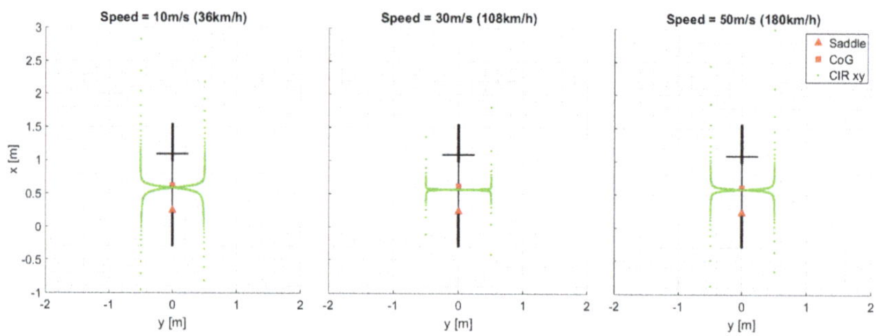

Figure 10. Velocity centres of wobble in a transverse plane through the centre of mass.

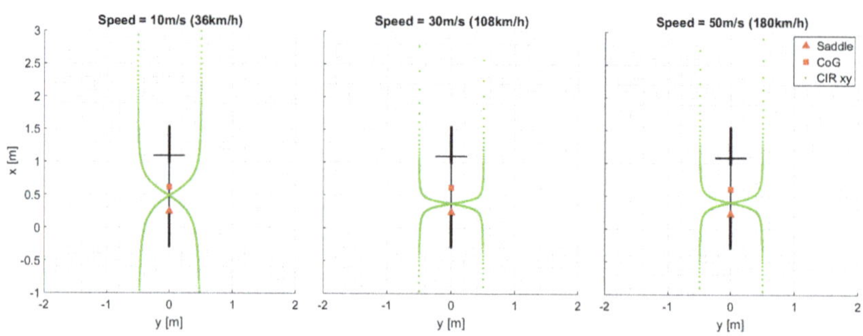

Figure 11. Velocity centres of wobble in the road plane.

5.2. Front-View

When the plane through the saddle point is considered, see Figure 12, the centrode intersection occurs below the road plane, with the distance from the road plane decreasing from 1.355 to 0.383 m as the speed increases from 10 to 30 m/s.At high speeds (>30 m/s) the intersection distance slightly increases up to 0.555 m at 50 m/s.

When the reference is a plane through the centre of mass, see Figure 13, then the centrode intersection moves above the road plane, decreasing the height as the speed increases, moving from −0.753 m to −0.715/−0.673 m at 10, 30 and 50 m/s, respectively, and getting closer to the centre or mass.

Finally, in the case the reference plane is through the rear contact point, see Figure 14, the diagrams get closer to those with a reference on the saddle and the cross point moves from 2.756 m to 1.099–1.322 m at 10, 30 and 50 m/s, respectively—again this is not surprising since the two point are close in the longitudinal direction.

The same considerations related to the selection of the reference plane and reported at the end of the weave section still hold.

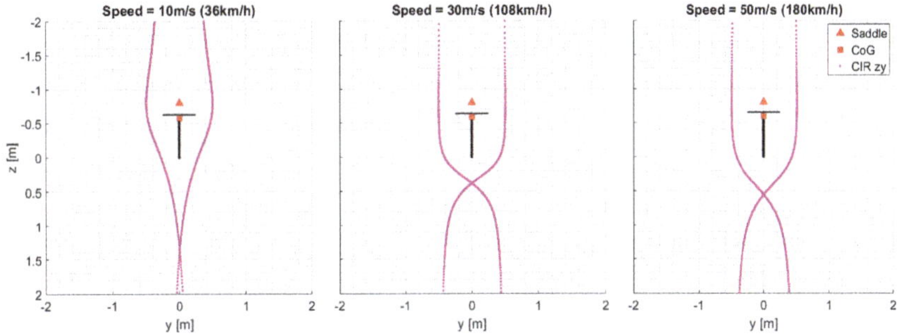

Figure 12. Velocity centres of wobble in a frontal plane through the saddle point.

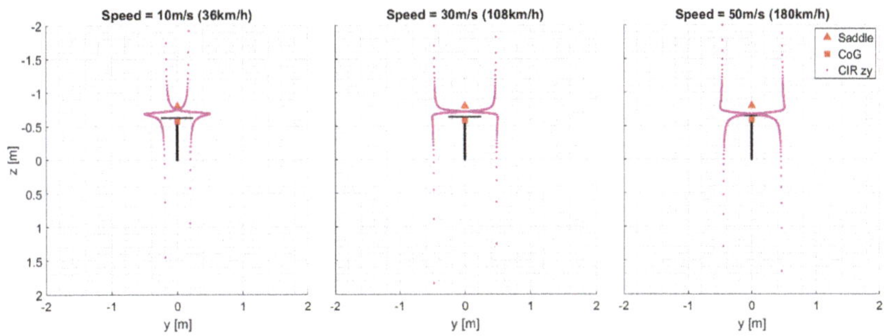

Figure 13. Velocity centres of wobble in a frontal plane through the centre of mass.

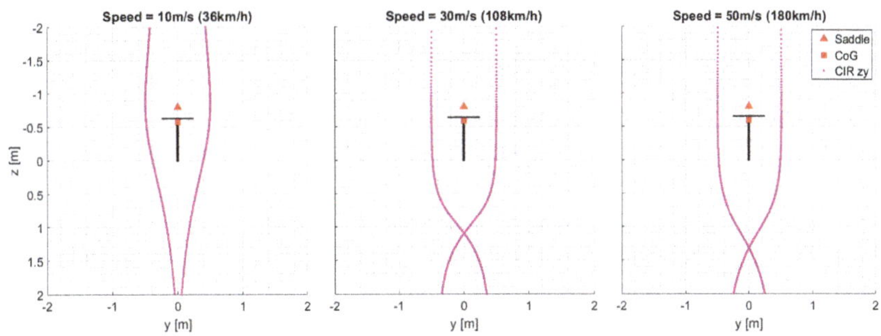

Figure 14. Velocity centres of wobble in a frontal plane through the rear contact point.

6. Conclusions

The mode shape of the most important lateral vibration modes of two-wheeled vehicles, namely weave and wobble, are analysed through a screw axis approach (instead of the classic compass diagram approach), which is an extension of the velocity centre approach recently proposed in the literature. The analyses reported in the literature have been extended to include also the frontal plane, in addition to the transverse plane, and wobble, in addition to weave. Finally, multiple transverse and frontal plane have been considered. It has been shown that different patterns can be obtained depending on the selected plane, and also different dynamic classifications, e.g., under- vs. over-steer behaviour.

This highlights the importance of choosing the appropriate reference plane to correlate the rider's sensations with the mode shapes: Saddle planes are closely related to the effective rider position, while centre of mass planes are related to the whole system dynamics. Both of these options are dependent on the motorcycle geometry, so another option is the use of the rear contact point planes, which do not change with the rider and bike.

Author Contributions: Conceptualization, M.B. and M.M.; methodology, M.B. and M.M.; software, M.B.; validation, M.B. and M.M.; formal analysis, M.B. and M.M.; investigation, M.B. and M.M.; resources, M.M.; writing—original draft preparation, M.B. and M.M.; writing—review and editing, M.B. and M.M.; supervision, M.M.; project administration, M.M. All authors have read and agreed to the published version of the manuscript.

Funding: This research received no external funding.

Conflicts of Interest: The authors declare no conflict of interest.

References

1. Sharp, R.S. The Stability and Control of Motorcycles. *J. Mech. Eng. Sci.* **1971**, *13*, 316–329. [CrossRef]
2. Sharp, R.S.; Evangelou, S.; Limebeer, D.J.N. Advances in the modelling of motorcycle dynamics. *Multibody Syst. Dyn.* **2004**, *12*, 251–283. [CrossRef]
3. Cossalter, V. *Motorcycle Dynamics*, 2nd ed.; Lulu.com: Morrisville, NC, USA, 2006.
4. Cossalter, V.; Lot, R.; Massaro, M. An advanced multibody code for handling and stability analysis of motorcycles. *Meccanica* **2011**, *46*, 943–958. [CrossRef]
5. Pacejka, H.; Besselink, I.J.M. *Tire and Vehicle Dynamics*, 3rd ed.; Butterworth-Heinemann: Oxford, UK, 2012.
6. Limebeer, D.J.N.; Massaro, M. *Dynamics and Optimal Control of Road Vehicles*; Oxford University Press: Oxford, UK, 2018.
7. Cossalter, V.; Lot, R.; Massaro, M. The influence of frame compliance and rider mobility on the scooter stability. *Veh. Syst. Dyn.* **2007**, *45*, 313–326. [CrossRef]
8. Passigato, F.; Eisele, A.; Wisselmann, D.; Gordner, A.; Diermeyer, F. Analysis of the Phenomena Causing Weave and Wobble in Two-Wheelers. *Appl. Sci.* **2020**, *10*, 6826. [CrossRef]
9. Bova, M.; Massaro, M.; Petrone, N. A Three-Dimensional Parametric Biomechanical Rider Model for Multibody Applications. *Appl. Sci.* **2020**, *10*, 4509. [CrossRef]
10. Nishimi, T.; Aoki, A.; Katayama, T. Analysis of Straight Running Stability of Motorcycles. *SAE Tech. Pap.* **1985**, *856124*, 1080–1094.
11. Katayama, T.; Aoki, A.; Nishimi, T.; Okayama, T. Measurements of structural properties of riders. *SAE Tech. Pap.* **1987** , *871229*, 1–9.
12. Doria, A.; Formentini, M.; Tognazzo, M. Experimental and numerical analysis of rider motion in weave conditions. *Veh. Syst. Dyn.* **2012**, *50*, 1247–1260. [CrossRef]
13. Uchiyama, H.; Tanaka, K.; Nakagawa, Y.; Kinbara, E.; Kageyama, I. Study on Weave Behavior Simulation of Motorcycles Considering Vibration Characteristics of Whole Body of Rider. 2018. Available online: https://saemobilus.sae.org/content/2018-32-0052/ (accessed on 8 August 2021).
14. Cossalter, V.; Doria, A.; Lot, R.; Massaro, M. The effect of rider's passive steering impedance on motorcycle stability: Identification and analysis. *Meccanica* **2011**, *46*, 279–292. [CrossRef]
15. Doria, A.; Marconi, E.; Massaro, M. Identification of rider's arms dynamic response and effects on bicycle stability. In Proceedings of the AMSE IDETC International Design Engineering Technical Conferences & Computers and Information in Engineering Conference, St. Louis, MO, USA, 17–19 August 2020.
16. Massaro, M.; Cole, D.J. Neuromuscular-Steering Dynamics: Motorcycle Riders vs. Car Drivers. In Proceedings of the ASME 2012 5th Annual Dynamic Systems and Control Conference, Fort Lauderdale, FL, USA, 31 October–2 November 2012; pp. 217–224.
17. Massaro, M.; Lot, R.; Cossalter, V.; Brendelson, J.; Sadauckas, J. Numerical and experimental investigation of passive rider effects on motorcycle weave. *Veh. Syst. Dyn.* **2012**, *50*, 215–227. [CrossRef]

18. Klinger, F.; Nusime, J.; Edelmann, J.; Plöchl, M. Wobble of a racing bicycle with a rider hands on and hands off the handlebar. *Veh. Syst. Dyn.* **2014**, *52*, 51–68. [CrossRef]
19. Marcolongo, R. Notizie Sul Discorso Matematico e Sulla vita di Giuio Mozzi. *Boll. Bibliogr. E Stor. Delle Sci. Mat.* **1905**, *8*, 1–8.
20. Shabana, A.A. *Dynamics of Multibody Systems*, 4th ed.; Cambridge University Press: Cambridge, UK, 2013.
21. Cossalter, V.; Doria, A. Analysis of motorcycle slalom manoeuvres using the Mozzi axis concept. *Veh. Syst. Dyn.* **2004**, *42*, 175–194. [CrossRef]
22. Cossalter, V.; Doria, A. Instantaneous screw axis of two-wheeled vehicles in typical manoeuvres. *Veh. Syst. Dyn.* **2006**, *44*, 669–678. [CrossRef]
23. Cossalter, V.; Bellati, A.; Doria, A.; Peretto, M. Analysis of racing motorcycle performance with additional considerations for the Mozzi axis. *Veh. Syst. Dyn.* **2008**, *46*, 815–826. [CrossRef]
24. Armanini, S.; Leo, E.; Pezzola, M.E.; Taroni, N.; Cheli, F. Riding velocity and payload conditions affecting weave modal shape: Subjective assessment on riding safety. In Proceedings of the Symposium on the Dynamics and Control of Single Track Vehicles, Padova, Italy, 9–11 September 2019.

applied sciences

MDPI

Article

Variability of Gravel Pavement Roughness: An Analysis of the Impact on Vehicle Dynamic Response and Driving Comfort

Vidas Žuraulis [1], Henrikas Sivilevičius [1], Eldar Šabanovič [1], Valentin Ivanov [2,*] and Viktor Skrickij [1]

1 Transport and Logistics Competence Centre, Vilnius Gediminas Technical University, Saulėtekio al. 11, LT-10223 Vilnius, Lithuania; vidas.zuraulis@vilniustech.lt (V.Ž.); henrikas.sivilevicius@vilniustech.lt (H.S.); eldar.sabanovic@vilniustech.lt (E.Š.); viktor.skrickij@vilniustech.lt (V.S.)

2 Automotive Engineering Group, Technische Universität Ilmenau, Ehrenbergstr. 15, 98693 Ilmenau, Germany

* Correspondence: valentin.ivanov@tu-ilmenau.de

Featured Application: In this article, we present a methodology for adapting pavement quality indexes for roads with gravel pavement. The findings presented in this article can be used to improve driving safety and gravel pavement maintenance.

Abstract: Gravel pavement has lower construction costs but poorer performance than asphalt surfaces on roads. It also emits dust and deforms under the impact of vehicle loads and ambient air factors; the resulting ripples and ruts constantly deepen, and therefore increase vehicle vibrations and fuel consumption, and reduce safe driving speed and comfort. In this study, existing pavement quality evaluation indexes are analysed, and a methodology for adapting them for roads with gravel pavement is proposed. We report the measured wave depth and length of gravel pavement profile using the straightedge method on a 160 m long road section at three stages of road utilization. The measured pavement elevation was processed according to ISO 8608, and the frequency response of a vehicle was investigated using simulations in MATLAB/Simulink. The international roughness index (IRI) analysis showed that a speed of 30–45 km/h instead of 80 km/h provided the objective results of the IRI calculation on the flexible pavement due to the decreasing velocity of a vehicle's unsprung mass on a more deteriorated road pavement state. The influence of the corrugation phenomenon of gravel pavement was explored, identifying specific driving safety and comfort cases. Finally, an increase in the dynamic load coefficient (DLC) at a low speed of 30 km/h on the most deteriorated pavement and a high speed of 90 km/h on the middle-quality pavement demonstrated the demand for timely gravel pavement maintenance and the complicated prediction of a safe driving speed for drivers. The main relevant objectives of this study are the adaptation of a road roughness indicator to gravel pavement, including the evaluation of vehicle dynamic responses at different speeds and pavement deterioration states.

Keywords: gravel pavement; roughness; straightedge; power spectral density; international roughness index; vehicle response; driving comfort

Citation: Žuraulis, V.; Sivilevičius, H.; Šabanovič, E.; Ivanov, V.; Skrickij, V. Variability of Gravel Pavement Roughness: An Analysis of the Impact on Vehicle Dynamic Response and Driving Comfort. *Appl. Sci.* **2021**, *11*, 7582. https://doi.org/10.3390/app11167582

Academic Editors: Flavio Farroni, Andrea Genovese and Aleksandr Sakhnevych

Received: 15 July 2021
Accepted: 15 August 2021
Published: 18 August 2021

1. Introduction

Pavement surfaces for roads and urban streets are mostly made of concrete, asphalt, cobblestone, and gravel. The share of gravel pavement, particularly on local and regional roads of low- and middle-income countries, remains relatively significant. Decreased handling and comfort, the impact on roadside areas, and other destructive indicators from this point of view make global sustainable transportation hardly achievable [1]. The costs of gravel pavement construction are the lowest; however, its strength, environmental impact, and driving conditions are the poorest [2–4]. Gravel pavement must be maintained continuously by employing the blading technique to reduce transverse ripples and the resulting changes in surface elevation. It is also necessary to add bulk material periodically to gravel pavement due to the deterioration process. Rough gravel pavement increases fuel consumption and gas emissions [5], which is incompatible with mobility and life quality needs and environmental principles in today's society. It is estimated that, based on a reference of 25 years, paving 1 km of gravel road results in 230,000 L of fuel saved and 580 tons of CO_2 emissions avoided [6]. Moreover, many road accidents occur on gravel roads with relatively high traffic volume [7]. Finally, safety, stability, and comfort systems including suspension development must also be adapted to operate efficiently under imperfect pavement with different degrees of roughness [8–14].

Currently, there are no straightforward methodologies for determining gravel road surface condition according to a driver's need to travel safely and comfortably. The existing studies on vehicle response analyses, safety, durability, and fatigue loads are mostly related to paved roads [15–18]. Different indexes are used for road pavement quality evaluation; however, they are mainly used for asphalt or concrete pavements. One of the goals of this study is to analyse the existing indexes and develop a methodology for adapting such indexes to roads with gravel pavement. Instead of developing an original index, we proposed modifying one of the widely used indexes, thus, increasing the possibility of its further usage by operators responsible for ensuring road quality and by other researchers.

The goal is to develop the methodology for gravel road quality evaluation concerning pavement deterioration and vehicle dynamic response, providing appropriate maintenance indication for this pavement type.

This paper is organized into six sections, which includes the Introduction. In Section 2, related studies on road roughness are analysed, and the specificity of gravel pavement is provided. The impact of a gravel road on vehicle dynamic responses considering comfort, handling, and vehicle components' degradation are taken into account. Road roughness measurement techniques and indexes used for quality evaluations are reviewed and the methodological gaps are identified. In Section 3, we provide the strategy for this study. In addition, we present the theoretical background for the selected index calculations. In Section 4, we describe the experimental procedure. In Section 5, the specificity of the international roughness index (IRI) approach for gravel pavement is analysed, and vehicle dynamic responses in the context of the excitation of sensitive frequencies at different driving speeds on gravel pavement are examined. In the Conclusions, we summarise the main results and explain how the IRI can be used for quality evaluation of gravel pavement and define future study guidelines.

2. Related Works

Three classes of grouped parameters of road pavement roughness, including geometric road parameters, statistical values, and performance-based indicators, are usually used in the literature [19]. Statistical parameters of geometric characteristics are directly related to the pavement texture profile's geometry, which is described in the international standard ISO 13473-2 [20]. The performance indicators describe tyre–surface interaction, and this area is related to a broad scope of vehicle dynamic response.

Rapid changes in surface characteristics are one aspect of gravel pavement's specific properties. The deterioration of gravel pavement is affected by climatic and meteorological factors such as wind, water, and temperature fluctuations. These factors alone, or in combination with vehicle loads, break down gravel pavement even faster. Thus, a five-level rating system has been introduced for evaluating the surface of gravel roads [21]. The system includes: (i) crown (shaping and grading quality); (ii) drainage (roadside ditches and culverts); (iii) gravel layer (thickness, gradation, particles shape and regularity, and gravel durability); (iv) surface deformation (corrugation, potholes, and ruts); (v) surface defects (dust and loose aggregate particles). Deteriorated pavement with specific waves of a narrow band spectrum is partly similar to rigid pavement that contains a high number of cosine shape obstacles. In such a case, the road profile is considered to be quasi-homogeneous, i.e., standard profile processing is not directly applicable [22].

The corrugation or 'washboard' phenomenon of gravel pavement occurs in small transverse ripples. The periodic vertical motion of a vehicle's wheels, which is caused by suspension characteristics and body bounce, affects harmonic loads and ripple formation. Very small and initially insignificant imperfections primarily cause corrugation on an unpaved road, such as small undulations, variance in surface flexibility, or even sporadic grains [23,24]. Following a study on the factors causing gravel road corrugation, a regression analysis enabled the researchers to define the extent of corrugation [25]. Surface deformation, especially corrugation, is usual for dirt roads under dry weather conditions; however, ripples also occur on paved surfaces where vehicle traffic involves intensive braking or acceleration, which are typical in vertical curves of roads near intersections.

The process of decomposition and overall deterioration of quality is more difficult to characterize for gravel pavement than asphalt, concrete, or cobble road pavements. During periods of dry weather, due to the interaction between wheels and road surface, pavement wear increases roads' dustiness, and thus, negatively impacts the comfort and health of vehicle occupants accessing and using this particular type of road [26,27]. On the contrary, rain washes away materials used for reducing dustiness as well as fine particles from the pavement surface, weakening the bedrock and resulting in an increased number of defects and deeper tracks in gravel pavements [28].

Defects in gravel pavement have been shown to result in a decrease in driving speed, comfort, and friction [29,30], as well as an increase in vehicle chassis damage [31]. Wavelength and the spatial frequency of road pavement characterise irregularities and a texture effect on vehicle dynamic response. Wear on vehicle components, rolling resistance, and discomfort properties are sensitive to longer wavelengths (unevenness and partly megatexture), while friction between tyres and the road interacts at mega-, macro-, and microtexture levels. Moreover, two types of pavement profile wave sizes have been specified in the tyre-asphalt contact area: short waves are 3–6 mm in length and long waves are 20–30 mm [32]. This particular study demonstrated that the maximum height of 1 mm micro-irregularities were found until adhesion tended to increase; however, this regularity was not applicable for rough surfaces.

Different maintenance procedures are applied to gravel pavement which mainly include light or hard blading, sand cushioning, reshaping, regravelling, etc. [33]. Bitumen emulsion spraying is used as an effective method for decreasing dustiness [34]. Jurkevičius et al. [35] defined a road maintenance management system based on a model for predicting pavement performance and deterioration. The model could be applied for gravel pavement and calculated the loss of gravel thickness [36], therefore, applicable for planning the thickness of gravel layer for road rehabilitation or a new road. Data on traffic, climate, and roughness variation; material quality; gradeability; and cost should be collected to select an optimal road maintenance strategy.

Pavement quality evaluations that include roughness measurements are necessary for planning road pavement maintenance. Various methods and devices have been used for evaluating the roughness and deterioration level of road pavement [37,38]. For profiling, various devices such as a straightedge, a dipstick, the Merlin machine, and an inertial profiler have commonly been used [39,40]. Response-based devices (displacement or acceleration sensors, equipped trailers, or vehicles) are widely applied due to fast processing and low-cost equipment availability [41–47]. Monitoring the state and quality of unpaved roads can be based on visual observations and in-person surveying, contrary to paved road evaluations in which data are collected automatically [48]. A digital image acquisition system has been applied utilizing an unmanned aerial vehicle (UAV); however, limited pavement distresses were captured (mainly potholes and rutting) and the flight range of the UAV made not widely applicable. A ride quality assessment of vehicle response is convenient enough, although the most characteristic driving speed is not the same for different quality pavements. Various identification principles are also known from studies in the literature, for example, a mechanical profilometer for three-dimensional terrain models [49], smartphone accelerometer processing [50], digital image analysis [51], and reconstruction methods using additional sensors [52,53].

Various ride quality assessment indicators have been defined to describe the impact of road roughness and, in the last 15 years, several new indexes have been introduced [54]. Loprencipe and Zoccali [55] compared the relations among Ride Number (RN), Michigan Ride Quality Index (RQI_{Mich}), Minnesota Ride Quality Index (RQI_{Mn}), and frequency-weighted vertical acceleration (a_{wz}) and their relationships with the most-used IRI. The RN, RQI_{Mich}, and RQI_{Mn} were developed considering road customer opinions, and their thresholds of ride quality level were not speed related. Therefore, the weighted acceleration a_{wz} provided by ISO 2631 [56] was applicable for lower speed roads (not exceeding 50 km/h). Gurmail and Kiss [57] analysed the response of towed vehicles employing the power spectral density (PSD) function to determine the effect of vehicle wheel–ground interaction on structural deterioration. Different types of roads with high- and low-quality surfaces were selected for the simulation analysis. The same frequency ranges with the elevated local maximum of the PSD characteristic increasing speeds from 3.6 to 18 km/h were determined for a rough gravel road. This study also stated that, under low speed, towed vehicle conditions (up to 20 km/h), micro-obstacles (humps and holes) occurred with shorter than 2.4 m of wavelength or greater than 0.4–1 m wave numbers on agricultural roads. A smooth exponential increase in root mean square (RMS) values of measured accelerations in a speed range from 3.6 to 25 km/h was determined. Lower-quality road surfaces with irregular forms of roughness are not suitable for the conventional IRI-based description, despite its broad applicability and simple enough computational implementation [58]. Different values of other indicators (for passengers in particular) were determined for roads with the same IRI values; therefore, alternative roughness or vehicle/passenger response indicators (such as the RMS value of acceleration) could be used for specific pavements. Some significant aspects of comfort evaluation in the cabins of different bus topologies were also highlighted as an IRI limitation [59]. First, it was argued that, in the case of similar IRI values, it was possible that there could be three different road profiles and that passengers would experience different vibration conditions. Then, insufficient sensitivity to lateral motions (acceleration) was proven based on experimental data. Due to the limited IRI evaluation of specific or unpaved road surfaces, dynamic response-based devices or road profile scanners have been used [45]. Devices added onto a vehicle have shown a high enough correlation with the IRI and allowed real-time measurements. However, evaluations of different roughness indicators have shown that nearly all the roughness measuring devices employed around the world could produce measurements on the same scale, as long as that scale was suitably selected; therefore, the IRI was defined [60,61]. The World Bank developed an IRI after conducting the International Road Roughness Experiment (IRRE) in Brazil in 1982. The aim of the experiment was to harmonise the existing indices for evaluating pavement roughness. Due to its stability over time and transferability over the world, it has become the most widely employed pavement index (not only for roughness assessment) in developed countries [62–65] and in developing countries [66–68]. Therefore, it is preferable to adopt an existing and known index to gravel roads instead of developing a new one.

Road infrastructure in different regions has been developed unevenly as there are various gravel coatings in suburban and rural areas. The majority of studies on vehicle–road interaction have focused on asphalt pavement. Evaluations of gravel road conditions have commonly been performed using methodologies developed several decades ago, and therefore new solutions are needed.

3. Indexes for Gravel Road Quality Assessment

The fundamental concept of road surface profiles at different mechanical vibration levels is described by the international standard ISO 8608 [69]. The PSD of vertical road profile displacement, $G_d(n)$, or its rate of change by velocity, $G_v(n)$, or acceleration, $G_a(n)$, conforming to spatial frequency, (n), are used for defining the degree of roughness:

$$G_d(n) = G_d(n_0)\left(\frac{n}{n_0}\right)^{-w}, \tag{1}$$

$$G_v(n) = (2\pi n)^2 \cdot G_d(n), \tag{2}$$

$$G_a(n) = (2\pi n)^4 \cdot G_d(n), \tag{3}$$

where n is the spatial frequency (cycles/m); n_0 is the reference spatial frequency (0.1 cycles/m); w is the pavement waviness indicator, or the exponent of the fitted PSD. For general use, the road roughness indicator usually takes the values of $1.75 < w < 2.25$ [70] and for ISO classification assuming constant velocity PSD, $w = 2$.

The profile of gravel pavement cannot be approximated with the waviness of $w = 2$ used in ISO 8608, which usually occurs on paved roads [46]. Múčka [71] reported that up to 40% of roads had differences in wavebands, and 20% of tested sections had short wave unevenness ($w > 2$). Another type of specific road usually used for evaluating vehicle dynamic response, known as Belgian paving, also has few brakes in its displacement spectral density characteristic [72]. There, the wavenumber of 6 cycles/meter corresponds with the size of cobbles forming the pavement [73].

Generally, PSD expresses power in a signal per unit frequency [74]. The PSD of road profile displacement can also be expressed by an unevenness index C [75]:

$$G_d(n) = C \cdot n^{-w}. \tag{4}$$

Wavelength λ (m) has an inverse value of spatial frequency. In the time domain, frequency is expressed as the ratio between wavelength and vehicle speed v (m/s), which generally satisfies conditions for light vehicles [76,77]:

$$0.5\ \text{Hz} < \frac{v}{\lambda} < 15\ \text{Hz} \tag{5}$$

In the context of vehicle dynamic response and vibration response analysis, it is more convenient to convert excitation from road pavement spatial frequency n (cycles per meter) to temporal frequency f (cycles per second or Hz). In this case, the supposed vehicle speed (v) is involved [78]:

$$f = n \cdot v. \tag{6}$$

Then, the PSD of the road pavement, in consonance to temporal frequency $G_d(f)$, is adapted to vehicle speed:

$$G_d(f) = \frac{G_d(n)}{v}. \tag{7}$$

The IRI is the most frequently used single-number indicator usually implemented in road maintenance management plans, ride comfort, or stability assessments. The IRI value is estimated from the reference quarter car model (Figure 1a), also known as the Golden car model [39], driving at 80 km/h. The highest sensitivity for IRI computation was found from 0.065 to 0.42 cycles/m (wavelength from 15 to 2.4 m) (Figure 1b), which also varied from the selected sample interval and integration method [79]. The general IRI computation is based on two components as the velocity of suspension displacement [79,80]:

$$\text{IRI} = \frac{1}{L}\int_0^{L/v} \left|\dot{z}_{sm} - \dot{z}_{usm}\right| dt, \tag{8}$$

where L is the length of the evaluated road profile (m), v is the longitudinal vehicle speed applied for the quarter car model (m/s), $\dot{z}_{sm}$ is the the velocity of sprung mass (SM) vertical displacement (m/s), $\dot{z}_{usm}$ is the the velocity of unsprung mass (USM) vertical displacement (m/s), and dt is the time increment (s).

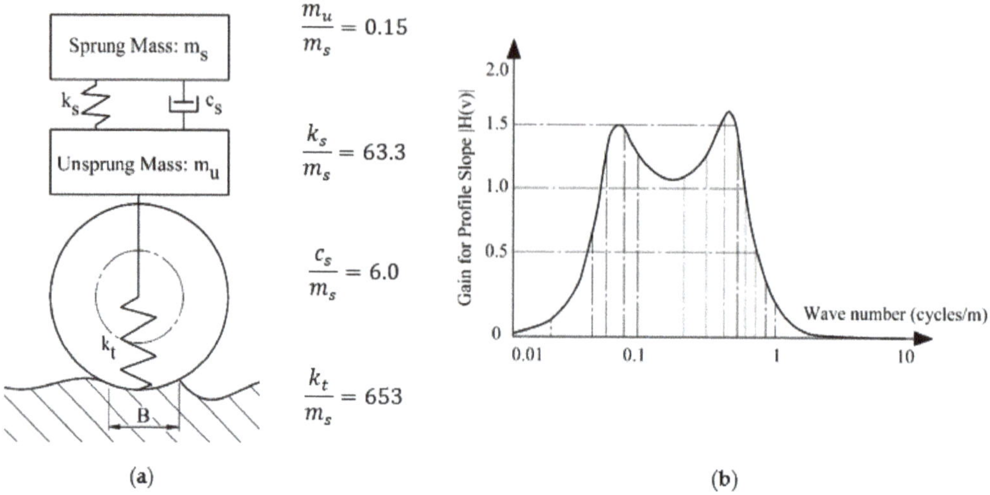

Figure 1. IRI computation and evaluation [39]: (**a**) Golden car model and its parameters; (**b**) IRI sensitivity to wave number.

The dynamic load coefficient (DLC) has been used for evaluating vehicle response to ride safety and driving stability. It shows variations in the vertical tyre force as compared with its static force [81,82]:

$$\text{DLC} = \frac{\text{RMS}_{Fz,\text{dyn}}}{F_{\text{stat}}}, \quad (9)$$

where F_{stat} is the static force and $\text{RMS}_{Fz,\text{dyn}}$ is the the root mean square of the dynamic vertical tyre force ($F_{z,\text{dyn}}$), determined by applying a vehicle model simulation or using the formula:

$$F_{z,\text{dyn}} = k_t(z_{usm} - h), \quad (10)$$

where k_t is the tyre radial stiffness (taken from the Golden car model), z_{usm} is the vertical displacement of USM, and h is the vertical roughness of the road pavement.

The DLC value usually ranges from 0.05 to 0.3, and ride safety is high when DLC < 0.1. In contrast, a value above 0.15 is related to shaking wheel interaction with the rough pavement and an increase in contact loss [83].

4. Roughness Measurement Using a Straightedge

The straightedge method for the pavement rut depth measurement [84] was selected for the data collection from the real gravel pavement. The proposed static method is based on measuring the gap between the road surface and a constant length (3.0 m) beam placed along the track corresponding to a vehicle's left-side wheels. In this study, the straightedge method for measuring gravel pavement longitudinal roughness was adopted, which is usually characterised by higher roughness. Therefore, a small correction in measurement was made by adding 20 mm height spacers (h_{spacer}) at the ends of the beam (Figure 2). An additional spacer was used in the cases when the end of the beam was over pavement dent.

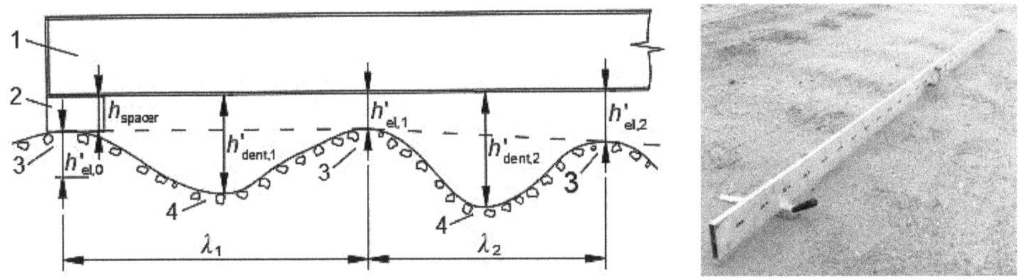

(a) (b)

Figure 2. The measurement of gravel road roughness applying a straightedge: (**a**) schematic view, 1—straightedge of 3 m in length; 2—added spacer; 3—pavement elevation; 4—pavement dent; (**b**) measurement of gravel pavement.

The actual height of elevation and dent is determined by subtracting added spacer height (20 mm) considering each measured vertical distance, which is the clearance between the bottom of the straightedge and the pavement surface at corresponding elevation and dent:

$$h_{\text{el}} = h'_{\text{el}} - 20, \tag{11}$$

$$h_{\text{dent}} = h'_{\text{dent}} - 20, \tag{12}$$

where h'_{el} and h'_{dent} are the measured height of elevation and dent, respectively (mm); h_{el} is the actual height of elevation or clearance between the bottom of the straightedge and the pavement at corresponding elevation (mm); h_{dent} is the actual depth of dent or clearance between the bottom of the straightedge and the pavement at corresponding dent (mm).

The first stage of data post-processing included the inversion of data on the measured pavement elevation. Thus, the negative values of pavement dent were obtained. The absolute depth (amplitude) of the first wave Δ_{λ_1} (Figure 2), meaning the difference between dent depth and the average of neighbouring elevations, is expressed by the formula:

$$\Delta_{\lambda_1} = h_{\text{dent},1} - \frac{h_{\text{el},0} + h_{\text{el},1}}{2}. \tag{13}$$

The absolute depth (amplitude) of the second and next waves are as follows:

$$\begin{aligned} \Delta_{\lambda_2} &= h_{\text{dent},2} - \tfrac{h_{\text{el},1} + h_{\text{el},2}}{2}, \\ &\vdots \\ \Delta_{\lambda_N} &= h_{\text{dent},N} - \tfrac{h_{\text{el},N-1} + h_{\text{el},N}}{2}. \end{aligned} \tag{14}$$

A straight section of rural road, 160 m in length, with gravel pavement was selected for this study. Pavement roughness was measured considering the different states of the pavement. The first state (Pavement State 1, PS-1) means that the gravel pavement measurement was carried out two days after pavement maintenance using the blading technique without adding gravel material to the existing road surface. This state corresponds to the best-expected quality and driving conditions. The second state (Pavement State 2, PS-2) is described as a middle-quality gravel road, three weeks after pavement maintenance and the start of intensive road deterioration. The third state (Pavement State 3, PS-3) corresponds to the worst pavement quality, when maintenance is expected.

After blading, large grains were observed on the gravel's flat surface; however, it did not have a significant influence on vehicle response because of tyre enveloping properties. The formed ripples were noticeable on the surface before maintenance (worst condition), and almost no individual free grains were found.

5. Results

The processed data on the measured profile elevation conforming to the pavement state of the road section are shown in Figure 3. The amplitudes do not exceed the elevation value of 20 mm in all pavement states; however, the profiles have noticeable differences which are described further.

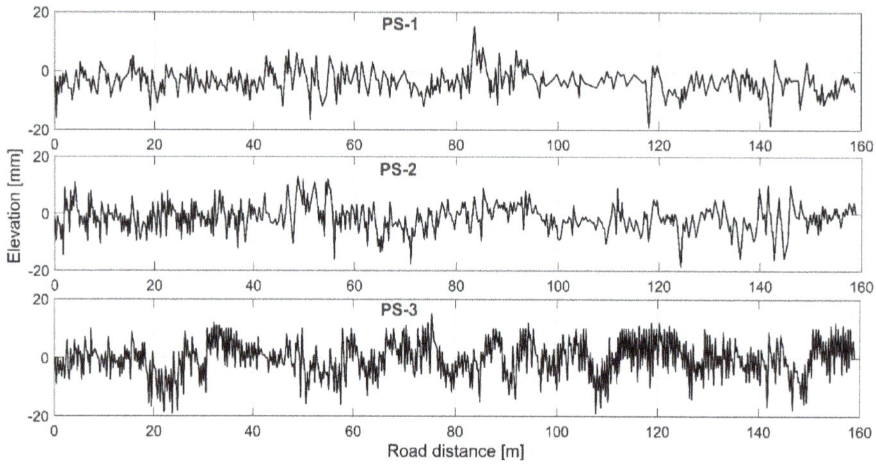

Figure 3. Pavement elevation of the same road section in different pavement states (PSs).

The determined surface roughness of the analysed road section is presented in line with ISO 8608 methodology in the form of displacement spectral density—spatial frequency (Figure 4). The pavement roughness description concerning the PSD log-log plot has a specific break from a smooth characteristic at spatial frequency $n = 2$ cycles/m (Figure 4).

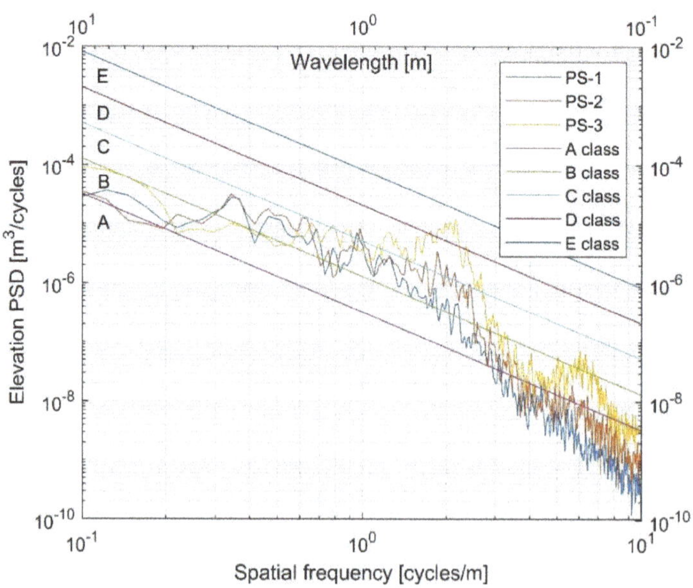

Figure 4. Roughness representation of the measured road section in its different states.

The characteristic of PS-3 has an additional break at around 6.5 cycles/m. The characteristic PSD slope changes show variations in profile waviness at different bands of waves (long, medium, or short). The corrugation or 'washboard' phenomenon is recognised at the investigated road section. It is the most characteristic for the PS-3 state, where the shortest waves with the smallest variance are established. Furthermore, the maximum number of the estimated waves represented PS-3, which confirmed the worst pavement quality. Such smooth and constant ripples at certain road sections make driving extremely uncomfortable and unsafe at a specific speed, which is also one of the reasons for pavement corrugation. The results of the performed analysis of the wavelengths formed on the gravel pavement were used to study the reliability of the calculated IRI of the investigated pavement.

5.1. The IRI Approach for Gravel Pavement

Methodically, the velocities of SM and USM in a vertical direction of the Golden car model are determined and used for calculating the IRI value (Equation (8)). The second column in Table 1 shows the IRI values of the different states of the gravel pavement at a driving speed of 80 km/h, as provided in the ASTM [80]. Under the deteriorating road pavement condition, the IRI values change inconsistently; the highest IRI value (7.28 m/km) is obtained under PS-2 condition, while the lowest IRI value (5.51 m/km) is obtained under PS-3 condition. This is not in line with the visually observed (Figure 3) trends of pavement roughness deterioration from PS-1 to PS-3. To analyse the resulting distribution of IRI values concerning pavement condition, the vertical velocities of SM and USM were calculated. Table 1 shows the calculated RMS values of the SM and USM velocities in vertical directions. The RMS is calculated as follows:

$$\mathrm{RMS}_{\dot{z}_i} = \left[\frac{1}{T}\int_0^T \dot{z}_i^{\,2}\,dt\right]^{1/2}, \tag{15}$$

where T is the duration of the evaluated road profile (s) and i is the indicator for the SM or USM depending on the calculated parameter ($\dot{z}_{sm}$ or $\dot{z}_{usm}$).

The values presented in Table 1 show that the RMS of the SM velocity increases along with consistent deterioration of the road pavement condition (from PS-1 to PS-3), but the RMS value of the USM velocity is unevenly distributed (such as the IRI values). The SM movement is inhibited by suspension elements (mainly dampers). At the same time, the USM interacts with the gravel pavement through the tyres, for which stiffness (k_t) is around ten times higher than that of suspension (k_s). As a result, the USM moves faster than the SM and has a more significant effect on calculating the IRI. These results indicate that the calculated IRI values are not suitable for analysing gravel pavements in which the wavelengths of the dominant length in the mega-texture area are formed. Similar results of higher sensitivity of wheel holder vibrations to the IRI value as compared with vehicle frame vibrations have been captured following measurements of reconstructed urban roads with cobblestone and asphalt pavements in [85].

Table 1. The estimated IRI values of the analysed gravel pavement.

Pavement State	IRI, m/km	Velocity RMS, m/s	
		SM	USM
PS-1	6.33	0.0283	0.1821
PS-2	7.28	0.0313	0.2303
PS-3	5.51	0.0378	0.1530

The IRI values of the gravel pavement are determined by simulating a quarter car model as a vehicle moving at a speed of 80 km/h. Due to comfort and sufficient stability, the tested road section of gravel pavement is characterised by lower driving speeds. Therefore, the IRI values were calculated for a broader (30–90 km/h) speed range (Figure 5a).

The received IRI values are significantly higher at lower speeds because the same road section is covered over a longer time duration, leading to higher accumulated suspension displacement in the IRI calculation. In this case, the distribution value at a given speed under a certain gravel pavement condition rather than the IRI absolute value is analysed. Figure 5a shows that, at lower speeds (30 and 40 km/h), the IRI values increase along with road pavement deterioration from PS-1 to PS-3. At speeds above 45 km/h, the distribution of the IRI values changes; the values of IRI_{PS-3} corresponding to the worst gravel pavement condition become lower than the values of IRI_{PS-2}, whereas, at speeds above 55 km/h, the values of $IR_{IPS\ 3}$ are the same as the worst gravel pavement condition and become lower than that of IR_{IPS-1} (under the best-quality gravel pavement state). Thus, for calculating the IRI value, a speed of 45–55 km/h meets the transition phase, the IRI value of which is incorrectly consistent with the pavement condition.

The RMS of the SM and USM velocities was used to analyse variations in the IRI values (Figure 5b). Fluctuations in the RMS values demonstrated that an increase in driving speed resulted in minor changes in the SM velocity (continuous lines). However, changes in the velocity of the USM were found to be significant (dotted lines). In addition, USM_{PS-3} is on a declining trend as speed increases while USM_{PS-1} and USM_{PS-2} are on an upward trend. These trends in the USM are repeated for IRI values and confirm that a speed of 30–45 km/h is compatible with the pavement state (PS) of the gravel road (Figure 5a).

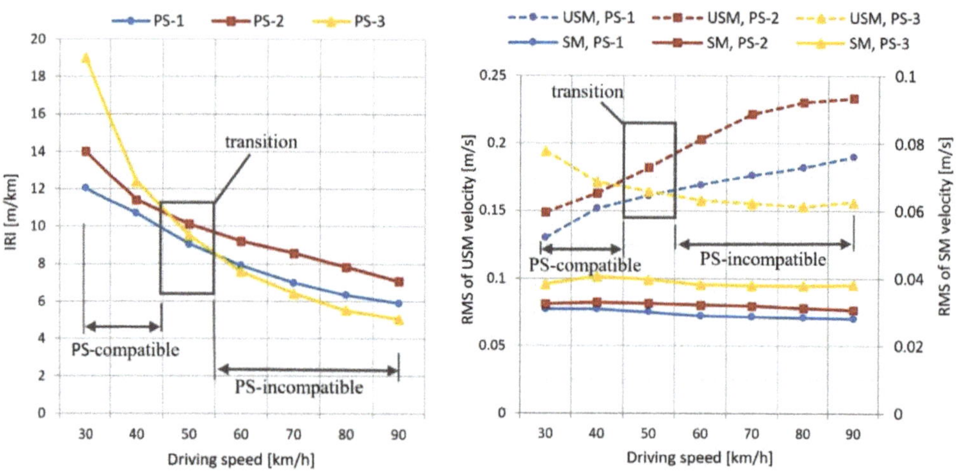

(a) (b)

Figure 5. The analysis of IRI calculation sensitivity in line with calculation-applicable speed: (**a**) IRI values; (**b**) RMS values of SM and USM displacement velocity.

5.2. Vehicle Response to the Pavement Surface State

In line with ISO 8608, the presented initial characteristic of the measured road roughness (Figure 4) shows a clear peak of displacement PSD at a spatial frequency of 2 cycles/m for the third pavement surface (PS-3) state. This peak is also repeated at six cycles/m of the frequency range. Such unevenness of the characteristic is considerably weakened in the second pavement surface (PS-2) state and is completely invisible under the best pavement surface (PS-1) state. Converting road roughness characteristics to temporal frequency (Equations (6) and (7)) reveals these characteristics' irregularities at different vehicle driving speeds (Figure 6). Speeds up to 70 km/h, shown as higher road speeds on the gravel pavement, are prohibited under national road traffic regulations [86]. The most pronounced fractures in the elevated PSD are observed for the third pavement surface PS-3 state (Figure 6): a peak for 10 km/h at 5–6 Hz, a peak for 30 km/h at 15–18 Hz, a peak for 50 km/h at 23–30 Hz, and a peak for 70 km/h at 33–42 Hz.

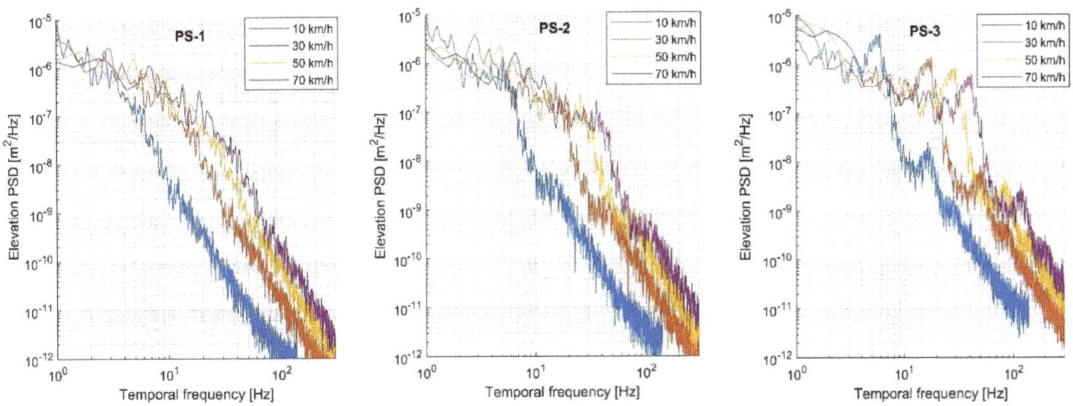

(a) (b) (c)

Figure 6. Roughness representation in temporal frequency measured at different driving speeds and pavement surface states: (**a**) PS-1; (**b**) PS-2; (**c**) PS-3.

Driving at lower speeds (10–30 km/h) through short wavelength roughness (PS-3) formed on the gravel pavement (macro-texture level) demonstrates that vibration frequency caused by waves approaches the natural frequency of vehicle USM (11–15 Hz). This order vibration tends to increase the variability in the normal load of the wheel described by the DLC [87,88]. The normal force of the tyre depends on suspension characteristics [89]; however, road pavement and driving speed have a direct influence. The calculated DLC values for the quarter car model working at a broad speed range (10–90 km/h) are shown in Figure 7. Here, the Golden car model simulation in the MATLAB/Simulink environment was used. Conforming to the pavement surface state, the obtained DLC values have the same distribution as the IRI (Table 1) due to the dominant movement of the USM (the highest RMS of the USM for PS-2). Both tendencies (IRI and DLC) prove that the second pavement surface (PS-2) state excites resonant motion for the analysed light car model.

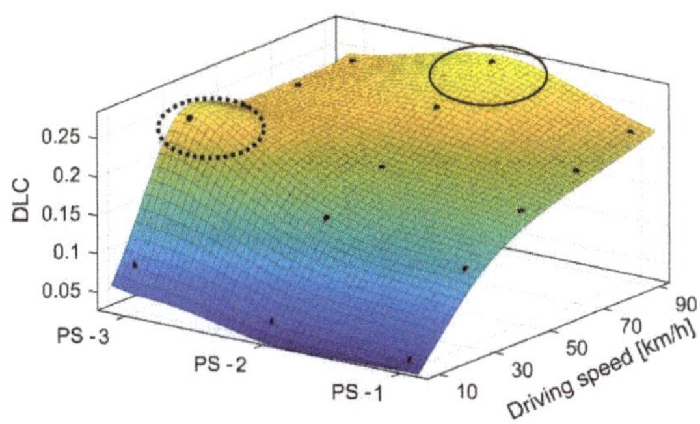

Figure 7. The estimated DLC at different driving speeds and investigated pavement states.

To analyse vehicle response changes in different gravel pavement surface states, the spectral density of pavement displacement at a vehicle vertical dynamics sensitive range of 0–20 Hz was selected (Figure 8). The excitation of the road pavement to the vehicle driven at a speed of 10 km/h corresponds to an increase in pavement deterioration level (Figure 8a). Although there is no significant rise in pavement excitation for the SM and

USM dynamics at 10 km/h (at 1–2 Hz and 10–14 Hz, respectively), the undesired excitation occurs at a human body sensitive frequency range of 4–8 Hz (ISO 2631-1) on PS-3 and PS-2 states. Since a speed of 10 km/h is more typical for a special purpose vehicle with different suspension and tyre properties, in this study on light vehicle application, we continued analysing higher speeds.

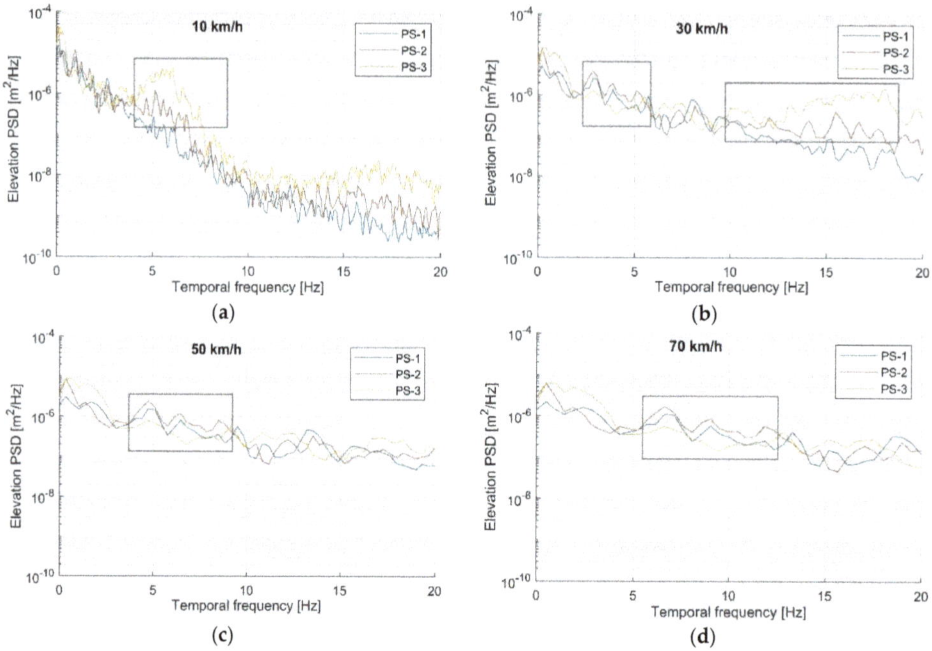

Figure 8. Roughness representation in the temporal frequency: (**a**) 10 km/h; (**b**) 30 km/h; (**c**) 50 km/h; (**d**) 70 km/h.

The characteristic of 30 km/h (Figure 8b) highlights two crucial cases. Excitation at a human body sensitive frequency range of 2.5–5.5 Hz is higher in the middle-quality pavement state (PS-2) and is the lowest in the PS-3 state. This also shows that a slightly higher driving speed (from 10 km/h) makes a milder impact on the human body. The second important point, in the case of 30 km/h, indicates the exclusively high excitation of the PS-3 state at a frequency range of 10–18 Hz. This frequency range covers the resonant frequency of the USM and increases variability in dynamic tyre force, contact loss, and vehicle chassis wear. The corresponding peak of the DLC characteristic is marked with a dotted line in Figure 7. The conversion of PS-3 characteristics from the lowest position at 2.5–5.5 Hz to the highest at 10–18 Hz (Figure 8b) shows possible erroneous driver's sensations about comfortable enough driving, but an imperceptible decrease in road holding.

The characteristics of 50 km/h and 70 km/h do not have distinguished peaks or rises (Figure 8c,d); however, the two characteristics have the same feature of higher PS-2 excitation in the middle-frequency range, i.e., 4–9 Hz for 50 km/h and 5.5–12.5 Hz for 70 km/h. The difference between these two cases is that excitation corresponding to 50 km/h affects the human body, while excitation corresponding to 70 km/h additionally affects USM dynamics (wheel contact). The latter is also recognised as a DLC characteristic (marked with the continuous line in Figure 7).

The excitation at a low-frequency range of 1–2 Hz sensitive for driving comfort is higher on the PS-3 state, but this is reasonable because it is the worst pavement quality.

In general, all PSD curves of pavement excitation tend to descend with each increase (Figure 8). This may lead to a false perception that a higher speed on the gravel pavement gives better driving comfort or safety. The individual analysis of different operating frequencies made it possible to distinguish vehicle stability, comfort, and human body sense. Lower pavement excitation for specific frequencies causes better comfort, but the individual cases of a high DLC warn about the potential of vehicle stability loss.

Natural frequencies utilising the quarter car model's damping ratios (Golden car parameters [39]) were determined from the Bode plot (Figure 9a). Natural frequencies were placed into the speed-wavelength-frequency characteristic (Figure 9b), showing sensitive frequencies overlapping pavement wavelengths at a specific driving speed. The characteristic shows that SM (vehicle body) vibrations are amplified by the pavement with the dominant wavelength of 2–30 m at a driving speed in the range of 10–130 km/h, while USM vibrations (wheels and axles) are amplified by the wavelengths of 0.25–3.5 m in length (Figure 9b).

For example, normal highway asphalt pavement with an unevenness of 10 m wavelength causes uncomfortable driving at around 30–50 km/h (speed curves cross the solid black line in Figure 9b). However, this speed is not characteristic of highway driving, and thus a hardly deteriorated pavement prevents discomfort. Vehicle instability caused by wheel resonance under such conditions is barely activated as the natural frequency of the USM (wheel hop) does not cross any speed curve at the above mentioned 10 m wavelength. On the contrary, the second example presents unpaved roads with gravel pavement where short wavelengths dominate. Considering the 1 m wavelength, the natural frequency of SM is not reached (solid line in Figure 9b). However, the USM resonances at a speed range of 30–50 km/h (crosses dotted line), thus, causing unstable wheel contact with the pavement. Shorter wavelengths activate USM resonance at a lower driving speed. Furthermore, the human body sensitive frequency range of 4–8 Hz falls into the short-middle wavelength band (0.3–5 m) for a wide range of speeds. The obtained vibrations correspond to uncomfortable and unstable driving at a specific road pavement waviness for a certain speed.

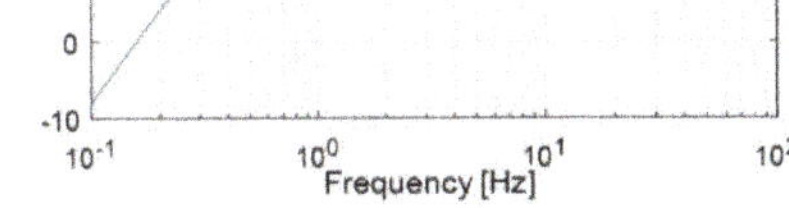

Vehicle mode shape	Natural frequency, ω_n, Hz	Damping ratio, ζ
SM bounce	1.17	0.33
USM vertical hop	10.75	0.30

(a)

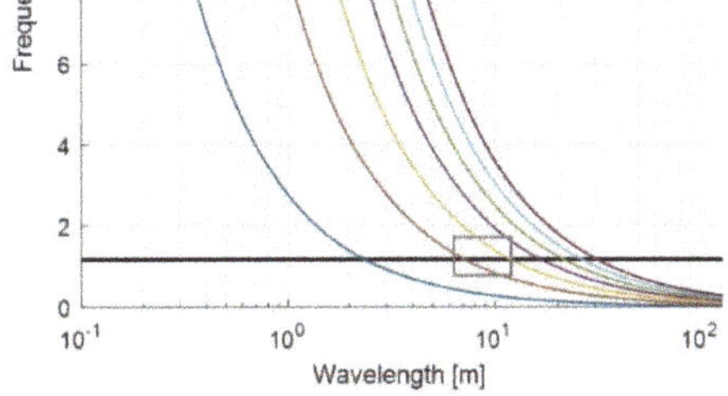

(b)

Figure 9. Sensitive vehicle frequencies of pavement wavelengths: (**a**) Natural frequencies and damping ratios of the Bode plot; (**b**) wavelength frequency characteristics conforming to driving speed.

In this Section, it was proven that the IRI index could be used to evaluate gravel road conditions; the significance of the proposal for calculating the index using the specific velocity interval from 30 to 45 km/h for gravel pavement was indicated. The PSD and

DLC indexes with frequency response characteristics provide valuable information about the impact of different road roughness levels on discomfort and safety at a certain driving speed.

6. Conclusions

In this study, we aimed to determine changes in the quality of gravel pavement overtime during the service life of a road, in line with the level of surface roughness and pavement excitation on vehicle response and driving comfort. The roughness of the experimental 160 m section of road with gravel pavement was measured continuously three times by adopting the 3 m straightedge method: (1) two days after pavement maintenance (blading), (2) three weeks after pavement maintenance, and (3) assessing the worst pavement quality before maintenance. The measurement results showed that the road quality continuously deteriorated with an increase in pavement roughness. Pavement elevation spectral density disclosed several fractures in the roughness log-log plot, which meant variations in pavement waviness. For the worst pavement quality (PS-3), a dominance of short waves of approximately 0.5 m in length was found (lowest standard deviation), which was compatible with the corrugation phenomenon of the transverse ripples formed. Furthermore, such a wavy pattern was found to cause unsprung mass (USM) resonant excitation of the vehicle before the worst road quality was achieved.

Spatial frequency conversion to temporal was done to analyse vehicle dynamic responses at different driving speeds. Although the specific fractures of surface power spectral density (PSD) characteristics remained, for the worst pavement quality state, in particular, the speed-affected frequencies shifted to higher bands. Speed is mostly related to driving safety, and therefore the dynamic load coefficient (DLC), as the indicator of tyre–pavement interaction and vehicle stability, was determined considering the analysed pavement states. Even quarter car simulations of driving at a low speed (<30 km/h) showed that the DLC reached values higher than 0.15 but did not exceed 0.3 when the proportion of wheel contact loss increased rapidly. However, the DLC undesirably increases at a low speed of 30 km/h on the worst pavement quality and a high speed of 90 km/h on the middle-quality pavement. Even though most countries have a speed limit of 70 km/h on gravel pavement, the potential wheel contact force loss draws attention to timely road maintenance. Variations in driving comfort sensation and the indirect prediction of wheel–pavement interaction complicate a driver's selection of safe driving speed. Therefore, vehicle active chassis systems should be continuously developed for better performance on a specific road such as variable roughness gravel pavement.

In consonance with the estimated vehicle suspension response to the investigated type of gravel pavement, IRI values have been found to rarely correspond to the actual level of pavement roughness. Therefore, this worldwide single-number indicator used for road maintenance planning is not appropriate for unpaved pavements with unsteady waviness. Absolute IRI values do not provide reliable results that can be used for safe speed estimation. However, calculating the IRI at a speed range of 30–45 km/h shows an increase in the values corresponding to the deteriorating pavement state. Thus, the IRI values calculated at a low speed (30 km/h) are suitable for assessing gravel pavement condition and road maintenance planning.

The achieved results can be used to assist operators responsible for ensuring road quality and for decision making about road maintenance, to analyse the effects of unpaved roads on vehicle driving stability and comfort, and to support the performance of active safety systems, drivers, road maintenance planners, and road supervisors.

Future experimental and theoretical studies should focus on vehicle dynamic response measurements and driver comfort evaluation in different driving modes under varying pavement states. Continuous research should also investigate higher-order frequency bands related to the issues of tyre dynamics and active vehicle safety control while driving on unpaved roads. The development of pavement roughness measurement utilising laser sensors and artificial intelligence-based image analysis for pavement state identification is planned. The results of direct roughness measurements and their analysis presented in this study could also be used for validating the results of further research.

Author Contributions: Conceptualisation, V.Ž. and H.S.; methodology, V.Ž. and H.S.; software, V.Ž. and E.Š.; validation, V.Ž., V.I. and V.S.; formal analysis, V.Ž.; investigation, all authors; resources, V.S.; data curation, E.Š. and V.Ž.; writing—original draft preparation, all auhors; writing—review and editing, V.Ž., E.Š. and V.I.; visualisation, V.Ž.; supervision, V.S. All authors have read and agreed to the published version of the manuscript.

Funding: This research was funded by the European Union Horizon 2020 Framework Program, Marie Skłodowska-Curie actions, under grant agreement no. 872907.

Institutional Review Board Statement: Not applicable.

Informed Consent Statement: Not applicable.

Data Availability Statement: Data available by request from vidas.zuraulis@vilniustech.lt.

Conflicts of Interest: The authors declare no conflict of interest.

References

1. Horiuchi, S.; Tsuda, A.; Kobayashi, H.; Redding, C.A.; Prochaska, J.O. Sustainable transportation pros, cons, and self-efficacy as predictors of 6-month stage transitions in a Chinese sample. *J. Transp. Health* **2017**, *6*, 481–489. [CrossRef]
2. Obeng, D.A.; Tuffour, Y.A. Prospects of alternative funding sourcing for maintenance of road networks in developing countries. *Transp. Res. Interdiscip. Perspect.* **2020**, *8*, 100225. [CrossRef]
3. Pasindu, H.R.; Gamage, D.E.; Bandara, J.M.S.J. Framework for selecting pavement type for low volume roads. *Transp. Res. Proc.* **2020**, *48*, 3924–3938. [CrossRef]
4. Petkevičius, K.; Maskeliūnaitė, L.; Sivilevičius, H. Determining travel conditions on motorways for automobile transport based on the case study for Lithuanian highways. *Transport* **2019**, *34*, 89–102. [CrossRef]
5. Ziyadi, M.; Ozer, H.; Kang, S.; Al-Qadi, I.L. Vehicle energy consumption and an environmental impact calculation model for the transportation infrastructure systems. *J. Clean Prod.* **2018**, *174*, 424–436. [CrossRef]
6. Gleave, S.D.; Frisoni, R.; Dionori, F.; Casullo, L.; Vollath, C.; Devenish, L.; Spano, F.; Sawicki, T.; Carl, S.; Lidia, R.; et al. EU Road Surfaces: Economic and Safety Impact of the Lack of Regular Road Maintenance. European Parliament—Directorate General for Internal Policies, Policy Department B: Structural and Cohesion Policies, Transport and Tourism. 2014. Available online: https://www.europarl.europa.eu/RegData/etudes/STUD/2014/529059/IPOL_STU(2014)529059_EN.pdf (accessed on 16 August 2021).
7. Mamčic, S.; Sivilevičius, H. The analysis of traffic accidents on Lithuanian regional gravel roads. *Transport* **2013**, *28*, 108–115. [CrossRef]
8. Pretagostini, F.; Ferranti, L.; Berardo, G.; Ivanov, V.; Shyrokau, B. Survey on Wheel Slip Control Design Strategies, Evaluation and Application to Antilock Braking Systems. *IEEE Access* **2020**, *8*, 10951–10970. [CrossRef]
9. Van der Merwe, N.A.; Els, P.S.; Žuraulis, V. ABS braking on rough terrain. *J. Terramech.* **2018**, *80*, 49–57. [CrossRef]
10. Cao, D.; Song, X.; Ahmadian, M. Editors' perspectives: Road vehicle suspension design, dynamics, and control. *Veh. Syst. Dyn.* **2011**, *49*, 3–28. [CrossRef]
11. Hamersma, H.A.; Els, P.S. Improving the braking performance of a vehicle with ABS and a semi-active suspension system on a rough road. *J. Terramech.* **2014**, *56*, 91–101. [CrossRef]
12. Šabanovič, E.; Žuraulis, V.; Prentkovskis, O.; Skrickij, V. Identification of Road-Surface Type Using Deep Neural Networks for Friction Coefficient Estimation. *Sensors* **2020**, *20*, 612. [CrossRef]
13. Žuraulis, V.; Surblys, V.; Šabanovič, E. Technological measures of forefront road identification for vehicle comfort and safety improvement. *Transport* **2019**, *34*, 363–372. [CrossRef]
14. Savitski, D.; Ivanov, V.; Shyrokau, B.; De Smet, J.; Theunissen, J. Experimental study on continuous ABS operation in pure regenerative mode for full electric vehicle. *SAE Int. J. Passeng. Cars Mech. Syst.* **2015**, *8*, 364–369. [CrossRef]
15. Gao, H.; Jézéquel, L.; Cabrol, E.; Vitry, B. Chassis durability and comfort trade-off at early stage of project by virtual proving ground simulation. *Veh. Syst. Dyn.* **2021**, 5–20. [CrossRef]
16. Fauriat, W.; Mattrand, C.; Gayton, N.; Bekou, A.; Cembrzynski, T. Estimation of road profile variability from measured vehicle responses. *Veh. Syst. Dyn.* **2016**, *54*, 585–605. [CrossRef]
17. Kerst, S.; Shyrokau, B.; Holweg, E. Anti-lock braking control based on bearing load sensing. In Proceedings of the EuroBrake, Dresden, Germany, 4–6 May 2015; pp. 4–6.
18. Vantsevich, V.V.; Shyrokau, B.N. Autonomously operated power-dividing unit for driveline modeling and AWD vehicle dynamics control. *Dyn. Syst. Control Conf.* **2008**, *43352*, 891–898. [CrossRef]
19. Bitelli, G.; Simone, A.; Girardi, F.; Lantieri, C. Laser Scanning on Road Pavements: A New Approach for Characterizing Surface Texture. *Sensors* **2012**, *12*, 9110–9128. [CrossRef] [PubMed]

20. ISO 13473-2: 2002. *Characterisation of Pavement Texture by Use of Surface Profiles–Part 2: Terminology and Basic Requirements Related to Pavement Texture Profile Analysis*; International Standardization Organization: Geneva, Switzerland, 2002.
21. Walker, D.; Entine, L.; Kummer, S. *Pavement Surface Evaluation and Rating (PASER) Manual*; Wisconsin Transportation Information Center, University of Wisconsin-Madison: Madison, WI, USA, 2002.
22. Kropáč, O.; Múčka, P. Deterioration Model of Longitudinal Road Unevenness Based on its Power Spectral Density Indices. *Road Mater. Pavement Des.* **2008**, *9*, 389–420. [CrossRef]
23. Taberlet, N.; Morris, S.W.; McElwaine, J.N. Washboard Road: The dynamics of granular ripples formed by rolling wheels. *Phys. Rev. Lett.* **2007**, *99*, 068003. [CrossRef]
24. Both, J.A.; Hong, D.C.; Kurtze, D.A. Corrugation of roads. *Phys. A Stat. Mech. Appl.* **2001**, *301*, 545–559. [CrossRef]
25. Mahgoub, H.; Bennett, C.; Selim, A. Analysis of factors causing corrugation of gravel roads. *Transp. Res. Rec.* **2011**, *2204*, 3–10. [CrossRef]
26. Edvardsson, K.; Magnusson, R. Monitoring of dust emission on gravel roads: Development of a mobile methodology and examination of horizontal diffusion. *Atmos. Environ.* **2009**, *43*, 889–896. [CrossRef]
27. Zhu, D.; Gillies, J.A.; Etyemezian, V.; Nikolich, G.; Shaw, W.J. Evaluation of the surface roughness effect on suspended particle deposition near unpaved roads. *Atmos. Environ.* **2015**, *122*, 541–551. [CrossRef]
28. McClelland, D.E.; Foltz, R.B.; Falter, M.C.; Wilson, W.D.; Cundy, T.; Schuster, R.L.; Saurbier, J.; Rabe, C.; Heinemann, R. Relative Effects on a Low-Volume Road System of Landslides Resulting from Episodic Storms in Northern Idaho. *Transp. Res. Rec. Transp. Res. Board* **1999**, *1652*, 235–243. [CrossRef]
29. Uys, P.E.; Els, P.S.; Thoresson, M. Suspension settings for optimal ride comfort of off-road vehicles travelling on roads with different roughness and speeds. *J. Terramech.* **2007**, *44*, 163–175. [CrossRef]
30. Scholtz, O.; Els, P.S. Tyre rubber friction on a rough road. *J. Terramech.* **2021**, *93*, 41–50. [CrossRef]
31. Farrahi, G.H.; Ahmadi, A.; Kasyzadeh, K.R. Simulation of vehicle body spot weld failures due to fatigue by considering road roughness and vehicle velocity. *Simul. Model. Pract. Theory* **2020**, *105*, 102168. [CrossRef]
32. Raslavičius, L.; Pakalnis, A.; Keršys, A.; Skvireckas, R.; Juodvalkis, D. Investigation of asphalt texture roughness on friction evolution for wheeled vehicles. *Transport* **2016**, *31*, 133–141. [CrossRef]
33. Van Zyl, G. Blading optimisation Reverting from Theory to Practice. *Transp. Res. Rec.* **2011**, *2204*, 11–20. [CrossRef]
34. Žilionienė, D.; Čygas, D.; Juzėnas, A.A.; Jurgaitis, A. Improvement of functional designation of low-volume roads by dust abatement in Lithuania. *Transp. Res. Rec. Transp. Res. Board* **2007**, *1989*, 293–298. [CrossRef]
35. Jurkevičius, M.; Puodžiukas, V.; Laurinavičius, L. Implementation of Road Performance Calculation Models Used in Strategic Planning Systems for Lithuania Conditions. *Balt. J. Road Bridge Eng.* **2020**, *15*, 146–165. [CrossRef]
36. Archondo-Callao, R. *HDM-4 Road User Cost Model Documentation*; Version 1.20; User's Guide; The World Bank: Washington, DC, USA, 2009.
37. Yunusov, A.; Eshkabilov, S.; Riskaliev, D.; Abdukarimov, N. Estimation and evaluation of road roughness via different tools and methods. In Proceedings of the Transport Problems 2019, XI International Scientific Conference, Silesian University of Technology Faculty of Transport, Silesia, Poland, 26–28 June 2019; pp. 770–784.
38. Tomiyama, K.; Kawamura, A. Application of lifting wavelet transform for pavement surface monitoring by use of a mobile profilometer. *Int. J. Pavement Res. Technol.* **2016**, *9*, 345–353. [CrossRef]
39. Sayers, M.W.; Karamihas, S.M. *The Little Book of Profiling*; University of Michigan: Ann Arbor, MI, USA, 1998.
40. Rajkamal, K.; Reddy, T.D.; Rohith, D.; Chowdary, V.; Prasad, C.S.R.K. Performance Evaluation of Gravel Road Sections Sealed with Surface Dressing. *Transp. Res. Proc.* **2019**, *17*, 81–89. [CrossRef]
41. Abulizi, N.; Kawamura, A.; Tomiyama, K.; Fujita, S. Measuring and evaluating of road roughness conditions with a compact road profiler and ArcGIS. *J. Traffic Transp. Eng.* **2016**, *3*, 398–411. [CrossRef]
42. Bidgoli, A.M.; Galroo, A.; Nadjar, H.S.; Rashidabad, A.G.; Ganji, M.R. Road Roughness measurement using a cost-effective sensor-based monitoring system. *Autom Constr.* **2019**, *104*, 140–152. [CrossRef]
43. Du, Y.; Liu, C.; Wu, D.; Jiang, S. Measurement of International Roughness Index by Using Z-Axis Accelerometers and GPS, Hindawi Publishing Corporation. *Math. Probl. Eng.* **2014**, *2014*, 928980. [CrossRef] [PubMed]
44. Eshkabilov, S.; Yunusov, A. Measuring and Assessing Road Profile by Employing Accelerometers and IRI Assessment Tools. *Am. J. Traffic Transp. Eng.* **2018**, *3*, 24–40. [CrossRef]
45. Leitner, B.; Decký, M.; Kováč, M. Road pavement longitudinal evenness quantification as stationary stochastic process. *Transport* **2019**, *34*, 195–203. [CrossRef]
46. Pawar, P.R.; Mathew, A.T.; Saraf, M.R. IRI (International Roughness Index): An Indicator of Vehicle Response. *Mater. Today Proc.* **2018**, *5*, 11738–11750. [CrossRef]
47. Žuraulis, V.; Levulytė, L.; Sokolovskij, E. The impact of road roughness on the duration of contact between a vehicle wheel and road surface. *Transport* **2014**, *29*, 431–439. [CrossRef]
48. Shtayat, A.; Moridpour, S.; Best, B.; Shroff, A.; Raol, D. A review of monitoring systems of pavement condition in paved and unpaved roads. *J. Traffic Transp. Eng.* **2020**, *7*, 629–638. [CrossRef]
49. Becker, C.M.; Els, P.S. Profiling of rough terrain. *Int. J. Veh. Des.* **2014**, *64*, 240–261. [CrossRef]
50. Souza, V.M.A. Asphalt pavement classification using smartphone accelerometer and Complexity Invariant Distance. *Eng. Appl. Artif. Intell.* **2018**, *74*, 198–211. [CrossRef]

51. Botha, T.R.; Els, P.S. Rough terrain profiling using digital image correlation. *J. Terramech.* **2015**, *59*, 1–17. [CrossRef]
52. Kerst, S.; Shyrokau, B.; Holweg, E. Reconstruction of wheel forces using an intelligent bearing. *SAE Int. J. Passeng. Cars Electron. Electr. Syst.* **2016**, *9*, 196–203. [CrossRef]
53. Kerst, S.; Shyrokau, B.; Holweg, E. A Model-based approach for the estimation of bearing forces and moments using outer ring deformation. *IEEE Trans. Ind. Electron.* **2019**, *67*, 461–470. [CrossRef]
54. Múčka, P. Current approaches to quantify the longitudinal road roughness. *Int. J. Pavement Eng.* **2016**, *17*, 659–679. [CrossRef]
55. Loprencipe, G.; Zoccali, P. Ride Quality Doe to Road Surface Irregularities: Comparison of Different Methods Applied on a Set of Real Road Profiles. *Coatings* **2017**, *7*, 59. [CrossRef]
56. ISO 2631-1: 1997. *Mechanical Vibration and Shock—Evaluation of Human Response to Whole-Body Vibration. Part I: General Requirements*; International Standardization Organization: Geneva, Switzerland, 1997.
57. Gurmail, L.; Kiss, P. A comparative study of destructive effects resulting from road profile acting on off-road towed vehicles. *J. Terramech.* **2019**, *81*, 57–65. [CrossRef]
58. Kropáč, O.; Múčka, P. Be careful when using the International Roughness Index as an indicator of road unevenness. *J. Sound Vib.* **2005**, *287*, 989–1003. [CrossRef]
59. Fichera, G.; Scionti, M.; Garescì, F. Experimental correlation between the road roughness and the comfort perceived in bus cabins. *SAE Tech. Pap.* **2007**, *116*, 39–49. [CrossRef]
60. Sayers, M.W.; Gillespie, T.D.; Queiroz, C.A.V. *International Road Roughness Experiment: Establishing Correlation and a Calibration Standard for Measurements*; Technical Report, World Bank Technical Paper 1986, No. WTP 45; World Bank Group: Washington, DC, USA.
61. Gillespie, T.D.; Paterson, W.D.O.; Sayers, M.W. *Guidelines for Conducting and Calibrating Road Roughness Measurements*; World Bank Technical Paper 1986, No. WTP 46; World Bank Group: Washington, DC, USA. (In English)
62. Sidess, A.; Ravina, A.; Oged, E. A model for predicting the deterioration of the international roughness index. *Int. J. Pavement Eng.* **2020**. [CrossRef]
63. Pérez-Acebo, H.; Gonzalo-Orden, H.; Findley, D.J.; Rojí, E. Modeling the international roughness index performance on semi-rigid pavements in single carriageway roads. *Constr. Build. Mater.* **2021**, *272*, 121665. [CrossRef]
64. Yamany, M.S.; Abraham, D.M. Hybrid approach to incorporate preventive maintenance effectiveness into probabilistic pavement performance models. *J. Transp. Eng. Part. B Pavements* **2021**, *147*, 04020077. [CrossRef]
65. Yamany, M.S.; Abraham, D.M.; Labi, S. Comparative analysis of Markovian methodologies for modeling infrastructure system performance. *J. Infrastruct. Syst.* **2021**, *27*, 04021003. [CrossRef]
66. Mirtabar, Z.; Golroo, A.; Mahmoudzadeh, A.; Barazandeh, F. Development of a crowdsourcing-based system for computing the international roughness index. *Int. J. Pavement Eng.* **2020**. [CrossRef]
67. Pérez-Acebo, H.; Mindra, N.; Railean, A.; Rojí, E. Rigid pavement performance models by means of Markov Chains with half-year step time. *Int. J. Pavement Eng.* **2019**, *20*, 830–843. [CrossRef]
68. Obunguta, F.; Matsushima, K. Optimal pavement management strategy development with a stochastic model and its practical application to Ugandan national roads. *Int. J. Pavement Eng.* **2020**, 1–15. [CrossRef]
69. ISO 8608: 2016. *Mechanical Vibrations—Road Surface Profiles—Reported of Measured Data*; International Standardization Organization: Geneva, Switzerland, 2016.
70. Goenaga, B.; Fuentes, L.; Mora, O. Evaluation of the methodologies used to generate random pavement profiles based on the power spectral density: An approach based on the International Roughness Index. *Ing. E Investig.* **2017**, *37*, 49–57. [CrossRef]
71. Múčka, P. Road waviness and the dynamic tyre force. *Int. J. Veh. Des.* **2004**, *36*, 216–232. [CrossRef]
72. Els, P.S.; Theron, N.J.; Uys, P.E.; Thoresson, M.J. The ride comfort vs. handling compromise for off-road vehicles. *J. Terramech.* **2007**, *44*, 303–317. [CrossRef]
73. Ngwangwa, H.M.; Heyns, P.S.; Breytenbach, H.G.A.; Els, S. Reconstruction of road defects and road roughness classification using Artificial Neural Networks simulation and vehicle dynamic responses: Application to experimental data. *J. Terramech.* **2014**, *53*, 1–18. [CrossRef]
74. ISO/TS 13473-4: 2008. *Characterisation of Pavement Texture by Use of Surface Profiles—Part 4: Spectral Analysis of Surface Profiles*; Technical Specification; International Standardization Organization: Geneva, Switzerland, 2008.
75. Múčka, P.; Stein, G.J.; Tobolka, P. Whole-body vibration and vertical road profile displacement power spectral density. *Veh. Syst. Dyn.* **2020**, *58*, 630–656. [CrossRef]
76. Rill, G. *Road Vehicle Dynamics: Fundamentals and Modeling*; CRC Press LLC: New York, NY, USA, 2020.
77. Tyan, F.; Hong, Y.F.; Tu, S.H.; Jeng, W.S. Generation of random road profiles. *J. Adv. Eng.* **2009**, *4*, 373–1378.
78. Reza-Kashyzadeh, R.; Ostad-Ahmad-Ghorabi, M.J.; Arghavan, A. Investigating the effect of road roughness on automotive component. *Eng. Fail. Anal.* **2014**, *41*, 96–107. [CrossRef]
79. Sayers, M.W. On the calculation of international roughness index from longitudinal road profile. *Transp. Res. Rec* **1995**, *1501*, 1–12.
80. ASTM E1926-98. *Standard Practice for Computing International Roughness Index for Roads from Longitudinal Profile Measurement*; ASTM E 1926–98 ASTM International: West Conshohocken, PA, USA, 1998. [CrossRef]
81. Cebon, D. *Handbook of Vehicle-Road Interaction*; CRC Press: Lisse, The Netherlands, 1999.
82. Múčka, P. Simulated Road Profiles According to ISO 8608 in Vibration Analysis. *J. Test. Eval.* **2018**, *46*, 405–418. [CrossRef]

83. Buhari, R.; Rohani, M.M.; Abdullah, M.E. Dynamic Load Coefficient of Tyre Forces from Truck Axles. *Appl. Mech. Mater.* **2013**, *405–408*, 1900–1911. [CrossRef]
84. Wang, D.; Falchetto, A.C.; Goeke, M.; Wang, W.; Li, T.; Wistuba, M.P. Influence of computation algorithm on the accuracy of rut depth measurement. *J. Traffic Transp. Eng.* **2017**, *4*, 156–164. [CrossRef]
85. Lakušić, S.; Brčić, D.; Tkalčević Lakušić, V. Analysis of Vehicle Vibrations—New Approach to Rating Pavement Condition of Urban Roads. *Promet Traffic Transp.* **2012**, *23*, 485–494. [CrossRef]
86. Seimas, L.R. Dėl Lietuvos Respublikos Vyriausybės 2002 m. Gruodžio 11 d. Nutarimo Nr. 1950 „Dėl Kelių Eismo Taisyklių Patvirtinimo" Pakeitimo. Available online: https://e-seimas.lrs.lt/portal/legalAct/lt/TAD/2a948a80506e11e485f39f55fd139d01 (accessed on 6 August 2021).
87. Múčka, P. Road Roughness Limit Values Based on Measured Vehicle Vibration. *J. Infrastruct. Syst.* **2016**, *23*, 04016029. [CrossRef]
88. Múčka, P. Proposal of Road Unevenness Classification Based on Road Elevation Spectrum Parameters. *J. Test. Eval.* **2016**, *44*, 930–944. [CrossRef]
89. Xu, Y.; Ahmadian, M. Improving the capacity of tire normal force via variable stiffness and damping suspension system. *J. Terramech.* **2013**, *50*, 121–132. [CrossRef]

applied sciences

MDPI

Communication

A Double Sky-Hook Algorithm for Improving Road-Holding Property in Semi-Active Suspension Systems for Application to In-Wheel Motor

Seunghoon Woo [1] and Donghoon Shin [2,*]

[1] R&D Division, Hyundai Motor Company, Onseokdong Hwaseongsi, Seoku-dong, Hwaseong-si 18280, Korea; boltra@naver.com
[2] Department of Mechanical Systems Engineering, Sookmyung Women's University, Seoul 04310, Korea
* Correspondence: dhshin@sookmyung.ac.kr; Tel.: +82-2-710-9154

Featured Application: Double Sky-Hook Algorithm, Vehicle Dynamics and Control, Semi-Active Suspension Systems, In-Wheel Motor.

Abstract: This paper presents a double sky-hook algorithm for controlling semi-active suspension systems in order to improve road-holding property for application in an in-wheel motor. The main disadvantage of the in-wheel motor is the increase in unsprung masses, which increases after shaking of the wheel, so it has poor road-holding that the conventional theoretical sky-hook algorithm cannot achieve. The double sky-hook algorithm uses a combination of damper coefficients, one from the chassis motion and the other from the wheel motion. Computer simulations using a quarter and full car dynamic models with the road conditions specified by ISO2631 showed the effectiveness of the algorithm. It was observed that the algorithm was the most effective in the vicinity of the wheel hop frequency. This paper also proposed the parameter set of the double sky-hook algorithm to differentiate the driving mode of vehicles under advanced development.

Keywords: sky-hook; in-wheel motor; semi-active suspension

Citation: Woo, S.; Shin, D. A Double Sky-Hook Algorithm for Improving Road-Holding Property in Semi-Active Suspension Systems for Application to In-Wheel Motor. *Appl. Sci.* **2021**, *11*, 8912. https://doi.org/10.3390/app11198912

Academic Editors: Flavio Farroni, Andrea Genovese and Aleksandr Sakhnevych

Received: 5 September 2021
Accepted: 21 September 2021
Published: 24 September 2021

1. Introduction

In the layout of electric vehicles, one of the most interesting aspect is represented by the use of four in-wheel electric motors. As the motor is directly inside the wheel, no powertrain is required, thus leading to the elimination of elements such as drive shafts, gearboxes, and differentials [1,2]. There are also other advantages: space inside the car can be enlarged, thus having more room for passengers and for the battery pack [2,3]. Since the in-wheel motor is installed at each wheel, it allows direct yaw moment control function using the active drive torque distribution to improve the handling and maneuverability of electric vehicles [4,5]. The main disadvantage of in-wheel motor is the increase in unsprung masses, which increases after wheel shake, so it has poor road-holding [2,3,6]. One of the most effective ways to reduce this resonance is controlling the damping characteristics. Among the many control algorithms for semi-active suspension systems, the sky-hook algorithm is known to be very effective in controlling the sprung mass resonance [7–13]. However, this algorithm is not effective in controlling the unsprung mass resonance, so that the road-holding property in the vicinity of wheel hop frequency is worse than that of the passive suspension [14–16]. The resonance of wheels is transferred with noise and vibration to the chassis through the suspension and steering systems [15]. In addition, the vehicle dynamic stability is degraded. An optimal control algorithm with a performance index in the frequency domain [17] may overcome this defect. However, an optimal control algorithm is not robust to the system parameter uncertainty and nonlinearities. Moreover, this algorithm needs information of every state, so the actual implementation of an optimal control algorithm is not practical.

In order to overcome the above-mentioned problems, a practical double sky-hook algorithm was proposed [18], which can effectively control the unsprung mass resonance as well as the sprung mass resonance. The proposed algorithm utilizes an additional sky-hook damper in the wheel combined with a conventional sky-hook damper that is attached to a sprung mass. In other words, the damper coefficient in a real physical semi-active damper is adjusted according to the principle of the sky-hook algorithm not only in the vicinity of the sprung mass resonant frequency, but also in the vicinity of the unsprung mass resonant frequency. In the algorithm, hardware components such as a semi-active damper and a chassis vertical accelerometer and a relative distance sensor for the proposed algorithm are the same as those of the conventional sky-hook algorithm. However, average damping force transmitted to the sprung mass can be increased due to larger value of an additional sky-hook damper at unsprung mass, which in turn, will increase the sprung mass acceleration levels. Therefore, there has been a great deal of research on determining the contribution of the sky-hook damping at unsprung mass for a semi-active suspension system [19–21].

This paper focused on the robust trade-off performance of ride comfort and handling by guaranteeing ride quality in every driving situation regardless of the value of the sky-hook damping coefficient at unsprung mass. For practical proof, trade-off performance analysis using cross-plot of RMS (root mean square) tire normal force variation, and RMS body acceleration using the proper model was used. The analysis was performed with the model of the in-wheel motored vehicle under development and with enough fidelity model for ride comfort simulation, which consists of actuator delay, suspension friction, and damper nonlinearity.

The rest of this paper is organized as follows. In Section 2, modeling of the suspension dynamics of a quarter and a full car is described. Section 3 explains the principle of the double sky-hook algorithm and suggests a design method to select suitable damping coefficients. In Section 4, computer simulations with a quarter and a full car suspension dynamics model are presented with a discussion of some the results for the purpose of comparison with a passive and the conventional sky-hook suspension system in the RMS of the suspension performance measures. Conclusions are made in the final section.

2. Modeling of Suspension Dynamics

For the purpose of comparison between the conventional sky-hook algorithm and the double sky-hook algorithm, this section briefly summarizes the conventional one as follows. The detailed representations of model have been stated in the Appendix A.

2.1. Quarter Car Model

A simple two degrees-of-freedom model was built to compare the frequency responses, as shown in Figure 1. The suspension force, f_s, is generated by spring, damper, and friction. The tire force, f_t, is calculated by tire stiffness and tire damping. The dynamic equations of the system [22] can be represented as:

$$m_u\ddot{z}_u - f_s + f_t = 0 \tag{1}$$

$$m_s\ddot{z}_s - f_s = f_z \tag{2}$$

2.2. Full Car Model

A full car dynamic model that had eight degrees-of-freedom including roll, pitch, heave, driver seat, and four wheel vertical motions was constructed in order to demonstrate a more realistic vehicle behavior, as shown in Figure 2. This dynamic model included the wheel-base filtering effect, roll resonance effect by anti-roll bar stiffness and chassis roll moment of inertia, suspension friction, and the driver seat suspension effect on the three measures of suspension performance. The dynamic equations of the system [22] are represented as

$$m_{ufl}\ddot{z}_{ufl} - f_{sfl} + f_{tfl} = 0 \tag{3}$$

$$m_{ufr}\ddot{z}_{ufr} - f_{sfr} + f_{tfr} = 0 \tag{4}$$

$$m_{url}\ddot{z}_{url} - f_{srl} + f_{trl} = 0 \tag{5}$$

$$m_{urr}\ddot{z}_{urr} - f_{srr} + f_{trr} = 0 \tag{6}$$

$$m_s\ddot{z}_s + f_{sfl} + f_{sfr} + f_{srl} + f_{srr} = f_z \tag{7}$$

$$J_\phi\ddot{\phi} + L_f\left(f_{sfl} + f_{sfr}\right) - L_r(f_{srl} + f_{srr}) = \tau_\phi \tag{8}$$

$$J_\theta\ddot{\theta} + W_l\left(f_{sfl} + f_{srl}\right) - W_r\left(f_{sfr} + f_{srr}\right) = \tau_\theta \tag{9}$$

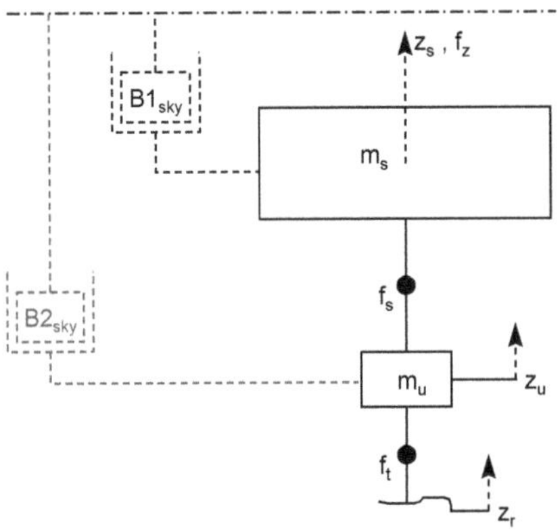

Figure 1. Quarter car model.

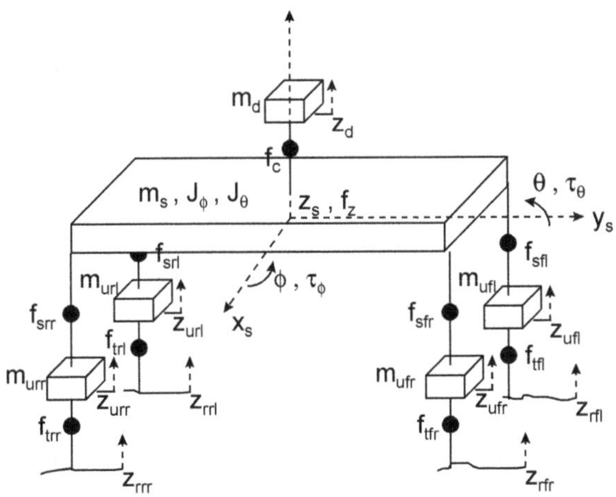

Figure 2. Full car model.

3. Double Sky-Hook Control Algorithm

3.1. Conventional Sky-Hook Algorithm

For the purpose of comparison between the conventional sky-hook algorithm and the double sky-hook algorithm, this section briefly summarizes the conventional one as follows: the force transmitted to the sprung mass through an ideal sky-hook damper is expressed as:

$$F_{sprung} = B1_{sky} \cdot \dot{x}_s \tag{10}$$

where $B1_{sky}$ is the sky-hook damping coefficient and $\ddot{x}_s$ is the sprung mass absolute vertical speed. In the real suspension configuration, however, the damping force is a function of a relative speed between a chassis and a wheel and is represented as

$$F_{sprung} = B1_d \cdot (\dot{x}_s - \dot{x}_u) \tag{11}$$

where $\ddot{x}_u$ is the unsprung mass absolute vertical speed. The desirable damping coefficient $B1_d$ that generates the same damping force as that of the ideal sky-hook damper is determined from Equations (1) and (2) as

$$B1_d = \frac{B1_{sky} \cdot \dot{x}_s}{\dot{x}_s - \dot{x}_u} \tag{12}$$

As discussed in the introduction, this theoretic sky-hook damper is effective only in the vicinity of the sprung mass resonance frequency and has a worse road-holding property than that of a passive damper.

3.2. Double Sky-Hook Algorithm

Following the basic idea of the ideal sky-hook damper, if an additional sky-hook damper is attached to the unsprung mass in order to reduce the wheel hop resonance, the force transmitted to the unsprung mass through an ideal sky-hook damper is expressed as

$$F_{unsprung} = B2_{sky} \cdot \dot{x}_u \tag{13}$$

where $B2_{sky}$ is the sky-hook damping coefficient. In the real suspension configuration, however, the damping force is a function of a relative speed between a chassis and a wheel and is represented as

$$F_{unsprung} = B2_d \cdot (\dot{x}_u - \dot{x}_s) \tag{14}$$

The desired damping coefficient $B2_d$ that generates the same damping force as that of the ideal sky-hook damper is determined from Equations (4) and (5) as

$$B2_d = \frac{B2_{sky} \cdot \dot{x}_u}{\dot{x}_u - \dot{x}_s} \tag{15}$$

If we use the sky-hook damper only to the wheel, in other words, a real damping coefficient is adjusted by $B2_d$, then the road-holding performance, which in turn, will greatly improve the vehicle dynamic stability.

Since a real semi-active suspension system is capable of controlling a single damping coefficient, however, we need to combine two damping coefficients $B1_d$ and $B2_d$ in such a way that both sprung and unsprung mass resonances are controlled effectively in each frequency region. For a simple combination, we propose the following combination as

$$B_d = B1_d + B2_d \tag{16}$$

Note that for a constant magnitude of relative velocity $(\dot{x}_u - \dot{x}_s)$, $B1_d$ in Equation (12) takes a large value in the sprung mass resonance frequency where the sprung mass vertical velocity is magnified dynamically while $B2_d$ in Equation (15) takes a large value in the unsprung mass resonance frequency where the unsprung mass vertical velocity is magnified. Therefore, each damping coefficient is considered to play an effective role in each frequency region of interest.

In the case of Equation (16), the damping force transmitted is represented as

$$F_d = B1_{sky} \cdot \dot{x}_s - B2_{sky} \cdot \dot{x}_u \tag{17}$$

Then, the following four cases should be noted:

Case I. If $B2_{sky} = 0$, the algorithm behaves like a conventional ideal sky-hook algorithm where a sky-hook damper with a damping coefficient $B1_{sky}$ is attached to a sprung mass.

Case II. If $0 < B2_{sky} < B1_{sky}$, as the value of $B2_{sky}$ increases from zero when the value of $B1_{sky}$ is fixed, we expect better unsprung mass resonance control over the wheel hop frequency region. This will result in a decrease in the tire contact load variation (thus tire deflection). In this case, however, average damping force transmitted to the sprung mass will be increased due to the larger value of B_d, which in turn, will increase the sprung mass acceleration levels. Therefore, the selection of $B2_{sky}$ for the fixed value of $B1_{sky}$ should be made based on the relative importance between the sprung mass acceleration (ride quality) and the tire contact load variation (road-holding property: stability).

Case III. If $B2_{sky} = B1_{sky}$, then $F_d = B1_{sky} \cdot (\dot{x}_s - \dot{x}_u)$ from Equation (17). This case is equivalent to a passive damper with damping coefficient of $B1_{sky}$.

Case IV. If $B1_{sky} = 0$ or if $B2_{sky} > B1_{sky}$, then too large a damping force will be generated due to the fact that the vertical speed of the unsprung mass is usually far larger than that of the sprung mass. These cases will therefore degrade the ride quality significantly.

Step I. For a fixed trial value of $B1_{sky}$, which may be determined according to the conventional sky-hook theory, adjust $B2_{sky}$ from zero up to $B1_{sky}$ until the tire contact load variation in the vicinity of the wheel hop frequency is similar to that of a passive suspension. This adjustment is based on the fact that the larger the $B2_{sky}$, the greater the sprung mass acceleration to that of a passive suspension in the vicinity of the wheel hop frequency.

Step II. If $B2_{sky}$ is determined according to Step I while $B1_{sky}$ is fixed, we will obtain a sub-optimal damper in the wheel hop frequency region. However, overall damping coefficient will be increased so that the tire force variation and chassis acceleration will tend to be increased in-between sprung and unsprung mass resonant frequencies. However, three performance measures still became better than those of the conventional sky-hook algorithm in the sprung mass resonant frequency region. Therefore, we can reduce the value of $B1_{sky}$, which controls the suspension characteristics in the sprung mass resonant frequency region to a value where three performance measures were still better than those of the conventional sky-hook algorithm.

It was also noted that the range of a real adjustable damping coefficient is limited by physical hardware limits such as $B_{min} \leq B_d \leq B_{max}$.

4. Simulation Studies

4.1. Simulation Conditions

Computer simulation was performed to show the effectiveness and design procedure of the double sky-hook algorithm with a quarter and a full car suspension dynamic model. In this simulation, in order to include nonlinear behavior of actual semi-active suspension systems, we considered the dead time from an electronic circuit and a variable damper valve action as well as the ramp up time from the damping force development after the valve was closed. Figure 3 shows these time delays and its model using MATLAB Simulink. In addition, we included nonlinear characteristics of the damper, as shown in Figure 4. In the simulation, we used irregular random road profiles classified by the ISO [22]. Two different vehicle speeds with different road conditions were considered as follows:

Case (a): Grade A (Very Good), $S_g(\Omega_0) = 4$ with 80 km/h
Case (b): Grade B (Very Poor), $S_g(\Omega_0) = 256$ with 10 km/h

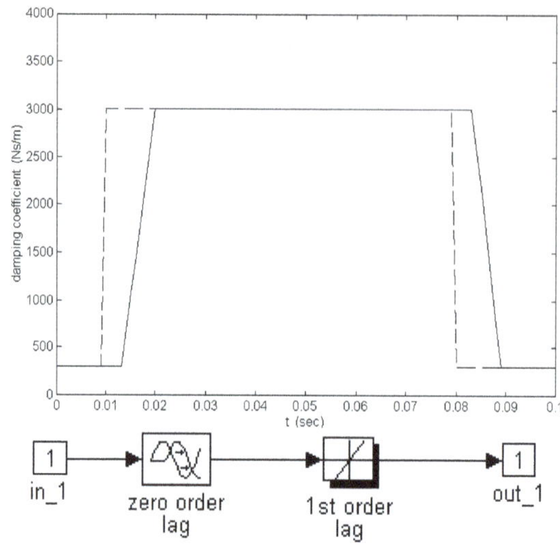

Figure 3. Time delay effect of variable damper (solid: with time delay, dashed: without time delay).

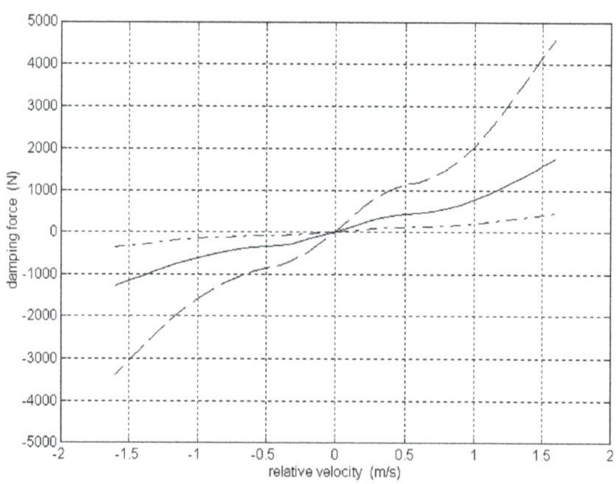

Figure 4. Nonlinear damping characteristics (solid: passive, dashed: max., dashdot: min.).

For the above simulation conditions, we simulated each vehicle dynamics model with a passive, conventional sky-hook, and double sky-hook damper.

4.2. Quarter Car Model Simulation

This quarter car model was selected to compare the frequency responses of three dampers. Table 1 shows the parameters of the quarter car model considering the in-wheel motor weight. The three suspension performance measures (chassis acceleration, tire force variation, suspension stroke) were evaluated by 1/3 octave band RMS values on the frequency axis in Figures 5–7. In these figures, the solid line denotes the double sky-hook case, dashed line denotes the conventional sky-hook, and dash-dot line for passive damper. In these figures, case (a) is for smooth road profile with high speed and case (b) is for rough road profile with slow speed.

Table 1. Model parameters for the quarter car.

Parameter	Value
Effective weight of wheel (m_u)	85 kg
Weight of sprung mass (m_s)	563 kg
spring stiffness	32,366 N/m
Passive damper coefficient (effective)	1577 Ns/m
Tire vertical stiffness	248,000 N/m
Tire vertical damping coefficient	322 Ns/m
Max. damping coefficient	3858 Ns/m
Min. damping coefficient	300 Ns/m
Sky-hook damper on sprung mass ($B1_{sky}$)	18,015 Ns/m

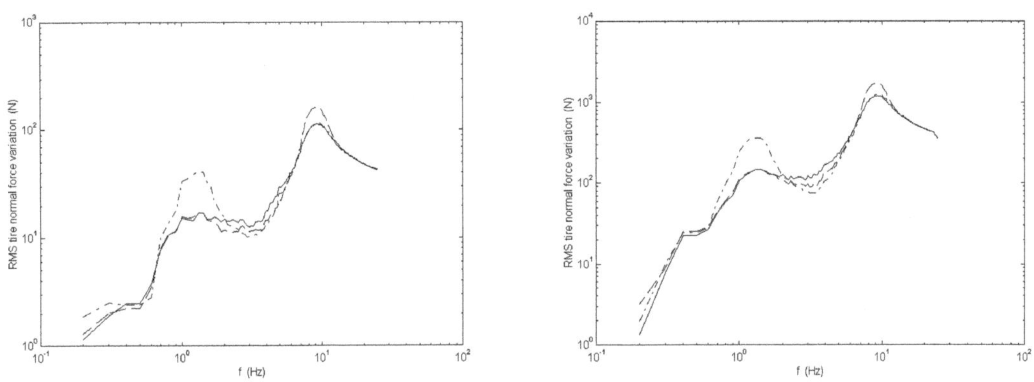

(a) **(b)**

Figure 5. RMS of the tire normal force variation. (**a**) smooth road; (**b**) rough road.

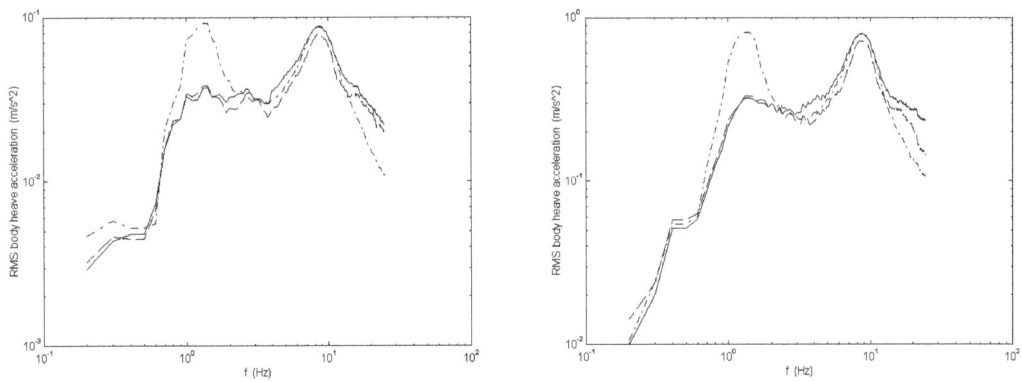

(a) **(b)**

Figure 6. RMS of chassis heave acceleration. (**a**) smooth road; (**b**) rough road.

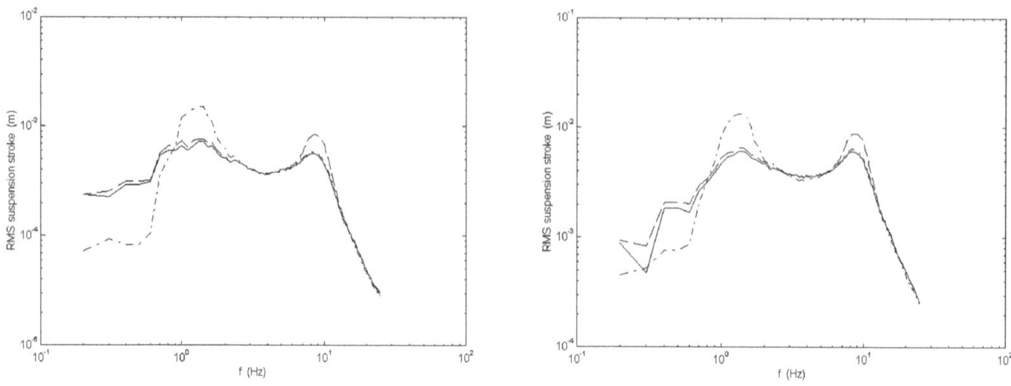

(**a**) (**b**)

Figure 7. RMS of suspension stroke. (**a**) smooth road; (**b**) rough road.

In this simulation, $B1_{sky}$ and $B2_{sky}$ have the values of 12,859 Ns/m and 1440 Ns/m, respectively. These values were determined by the trial-and-error steps described in Section 2 to have a lower chassis acceleration, tire force variation, and suspension stroke with the double sky-hook damper than those with the conventional sky-hook and passive dampers in the sprung and unsprung mass resonant frequency regions. Figures 5–7 show the following characteristics by using the double sky-hook algorithm:

(1) The double sky-hook damper generated better suspension performance than the passive damper for all three performance measures except for slight degradation between the sprung and unsprung mass resonant frequencies, even with increased unsprung mass by the in-wheel motors.
(2) The double sky-hook damper had better road-holding property, as shown in Figure 5, and smaller suspension stroke, as shown in Figure 7, than the conventional sky-hook damper not only in the unsprung mass resonant frequency region, but also in the sprung mass resonant.
(3) The double sky-hook damper had a similar body acceleration as shown in Figure 6 to the conventional sky-hook damper in all frequency regions.

In conclusion, the double sky-hook damper can achieve better a road-holding property, as intended.

4.3. *Full Car Model Simulation*

A full car dynamic model that had eight degrees-of-freedom including roll, pitch, heave, driver seat, and four wheel vertical motions was simulated for more realistic vehicle behavior. Table 2 shows the model parameters of the vehicle under advanced development with in-wheel motors on the rear axle. In order to evaluate the performance variation for the change in damping coefficient, a simulation was performed for the damping coefficients listed in Table 3. Note that the fourth passive damping coefficient in Table 3 was the same as the passive damping coefficient in Table 2.

Table 2. Model parameters for a full car.

Parameter	Value
Effective weight of front wheel (m_{ufl}, m_{ufr})	54 kg
Effective weight of rear wheel (m_{url}, m_{urr})	68 kg
Weight of sprung mass (m_s)	1754 kg
Pitch moment of inertia (J_ϕ)	3012 kg m^2
Roll moment inertia (J_θ)	598 kg m^2
Front axle location from C.G. (L_f)	1.44 m
Rear axle location from C.G. (L_r)	1.56 m
Left/Right side from C.G. (W_l, W_r)	0.82 m
Front spring stiffness	24,426 N/m
Rear spring stiffness	31,954 N/m
Front passive damper coefficient (effective)	1400 Ns/m
Rear passive damper coefficient (effective)	2000 Ns/m
Suspension friction	54 N
Front stabilizer bar stiffness	19,984 N/m
Rear stabilizer bar stiffness	10,204 N/m
Tire vertical stiffness	248,000 N/m
Tire vertical damping coefficient	322 Ns/m
Weight of driver (m_d)	65 kg
Driver's seat stiffness	10,507.1 N/m
Driver's seat damping coefficient	875.6 Ns/m

Table 3. Damping coefficients.

No.	Passive Damping (Ns/m)		Sky-Hook Algorithm $B1_{sky}$ (Ns/m)	Double Sky-Hook Algorithm $B2_{sky}$ (Ns/m)
	Front	Rear		
1	560	800	5000	500
2	840	1200	10,000	1000
3	1120	1600	40,000	1500
4	1400	2000	100,000	2000
5	1680	2400	1,000,000	2500
6	1960	2800	-	-

Figures 8–10 show the cross plots for the RMS driver seat heave, chassis pitch, and chassis roll accelerations against RMS tire normal force variation (road-holding performance), respectively. The performance measures were frequency-weighted performance indices computed by multiplication of frequency-weighted curves (Figures 11 and 12) as suggested by ISO2631 [7] with RMS heave, pitch, and the value of roll accelerations and RMS tire normal force variation. The RMS values were calculated using 1/3 octave band frequencies of frequency responses, which were obtained from the time-domain simulation of the full car dynamics model with nonlinearities over the road conditions described in the simulation conditions. In the figures, the solid line denotes the double sky-hook case, the dashed line denotes the conventional sky-hook, and the dash-dot line represents a passive damper. The numbers in each plot correspond to the different damping coefficients listed in Table 3. Note that in the case of the double sky-hook algorithm, the value of $B1_{sky}$ was fixed to the value of 10,000 Ns/m and the simulation was performed for the values of $B2_{sky}$ from zero (this point in Figures 8–10 is point 2 in the conventional sky-hook algorithm) to the values in Table 3.

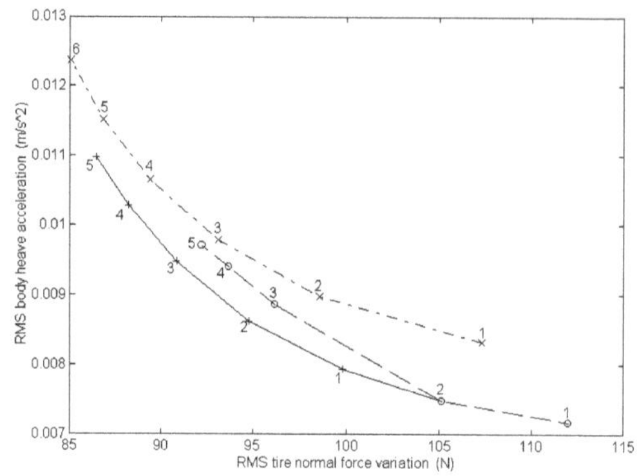

Figure 8. Tire normal force variation vs. heave acceleration.

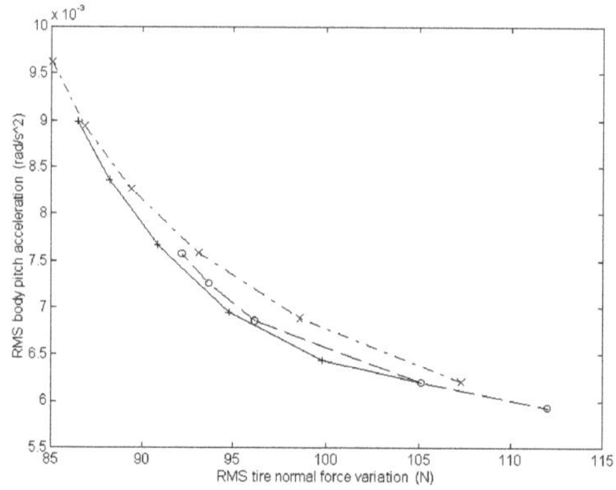

Figure 9. Tire normal force variation vs. pitch acceleration.

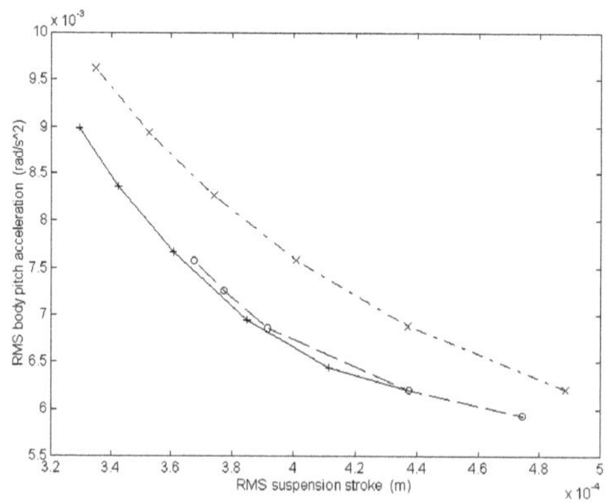

Figure 10. Tire normal force variation vs. roll acceleration.

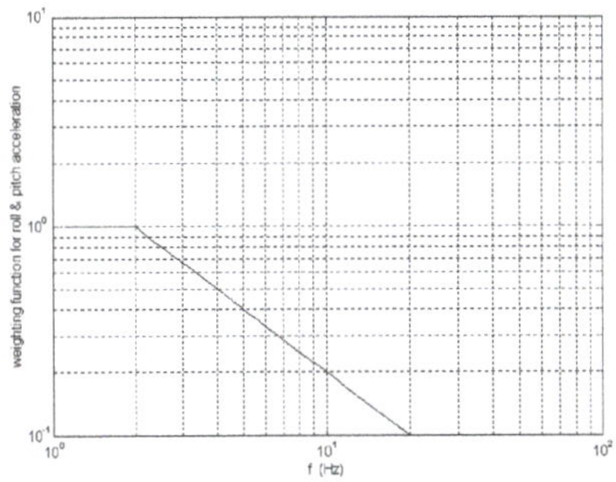

Figure 11. Frequency weighting function for roll and pitch acceleration.

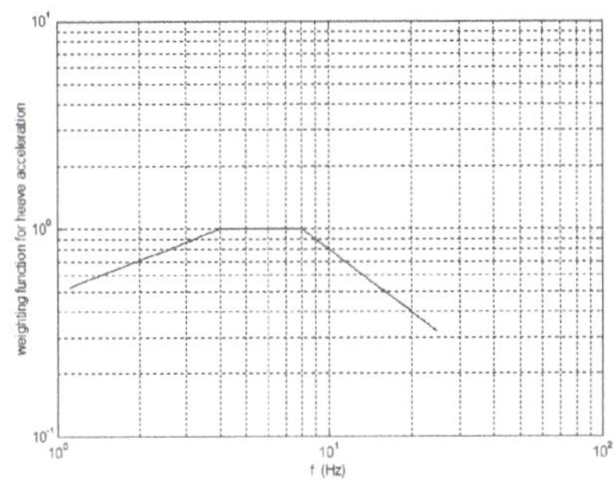

Figure 12. Frequency weighting function for heave acceleration.

Figures 8–10 show the following characteristics:

(1) The double sky-hook damper had smaller RMS seat accelerations and smaller RMS tire normal force variation than the passive and conventional sky-hook dampers, even in the development vehicle with in-wheel motors.

(2) In the case of the conventional sky-hook damper, the performance indices converged to the vicinity of a certain point (e.g., number 5 in the plots) as $B1_{sky}$ was increased close to infinity. This means that the conventional sky-hook damper cannot generate better road-holding performance than the passive damper. In the meantime, the double sky-hook damper could achieve better road-holding performance and ride quality than a passive damper.

(3) The sky-hook damping parameter at unsprung mass of double sky-hook damper provided tuning freedom to improve tire normal force variation while minimizing deterioration in seat acceleration. Therefore, the parameter was used to differentiate the driving mode of the vehicle under advanced development, as shown in Table 4.

Table 4. Drive mode differentiation of the development vehicle.

Driving Mode	$B2_{sky}$ (Ns/m)
Comfort mode	0
Normal mode	500
Sports mode	1500
Racing track mode (Hyundai *N* mode)	2000

5. Conclusions and Discussion

A double sky-hook algorithm for application to in-wheel motored vehicle has been presented. It was proven that the double sky-hook algorithm could achieve a higher level of compromise for road-holding and ride comfort compared to not only passive suspension, but also conventional sky-hook suspension.

Even in the development vehicle with in-wheel motors, the double sky-hook damper for a semi-active suspension system could improve the vehicle dynamic stability and reduce suspension travel distance while maintaining a better ride comfort than the passive and the conventional sky-hook dampers could generate. The smaller suspension travel also reduced the noise and vibration transmitted to the chassis.

The sky-hook damping parameter at unsprung mass of double sky-hook damper provided tuning freedom to improve vehicle dynamic stability while minimizing deterioration in ride quality. This paper also proposed the parameter set of the double sky-hook algorithm to differentiate the driving mode of the in-wheel motored vehicle under development.

The advanced development vehicle using the proposed algorithm is currently in the virtual development stage. Therefore, implementation, tests, and evaluation of the double sky-hook algorithm for in-wheel motored vehicles are the topics of our future research. For comparison with the simulation results, it is necessary to secure the reliability and repeatability of the test, so not only the actual vehicle test, but also the equipment test will be conducted as shown in Figure 13. The Hyundai Motor Company is expanding the scope of virtual development by performing these tasks.

Figure 13. Roller bench test rig to secure the reliability and repeatability of the ride comfort test.

For electric vehicles currently under development, due to the weight of the motor and battery, not only the unsprung mass but also the sprung mass is large, therefore algorithms such as the acceleration driven damper, which functions similar to the technology mainly applied to light racing cars, were not considered. In the future, in preparation for a weight reduction in electric vehicles, we plan to conduct mass-production application studies on algorithms to reduce the vibration acceleration of light vehicles.

Author Contributions: Conceptualization, D.S. and S.W.; Methodology, D.S.; Software, D.S.; Validation, D.S. and S.W.; Formal analysis, D.S.; Investigation, S.W.; Resources, D.S.; Data curation, S.W.; Writing—original draft preparation, D.S.; Writing—review and editing, S.W.; Visualization, D.S.; Supervision, S.W.; Project administration, S.W.; Funding acquisition, S.W. Both authors have read and agreed to the published version of the manuscript.

Funding: This paper was supported by a National Research Foundation of Korea (NRF) grant funded by the Korean government (MSIT; No. 2021R1F1A1048133) and by the Sookmyung Women's University Research Grants (1-2103-1077).

Institutional Review Board Statement: Not applicable.

Informed Consent Statement: Not applicable.

Data Availability Statement: Not applicable.

Conflicts of Interest: The authors declare no conflict of interest.

Appendix A

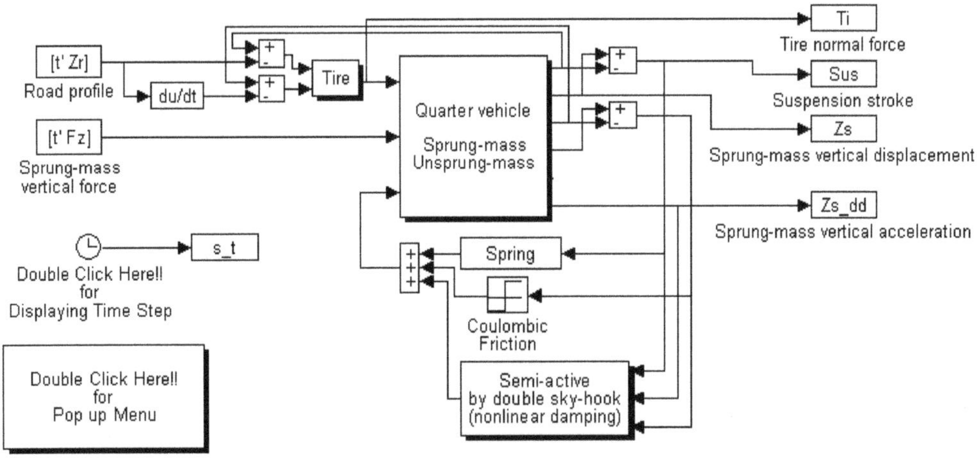

Figure A1. Quarter car model using the simulation modules.

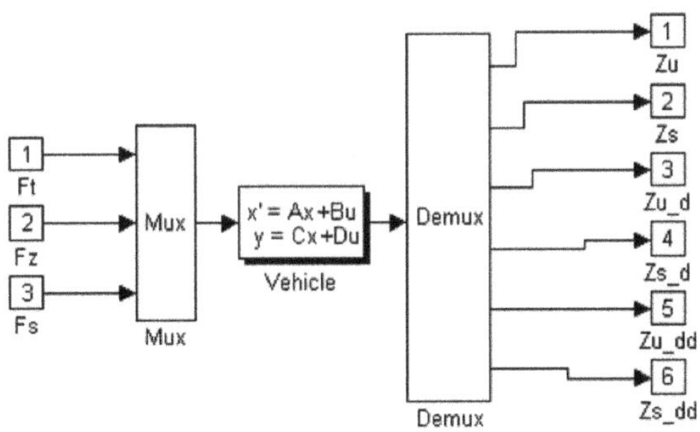

$$m_u \ddot{z}_u - f_s + f_t = 0$$
$$m_s \ddot{z}_s + f_s = f_z$$

$$\mathbf{u} = \begin{bmatrix} f_t \\ f_z \\ f_s \end{bmatrix}, \quad \mathbf{x} = \begin{bmatrix} z_u \\ z_s \\ \dot{z}_u \\ \dot{z}_s \end{bmatrix}, \quad \mathbf{y} = \begin{bmatrix} z_u \\ z_s \\ \dot{z}_u \\ \dot{z}_s \\ \ddot{z}_u \\ \ddot{z}_s \end{bmatrix}$$

$$\dot{\mathbf{x}} = \mathbf{Ax} + \mathbf{Bu}$$
$$\mathbf{y} = \mathbf{Cx} + \mathbf{Du}$$

$$\mathbf{A} = \begin{bmatrix} \mathbf{0}_{(2\times2)} & \mathbf{I}_{(2\times2)} \\ \mathbf{0}_{(2\times2)} & \mathbf{0}_{(2\times2)} \end{bmatrix}, \quad \mathbf{B} = \begin{bmatrix} \mathbf{0}_{(2\times3)} \\ \mathbf{B}_{\mathbf{L}(2\times3)} \end{bmatrix}$$

$$\mathbf{C} = \begin{bmatrix} \mathbf{I}_{(4\times4)} \\ \mathbf{0}_{(2\times4)} \end{bmatrix}, \quad \mathbf{D} = \begin{bmatrix} \mathbf{0}_{(4\times3)} \\ \mathbf{D}_{\mathbf{L}(2\times3)} \end{bmatrix}$$

$$\mathbf{B_L} = \begin{bmatrix} \frac{-1}{M_u} & 0 & \frac{1}{M_u} \\ 0 & \frac{-1}{M_s} & \frac{1}{M_s} \end{bmatrix}, \quad \mathbf{D_L} = \mathbf{B_L}$$

Figure A2. Quarter car body module.

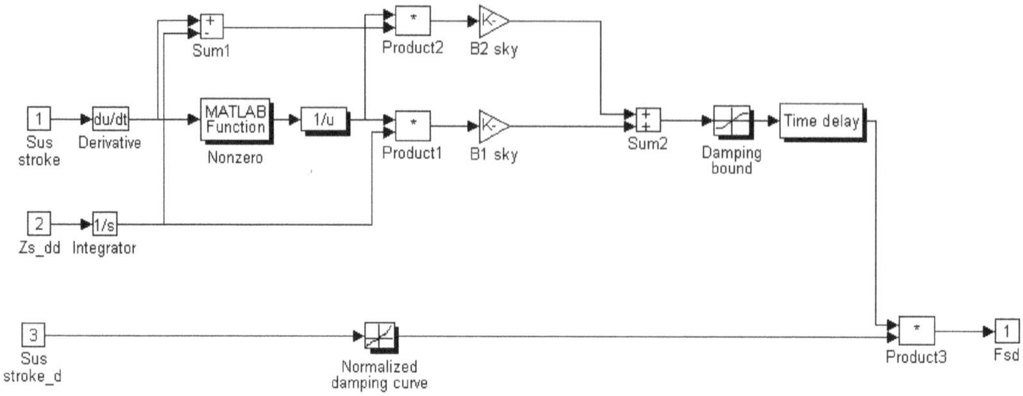

Figure A3. Semi-active control module using the continuous double sky-hook algorithm with a nonlinear damper.

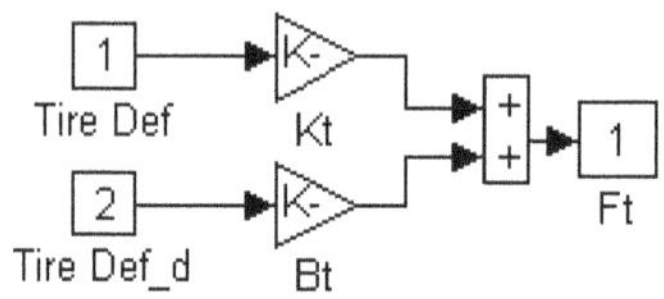

Figure A4. Tire module.

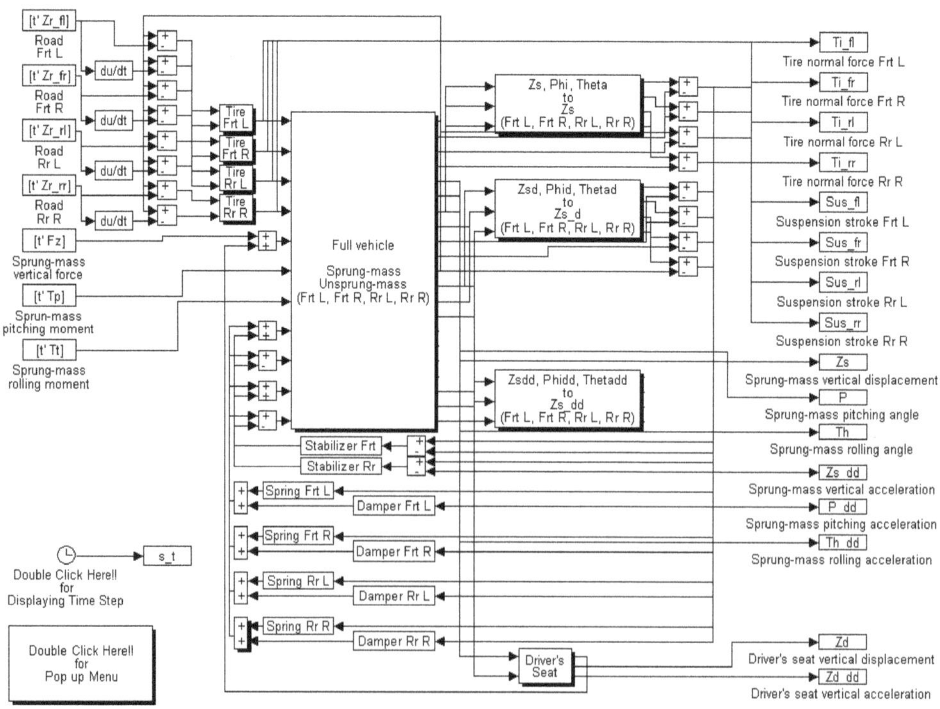

Figure A5. Full car model using the simulation modules.

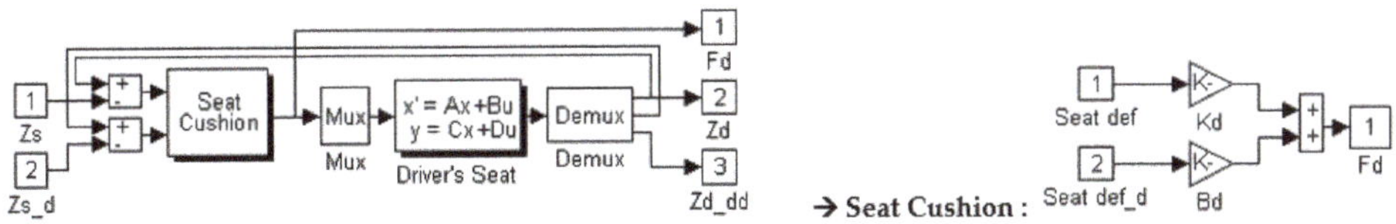

$$m_d \ddot{z}_d + f_c = 0$$

$$u = f_c, \quad x = \begin{bmatrix} z_d \\ \dot{z}_d \end{bmatrix} \quad y = \begin{bmatrix} z_d \\ \dot{z}_d \\ \ddot{z}_d \end{bmatrix} \qquad A = \begin{bmatrix} 0 & 1 \\ 0 & 0 \end{bmatrix}, \quad B = \begin{bmatrix} 0 \\ \frac{-1}{m_d} \end{bmatrix}$$

$$C = \begin{bmatrix} \mathbf{I}_{(2\times 2)} \\ \mathbf{0}_{(1\times 2)} \end{bmatrix}, \quad D = \begin{bmatrix} \mathbf{0}_{(2\times 1)} \\ \frac{-1}{m_d} \end{bmatrix}$$

$$\dot{\mathbf{x}} = \mathbf{A}\mathbf{x} + \mathbf{B}\mathbf{u}$$
$$\mathbf{y} = \mathbf{C}\mathbf{x} + \mathbf{D}\mathbf{u} \qquad f_c = k_d(z_d - z_s) + k_d(\dot{z}_d - \dot{z}_s)$$

Figure A6. Driver's seat module.

References

1. Wu, F.-K.; Yeh, T.-J.; Huang, C.-F. Motor Control and Torque Coordination of an Electric Vehicle Actuated by Two In-Wheel Motors. *Mechatronics* **2013**, *23*, 46–60. [CrossRef]
2. Murata, S. Innovation by In-Wheel-Motor Drive Unit. *Veh. Syst. Dyn.* **2012**, *50*, 807–830. [CrossRef]
3. Crolla, D. *Automotive Engineering: Powertrain, Chassis System and Vehicle Body*; ELSEVIER: Amsterdam, The Netherlands, 2009.
4. Kobayashi, T.; Katsuyama, E.; Sugiura, H.; Ono, E.; Yamamoto, M. Direct Yaw Moment Control and Power Consumption of In-Wheel Motor Vehicle in Steady-State Turning. *Veh. Syst. Dyn.* **2017**, *55*, 104–120. [CrossRef]
5. Wang, Q.; Zhao, Y.; Deng, Y.; Xu, H.; Deng, H.; Lin, F. Optimal Coordinated Control of ARS and DYC for Four-Wheel Steer and in-Wheel Motor Driven Electric Vehicle with Unknown Tire Model. *IEEE Trans. Veh. Technol.* **2020**, *69*, 10809–10819. [CrossRef]
6. Anderson, M. Unsprung Mass with In-Wheel Motors-Myths and Realities. *Proc. AVEC* **2010**, *10*, 261–266.
7. Du, H.; Sze, K.Y.; Lam, J. Semi-Active H∞ Control of Vehicle Suspension with Magneto-Rheological Dampers. *J. Sound Vib.* **2005**, *283*, 981–996. [CrossRef]
8. Gu, X.; Yu, Y.; Li, J.; Li, Y. Semi-Active Control of Magnetorheological Elastomer Base Isolation System Utilising Learning-Based Inverse Model. *J. Sound Vib.* **2017**, *406*, 346–362. [CrossRef]
9. Jansen, L.M.; Dyke, S.J. Semiactive Control Strategies for MR Dampers: Comparative Study. *J. Eng. Mech.* **2000**, *126*, 795–803. [CrossRef]
10. Karnopp, D.; Crosby, M.J.; Harwood, R.A. Vibration Control Using Semi-Active Force Generators. *J. Manuf. Sci. Eng.* **1974**, *96*, 619–626. [CrossRef]
11. Anaya-Martinez, M.; Lozoya-Santos, J.-J.; Félix-Herrán, L.C.; Tudon-Martinez, J.-C.; Ramirez-Mendoza, R.-A.; Morales-Menendez, R. Control of Automotive Semi-Active MR Suspensions for in-Wheel Electric Vehicles. *Appl. Sci.* **2020**, *10*, 4522. [CrossRef]
12. Basargan, H.; Mihály, A.; Gáspár, P.; Sename, O. Adaptive Semi-Active Suspension and Cruise Control through LPV Technique. *Appl. Sci.* **2021**, *11*, 290. [CrossRef]
13. Emura, J.; Kakizaki, S.; Yamaoka, F.; Nakamura, M. Development of the Semi-Active Suspension System Based on the Sky-Hook Damper Theory. *SAE Trans.* **1994**, *103*, 1110–1119.
14. Venhovens, P.T. The Development and Implementation of Adaptive Semi-Active Suspension Control. *Veh. Syst. Dyn.* **1994**, *23*, 211–235. [CrossRef]
15. Williams, R.A. Electronically Controlled Automotive Suspensions. *Comput. Control Eng. J.* **1994**, *5*, 143–148. [CrossRef]
16. Palkovics, L.; Bokor, J.; Venhovens, P. Design Problems of the Semi-Active Wheel Suspension System and a Possible Way of Their Elimination. In Proceedings of the FISITA'94 World Congress, Beijing, China, 17–21 October 1994.
17. Kim, H. A Study for the Control of Passenger Car Suspension for Ride Comfort. Ph.D. Thesis, KAIST, Daejeon, Korea, 1994.
18. Woo, S.; Ryu, J.H. Design of Double Sky-hook Algorithm for Semi-active Suspension System. In Proceedings of the KSAE 1996 Annual Autumn Conference, Seoul, Korea, 30 November 1996.
19. Goncalves, F.D.; Ahmadian, M. A Hybrid Control Policy for Semi-Active Vehicle Suspensions. *Shock Vib.* **2003**, *10*, 59–69. [CrossRef]
20. Ren, H.; Zhao, Y.; Chen, S.; Liu, G. State Observer Based Adaptive Sliding Mode Control for Semi-Active Suspension Systems. *J. Vibroeng.* **2015**, *17*, 1464–1475.
21. Geng, G.; Yu, Y.; Sun, L.; Li, H. Research on Ride Comfort and Driving Safety under Hybrid Damping Extension Control for Suspension Systems. *Appl. Sci.* **2020**, *10*, 1442. [CrossRef]
22. Wong, J.Y. *Theory of Ground Vehicles*; John Wiley & Sons: Hoboken, NJ, USA, 2008.

applied sciences

MDPI

Article

Investigation of Vehicle Stability with Consideration of Suspension Performance

Vaidas Lukoševičius [1,*], Rolandas Makaras [1], Arūnas Rutka [2], Robertas Keršys [1], Andrius Dargužis [1] and Ramūnas Skvireckas [1]

1 Faculty of Mechanical Engineering and Design, Kaunas University of Technology, Studentų Str. 56, 51424 Kaunas, Lithuania; rolandas.makaras@ktu.lt (R.M.); robertas.kersys@ktu.lt (R.K.); andrius.darguzis@ktu.lt (A.D.); ramunas.skvireckas@ktu.lt (R.S.)
2 Lithuanian Road Administration, J. Basanavičiaus Str. 36, 03109 Vilnius, Lithuania; arunas.rutka@lakd.lt
* Correspondence: vaidas.lukosevicius@ktu.lt

Abstract: The issue of movement stability remains highly relevant considering increasing vehicle speeds. The evaluation of vehicle stability parameters and the modeling of specific movement modes is a complex task, as no universal evaluation criteria have been established. The main task in modeling car stability is an integrated assessment of the vehicle's road interactions and identification of relationships. The main system affecting the vehicle's road interaction is the suspension of the vehicle. Vehicle suspension is required to provide constant wheel to road surface contact, thus creating the preconditions for stability of vehicle movement. At the same time, it must provide the maximum possible body insulation against the effect of unevennesses on the road surface. Combining the two marginal prerequisites is challenging, and the issue has not been definitively solved to this day. Inaccurate alignment of the suspension and damping characteristics of the vehicle suspension impairs the stability of the vehicle, and passengers feel discomfort due to increased vibrations of the vehicle body. As a result, the driving speed is artificially restricted, the durability of the vehicle body is reduced, and the transported cargo is affected. In the study, analytical computational and experimental research methods were used. Specialized vehicle-road interaction assessment programs were developed for theoretical investigation. The methodology developed for assessing vehicle movement stability may be used for the following purposes: design and improvement of vehicle suspension and other mechanisms that determine vehicle stability; analysis of road spans assigned with characteristic vehicle movement settings; road accident situation analysis; design of road structures and establishment of certain operational restrictions on the road structures. A vehicle suspension test bench that included original structure mechanisms that simulate the effect of the road surface was designed and manufactured to test the results of theoretical calculations describing the work of the vehicle suspension and to study various suspension parameters. Experimental investigations were carried out by examining the vibrations of vehicle suspension elements caused by unevenness on the road surface.

Keywords: quarter-car model; road unevenness; suspension performance; suspension test bench; vehicle stability

Citation: Lukoševičius, V.; Makaras, R.; Rutka, A.; Keršys, R.; Dargužis, A.; Skvireckas, R. Investigation of Vehicle Stability with Consideration of Suspension Performance. *Appl. Sci.* **2021**, *11*, 9778. https://doi.org/10.3390/app11209778

Academic Editors: Flavio Farroni, Andrea Genovese and Aleksandr Sakhnevych

Received: 24 September 2021
Accepted: 11 October 2021
Published: 19 October 2021

1. Introduction

Vehicle stability depends on a number of factors, and a general investigation of stability would not accurately reflect the effect of individual systems. Vehicle-road interaction and the factors that affect this relationship have the greatest impact on vehicle stability. The key elements of the interaction are the following: the structure of the vehicle suspension and the changes in wheel position influenced by it, the nature of the road surface, the structure, and the principles of operation of the elastic and damping elements. Most applied vehicle stability research is dedicated to the study and improvement of individual elements of vehicle-road interaction.

The main assembly affecting vehicle-road interaction is the vehicle suspension. Designing suspension of a completely new type would be a highly difficult task, so the maximum scientific capacity and research base must be dedicated to improving conventional structures. Fewer cars feature classic passive suspensions. Various control systems with active elements that adjust the suspension properties as necessary are becoming increasingly popular. Anti-lock braking, anti-skid systems, and four-wheel steering systems have become popular in recent decades. All of the above factors determine the stability of vehicle movement.

The unevenness and roughness of the road surface are the key environmental factors affecting a moving vehicle. Therefore, it is crucial to accurately assess the dynamic response of the vehicle that is subject to the most realistic road model possible. However, the development of the latter presents a great challenge. Road models are usually expressed as a function of spectral density that accounts for the rises and falls of a road profile. In road unevenness models, the idealized road profile is considered to be a random process, and random deviations are not analyzed.

The work of M.W. Sayers, T.D. Gillespie, and S.M. Karamihas at the University of Michigan Transportation Research Institute should be emphasized in an analysis of road surface assessment. This research work [1–3] is among the key works that have contributed to the establishment of globally recognized principles of road surface measurement and data processing criteria. In this work, the researchers investigated methods and equipment for measuring various parameters of the road surface and provided comprehensive criteria for the determination of IRI (International Roughness Index). IRI measurement is based on an assessment of the effect of the road surface on the parameters established for the quarter-car model. Other studies [4–9] can also be considered, since they evaluate the effect of the road surface on the vehicle.

The tire is another important element that influences vehicle stability. Pneumatic tires are essentially complex force generators. This means that only complex algebraic expressions enable the exact modeling of the tires. In their paper [10], M. EI-Gindy and H. Lewis present an extensive investigation of the contact area of the tire-road surface. Several other researchers have also evaluated the effect of the tire [11–18]. H.B. Pacejka from TU Delft performed important research in the field of vehicle stability, performance variables, and tyre modelling. He investigated the nonlinear motion of the car during sudden manoeuvres, modelling the behaviour of the car during sudden acceleration or braking. This research works dealt with various wheel-slip situations and provided model simulations of longitudinal and lateral slip [19].

Research articles have primarily investigated the design and improvement of active suspension. The use of an active suspension system enables the solution of many complex suspension design issues. Designers of active systems have provided significant contribution to the development of suspensions with different levels of activity: semi-active, adaptive, active damping, and other types. In individual cases, these suspensions can be compared to fully active ones, in particular when considering a simpler design for reliability. The main research objects considered when dealing with suspension work are single or two-mass quarter-car models. The synthesis criteria and optimization methods of control theory are applied to the analysis of these models. Accurate assessment of an active car suspension is possible only with the use of more complex two or three-dimensional car models. The accuracy of these models is further enhanced by the lateral tilt and turns of the entire vehicle. The most common research works focus on the separate analysis of dependencies: the delay of the excitation effect between the front and rear wheels, differences between parallel rolling wheels on one axle and deformability of the car frame, among others.

An article published by J. Wattan, K.M. Holford, P. Surwattanawan [20] is insightful as it deals with the issues of active suspension modeling and employs a real-life quarter-car model described in the article. Other articles deal with active suspension control problems [21–24] or analyze the performance of specific suspensions [25–27].

Some research investigates the active suspension anticipation function. Two main control schemes are used most often. The one located in front of the data vehicle uses the obtained data in the control system that is designed to eliminate the excitatory effect of the road surface. The performance of another system is based on an assessment of the road surface by examining the dynamics of the front axle. Studies have shown that the control system is highly dependent on the quality of the road surface. Theoretical calculations provide for improvements of the active suspension and account for the applicability of the anticipation function.

Research by D. Horvat [28,29] analyzed the methods of road surface assessment and model the vehicle movement. The research was performed using quarter-car and spatial models of a vehicle with a different number of degrees of freedom. The main criterion for assessing the impact of the road surface is the acceleration of the sprung mass of the vehicle. The properties of active suspensions may be provided to semiactive suspension systems by using components similar to passive suspensions. The main idea is to use an additional elastic element mounted parallel to the damper, the properties of which can be adjusted to high-frequency performance. A semi-active damper is an externally controlled force generator. Its capacities are limited, as the stored energy must be dissipated.

A separate research field is dedicated to the investigation of the design of various suspension elements and possibilities for their improvement. Considerable research efforts have been made to improve suspension elements and their properties [30–35]. Certain articles deal with the vibrations of various vehicle elements and the methods of their measurement [36–38]. Very few research works provide a generalized analysis of different types of car suspension design. The German researcher J. Reimpell is one of the most prominent world-class experts in this field. His research papers have been published all over the world and translated into different national languages [39–43]. His work provides a comprehensive investigation of vehicle suspension structures and discussion of the principles and methods of calculation of the characteristic parameters of suspension.

The designs of suspension systems and their main characteristic parameters have been investigated by T. D. Gillespie [44], W. Matschinsky [45] and D. Bastow [46]. In their studies, they analyze different types of suspension and discuss key parameters that affect the movement and stability of a vehicle.

Reviews have confirmed that efforts are being made to make use of the rigid and damping elements of suspensions with variable characteristics in an attempt to combine conflicting requirements such as driving comfort and directional stability of the vehicle. However, there are no comprehensive data on the performance of these systems for poorer quality roads.

This article reviews the road surface and vehicle tire interaction models that include real road surface properties. Classification of vehicle suspensions according to summarized kinematic suspension properties that provide the prerequisites for the development of universal models that account for the performance of suspensions with diverse kinematic properties are proposed. Models that account for the effect of the suspension on the vehicle movement stability were developed. A vehicle suspension test bench that includes the mechanisms of the original design to simulate the effect of the road surface was designed and manufactured to verify the results of the theoretical calculations describing the performance of the vehicle suspension and to study various suspension parameters.

Based on the topics above, the main contributions of this paper are: (1) a description of the road surface is revised and provides specifications of the interaction of road surface-tire contact areas and the equalizing properties of the tire; (2) a dynamic quarter-car model is developed and accounts for the effect of suspension, and the effect of the individual suspension elements on the characteristics that determine movement stability is determined, with the friction force originating from the transverse displacement of the wheel in the damper frame and the variation thereof being assessed; (3) the applicability of a vertical dynamics model is verified by experiments, with experimental investigations being conducted and the vibrations of the vehicle suspension elements caused by the road surface unevennesses explored using the developed vehicle suspension test bench that simulates driving conditions and replicating the real road surface effect; (4) the effect of geometric position changes of the wheel on the parameter of directional stability of the vehicle, i.e., on the slip angle, is determined using kinematic models of the vehicle suspension. The developed software application is used to calculate the effect of the above factors on the dependence of the vehicle limit turning radius on the speed.

2. Vehicle Stability Modeling Methods

Trend analysis of the traffic and vehicle design patterns suggests that handling, i.e., the ability to move in the direction intended by the driver, is one of the key performance properties of vehicles that influence their safety. The driver sets the intended direction using the steering wheel, and the handling tasks are targeted at the steering mechanism. One of the first tasks addressed was the interrelation between the front wheel angles. The Ackermann-Jeantaud scheme (Figure 1) was used for this purpose.

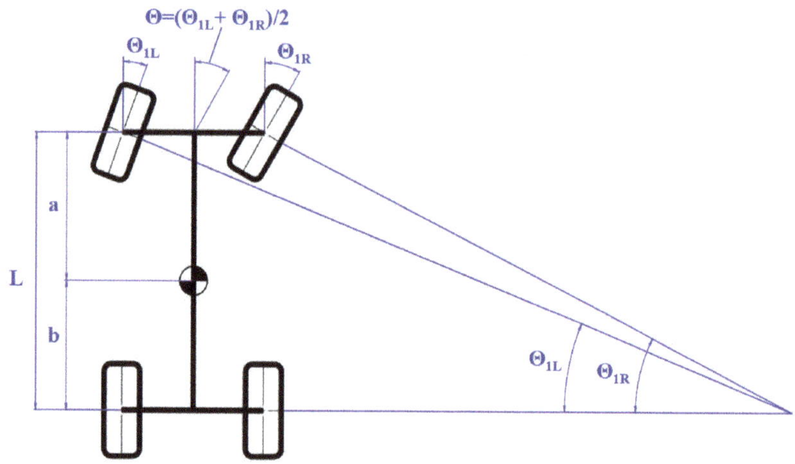

Figure 1. Ackermann—Jeantaud schema [47].

Following the introduction of pneumatic tires into modeling, the dependencies of the tire-road interrelation had to be analyzed in greater detail. The area of contact between the tire and the road is subject to deformation under the action of a lateral force. As a result, the contact patch axis is no longer parallel to the wheel symmetry plane, but forms angle δ, the former being referred to as the tire slip. In most tasks, the slip angle is determined as the angle between the velocities longitudinally and transversely to the wheel symmetry plane. The lateral force F_y that the wheel has the capacity to withstand depends on the radial load and slip angle. Tire slip angle (Figure 2) causes a considerable distortion of the cornering scheme [48]. There are several theories that can be used to determine the angle of tire-slip. The theory of M. V. Keldysh [48] is most convenient. Under M. V. Keldysh's theory, lateral tire deformation ξ and trajectory curvature ρ are described by the following Equations [47,48]:

$$F_y = C_y\xi, \frac{1}{\rho} = \alpha\xi - \beta\varphi_t \tag{1}$$

where C_y is the lateral stiffness of the tire, ρ is the trajectory curvature radius, α and β are the constant coefficients associated with the tire structure, and φ_t is the angle between the wheel symmetry plane and wheel trajectory tangent (Figure 2).

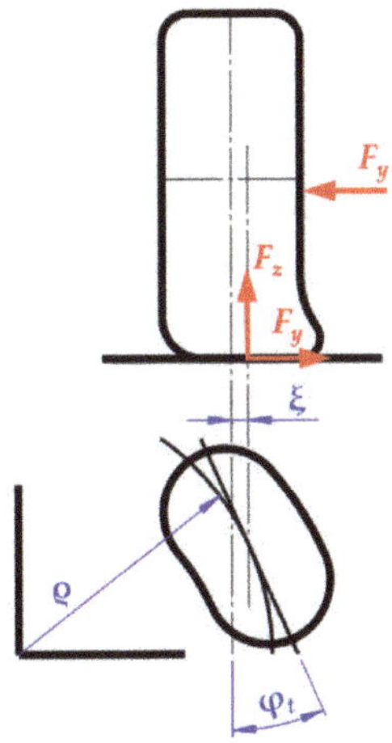

Figure 2. Tire deformation scheme in the presence of slip [48].

If $F_y \neq$const., then the wheel center movement velocity in the transverse direction equals is given by:

$$v_y = v\varphi_t + \dot{\xi} \tag{2}$$

where $v_y/v = \delta$, then $\delta = \varphi_t + \dot{\xi}/v$, from which it follows:

$$\varphi_t = \delta - \dot{\xi}/v \tag{3}$$

This would lead to the condition that only the tire slip angle would affect the vehicle movement direction, the prerequisite being $\dot{\xi} = 0$, i.e., F_y = const.

For low φ_t, the wheel trajectory is straight, and the contact patch area is positioned at angle $\varphi_t = \delta$ relative to the symmetry plane of the wheel. With the available inertia forces, the longitudinal and transverse movement of the vehicle can be described. Considering that θ, δ_1 and δ_2 are low, and their cosines are equal to each other, while sines are equal to the angle values (Figure 3), and forming the expression $F_{y1} = k_{y1}\delta_1$ and $F_{y2} = k_{y2}\delta_2$, the transverse force equilibrium is described by Equation (4) [48]:

$$F_{iy} = m_a(v\omega_a + \dot{v}_y) = k_{y1}\delta_1 + k_{y2}\delta_2 - F_{x1}\Theta \tag{4}$$

where m_a is the vehicle center of mass; ω_a is the vehicle angular velocity when cornering; k_{y1} and k_{y2} are the slip angle drag coefficients; F_{x1} and F_{y1}, F_{x2} and F_{y2} are drag forces for the front and rear wheels respectively and Θ is the wheel turning angle.

Following expression of δ_1 and δ_2 by using v_y and ω_a:

$$\delta_1 = \Theta - (a\omega_a + v_y)/v,\ \delta_2 = (b\omega_a - v_y)/v \tag{5}$$

Following input of the δ_1 and δ_2 values into Equation (5), the following is obtained:

$$\dot{v}_y + v_y(k_{y1} + k_{y2})/(m_a v) + \omega_a[v + (k_{y1}a - k_{y2}b)/(m_a v)] - k_{y1}\Theta/m_a = 0 \tag{6}$$

By applying the equation of the moments at the center of mass $I_z\varpi_z = k_{y1}a - k_{y2}b$ and identifying variables v and ω, the following is obtained:

$$\ddot{v}_y + m\dot{v}_y + pv_y = q_{y1}\dot{\Theta} + q_{y2}\Theta,\ \ddot{\omega}_a + m\dot{\omega}_a + p\omega = q_{\omega1}\dot{\Theta} + q_{\omega2}\Theta \tag{7}$$

where m, p, q_{y1}, q_{y2}, $q_{\omega1}$, $q_{\omega2}$ are the relative drag coefficients of the front and rear axles.

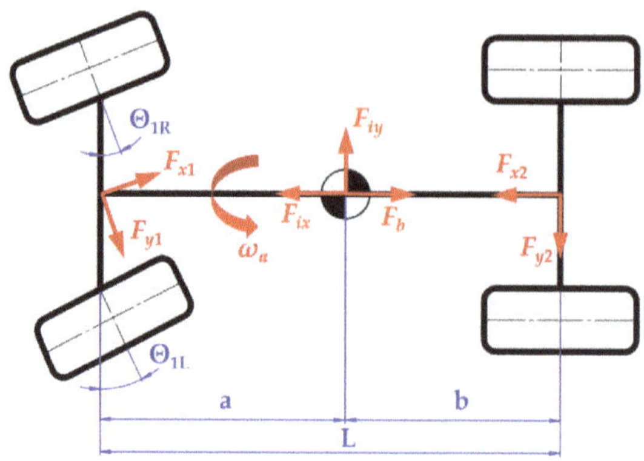

Figure 3. Forces acting at vehicle cornering [48].

The dependence of the longitudinal and lateral parameters of the vehicle on the steerable wheel turn angle can be determined by solving Equations (6) and (7), and their dependence on the steering wheel turn angle can be determined if the steering ratio value is available. Equations (6) and (7) can be expressed in a different form [49]:

$$I_z\dddot{\psi} + D\ddot{\psi} + C\dot{\psi} = \dot{M}_z + HM_z + k_{y1}a\dot{\delta}_1 + E\delta_1 + k_{y2}b\dot{\delta}_2 + E\delta \tag{8}$$

where I_z is the moment of inertia about the vertical axis; M_z is the external moment about the vertical axis and ψ is the angle of the turn about the moment center. Coefficients D, H, C, and E are calculated as follows:

$$D = \frac{I_z(k_{y1}+k_{y2})}{m_a v_c} + \frac{k_{y1}a^2+k_{y2}b^2}{v_c},\ H = \frac{k_{y1}+k_{y2}}{m_a v_c},\ C = \frac{L^2 k_{y1}k_{y2}}{m_a v_c} - k_{y1}a + k_{y2}b, \quad E = \frac{Lk_{y1}k_{y2}}{m_a v_c} \tag{9}$$

where v_c is the velocity of the center of mass (its direction does not correspond to the longitudinal axis of the vehicle).

The following is obtained:

$$I_z\dddot{\psi} + D\ddot{\psi} + C\dot{\psi} = \Phi(M_z, \delta_1, \delta_2) \tag{10}$$

where Φ is the function of effects.

The stability of the solutions of Equation (10) depends on the coefficients of the equation and the effect function Φ. If the function of effects were not considered, the vehicle movement would be linear, and its stability would depend on coefficients D and C.

Although designers may solve the problem and implement respective measures, the spatial position of the wheel changes at excitation, for example, due to road unevenness, and this may lead to a change in vehicle direction. More comprehensive studies have shown that slip angle characteristics are not enough, as the direction vector of the wheel is affected by the spatial position of the wheel, i.e., camber, alignment, etc. Direct application of dynamic equations is impossible due to the specifics of the tasks. Furthermore, the characteristics of individual elements are clearly nonlinear. Semiempirical formulas are used to describe the effect of the forces acting on the wheel and the spatial position of the wheel on the direction vector. Modeling requires dependences of the interrelation between the slip angle and lateral force, between the lateral force and relative wheel slip, between the slip angle and stabilizing moment. The analysis carried out has suggested that the currently available methodologies applicable to the factors affecting directional stability employ simplified models that individually address steering mechanism kinematics, suspension kinematics, suspension element stiffness, and forces acting in vehicle cornering. This kind of analysis does not enable the identification of key factors. Therefore, to perform an integrated analysis, models consisting of multiple elements must be used. However, the characteristics of the elements can only be identified if detailed information is available on the specific vehicle design. This creates difficulties when performing an integrated analysis and drawing general conclusions in an investigation of directional stability as a criterion for the assessment of the technical condition of a vehicle and the condition of the road surface.

3. Investigation of the Factors Affecting the Vehicle Stability

3.1. Road Description

Records of the actual Lithuanian road microprofile were used in modeling the vehicle stability tasks. Records were generated by road surface measurements performed using the DYNATEST 5051 RSP (Dynatest A/S, Ballerup, Denmark) profilometer mounted on the VW Transporter [18,49]. Whereas road profile records are characterized by exception and span-specific properties, the possibilities for formalization of the road microprofile description were analyzed by identifying the typical features of road surfaces of different quality. During formalization of the road description, the discrete nature and large interval of the road microprofile records, was considered. For example, the information on the micro-irregularity height within a single established line is recorded every 0.147 m under the IRI calculation methodology and every 1 m when spatial road microprofile is recorded. Road microprofile records contain road surface data on the two longitudinal lines that correspond to the vehicle wheel rolling surface, and the longitudinal distance between the data is 0.147 m. Analysis of available records confirmed that essential differences can be assessed using generalized data of the road span microprofile for an even asphalt pavement, uneven asphalt pavement (in service), and gravel road surfaces. The microprofile records of the road spans were analyzed using classic methodology by calculating the dispersion of unevennesses σ_q^2 and correlation function R_q:

$$\sigma_q{}^2 = \lim_{L_q \to \infty} \frac{1}{L_q} \int_0^{L_q} q_0{}^2(x)dx, \; R_q(x_s) = \frac{1}{L_q \sigma_q{}^2} \int_0^{L_q} q_0{}^2(x) q_0{}^2(x - x_s)dx \tag{11}$$

where L_q is the span length; q_0 is the microprofile height with the y–coordinate of the midline equal to zero interval x_s.

Road surface roughness data were used for an accurate assessment of the effect of the road surface and investigation of the equalization function of the tire. The data reflect the roughness of the road surface on a single line every 2.5 mm. Data from three different road surfaces were used for the investigation: relatively even (I), medium (II) and very rough (III). These microprofile records are fairly accurate, and the effect of roughness on the road surface was considered in addition. The road surface roughness span data was distributed in the microprofile according to the principle of a random number generator. These measures enabled the authors to fairly accurately reflect the key characteristics of the road surface. The data on the microprofile and road surface roughness characteristics of the investigated road spans are presented in Table 1. The characteristic wavelengths were revised by harmonic analysis involving identification of the lengths of maximum amplitude waves $R_i = \sqrt{a_i^2 + b_i^2}$ for Equation (12):

$$q(x) = \frac{x_0}{2} + \sum_{i=1}^{\infty} a_i \cos(i\omega x) + \sum_{i=1}^{\infty} b_i \cos(i\omega x) \tag{12}$$

Table 1. Microprofile and roughness characteristics of the investigated road surfaces.

Road Surface Type	Dispersion σ_q, mm	Characteristic Wavelength, m
Even asphalt pavement	3.492	50–52
Uneven asphalt pavement	12.437	75–77
Gravel road	13.667	45–47
Roughness I (relatively even)	0.276	0.04–0.1
Roughness II (medium)	0.695	–
Roughness II (very rough)	1.667	0.06–0.14

The waviness of the road surface and its amplitudes were determined following harmonic analysis of typical road microprofile fragments. The maximum amplitude waves were longer than 12 m in the cases analyzed. Only a few longer waves were found. The number of shorter waves was higher, but the amplitude of the shorter waves was very small at 0.3 to 1 mm depending on the microprofile evenness. It should be emphasized that short wavelength unevenness amplitudes were clearly dependent on the road surface type. For modeling using a flat model, the effect of the road surface was assessed by analyzing the microprofile points on a single longitudinal line.

The correlation functions of the road surface roughness are presented in Figure 4. It was noticed that waviness with waves shorter than those of the microprofile was characteristic of rough road surfaces.

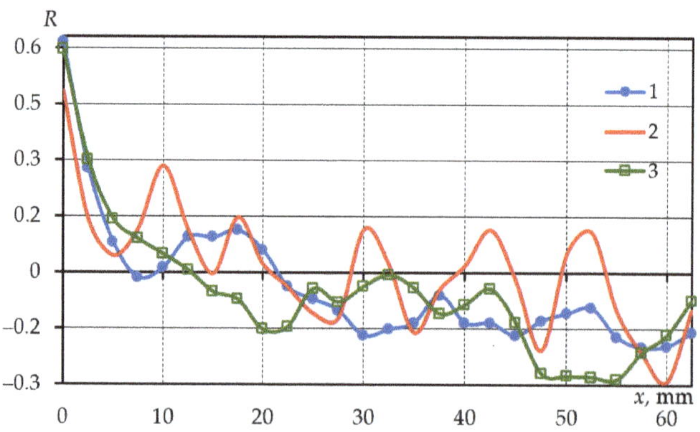

Figure 4. Correlation functions of road surfaces of different roughness: 1 = even, 2 = medium, 3 = very rough.

The spatial record of the road microprofile was used to determine the characteristics of the transverse road and to accurately model vehicle movement. Data from the 3 m wide road lane microprofile were recorded at nine points in the transverse, longitudinal direction, with a 1 m interval between the data. The description of the road span microprofiles was facilitated by the fact that the microprofile records generated by the DYNATEST 5051 RPS profilometer were densified in the spatial (3D) mode in the ruts to reproduce the microprofile specifics of the tire-road contact area more accurately.

The spatial microprofile record analysis showed waviness (Figure 5) that was characterized by perpendicularity, or a slight tilt relative to the axis, and relatively low transverse waves. The analysis of the spatial road surface roughness fragment (Figure 6) suggested that the roughness was a random set of unevennesses, having a certain prevailing shape of unevennesses.

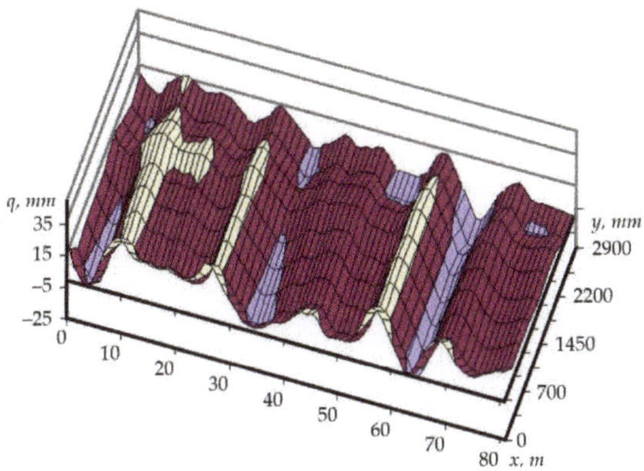

Figure 5. Spatial fragment of a road surface microprofile span.

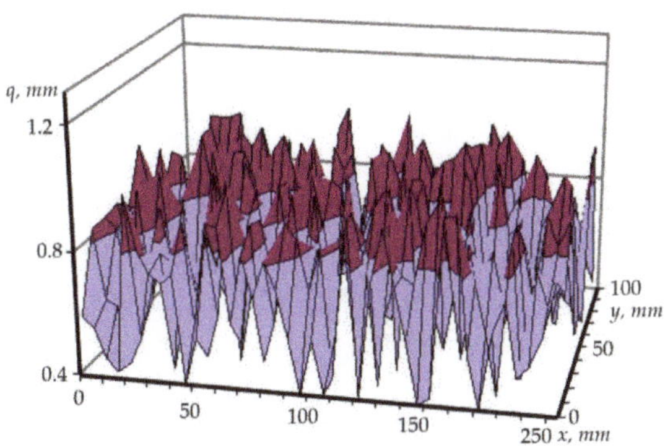

Figure 6. Fragment of a spatial road surface roughness.

In the case of spatial modeling, it is necessary to assess the nature of the road surface in the transverse plane relative to the vehicle movement. The spatial road microprofile record was used as the baseline in the calculation application and accounted for the nature of the variation of the microprofile in the longitudinal and transverse directions. These data provided accurate information on the specifics of the road surface in the transverse directions (ruts, camber, banks). A microprofile of a smaller interval was reproduced according to more detailed records. Road surface roughness was also assessed during modeling. The calculation application includes the option to assess random damages to the road surface, e.g., cracks present in the road surface.

3.2. *Investigation of the Equalizing Function of the Tire*

During modeling of vehicle wheel rolling on a road span with determined characteristics, assessment of road-tire interaction is important even where the road microprofile record is available. Two aspects were analyzed during the study. The first was related to the fact that a rolling wheel would not copy the unevenness of the road surface, but would rather be characterized by an equalizing effect. This had to be taken into account in the case of description of movement on a very uneven road. The second aspect was related to tire deformation. Theoretically, assuming that the tire deformability is linear, it needs to be described in the tire description as a nonlinear element. This is due to the constant variation of the tire-road contact area and nonlinear dependence between the contact dimensions (e.g., contact length) and vertical deformation during deformation of the tire.

During the study, the effect of the tire on vehicle-road interaction was investigated to select the simplest possible tire model that would be characterized by sufficient precision. The study was limited to linear and flat tire-road contact assessment irrespective of the tire structure (quasi-static model). Flexible narrow ring (2D) and flexible band (3D) models were chosen for a more detailed investigation of the equalizing function of the tire. The former (Figure 7a) assesses the linear tire-road contact and is used where a simplified microprofile record is available, i.e., the road microprofile on a single line. The latter (Figure 7b) assesses the spatial contact between the tires and the road, but requires a road microprofile record that reflects the condition of the entire traffic lane. During the application of the models referred to above, the contact length of the tire was determined according to the vertical deformation of the tire. Two marginal cases were analyzed: (i) tire deformation in the circular direction, which was absolutely absent with contact length equal to arc length (this assumption was inaccurate, as experimental data confirmed that the tire was subject to compression in the contact area), and (ii) compression of the tire tread up to the chord length. The calculations suggested that the methodology of calculation of the contact length did not have a significant effect on the results, where the rolling of contemporary tires on roads in good condition was analyzed. In the calculations, a deterministic research approach was employed. For this purpose, real road microprofile records were formed on a single line for the 2D tire model, and spatial road microprofile records were formed for the 3D tire model. The roughness of the road surface was assessed by distributing the road surface data on the road microprofile. The road surface roughness data (control points located every 2.5 mm) were selected from the baseline data file according to the random number generator principle. This provided the calculated road microprofile data $q(x)$ or $q(x,y)$.

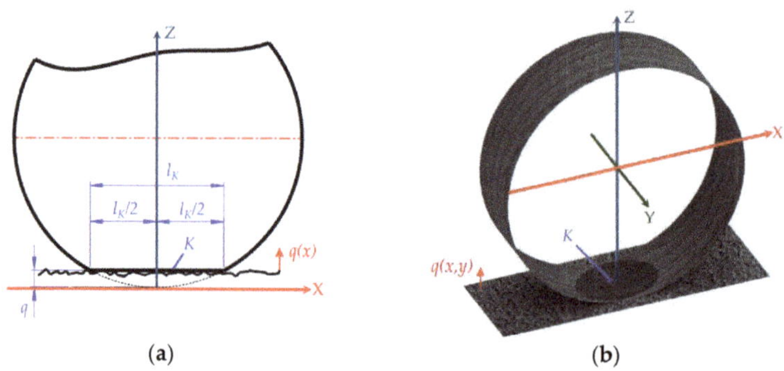

Figure 7. Tire models: (**a**) flexible ring; (**b**) flexible band.

The quarter-car model corresponding to the parameters of a city car was used in the calculations [18]. By using this model, the limits of variation in tire contact length were determined for roads of different quality (Figure 8). For roads with asphalt pavement, the length of the tire-road contact varied by about 16%, and about 15% of the length of the road contact was close to the length under static load. For an uneven road (Figure 8), the contact length curve was distributed over a much wider range. During the investigations, the cases where the tire became detached from the road, and the maximum allowable deformation of the tire was reached, were recorded. The analysis found that the role of the tire was more pronounced when modeling driving on very uneven roads. If necessary, the radii of the curvature of the road surface may be estimated at individual points of the profile. This kind of data processing significantly reduces the size of the initial data files and the duration of further calculations. Investigations showed that it is unreasonable to use the 3D model where there is no need to have data on vehicle wheel tilt or contact curvature in the transverse plane relative to wheel rolling.

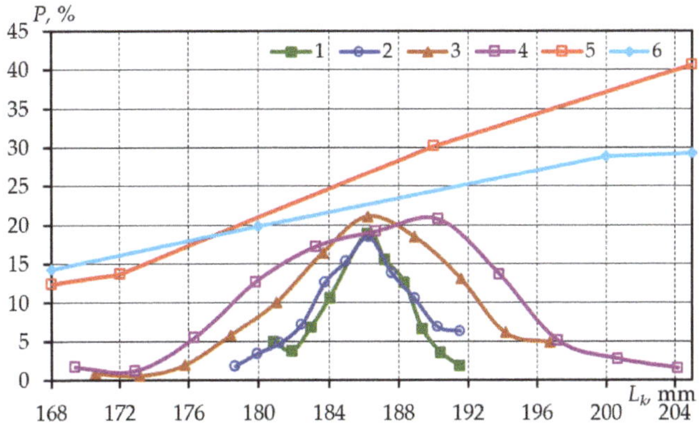

Figure 8. Tire contact length distribution for different road quality and tire models: 1—smooth asphalt pavement 2D; 2—smooth asphalt pavement 3D; 3—low quality asphalt pavement 2D; 4—low quality asphalt pavement 3D; 5—gravel road 2D; 6—gravel road 3D.

The vertical wheel displacement required for the evaluation of suspension performance is determined by the tire model with point contact. The results of the analysis of the microprofile records for roads of different quality suggested that the IRI measurement methodology proposed by M. Sayers [1,2] is solid when choosing the profile measurement criteria and the unevenness assessment interval.

3.3. Dynamic Quarter-Car Model

The quarter-car model was used for the tire-road interaction analysis. The model corresponded to the operating conditions of a single vehicle wheel and enabled researchers to simplify the analysis of vehicle dynamics. The quarter-car model is usually used to identify key vehicle suspension parameters and investigate tire performance.

Although the quarter-car model only covered the vertical elastic and damping suspension elements, the results enabled analysis of the directional stability of the vehicle at vertical excitation. The data obtained were used as the baseline data to identify the exact spatial position of the wheel (Section 5).

A revised quarter-car model (Figure 9) was used for the analysis of vehicle movement on road surfaces of different quality. In contrast to the two-mass model, the revised model also accounted for the tensile properties of the tire tread and the friction in the suspension elements. This model could provide more accurate structural parameters of the vehicle analyzed.

Model movement Equations for the vertical direction:

$$\begin{aligned} &m_1\ddot{z}_1 + k_2(\dot{z}_1 - \dot{z}_2) + k_1(\dot{z}_1 - \dot{q}) + c_2(z_1 - z_2) + c_1(z_1 - q) = 0, \\ &m_2\ddot{z}_2 + k_3(\dot{z}_2 - \dot{z}_3) + k_2(\dot{z}_2 - \dot{z}_1) + c_3(z_2 - z_3) + c_2(z_2 - z_1) + F_\mu \mathrm{sgn}(\dot{z}_2 - \dot{z}_3) = 0, \\ &m_3\ddot{z}_3 + k_3(\dot{z}_3 - \dot{z}_2) + c_3(z_3 - z_2) + F_\mu \mathrm{sgn}(\dot{z}_3 - \dot{z}_2) = 0 \end{aligned} \quad (13)$$

where m_1, m_2, m_3 are the tread part of the tire, unsprung, and sprung mass, respectively; z_1, z_2, z_3 are the tire, suspension and body displacements, respectively; k_1, k_2, k_3 are the tread, tire, and suspension damping factors, respectively; c_1, c_2, c_3 are the stiffness of the tread, tire, and suspension, respectively; F_μ is the friction force at the shock absorber rod and q is the height of road profile unevenness.

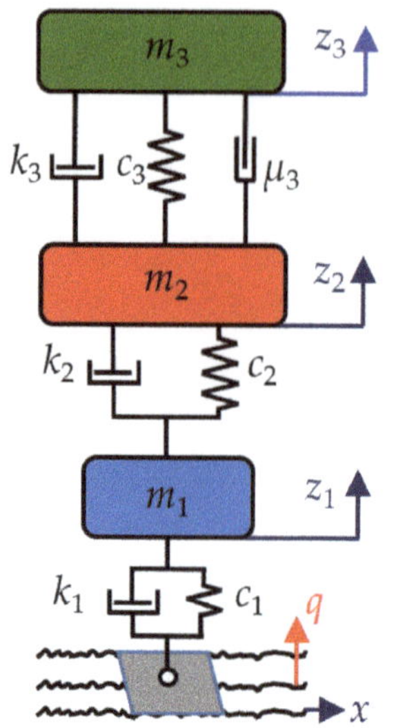

Figure 9. Quarter-car model.

When refining the quarter-car model, it is necessary to consider that the elastic elements and dampers of the suspension have distinctly nonlinear characteristics. In the investigated case, the characteristic of the front suspension spring of the modeled VW Golf was linear, as refined by the experiment. However, it must be taken into account that elastic travel stops block with nonlinear characteristics and start to act in extreme positions with respect to the car's suspension recoil and compression travel. Therefore, the overall characteristic of all the elastic elements of the suspension is close to typical (Figure 10a). In the model, the suspension characteristics are described by three lines. For the avoidance of uncertainty, additional refinements were used in the inflection zones of the characteristics. The damping characteristics of the damper were described in a similar way (Figure 10b).

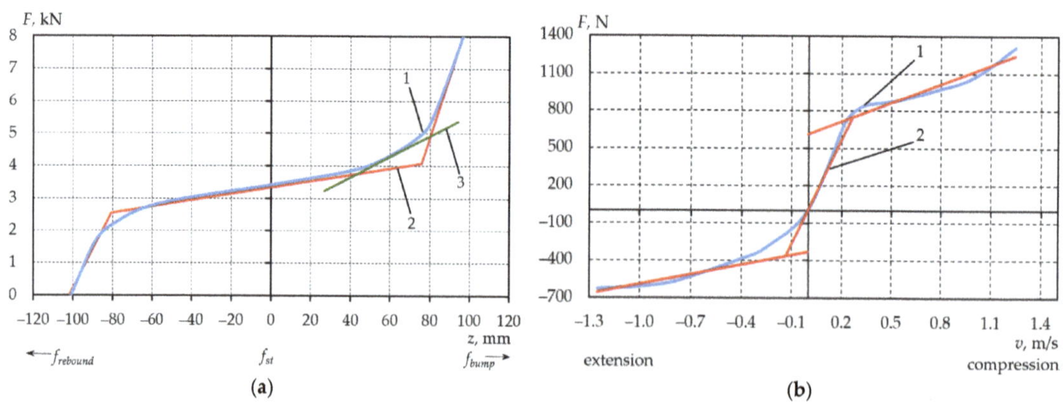

Figure 10. Modeling of suspension characteristics: (**a**)—elastic characteristic; (**b**)—damping characteristic of the damper; 1—typical characteristic; 2—simulated characteristic; 3—inflection zone refinement.

The frequency characteristic of the vehicle was determined by numerical modeling using calculation applications, with the road profile being sinusoidal in shape. The developed computer application enabled the analysis of vehicle movement with the set parameters related to the road surface at the respective frequencies. The movement of the model corresponding to the city car parameters was investigated to determine its frequency characteristics (Figure 11). When compared to the theoretical frequency characteristic, low-frequency resonance oscillations of higher amplitude sprung masses of the body were observed. The frequency of resonance oscillations of the unsprung masses was lower than that determined by the theoretical calculations.

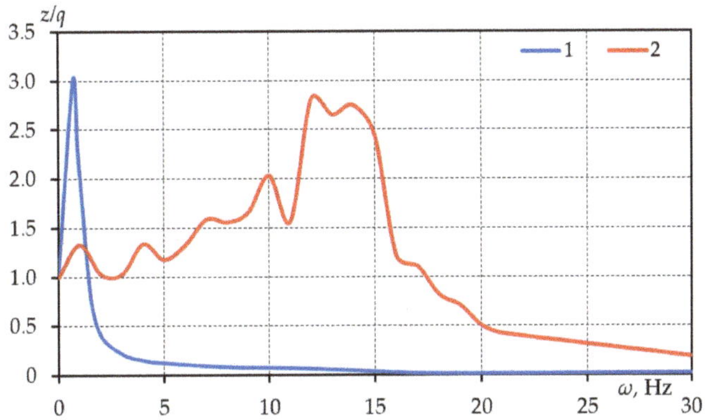

Figure 11. The frequency characteristic of the vehicle was determined with a numerical experiment: 1—sprung mass; 2—unsprung mass.

Calculations were performed to evaluate the uneven excitation effect of the road surface unevenness at various amplitudes on the frequency characteristic of the vehicle. During the evaluation of the effect of excitation amplitude, high-frequency resonance oscillation amplitudes of the unsprung mass were observed to decrease as the unevenness amplitude increased (Figure 12). However, the amplitudes of the low-frequency resonance oscillations increased in the latter case.

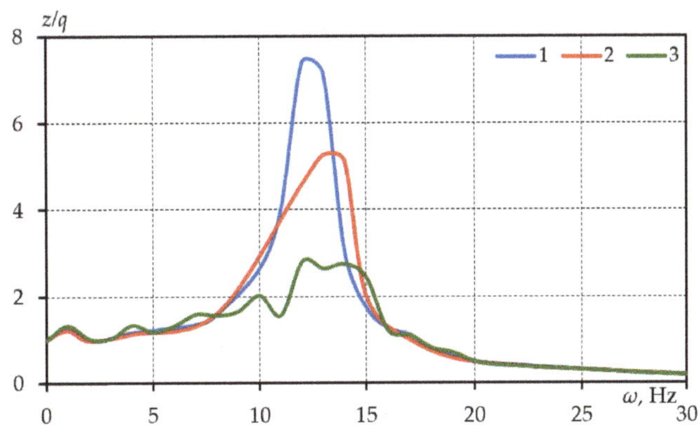

Figure 12. Frequency characteristic of the unsprung mass of the vehicle for the following road surface excitation amplitudes: 1—10 mm; 2—20 mm; 3—40 mm.

The effect of suspension characteristics on vehicle vertical displacements was determined by investigating the movement of the microprofile of the quarter-car model for the road surface microprofile of various parameters. A computer application was developed for the investigations. The application analyzed the vertical displacements of a three-mass quarter-car model and the effect of the nonlinear characteristics of the suspension.

Driving on one line was simulated without taking into account the transverse displacement of the wheel. The nonlinear characteristics of the suspension were found to mainly influence the movements of the sprung mass of the vehicle body (Figure 13). Simulation of driving on different road surfaces showed that the evaluation of the stiffness of the tire tread did not have a significant effect in most cases at low-frequency vibrations.

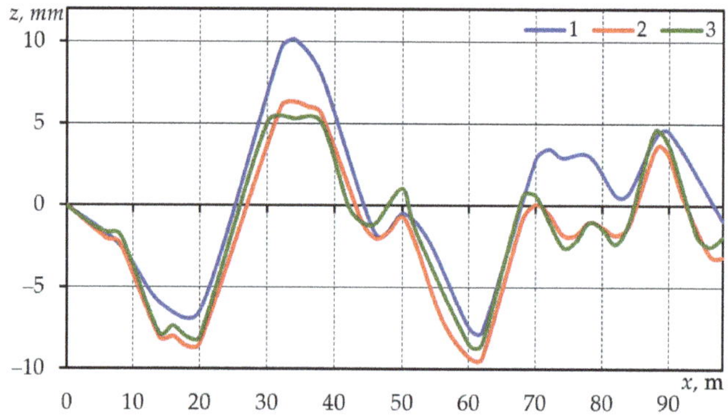

Figure 13. Vertical displacements of the sprung mass of the vehicle for driving on a smooth asphalt pavement road surface taking into account the effect of nonlinear elements: 1—linear elements; 2—nonlinear suspension characteristics; 3—effect of friction on the damper frame.

4. Experimental Investigations of Suspension Properties

4.1. Suspension Test Bench and Road Effect Simulation Equipment

A real quarter-car model was produced for the analysis of the theoretical calculations describing the performance of the vehicle suspension and for the investigation of the suspension parameters. This model enables investigation of vehicle suspension performance under laboratory conditions, different modes of movement, varying load conditions, and the excitation effect of different road surfaces.

The option of producing the car suspension test bench was chosen for the following main reasons:

- The car suspension test bench is not affected by random environmental conditions acting on the moving vehicle and the oscillations caused by the engine. This enables multiple experiments to be conducted under identical conditions. Therefore, this method provides more benefits compared to in-situ investigations of a real vehicle, where it is difficult to provide identical environmental conditions. Laboratory conditions allow for the use of stationary test equipment, thus simplifying the investigation process and reducing the duration of experiments.
- The test bench consists of the quarter car, i.e., the suspension of only one wheel is investigated. This makes it possible to investigate the performance of an individual wheel and its suspension elements without considering the remaining wheels or the movement of the entire vehicle body. This simplifies the research considerably. The results of the quarter car investigation revise the flat and spatial vehicle models.
- With the performance of the vehicle suspension elements is evaluated accurately, the test bench may be used for suspension improvement, for example, the development of a new active suspension.
- The vehicle suspension test bench can be used to investigate the performance of individual suspension elements by simulating various modes of vehicle movement and changing the load conditions.
- After assessing the lessons learned and installing a suspension of a different design, the test bench can be used to investigate other types of vehicle suspensions.

The test bench consists of the following main structural elements and assemblies.

- Main frame with road surface simulation equipment.
- Vehicle body part with suspension elements.
- Wheel drive with rotational speed changers.
- Equipment for recording the parameters describing the movement of the vehicle body.

The front part of the VW Golf-II vehicle with suspension elements was installed on the experimental test bench. During the investigation, the performance of the MacPherson suspension was analyzed. Various operating modes of the vehicle suspension were modeled by changing wheel rotational speed (four modes available); sprung mass of the body part, and the frequency of movement of the device simulating the effects of the road surface. The effect of the road surface was simulated by a pneumatic diaphragm chamber controlled by an electro-pneumatic distributor (Festo CPE 1 8-M1H-5/3G-1/4) that controlled the compressed air feed from the compressor. In the initial phase of the investigations, the road profile was modeled as a periodic function, and a discrete programmable controller (Festo FEC-FC30-DC) was used to control the distributor. For further investigation, microprocessor control of the distributor was provided according to real road surface microprofile records. The scheme of operation of the mechanisms that simulate the effects of the road surface and the quarter-car test bench is presented in Figure 14.

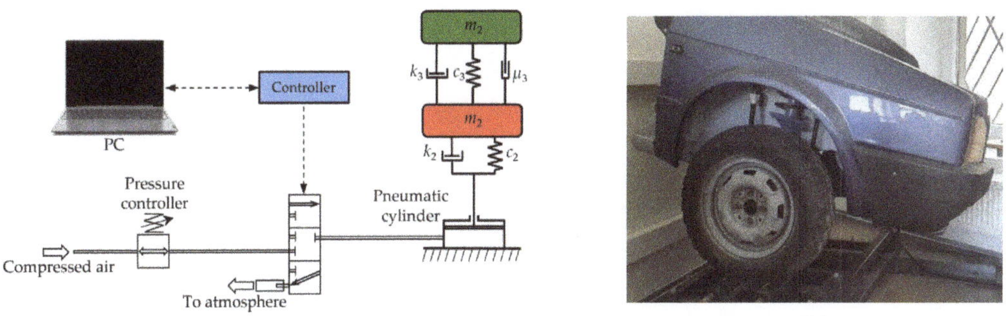

(a) (b)

Figure 14. Equipment simulating the road surface effects: (**a**) scheme; (**b**) quarter-car test bench.

Oscillation analysis of individual assemblies of test bench assemblies was performed to investigate the movement of the vehicle body and suspension and to determine the main characteristics of the model. Oscillation measurement equipment was used to record the oscillation parameters (displacement, speed, and acceleration). To measure the oscillation signal, a two-channel oscillation recording converter ADC-212 'Pico' (Pico Technology, St. Neots, UK) was used together with software installed on the computer. For additional tests, and control of the results of previous tests, a portable oscillation recording device 'Multiviber' by VMI (VMI International AB, Linköping, Sweden) was used. The results of the additional investigations allowed us to compare the performance of a real vehicle suspension with the operating conditions simulated by the vehicle suspension test bench.

Movement of the vehicle body and suspension elements was investigated by measuring the vertical displacements and the nature of their change. Movements of the equipment that simulate excitation of the road surface were recorded by a sensor located on a moving frame arm. The effect of excitation on the movement of the unsprung mass of the suspension was recorded by a sensor located on the lower arm of the suspension of the wheel. Vehicle body movements were measured by a sensor located on the body next to the mounting joint of the MacPherson damper frame. This arrangement of the sensors allowed the researchers to register changes in the movements between the individual assemblies of the system determined by the elastic and damping properties of the suspension elements and the tire.

4.2. Analysis of Experimental Investigation Results

During the tests, the characteristics of free oscillations of the suspension of the vehicle were determined by applying a load on the body using an additional mass and abruptly removing the load (Figure 15). The relative damping coefficient of the suspension was determined on the basis of these characteristics in relation to the theoretical relative damping coefficient corresponding to the front suspension of the VW Golf vehicle. Comparison of the relative damping coefficient of the suspension determined on the test bench with the theoretically calculated one showed that the actually measured coefficient was slightly (15%) higher. This change may have been determined by frictional forces in the suspension elements that had not been considered.

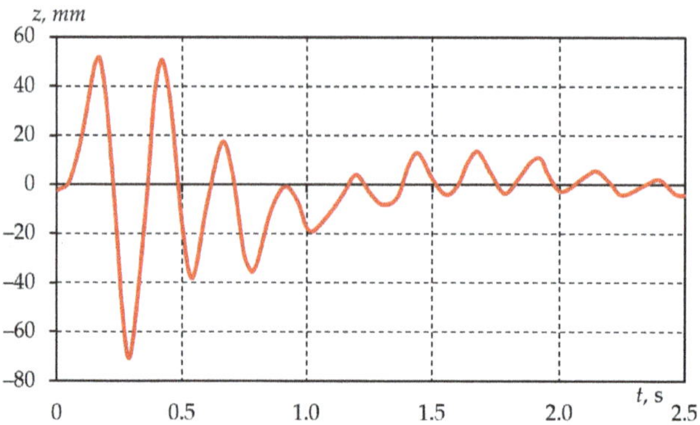

Figure 15. The characteristic of the free oscillations of the vehicle suspension was determined on the car suspension test bench.

Following measurement of the parameters describing the movement of the vehicle suspension elements, it was decided to use spectral records of oscillation displacement for the analysis of the results, as these records more accurately reflect the nature of low-frequency oscillations prevailing in the vehicle body and suspension elements. The spectral analysis of oscillations enabled a more accurate assessment of the components of oscillations of individual frequencies and the determination of their origin and the nature of the change. Investigation of vehicle body oscillations caused by wheel rotation (Figure 16) allowed the observation of the maximum values of the vertical displacements of the vehicle body corresponding to the wheel rotation speed.

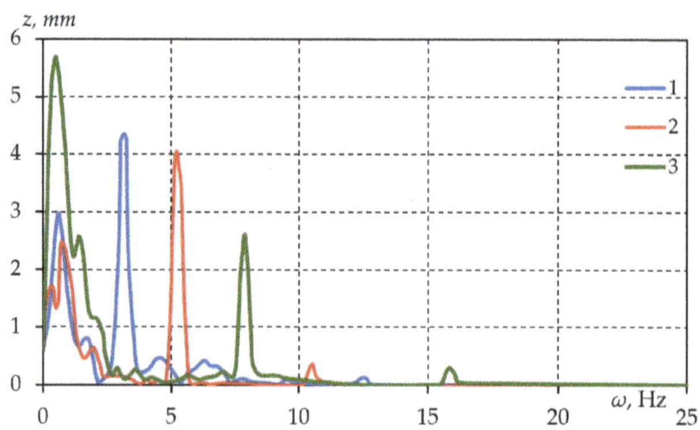

Figure 16. Spectrum of vehicle body displacements at different wheel rotational speeds: 1—gear I; 2—gear II; 3—gear III.

The frequency of these values corresponded to the rotational speed of the wheel set by the respective gear. The spectral record showed an evident increase in the vertical displacement of the 1 Hz frequency, which corresponded to the lower natural frequency of the resonance oscillations of the sprung mass. The amplitudes of the body resonance oscillations increased with increasing wheel rotational speed.

To verify the reality of the vertical displacements of the vehicle body measured on the test bench, real displacements of the vehicle body were measured when driving on a smooth asphalt pavement surface. Different low-frequency resonance displacements of the vehicle body were observed in the vertical displacement spectrum (Figure 17), and their amplitudes corresponded to the amplitudes measured on the test bench.

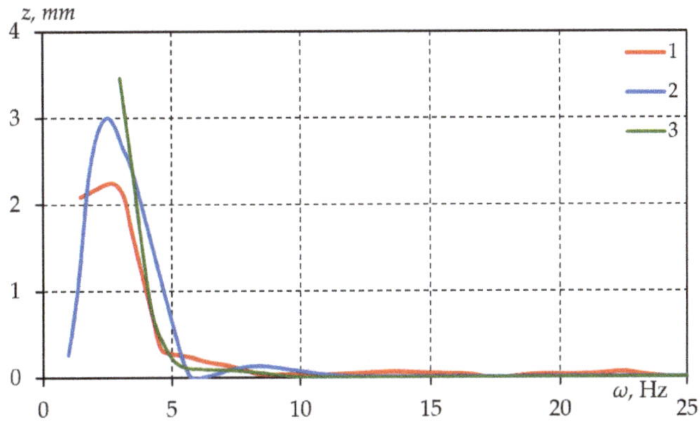

Figure 17. Spectrum of vertical displacements of the real vehicle body at different speeds: 1—23 km/h; 2—31 km/h, 3—56 km/h.

During the vehicle suspension investigation, the excitation of a rolling wheel of different frequencies was investigated, thereby simulating the effect of the road surface. Different peak displacement values were observed in the vertical displacement spectrum of the vehicle body (Figure 18).

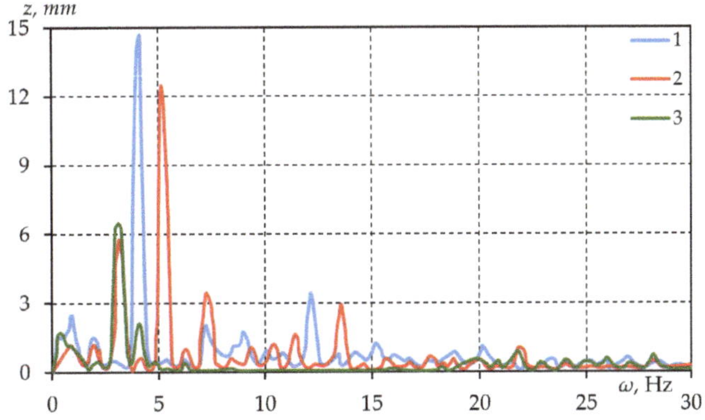

Figure 18. Spectrum of vertical displacements of the vehicle body determined by different excitation frequencies: 1—4.17 Hz; 2—5.55 Hz; 3—8.33 Hz; 4—25 Hz.

The frequency of the peak displacement values corresponded to the excitation frequency. The value of the vertical displacement determined by an excitation frequency of 25 Hz was not distinct. This may be due to the slow action of the pneumatic chamber that simulates the effect of the road surface. Therefore, it can be concluded that it is reasonable to simulate only the low-frequency (<20 Hz) excitation effect of the road surface on the test bench. An experiment was performed to compare the vertical displacements of the suspension on the test bench and of the real vehicle body to verify the reliability of the test bench results. Spectral analysis of the vertical displacements (Figure 19) suggested that the values of the displacements and the nature of the change corresponded to each other. Existing inaccuracies may have been affected by an inaccurately simulated road surface, suspension elements, or the tire on the test bench. Fixing the car body to the test bench is another problem. A sufficiently rigid mounting can dampen the movement of the sprung mass of the model.

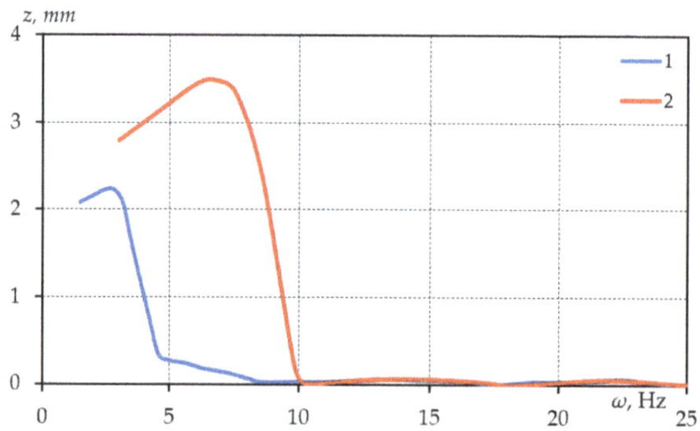

Figure 19. Comparison of vertical displacements of the vehicle body: 1—real vehicle; 2—suspension test bench.

5. The Effect of Vehicle Suspensions on Stability

5.1. Kinematic Model of Vehicle Suspensions

Three variants of vehicle suspension were modeled in the work: 1—front suspension: double arm Mercedes Benz type suspension (Figure 20a); 2—front suspension: double arm Honda type suspension (Figure 20a); 3—front suspension: MacPherson VW Golf type suspension (Figure 20b); 4—rear suspension: single-arm BMW type suspension (Figure 20c); 5—rear suspension: semi-dependent single-arm VW Golf type suspension (Figure 20c); 6—rear suspension—MacPherson Audi type suspension (Figure 20b).

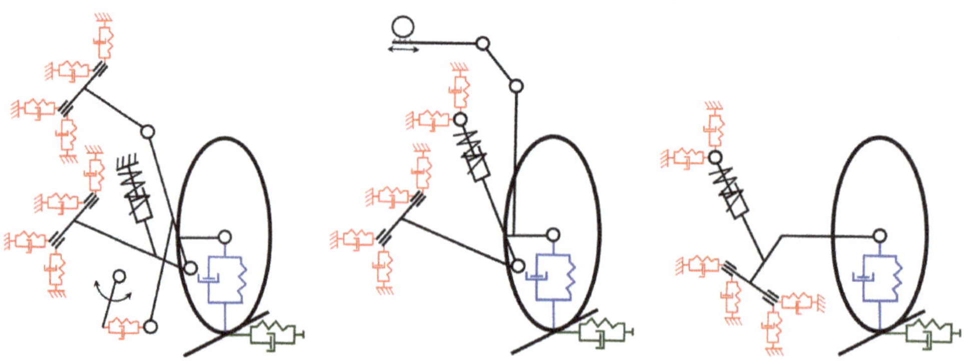

Figure 20. Kinematic schemes of the suspensions: (**a**)—double-arm suspension with lever steering mechanism; (**b**)—MacPherson suspension with gear-rack type steering mechanism; (**c**)—semidependent single-arm suspension.

Driving of the vehicle with each of the suspension combinations was simulated on three different road profiles (low quality pavement, even pavement, and gravel road) during the investigation. For this purpose, programs developed in Matlab were used. The main version was designed to study the kinematics of suspensions. The programs were developed on the basis of kinematic analysis of multiassembly mechanisms. In preparation for the use of the provided data for stability calculations, a calculation of the instantaneous centers of suspension on the transverse and longitudinal planes of the vehicle was introduced, and auxiliary values for the evaluation of the angular stiffness of the suspensions were determined. The parameters required for the calculation of the angular suspension stiffness values and the calculation results were transferred to the files and used directly for stability assessment programs. It was assumed that the vehicle was traveling at a constant speed of 90 km/h. The following characteristics of suspension performance were determined: wheel travel in the vertical direction Δz; wheel turning angle $\Delta\varphi$; wheel displacement in the transverse direction Δy and wheel camber ξ. The effect of the variation of these parameters on the stability was evaluated by using the generalized slip angle δ:

$$\delta = \Delta\varphi + \delta_{\xi} + \delta_y \tag{14}$$

where $\Delta\varphi$ is the change in wheel turning angle; δ_{ξ} is the tire slip angle due to wheel camber and δ_y is the tire tread slip angle due to the transverse displacement of the wheel.

Tire slip angle δ_{ξ} was determined as follows:

$$\delta_{\xi} = \frac{\xi}{k_{\xi}} \tag{15}$$

where ξ is the wheel camber and k_{ξ} = 4–6.

Additional tire slip angles δ_y were calculated using the Equations:

$$\delta_y = \frac{F_y}{k_y} = \frac{\Delta y C_y}{k_y} \tag{16}$$

where F_y is the load acting on the wheel in the transverse direction; Δy is the transverse displacement of the wheel, C_y is the transverse stiffness of the tire and k_y is the slip angle drag coefficient.

Table 2 shows the vertical travel of the wheel Δz with the car traveling on different road profiles. Based on these results, the suspension travel limits were set to calculate the changes in wheel position due to vertical suspension excitation.

Table 2. Vertical travel of the wheel (mm).

	1 (Mercedes Benz Front)	2 Honda (Front)	3 VW (Front)	4 BMW (Rear)	5 VW (Rear)	6 Audi (Rear)
Low Quality Pavement	42.4 −56.8	42.0 −55.2	41.7 −56.7	41.4 −54.9	41.2 −55.7	40.9 −55.0
Even Pavement	14.1 −14.1	14.0 −13.9	13.8 −25.7	13.0 −13.8	13.0 −13.8	12.8 −13.4
Gravel Road	65.5 −66.3	64.5 −64.2	63.4 −65.9	68.1 −65.5	68.4 −65.7	62.2 −65.1

Note: "+" and "−" signs indicate the wheel travel up and down relative to the static position of the wheel, respectively.

Investigation of the kinematic characteristics of the suspensions showed that different road profiles had significant effects on suspension performance. When driving on low quality asphalt, the total suspension angles of the front suspension for the Mercedes Benz suspension ranged from −0.5 to 0.4°, for Honda—from −0.5 to 0.8°, and for VW for front suspension from −0.7 to 1.3°. For the rear suspensions, the slip angles were much smaller. For example, in the case of the BMW they varied from −0.005 to 0.03°, and for VW from

−0.01 to 0°. The front suspension of the Mercedes Benz and the rear suspension of the BMW were found to provide the best kinematic stability. The modeling results are presented in Figures 21 and 22.

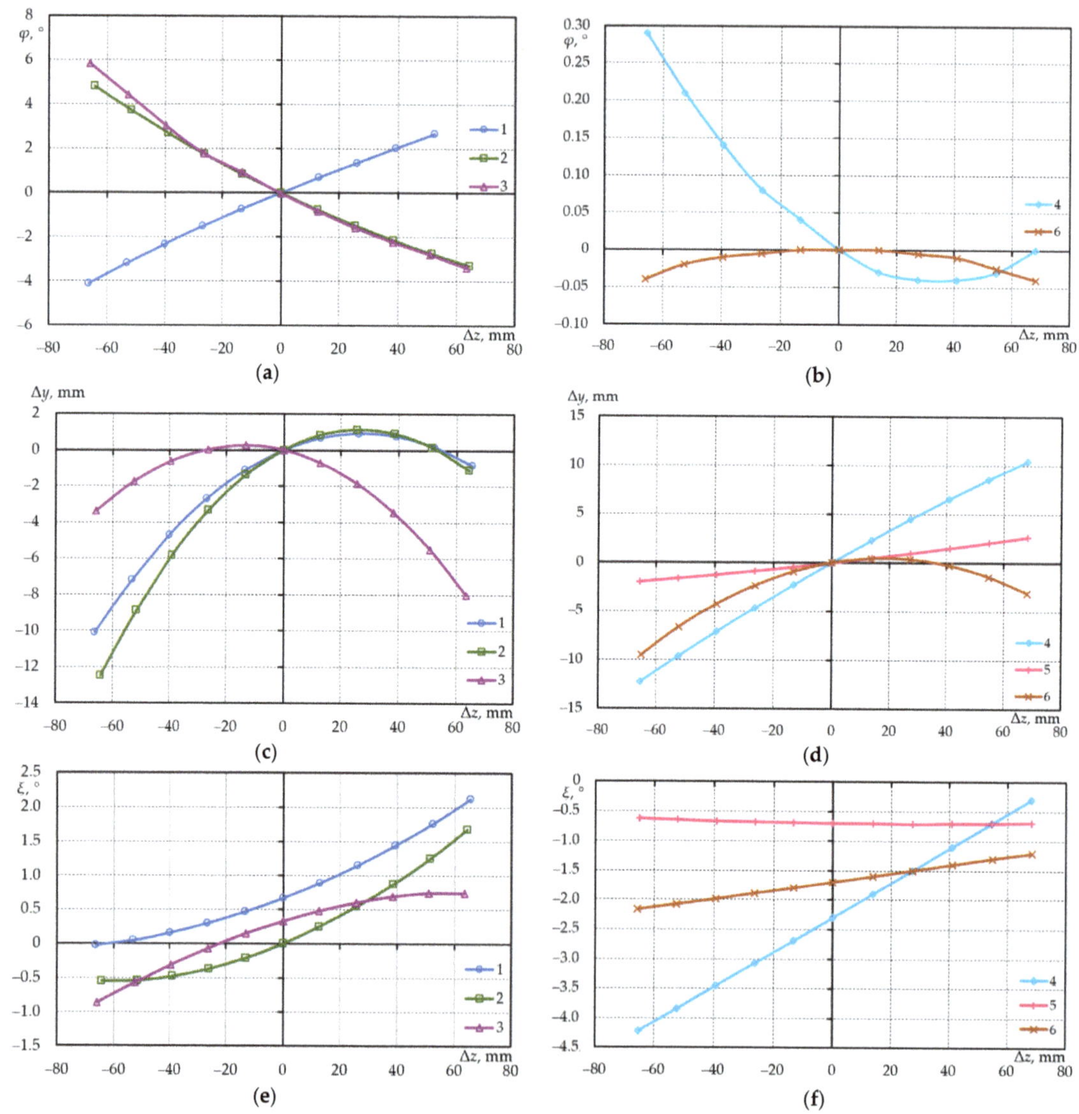

Figure 21. Dependence of front (**a**,**c**,**e**) and rear (**b**,**d**,**f**) suspensions on wheel position as a function of the vertical travel of the wheel: the turning angle of the wheels (**a**,**b**), transverse displacement of the wheels (**c**,**d**), wheel camber (**e**,**f**) (marking according to Table 2).

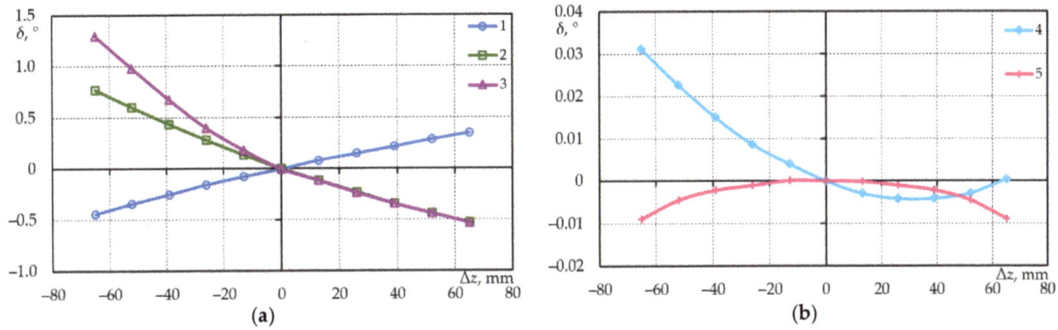

Figure 22. Dependence of the slip angle of the front (**a**) suspension and rear (**b**) suspension wheels on the vertical wheel travel (marking according to Table 2).

5.2. Investigation of the Dynamics of Change in Vehicle Direction

Following assessment of the kinematic properties of the suspensions, the suspension model was improved by introducing deformable elements at the characteristic locations of the suspension and steering mechanism. The deformability of the arms was evaluated in the suspension, and a nonrigid steering rod was used in the steering mechanism. All deformable elements of the suspension and steering mechanism were reduced to these elements (Figure 20a,b). To determine the spatial position of the wheels during variation of the vertical displacement, *z* of the wheel, a coordinate system was used (Figure 23).

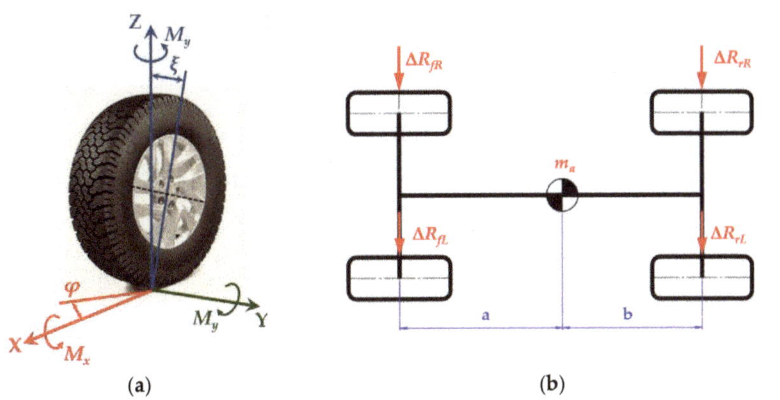

Figure 23. Wheel coordinate system of the 3D model (**a**); lateral forces acting on the wheels (**b**).

If the vehicle body is subjected to a lateral force, then the reaction resulting from transverse wheel displacements Δy is:

$$R_{iy} = C_y \cdot \Delta y \tag{17}$$

where C_y is the transverse stiffness of the tire and Δy is the transverse displacement of the wheel. Then:

$$m_a \cdot \ddot{y}_c = \Sigma \Delta R_{ih},\ I_{za} \cdot \ddot{\varphi}_c = \left(\Delta R_{fL} + \Delta R_{fR}\right) \cdot a + (\Delta R_{rL} + \Delta R_{rR}) \cdot b \tag{18}$$

When modeling vehicle stability, the effect of nonlinearity of the suspension on the vertical displacements of the body and wheels was investigated in view of the excitation frequency by combining different quality road profiles and vehicle speeds. The quarter-car model was used (Figure 9). The driving of the vehicle on three different road profiles (low quality pavement, even asphalt pavement, and gravel road) was simulated during the investigation. The VW Golf vehicle was assumed to travel at a constant speed of 10, 20 and 30 m/s (36, 72, 108 km/h).

The following suspension performance characteristics were determined: wheel travel in the vertical direction Δz; wheel turning angle change $\Delta \varphi$; wheel displacement in the transverse direction Δy; wheel camber ξ and wheel slip angle δ (Figures 24 and 25).

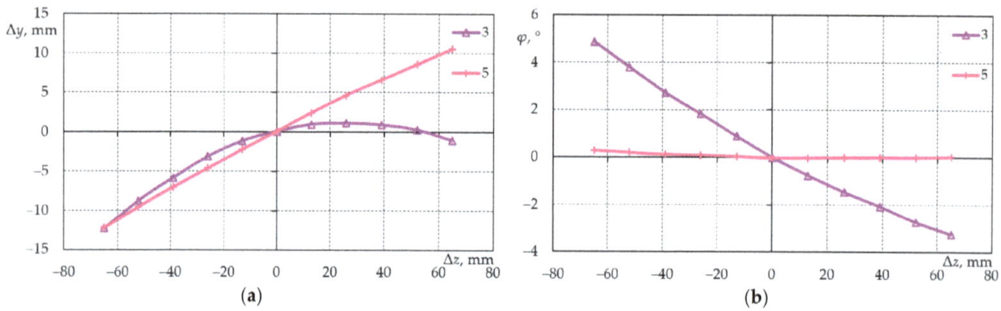

Figure 24. Dependence of transverse wheel displacement (**a**) and wheel turning angle (**b**) on vertical wheel travel (marking according to Table 2).

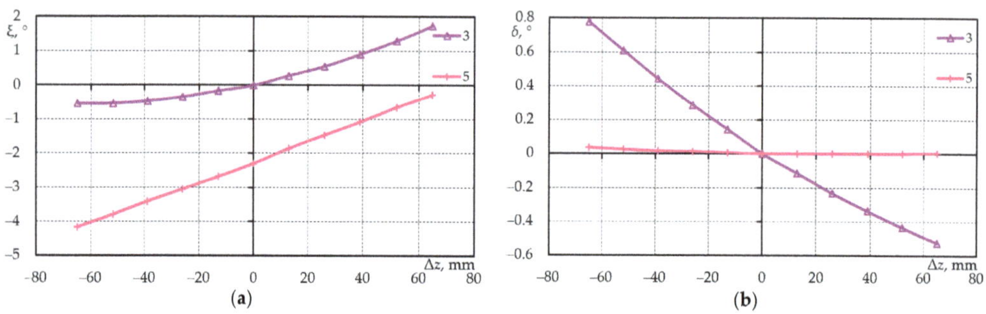

Figure 25. Dependence of wheel camber (**a**) and slip angle (**b**) on vertical wheel travel (marking according to Table 2).

The Matlab program for the investigation of suspension kinematics was used for description of values $\Delta \varphi$, Δy, ξ. The results obtained that describe the dependence of the parameters of the mentioned parameters on the vertical travel of the wheel are presented in Figure 26.

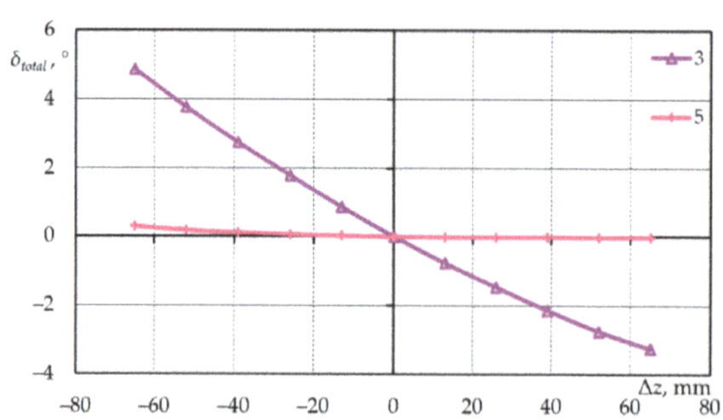

Figure 26. Dependence of total slip angle on vertical wheel travel (marking according to Table 2).

The results obtained show that additional investigations of vehicle stability are needed in the dynamic mode, i.e., when the car is moving, as the total slip angles obtained exceed 4 degrees in borderline cases. Previous vehicle stability calculations show that these angles can be critical under transverse stability conditions even at relatively low speeds (up to 100 km/h). To this end, additional investigations on the vertical dynamics of the vehicle were performed to evaluate the influence of nonlinear suspension elements on the tire tread and the dynamics coefficients of unsprung mass and sprung mass:

$$k_d = \frac{z_i}{q} \tag{19}$$

where z_i is the i-th mass displacement; q is the height of the unevennesses (q is assumed to be equal to the average height of the unevenness corresponding to the even asphalt pavement road surface).

5.3. Cornering Stability of the Vehicle

Modeling of the linear motion according to the equations provided in Section 5.1, including additional evaluation of the changes in the spatial position of the wheel, showed the necessity to input the driver's reaction. Without the driver, such modeling is not effective because the car changes direction due to accidental effects and requires frequent correction to simulate direction adjustment using the steering mechanism. However, in this work, the technical specifications for the model with a driver were prepared, while the excitation parameters when driving on roads of different quality could be established using this model. The developed Matlab programs enabled the determination of suspension deformations (and tire-road contact geometry for more accurate modeling).

The appropriateness of the solutions was evaluated using a more defined model that examined vehicle behavior under stabilized conditions, i.e., when moving in circles.

Vehicle cornering stability was investigated but the studies did not take into account the change in the position of the wheels during vehicle tilt, which may affect vehicle stability. In this work, the vehicle behavior model was refined by evaluating the kinematics of the vehicle's front and rear axles and their influence on the vehicle's behavior.

When examining vehicle cornering stability, two criteria are usually evaluated [50]: loss of stability when the car slips and loss of stability when the car rolls over. Our study additionally specified the extent of the change in the vehicle at cornering due to suspension deformations and changes in the position of the wheels at car rollover.

The main problem encountered in the evaluation of vehicle stability at corners was that the lateral inertia forces redistributed the vertical wheel loads. The solution to the problem was initiated by investigating the problem of car tilt in cornering. In the programs for the kinematic analysis of the suspension, the position of the center of instantaneous tilting was additionally determined for the front and rear axles. The application with the subroutine provides additional data required to determine the tilt angle, the instantaneous tilt centers, the angular stiffness coefficients of the suspension, and stabilizers.

In determining the distribution of vertical reactions, the assumption was made that under the action of a lateral force, the sprung part of the vehicle would rotate about the axis connecting the tilt centers of the two axles. When calculating vehicle tilt, simplified tilt schemes (Figure 27) are usually chosen, which do not take into account body (frame) deformations and changes in the center of position of the tilt center during vehicle tilt.

With acceleration in the lateral direction equal to μ, the damping mass m_a of the vehicle would be subject to force $m_a\mu$. If the designations in Figure 27 are used, the distance from the road surface to the tilt axis at the center of mass is equal to:

$$h_p = h_{1L} + (h_{2L} - h_{1L})/L = (h_{2L}a_1 + h_{1L}a_2)/L \tag{20}$$

The body tilt is equal to:

$$\beta = \frac{\Delta R_1 + \Delta R_2}{\frac{(c_{1k}+c_{1sk})c_{kpa}}{c_{1k}+c_{1sk}+c_{kpa}} + \frac{(c_{2k}+c_{2sk})c_{kpa}}{c_{2k}+c_{2sk}+c_{kpa}}} t \tag{21}$$

where ΔR_1, ΔR_2 are the changes in radial loads on the front and rear axles, respectively, due to the lateral force, and c_{ik}, c_{isk}, c_{kpa} are the coefficients of angular stiffness of the suspension, angular stabilizer and the tires, respectively.

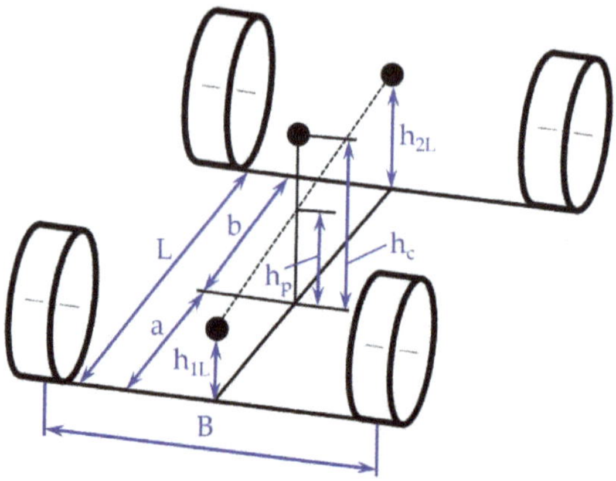

Figure 27. Wheel load recalculation scheme when the vehicle is subjected to lateral force.

The tilt of the front and rear wheels is equal to:

$$\beta_1 = \beta \frac{c_{kpa}}{c_{1k} + c_{1sk} + c_{kpa}}, \ \beta_2 = \beta \frac{c_{kpa}}{c_{2k} + c_{2sk} + c_{kpa}} \tag{22}$$

The angle of slipping of the wheel may be calculated after determining the tilt of the wheels. As already mentioned, changes in the camber angle of the wheel and the turning angle of the wheel about the vertical axis were evaluated by adjusting the kinematically obtained turning angle of the wheel, which was assumed to be the change in the angle of slip.

Calculations were performed by estimating the limit turning radius R_{rib} for different speeds. This was calculated according to the grip conditions (turning radius limited by the car slip or camber) with nondeformable and deformable suspension (grip coefficient μ = 0.8). The results are presented in Table 3. The additional body rollover angle can be observed to undergo no significant change due to suspension deformations under these conditions.

Table 3. Vehicle boundary turning radius.

Speed, km/val	Turning Radius at Slip, R_{ribS}, m	Turning Radius at Rollover, R_{ribV}, m	
		Deformable Suspension	Non-Deformable Suspension
20	3.50	1.80	1.80
40	13.98	7.20	7.21
60	31.46	16.20	16.23
80	55.93	28.81	28.86
100	87.39	45.01	45.09
120	125.85	64.82	64.92
140	171.29	88.23	88.37
160	223.73	115.23	115.42

In further investigations, a change of direction was noted to cause a transverse inertia force, which led to the redistribution of radial reactions. The kinematics of the suspension, in turn, led to possible changes in the spatial position of the wheels.

Since the slip angle depends on the radial loads, in particular under boundary conditions, the redistribution of the radial loads must be evaluated when solving stability problems. For this, the algorithm provided the determination of the reaction R_y of each tire according to it load.

The AUTOSTAB program was developed for the implementation of the model. The algorithm is shown in Figure 28.

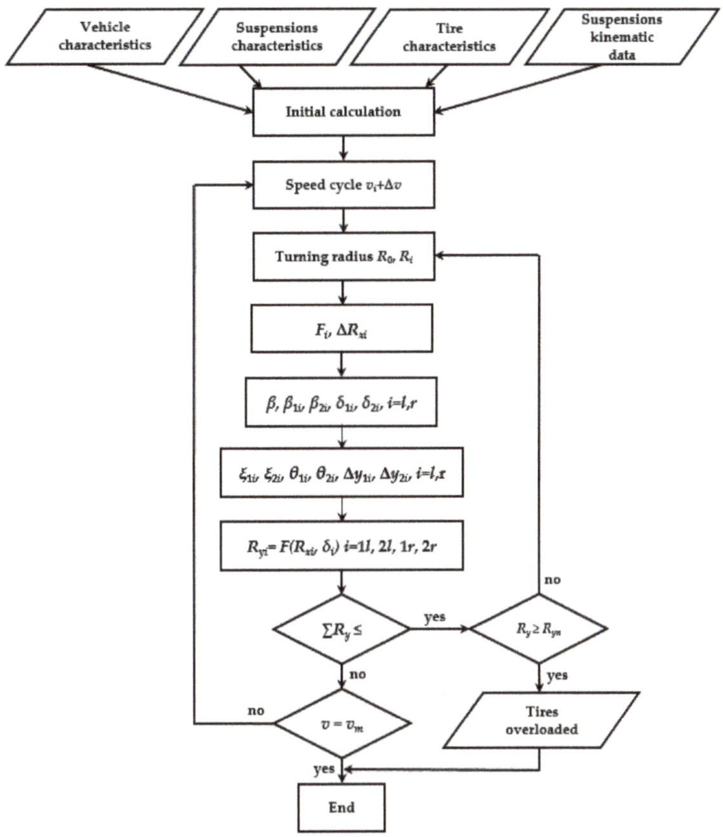

Figure 28. Vehicle boundary turning radius calculation program AUTOSTAB algorithm.

The simulation results show that in modeling the directional stability of the vehicle, it is necessary to evaluate the specifics of the suspension, the redistribution of radial loads under the action of lateral force and changes in the geometric position of the wheel due to the kinematic properties of the suspension.

The results obtained by modeling the movement of the car under stabilized conditions are presented in Figure 29.

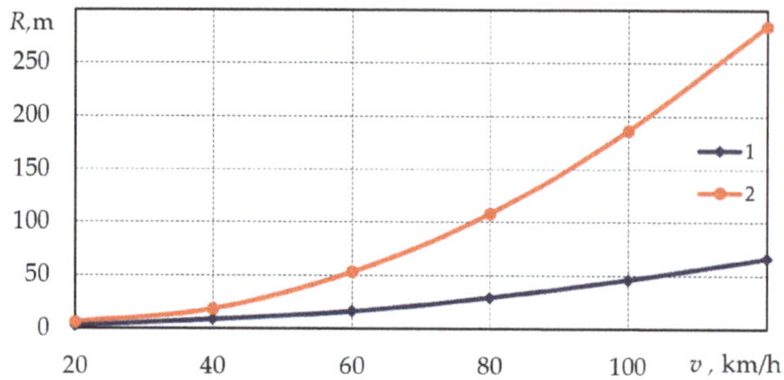

Figure 29. Dependence of the boundary turning radius on the vehicle speed with (1) and without (2) the distribution of radial loads, and the change in wheel position during deformation of the suspension.

Figure 29 shows that by estimating the distribution of radial loads and changes in wheel position, the vehicle boundary turning radius increased up to 4.39 times at 120 km/h speed.

The developed methodology enabled an evaluation of the influence of road condition on directional stability under vertical excitation, investigation of the sensitivity of individual suspension types to vertical excitation according to directional stability criteria, and an assessment of the technical condition of the vehicle if patterns of change in suspension parameters are available. The technical condition was assessed in the work by changing the stiffness of the suspension elements.

Application of the numerical model to linear movement showed that this kind of modeling would not be effective without a driver's model. This was due to the fact that the vehicle would change direction under the action of random effects, and frequent correction would be required to simulate direction adjustment using the steering mechanism. Technical specifications for this kind of model were prepared in this work, and the excitation parameters when driving on roads of different quality may be established using this model.

6. Discussion

The investigation showed that when modeling the effect of the road surface using a flat model, the effect of the optimum evaluation of the road surface effect could be provided using microprofile records on one longitudinal line. The spatial records of the road surface parameters reflect the properties of the road surface in the transverse direction. However, for a thorough evaluation of the interaction, small-interval microprofile data must be recovered from more accurate records. In the above cases, the roughness of the road surface could be assessed. Investigations have suggested that the IRI measurement methodology proposed by M. W. Sayers for measuring the microprofile every 0.147 m provides an adequate reflection of the road-vehicle interaction characteristics, as it corresponds to the average length of the passenger car tire-road contact, which allows accurate prediction of road surface effects. On the basis of the results of the tire comparison function investigations, it can be concluded that the description of a tire with an elementary filter that compares waves of a certain length is not efficient. Rough surface unevennesses with amplitudes greater than 4 mm were found to be not always absorbed by tires. The tire can absorb a single random unevenness, or a repetitive set of them, otherwise the roughness is registered as micro-irregularity and causes additional wheel loads. Therefore, in more accurate models, not only the road microprofile, but also the roughness of the road surface needs to be assessed. After examining various methods of evaluation of the tire comparison function, it was determined that when calculating the uniformity of vehicle travel, the data from spatial road microprofile records can be recalculated by comparing properties of the tire. These data are then used to investigate suspension performance, in which case the tire-road interaction is described by the point contact model. It was determined that where data on wheel tilt or contact curve on the plane transverse relative to the wheel roll are unnecessary, the use of the 3D tire model would also be unreasonable. The vertical wheel displacement required for the evaluation of suspension performance may be determined by the tire model with point contact. Programs were developed that analyze the movement of the quarter-car model on road surfaces of various characteristics. In order to increase the accuracy of calculations, it was necessary to evaluate the nonlinearity of vehicle suspension elements, friction in suspension elements, and the effect of tire tread properties on suspension performance, in particular when modeling driving on low quality roads or when assessing gross, random road surface damage.

To verify the theoretical calculations characterizing vehicle suspension, and to investigate various suspension parameters, a vehicle suspension test bench corresponding to the quarter-car model was designed and manufactured. A mechanism of the original design was developed to simulate the effect of the road surface, which allowed the modeling of various rolling conditions of the wheels. After evaluating the results of the investigation carried out on the experimental test bench, comparing theoretical and practical data, it can be concluded that the results obtained corresponded with reality. This was confirmed by analysis of the free oscillation characteristics of the vehicle suspension measured on the test bench and the correspondence of the relative damping coefficient of the suspension with that determined by theoretical calculations (the results obtained differed by 15%). Spectral analysis of the vertical displacement of the vehicle body and other suspension mechanisms measured on the suspension test bench confirmed correspondence of the properties of the quarter-car model with corresponding characteristics of the real vehicle. Therefore, it can be stated that the real quarter-car model corresponded to the theoretical model, as well as to real car movement. The developed model is appropriate for further investigation and modeling of the suspension properties of cars. Minor inaccuracies were also observed, which may have affected the reliability of the results. The limited capacities and slowness of the pneumatic chamber simulating the excitation of the road surface, which allowed testing only at a low-frequency (<20 Hz) of excitation should be mentioned. Another issue was the attachment of the vehicle body to the test bench frame, as a sufficiently rigid attachment may inhibit the movement of the model sprung mass. However, the conditions simulated by the test bench offered a fairly accurate reflection of the real conditions; therefore, it is reasonable to further investigate the characteristics of the real quarter-car module, as well as possibilities for its improvement.

A vehicle stability model was developed for the assessment of vehicle-road interaction. The model comprehensively evaluated the dynamic characteristics of the suspension, suspension kinematics, and road conditions, and enabled the development of methodology for the determination of the safe speed of the vehicle on roads of known quality using numerical models.

As a result of the combination of the vertical displacement of the quarter car and the kinematic models, a suspension model was developed and described changes in the spatial position of the wheel during suspension operation. When driving on roads of different quality, the data obtained allowed estimation of the directional stability of the vehicle, including the lateral displacement of the wheel, camber, and the turn about the vertical axis. To simplify the analysis, all these changes were reduced to one parameter, i.e., the calculated tire slip angle.

Simulations showed that the kinematic properties of the front suspension needed to be refined in the directional simulation application, as the boundary travel of the suspension resulted in large total slip angles (in particular, the front wheels). This causes the risk that the vehicle may lose its transverse stability at higher speeds.

The effect of changes in the geometric position of the wheel on the directional stability parameters of the vehicle was determined using kinematic models. The total slip angle was used for this. The software application AUTOSTAB was developed under this methodology taking into account the effect of the above factors on the dependence of the vehicle's limit turning radius on speed.

Following the example of the VW Golf, changes in the geometric position of the wheel during suspension operation were found to have a significantly greater effect on directional stability than body tilt due to centrifugal forces.

7. Conclusions

1. A comprehensive analysis of vehicle-road interactions shoed that in order to determine the stability of vehicle movement it is necessary to accurately assess the characteristics of the interacting subsystems. Factors influencing vehicle-road interactions were specified during the investigation, including road surface characteristics; nature of road surface, tire contact area and compared properties of the tire, suspension design; kinematic and dynamic models of the suspension, all of which affect stability of the vehicle movement.
2. A revised model of tire equalizing function was developed. The coordinates of relative measures were found to cause no essential differences in the assessment of tire deformations of different vehicle tires. This means that the number of tire groups may be reduced to evaluate their impact on stability.
3. A dynamic quarter-car model was developed for a comprehensive integrated assessment of suspension characteristics, suspension kinematics, and road surface condition. The model enabled the development of a methodology for the determination of the safe speed of the car on roads of known quality using numerical models.
4. Analysis of the revised quarter-car model showed that the nonlinearity of the suspension elements could be ignored only with examination of movement on a good quality asphalt pavement, because the suspension elements work in the linear part of the suspension characteristic. The effects of nonlinear elements occurred when driving on low quality roads and in case of severe, random damage to the road surface. During large suspension movements, it was necessary to account for friction forces in the damper frame; therefore, methodology for formalizing friction force was considered.
5. The adequacy of the theoretical characteristics of the quarter-car model in relation to the real characteristics was investigated using an experimental test bench. The relative damping coefficient of the vehicle suspension measured on the test bench was approximately 15% higher than the calculated coefficient. The frequency characteristics of the vertical displacements of the real vehicle and vehicle body measured on the test bench were similar to those predicted by the quarter-car model in terms of the magnitude and nature of the change. The developed model is appropriate for further investigations and improvement of vehicle suspension properties.
6. The revised kinematic model of suspension confirmed that the relative real camber changes by up to 2° when modeling an asphalt road surface compared to an ideally horizontal road surface. This causes the alignment angle of the wheel to change by up to 1.5°.
7. The results of the experiment on the asphalt pavement road surface were compared with the model. Certain issues were observed in the analysis of experimental results due to additional oscillations and vibrations occurring in the body and suspension elements, which could be avoided by modeling movement after recording the road microprofile with a profilograph. The parameters that had the greatest impact on the stability of the direction—the difference between the displacement of the sprung and unsprung masses in the model versus the field experiment—did not exceed 15%.
8. Changes in car behavior at changing speeds are related not only to the slip effect but also to changes in wheel spatial position due to suspension kinematics and suspension incompatibility with the steering mechanism. This effect can be particularly pronounced when road defects (pits, ruts) lead to an increase in suspension movement.

Author Contributions: Conceptualization, V.L., R.M., A.R., R.K., A.D. and R.S.; methodology, V.L., R.M., A.R., R.K., A.D. and R.S.; software, V.L., R.M., A.R., R.K., A.D. and R.S.; validation, V.L., R.M., A.R., R.K., A.D. and R.S.; formal analysis, V.L., R.M., A.R., R.K., A.D. and R.S.; investigation, V.L., R.M., A.R., R.K., A.D. and R.S.; resources, V.L., R.M., A.R., R.K., A.D. and R.S.; data curation, V.L., R.M., A.R., R.K., A.D. and R.S.; writing—original draft preparation, V.L., R.M., A.R., R.K., A.D. and R.S.; writing—review and editing, V.L., R.M., A.R., R.K., A.D. and R.S.; visualization, V.L., R.M., A.R., R.K., A.D. and R.S.; supervision, V.L., R.M., A.R., R.K., A.D. and R.S.; project administration, V.L., R.M., A.R., R.K., A.D. and R.S.; funding acquisition, V.L., R.M., A.R., R.K., A.D. and R.S. All authors have read and agreed to the published version of the manuscript.

Funding: This research received no external funding.

Institutional Review Board Statement: Not applicable.

Informed Consent Statement: Not applicable.

Data Availability Statement: Not applicable.

Conflicts of Interest: The authors declare no conflict of interest.

References

1. Sayers, M.W.; Karamihas, S.M. *The Little Book of Profiling*; University of Michigan: Ann Arbor, MI, USA, 1998.
2. Sayers, M.W.; Karamihas, S.M. Estimation of Rideability by Analyzing Longitudinal Road Profile. *Transp. Res. Rec.* **1996**, *1536*, 110–116. [CrossRef]
3. Gillespie, T.D.; Paterson, W.D.O.; Sayers, M.W. *Guidelines for Conducting and Calibrating Road Roughness Measurements*; World Bank Technical Paper 1986, No. WTP 46; World Bank Group: Washington, DC, USA, 1986. (In English)
4. Žuraulis, V.; Sivilevičius, H.; Šabanovič, E.; Ivanov, V.; Skrickij, V. Variability of Gravel Pavement Roughness: An Analysis of the Impact on Vehicle Dynamic Response and Driving Comfort. *Appl. Sci.* **2021**, *11*, 7582. [CrossRef]
5. Leitner, B.; Decký, M.; Kováč, M. Road pavement longitudinal evenness quantification as stationary stochastic process. *Transport* **2019**, *34*, 195–203. [CrossRef]
6. Misaghi, S.; Tirado, C.; Nazarian, S.; Carrasco, C. Impact of pavement roughness and suspension systems on vehicle dynamic loads on flexible pavements. *Transp. Eng.* **2021**, *3*, 100045. [CrossRef]
7. Mahajan, G.R.; Radhika, B.; Biligiri, K.P. A critical review of vehicle-pavement interaction mechanism in evaluating flexible pavement performance characteristics. *Road Mater. Pavement Des.* **2020**, 1–35. [CrossRef]
8. Yechen, Q.; Hong, W.; Huang, Y.; Xiaolin, T. *Real-Time Road Profile Identification and Monitoring: Theory and Application*; Morgan & Claypool Publishers: San Rafael, CA, USA, 2019. [CrossRef]
9. Yang, S.; Chen, L.; Li, S. *Dynamics of Vehicle-Road Coupled System*; Springer: Beijing, China, 2015.
10. El-Gindy, M.; Lewis, H.L.; Lewis, A.S. Development of a tire/pavement contact-stress model based on an artificial neural network. In *American Society of Mechanical Engineers, Design Engineering Division (Publication) DE*; ASME: Nashville, TN, USA, 1999; Volume 101, pp. 25–34. Available online: https://pennstate.pure.elsevier.com/en/publications/development-of-a-tirepavement-contact-stress-model-based-on-an-ar (accessed on 20 September 2021).
11. Jeong, D.; Choi, S.B.; Lee, J.; Kim, M.; Lee, H. Tire dimensionless numbers for analysis of tire characteristics and intelligent tire signals. *Mech. Syst. Signal Process.* **2021**, *161*, 107927. [CrossRef]
12. Jeong, D.; Kim, S.; Lee, J.; Choi, S.B.; Kim, M.; Lee, H. Estimation of Tire Load and Vehicle Parameters Using Intelligent Tires Combined with Vehicle Dynamics. *IEEE Trans. Instrum. Meas.* **2021**, *70*, 1–12. [CrossRef]
13. Gao, X.; Xiong, Y.; Liu, W.; Zhuang, Y. Modeling and experimental study of tire deformation characteristics under high-speed rolling condition. *Polym. Test.* **2021**, *99*, 107052. [CrossRef]
14. Swami, A.; Pandey, A.K. Vibration Analysis of a Tire Under Static Loading Using Flexible Ring-Based Model. *J. Vib. Acoust.* **2021**, *143*, 011007. [CrossRef]
15. Pang, Y.; Prokop, G. *Modeling of Tire–Obstacle Contact with Focus on Calculation Accuracy and Efficiency*; SAE Technical Paper 2021-01-0326; SAE International: Warrendale, PA, USA, 2021. [CrossRef]
16. Cuong, D.M.; Sihong, Z. Natural frequency analysis of tractor tire with different ground contacts and inflation pressures. *Coupled Syst. Mech.* **2020**, *9*, 455–471. [CrossRef]
17. Zhou, H.; Li, H.; Liang, C.; Zhang, L.; Wang, G. Relationship between Tire Ground Characteristics and Vibration Noise. *Stroj. Vestn.-J. Mech. Eng.* **2021**, *67*, 11–26. [CrossRef]
18. Lukoševičius, V.; Makaras, R.; Dargužis, A. Assessment of Tire Features for Modeling Vehicle Stability in Case of Vertical Road Excitation. *Appl. Sci.* **2021**, *11*, 6608. [CrossRef]
19. Pacejka, H.B.; Besselink, I. *Tire and Vehicle Dynamics*, 3rd ed.; Butterworth-Heinemann: Oxford, UK, 2012.

20. Watton, J.; Holford, K.M.; Surawattanawan, P. Electrohydraulic effects on the modelling of a vehicle active suspension. *Proc. Inst. Mech. Eng. Part D J. Automob. Eng.* **2001**, *215*, 1077–1092. [CrossRef]
21. Basargan, H.; Mihály, A.; Kisari, Á.; Gáspár, P.; Sename, O. Vehicle Semi-active Suspension Control with Cloud-based Road Information. *Period. Polytech. Transp. Eng.* **2021**, *49*, 242–249. [CrossRef]
22. Basargan, H.; Mihály, A.; Gáspár, P.; Sename, O. Road Quality Information Based Adaptive Semi-active Suspension Control. *Period. Polytech. Transp. Eng.* **2021**, *49*, 210–217. [CrossRef]
23. Meng, Q.; Chen, C.-C.; Wang, P.; Sun, Z.-Y.; Li, B. Study on vehicle active suspension system control method based on homogeneous domination approach. *Asian J. Control.* **2021**, *23*, 561–571. [CrossRef]
24. Basargan, H.; Mihály, A.; Gáspár, P.; Sename, O. Adaptive Semi-Active Suspension and Cruise Control through LPV Technique. *Appl. Sci.* **2021**, *11*, 290. [CrossRef]
25. Lv, T.; Zhang, Y.; Duan, Y.; Yang, J. Kinematics & compliance analysis of double wishbone air suspension with frictions and joint clearances. *Mech. Mach. Theory* **2021**, *156*, 104127. [CrossRef]
26. Von Wysocki, T.; Chahkar, J.; Gauterin, F. Small Changes in Vehicle Suspension Layouts Could Reduce Interior Road Noise. *Vehicles* **2020**, *2*, 2. [CrossRef]
27. Ciampaglia, A.; Santini, A.; Belingardi, G. Design and analysis of automotive lightweight materials suspension based on finite element analysis. *Proc. Inst. Mech. Eng. Part C J. Mech. Eng. Sci.* **2021**, *235*, 1501–1511. [CrossRef]
28. Hrovat, D. Optimal suspension performance for 2-D vehicle models. *J. Sound Vib.* **1991**, *146*, 93–110. [CrossRef]
29. Hrovat, D. Survey of Advanced Suspension Developments and Related Optimal Control Applications. *Automatica* **1997**, *33*, 1781–1817. [CrossRef]
30. Gao, J.; Wu, F. Analysis and optimization of the vehicle handling stability with considering suspension kinematics and compliance characteristics. *Adv. Mech. Eng.* **2021**, *13*, 16878140211015523. [CrossRef]
31. Dubrovskiy, A.; Aliukov, S.; Osintsev, K. On the Possibility of Generalizing the Results of Studies of Dynamical Systems on the Example of a Vehicle Suspension. *Machines* **2021**, *9*, 68. [CrossRef]
32. Dinesh, T. *Study of Influence of Tie Rod Hardpoints in Suspension Kinematics and Handling Behavior of a Vehicle*; SAE Technical Paper 2021-01-5013; SAE International: Warrendale, PA, USA, 2021. [CrossRef]
33. Akhmedov, D.A.; Aliboev, B.A. Kinematic Optimization of the Steering and Suspension Systems of a Modernized Car. *Russ. Eng. Res.* **2021**, *41*, 484–488. [CrossRef]
34. Avi, A.; Carboni, A.; Stella Costa, P. *Multi-Objective Optimization of the Kinematic Behaviour in Double Wishbone Suspension Systems Using Genetic Algorithm*; SAE Technical Paper 2020-36-0154; SAE International: Warrendale, PA, USA, 2021. [CrossRef]
35. Vahedi, A.; Jamali, A. Constraint optimization of nonlinear McPherson suspension system using genetic algorithm and ADAMS software. *J. Vib. Control* **2021**, 10775463211026036. [CrossRef]
36. Jiao, R.; Nguyen, V. Studies on the low frequency vibration of the suspension system for heavy trucks under different operation conditions. *Noise Vib. Worldw.* **2021**, *52*, 127–136. [CrossRef]
37. Yoon, D.-S.; Kim, G.-W.; Choi, S.-B. Response time of magnetorheological dampers to current inputs in a semi-active suspension system: Modeling, control and sensitivity analysis. *Mech. Syst. Signal Process.* **2021**, *146*. [CrossRef]
38. Swethamarai, P.; Lakshmi, P.; Gokul Prassad, S. Whale-optimized fuzzy-fractional order controller-based automobile suspension model. *Eng. Optim.* **2021**, 1–16. [CrossRef]
39. Reimpell, J. *Fahrwerktechnik: Lenkung*; Vogel-Buchverlag: Würzburg, Germany, 1978. (In German)
40. Reimpell, J. *Shassi Avtomobilya: Konstrukciji Podvesok*; Mashinostroenie: Moscow, Russia, 1989. (In Russian)
41. Reimpell, J. *Shassi Avtomobilya: Rulevoe Upravlenie*; Mashinostroenie: Moscow, Russia, 1987. (In Russian)
42. Reimpell, J.; Betzer, J.W. *Podwozia Samochodów. Podstawy Konstrukcji*; WKiŁ: Warszawa, Poland, 2008. (In Polish)
43. Reimpell, J. *The Automotive Chassis: Engineering Principles*; Butterworth Heinemann: Oxford, UK, 2001. Available online: https://books.google.com/books?hl=en&lr=&id=fuXf3wmahM8C&oi=fnd&pg=PP1&dq=46.%09Reimpell,+J.+The+automotive+chassis:+engineering+principles.+Butterworth+Heineman&ots=pdBnI9A3OS&sig=ug-Xlm6zSyxh4eV1-bx5qHkWLMc (accessed on 20 September 2021).
44. Gillespie, T. *Fundamentals of Vehicle Dynamics*; Society of Automotive Engineers, Inc.: Warrendale, PA, USA, 1992. Available online: https://scholar.google.com/scholar_lookup?title=Fundamentals+of+Vehicle+Dynamics&author=Gillespie,+T.&publication_year=1992 (accessed on 20 September 2021).
45. Matschinsky, W. *Radführungen der Straßenfahrzeuge: Kinematik, Elasto-Kinematik und Konstruktion*; Springer: Berlin, Germany, 2007. Available online: https://scholar.google.com/scholar?hl=en&as_sdt=0%2C5&q=48.%09W.+Matschinsky.+Radf%C3%BChrungen+der+Stra%C3%9Fenfahrzeuge%3A+Kinematik%2C+Elasto-Kinematik+und+Konstruktion.+&btnG= (accessed on 20 September 2021).
46. Bastow, D.; Howard, G.; Whitehead, J.P. *Car Suspension and Handling*, 4th ed.; Society of Automotive Engineers, Inc.: New York, NY, USA, 2004. Available online: https://scholar.google.com/scholar?hl=en&as_sdt=0,5&q=Bastow,+D%3B+Car+Suspension+and+Handling,+Fourth+Edition (accessed on 20 September 2021).
47. Andrzejewski, R.; Awrejcewicz, J. *Nonlinear Dynamics of a Wheeled Vehicle*; Springer Science+Business Media: New York, NY, USA, 2006. Available online: https://scholar.google.com/scholar?hl=en&as_sdt=0%2C5&q=Andrzejewski%2C+R.%3B+Awrejcewicz%2C+J.+Nonlinear+Dynamics+of+a+Wheeled+Vehicle+&btnG (accessed on 20 September 2021).

48. Litvinov, A.S.; Farobin, J.E. *Automobile: Theory of Performance Characteristics*; Mashinostroenije: Moscow, Russia, 1989. (In Russian)
49. Dimaitis, M. *Modern Technologies for Evaluating the Condition of Road Surfaces*; Vilnius: Ex Arte, Lithuania, 2008; pp. 32–35. (In Lithuanian)
50. Dixon, J.C. *Tires, Suspension and Handling*, 2nd ed.; Society of Automotive Engineers, Inc.: New York, NY, USA, 1996. Available online: https://scholar.google.com/scholar?hl=en&as_sdt=0%2C5&q=Tires%2C+Suspension+and+Handling%2C+Second+Edition&btnG (accessed on 20 September 2021).

applied sciences

MDPI

Article

Speed Oscillations of a Vehicle Rolling on a Wavy Road

Walter V. Wedig

Institut für Technische Mechanik, KIT—Karlsruher Institut für Technologie, 76131 Karlsruhe, Germany; wedigwalter08@googlemail.com

Abstract: Every driver knows that his car is slowing down or accelerating when driving up or down, respectively. The same happens on uneven roads with plastic wave deformations, e.g., in front of traffic lights or on nonpaved desert roads. This paper investigates the resulting travel speed oscillations of a quarter car model rolling in contact on a sinusoidal and stochastic road surface. The nonlinear equations of motion of the vehicle road system leads to ill-conditioned differential-algebraic equations. They are solved introducing polar coordinates into the sinusoidal road model. Numerical simulations show the Sommerfeld effect, in which the vehicle becomes stuck before the resonance speed, exhibiting limit cycles of oscillating acceleration and speed, which bifurcate from one-periodic limit cycle to one that is double periodic. Analytical approximations are derived by means of nonlinear Fourier expansions. Extensions to more realistic road models by means of noise perturbation show limit flows as bundles of nonperiodic trajectories with periodic side limits. Vehicles with higher degrees of freedom become stuck before the first speed resonance, as well as in between further resonance speeds with strong vertical vibrations and longitudinal speed oscillations. They need more power supply in order to overcome the resonance peak. For small damping, the speeds after resonance are unstable. They migrate to lower or supercritical speeds of operation. Stability in mean is investigated.

Keywords: road models; quarter car models; limit cycles; acceleration speed portraits; speed oscillations; velocity bifurcations; noisy limit cycles; limit flows of trajectories; Sommerfeld effects; differential-algebraic systems; polar coordinates of roads; covariance equations; stability in mean; supercritical speeds; analytical travel speed amplitudes; Floquet theory applied to limit cycles

Citation: Wedig, W.V. Speed Oscillations of a Vehicle Rolling on a Wavy Road. *Appl. Sci.* **2021**, *11*, 10431. https://doi.org/10.3390/app112110431

Academic Editors: Flavio Farroni, Andrea Genovese and Aleksandr Sakhnevych

Received: 19 September 2021
Accepted: 29 October 2021
Published: 5 November 2021

1. Introduction to the Problem

Vertical vibrations of a vehicle driven by a constant force and rolling on a sinusoidal road surface are coupled with its horizontal travel motion, affecting the vehicle speed which fluctuates around a mean value. The coupling between both planar motions is caused by the permanent direction change of the contact force to ground along the contour of the road profile. This paper explains the nonlinear model of this dynamic problem applying averaging methods to calculate stationary solutions before and after the resonance speed. Numerical integrations are applied to obtain limit cycles around the averaged solutions, plotting the fluctuating car acceleration against the true velocity. Stationary solutions are stable in mean when the slope of the driving force speed characteristic is positive. Vice versa, they are unstable for negative slopes. This leads to the so-called Sommerfeld effect [1] that for a given driving force the car becomes stuck before the resonance speed and can only pass over the resonance and the unstable velocity range after the resonance by considerably increasing the driving force [2]. First investigations of velocity jumps and turbulent speeds in nonlinear vehicle road dynamics are given by Wedig in [3–7] applying sinusoidal and random road models introduced by Robson et al. [8–10]. The first order road model in [11,12] is extended in [3,4] to a second order one which includes sinusoidal models. Blekhman and Kremer studied the same vehicle road system in [13,14] for the special case of small road excitations to calculate only the average response of driving cars (see also [15]). In [2,7], these investigations are extended to large sinusoidal road surfaces and double-periodic limit cycles in the phase plane of longitudinal acceleration and oscillating speed. In the present paper, new results for mean values and amplitudes of speed oscillations are calculated by means of Fourier expansions. Stability investigations by means of Floquet theory are proposed. Extensions to quarter car models with two degrees of freedom are made. The new speed amplitudes calculated in this paper show that the longitudinal speed oscillations of the vehicle are stable in the lower speed range before the resonance speed and in the upper higher speed range. In the middle range immediately after the resonance peak, stationary speeds are unstable and therefore physically not realizable. These stability properties correspond to the Duffing problem where vertical displacement vibrations of the vehicle possess three different amplitudes in the resonant speed range: the upper and lower displacement vibrations are stable and the middle range vibrations are unstable, as well.

2. Coupled Vertical and Longitudinal Vehicle Road Dynamics

Figure 1a shows the applied quarter car model rolling on a wavy road with vertical displacement z and derivative u taken along the travel way s. In the following, the quantities z and u are called road level and slope, respectively. They generate vertical car vibration displacement y and velocity $\dot{y}$, which are coupled by the vehicle speed $v = \dot{s}$ and described by the two equations of motion

$$\dot{v} = \left[\omega_1^2(y - z) + 2D\omega_1(\,\dot{y} - \dot{z})\right] \tan\alpha + f/m\,, \tag{1}$$

$$\ddot{y} + 2D\omega_1(\dot{y} - \dot{z}) + \omega_1^2(y - z) = 0,\ \dot{y} = \omega_1 x \tag{2}$$

where s is the travel displacement of the vehicle and $x = \dot{y}/\omega_1$ denotes the coordinate of the vertical vibration velocity. In Equations (1) and (2), dots denote derivatives with respect to time t. The parameter $\omega_1^2 = c/m$ determines the natural frequency ω_1 of the vertical vehicle vibrations, $2D\omega_1 = b/m$ denotes the damping, and f is the driving force which is constant or slightly decreasing with growing speed. In Figure 1b, both force characteristics are plotted in yellow-black. In the following, constant driving force is applied only. The nonlinear term in Equation (1) represents the damper and spring force multiplied by $\tan\alpha$ that takes the horizontal component of the contact force N by means of $\tan\alpha = dz/ds$ One finds Equations (1) and (2) already in the literature in [13,14] and in [2–7].

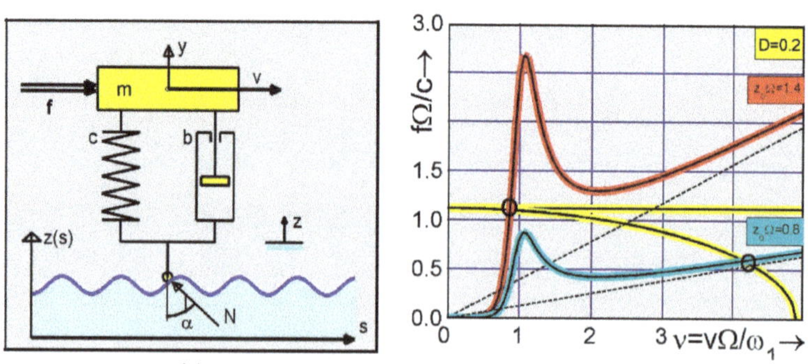

(a) (b)

Figure 1. (**a**) Quarter car model rolling on sinusoidal road surface driven by the constant force f. (**b**) Driving force speed characteristic for two parameters $z_o\Omega$ of the road unevenness marked by red and cyan color, respectively, and two driving force characteristics marked by yellow-black lines.

In first investigations, the road level and slope are assumed to be sinusoidal as

$$z(s) = z_0 \cos(\Omega s) \quad \text{and} \quad u(s) = -z_0 \sin(\Omega s), \tag{3}$$

where z_0 is the amplitude of the road surface and Ω is the angular spatial frequency determined by the wave length $L = 2\pi/\Omega$ of the road. Equations (1)–(3) represent a DA equation system where the role of algebraic terms is taken by sinusoidal terms. To eliminate these terms, the road level z and slope u in Equation (3) are differentiated with respect to the way coordinate s in order to obtain the increments $dz = -z_0\Omega\sin(\Omega s)ds$ and $du = -z_0\Omega\cos(\Omega s)ds$ that leads to the homogeneous nonlinear oscillator equations

$$\dot{z} = \mathrm{v}\Omega u\,, \qquad \dot{u} = -\mathrm{v}\Omega z\,, \tag{4}$$

which are obtained when both increments above are divided by dt and ds/dt is replaced by the speed v. Furthermore, $dz/ds = -z_o\Omega\sin(\Omega s) = \Omega u$ holds so that Equation (1) reads as

$$\dot{\mathrm{v}} = \left[\omega_1^2(y-z) + 2D\omega_1(\,\dot{y}-\dot{z})\right]\Omega u + f/m. \tag{5}$$

Equations (2), (4) and (5) describe a five-dimensional problem with five unknowns [4,7]: the horizontal travel speed v(t) of the vehicle, its vertical vibration by displacement $y(t)$ and velocity $\dot{y}(t) = \omega_1 x(t)$, and the road level $z(t)$ and slope $u(t)$. For analytical and numerical investigations, it is appropriate to introduce the dimensionless time $\tau = \omega_1 t$ and the related speed $v = \mathrm{v}\Omega/\omega_1$, as well as the related coordinates $(\overline{z}, \overline{u}) = (z, u)z_0$ and $(\overline{y}, \overline{x}) = \Omega(y, x)$. Their insertion into Equations (2), (4) and (5) leads to

$$v' + 2D(z_0\Omega)^2\overline{u}^2 v = z_0\Omega(\overline{uy} + 2D\overline{ux}) - (z_0\Omega)^2\overline{zu} + f\Omega/c, \tag{6}$$

$$\overline{z}' = v\overline{u}, \qquad \overline{u}' = -v\overline{z}, \qquad \text{IC.}:\ \overline{z}(0) = 1,\ \overline{u}(0) = 0, \tag{7}$$

$$\overline{y}' = x, \qquad \overline{x}' = -(\overline{y} + 2D\overline{x}) + z_o\Omega(\overline{z} + 2Dv\overline{u})\,. \qquad \tau = \omega_1 t, \tag{8}$$

where prime denotes differentiation with respect to the dimensionless time $\tau = \omega_1 t$. To improve numerical integration [2,7] in Equations (6)–(8), the polar coordinates

$$\overline{z} = \overline{r}\cos\varphi \qquad \text{and} \qquad \overline{u} = \overline{r}\sin\varphi \tag{9}$$

are introduced into the road Equation (7) that leads to the transformation equations

$$\overline{r}'\cos\varphi - \overline{r}\varphi'\sin\varphi = v\overline{r}\,\sin\varphi,$$

$$\overline{r}'\sin\varphi + \overline{r}\varphi'\cos\varphi = -v\overline{r}\cos\varphi,$$

which are solved by means of the determinant $\Delta = \cos^2\varphi + \sin^2\varphi = 1$ and Cramer's rule to

$$\overline{r}\,' = 0, \qquad \varphi' = -v.$$

Without loss of generality, the related polar radius is integrated to $\overline{r} = 1$. Note that the derivative of the polar angle is equal to the negative speed of the vehicle. According to the definition of polar coordinates, the polar angle turns counterclockwise into the mathematically positive direction. The applied oscillator, however, rotates clockwise. This is the reason why both quantities have opposite signs.

3. Speed Driving Force Characteristic of Traveling Vehicles

In order to derive the driving force speed characteristic, shown in Figure 2, the equations of motion are approximately investigated assuming that the oscillating speed of the vehicle can be averaged by $v(\tau) = v =$ const. In this case, the sinusoidal solutions

$$\overline{y}(\tau) = y_c\cos(v\tau) + y_s\sin(v\tau), \qquad \overline{z}(\tau) = \cos(v\tau),$$

$$\overline{x}(\tau) = x_c\cos(v\tau) + x_s\sin(v\tau), \qquad \overline{u}(\tau) = -\sin(v\tau)$$

are applicable and inserted into Equation (8) in order to obtain the matrix equation

$$\begin{bmatrix} 0 & -v & -1 & 0 \\ v & 0 & 0 & -1 \\ 1 & 0 & 2D & -v \\ 0 & 1 & v & 2D \end{bmatrix} \begin{bmatrix} y_s \\ y_c \\ x_s \\ x_c \end{bmatrix} = z_0\Omega \begin{bmatrix} 0 \\ 0 \\ 2Dv \\ 1 \end{bmatrix}$$

applying the coefficient comparison. The first two rows of this matrix equation yield $x_c = vy_s$ and $y_c = -x_s/v$. Both relations are inserted into the last two rows, obtaining

$$\left(v^2-1\right)x_s + 2Dv^2 y_s = z_0\Omega v,$$

$$-2Dx_s + \left(v^2-1\right)y_s = z_0\Omega\, 2Dv.$$

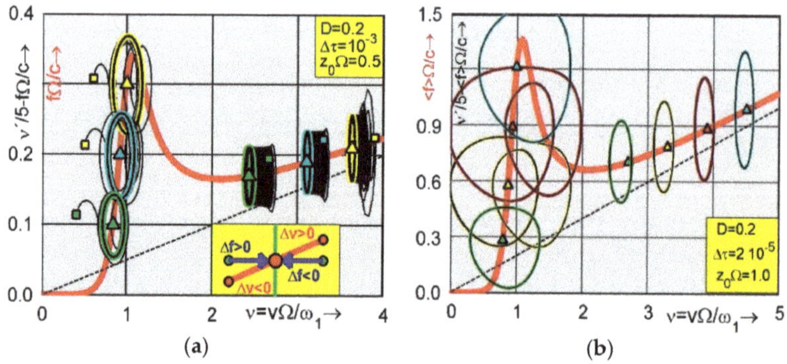

Figure 2. (**a**) Force speed characteristic (thick red) against speed. Transients of limit cycles for driving force values marked on the thick red characteristic by green, cyan, and yellow triangles. (**b**) Bifurcations of limit cycles of scaled and shifted accelerations against travel speed when the vehicle becomes stuck before the resonance speed. One-periodic limit cycles in the super-critical speed range. The applied driving forces are marked by green, yellow, red, and cyan triangles.

This reduced equation system possesses the determinant $\Delta = \left(v^2-1\right)^2 + (2Dv)^2$. It is solved by Cramer's rule in order to obtain first x_s and y_s and then x_c and y_c, as follows:

$$x_s = z_0\Omega v\left[v^2 - 1 - (2Dv)^2\right]/\Delta, \qquad x_c = z_0\Omega 2Dv^4/\Delta, \tag{10}$$

$$y_s = z_0\Omega 2Dv^3/\Delta, \qquad y_c = z_0\Omega[1 + v^2\left(4D^2-1\right)]/\Delta. \tag{11}$$

These results coincide with the linear theory of a constant vehicle speed. They are inserted into the velocity Equation (6) to calculate the time dependent driving force

$$f\Omega/c = 2D(z_0\Omega)^2 v\sin^2 v\tau + z_0\Omega[(\overline{y} + 2D\overline{x}) - z_0\Omega\cos(v\tau)\,]\sin(v\tau)$$

which is needed to keep the speed v constant. Its mean value is calculated as

$$f\Omega/c = (z_0\Omega)^2 Dv^5/[\left(v^2-1\right)^2 + (2Dv)^2], \qquad \text{with } \mathrm{v} = \mathrm{v}\Omega/\omega_1. \tag{12}$$

The approximated force speed characteristic in Equation (12) is plotted in Figure 1b for the two road surface parameters $z_0\Omega = 1.4$ and 0.8, marked by red and cyan lines, respectively. The dashed line represents the asymptote $f\Omega/c = (z_0\Omega)^2 Dv$, which is proportional to the speed, indicating that one needs a linearly growing driving force to reach higher speeds of operation. Obviously, the increasing driving force is needed to compensate the energy loss in the damper, which is growing with higher speeds, as well.

The force speed characteristic (12) represents a linear approximation obtained by averaging the oscillating driving speed. In order to check the validity and stability of the result in Equation (12), numerical simulations are performed by applying the Euler scheme to

$$v' + 2D(z_0\Omega)^2\overline{u}^2 v = z_0\Omega(\overline{uy} + 2D\overline{ux}) - (z_0\Omega)^2\,\overline{zu} + f\Omega/c, \tag{13}$$

$$\overline{y}' = x, \quad \overline{x}' = -(\overline{y} + 2D\overline{x}) + z_o\Omega(\overline{z} + 2Dv\overline{u})\,. \qquad \varphi' = -v, \tag{14}$$

where the related level and slope of the road surface are given by $\overline{z} = \cos\varphi$ and $\overline{u} = \sin\varphi$ dependent on the polar angle φ, the derivative of which is equal to the negative speed of the vehicle. The numerical results obtained are presented in Figure 2a by plotting the scaled and shifted acceleration of the vehicle against the true speed where squares denote initial values of acceleration and speed; triangles are mean values of acceleration and speed calculated after the initially transient period, providing a sufficiently long time for the averaging procedure. In Figure 2b, stationary limit cycles are shown for a stronger road excitation given, e.g., by $z_0\Omega = 1.0$, which leads to the road amplitude $z_o \approx 3.2$ mm, e.g., for the wave length $L = 20$ mm. Note that phase portraits of velocity over displacement are not applicable for the travel kinematics since the horizontal displacement of the traveling vehicle is growing infinitely. Instead of phase portraits of displacement and velocity, Figure 2b shows limit cycles of velocity against acceleration. Obviously, there are two speed regions where the limit cycles are stable; the first is in the under-critical speed range before the resonance speed $v = 1$. The second is far beyond the resonance in the higher speed range of operation. In between both stable speed ranges, the slope of the speed driving force characteristic is negative. In this range, limit cycles are not stable and therefore not realizable. This instability is plausible and physically explained inside the yellow area shown in Figure 2a. Accordingly, a speed perturbation by means of $\Delta v < 0$ into the negative speed direction on the left side of the dynamic equilibrium generates an acceleration back to the equilibrium since the applied force $\Delta f > 0$ is larger than that one being necessary in the new perturbed situation. In this case the positive driving force difference is equal to the vertical distance between the green and red circle. However, the vehicle is braked if the speed perturbation goes into the positive speed direction on the right side of the dynamical equilibrium. In this case, the applied driving force is smaller than the one necessary to maintain the new perturbed dynamic equilibrium, marked by the right red circle. Vice versa, a speed driving force characteristic with negative slope leads to monotonous instability with the effect that speed leaves the unstable branch. In Section 5, it is shown that the negative slope condition coincides with the instability in mean, which is derived by applying the Hurwitz criterion to the variational equations of the averaged equations of motion.

Figure 2a shows three limit cycles in the stable under-critical speed range for the driving forces: $f\Omega/c = 0.1,\ 0.2,$ and 0.3 marked by green, cyan, and yellow triangles, respectively. The limit cycles are obtained by plotting the scaled and shifted acceleration against the true travel speed. The selected initials are marked by colored squares. Thin black lines are transients which start in the squares and end in the thick colored lines of the one-periodic limit cycles. After this initial period, the simulation is continued a sufficiently long time in order to calculate the mean values of speed and acceleration, which are plotted marked by colored triangles. Obviously, the triangles showing speed and acceleration coincide with the initial values of the applied speed driving force characteristic. Since the accelerations are shifted upward by the applied value of the speed driving force characteristic, the acceleration mean value is vanishing when both triangles coincide. The same property is obtained in the stable region of super-critical speeds $2 < v < 4$, where three one-periodic limit cycles are shown on the right side in Figure 2a. Again, the transients are marked by thin black lines. After a sufficiently long initial time, the transients go over to the stationary limit cycles marked by green, cyan, and yellow.

In Figure 2b, the applied wave parameter is doubled by $z_0\Omega = 1$ with the consequence that the amplitudes of the speed oscillations become much broader. Moreover, the one-periodical limit cycles bifurcate into double-periodic ones when for growing driving force the vehicle becomes stuck in the under-critical speed range. This bifurcation scenario is demonstrated by the green limit cycle obtained (left side, Figure 2b) for the related driving force $f\Omega/c = 0.3$, which goes over to the yellow double-periodic one if the related force is increased to $f\Omega/c = 0.6$. For the further increased driving force $f\Omega/c = 0.9$, the calculated limit cycle is still double-periodic marked by red and bifurcates back into the one-periodic limit cycle marked by cyan if the driving force is again increased to $f\Omega/c = 1.2$. For further growing driving force, the resonance speed $v = 1$ is reached with strong oscillations in the vertical and horizontal direction and then passed up to high speed ranges where the driving force can be again decreased. Associated limit circles are single periodic as shown in the right side of Figure 2b. They are obtained for $f\Omega/c = 2.7$, 3.3, 3.9, and 4.5, marked by green, yellow, red, and cyan, respectively. Correspondingly, colored triangles represent the averaged travel speeds together with the applied driving force. The applied vehicle damping is $D = 0.2$ and the time step of integration is $\Delta\tau = 2 \times 10^{-5}$. Note that for $f\Omega/c = 0.9$ two different limit cycles are obtained: the first in the under-critical speed range and the second one in the upper-critical speed range. Both are marked by red. In between both stable limit cycles, there is a middle one which is unstable und therefore not realizable. Note that one finds the same behavior in rotor dynamics, where the Sommerfeld rotor leads to the same driving characteristic [2] when in Equation (12) the road factor $z_0\Omega$ is replaced by the mass ratio of the unbalanced rotor, the translation velocity by the rotation speed ratio, and the driving force by the moment. More details on Sommerfeld effects in rotor dynamics are given in [16–20].

4. Stable and Unstable Oscillation Amplitudes of Vehicle Speeds

In Figure 3a, the vibration amplitudes of the vertical displacement (red) and velocity (pink) of the vehicle are plotted against the travel speed together with the speed driving force characteristic shown in blue. The displacement and velocity amplitude are

$$A_y = \sqrt{y_s^2 + y_c^2}, \qquad \text{and} \qquad A_x = \sqrt{x_s^2 + x_c^2},$$

respectively, where the coefficients y_s, y_c and x_s, x_c are determined by Equations (10) and (11). The blue curve in Figure 3a represents the speed driving force characteristic dependent on speed. As already derived, stable travel speeds are only realizable in those speed regions where the slope of the characteristic is positive marked by thick blue lines in the under- and super-critical range. In the middle range, where the slope is negative, the blue curve is replaced by a thin black line, indicating that the calculated stationary speed is unstable and therefore not realizable. From this it follows that the calculated amplitudes A_y and A_x of the vertical vibrations of the vehicle are not realizable in the middle speed range, as well. In the super-critical speed range, however, all stationary vibration amplitudes are stabilized with growing speed and end into asymptotes marked by a dashed line, which are linearly growing, constant without increasing and decreasing for the force characteristic, the velocity, and the displacement amplitude, respectively. Obviously, the asymptote of the driving force corresponds to acceleration. Both quantities, force and acceleration, are linearly increasing with growing speed of the vehicle.

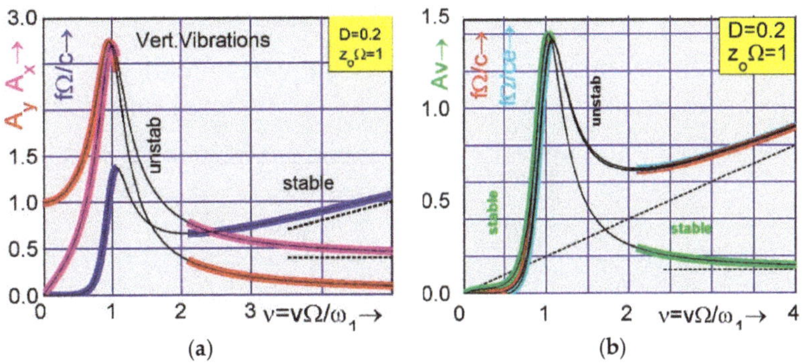

Figure 3. (**a**) Vibration amplitudes of the vertical velocity (pink) and displacement (red) of the vehicle in comparison with the force speed characteristic (blue) in the under- and supercritical speed range. (**b**) Travel speed oscillation amplitudes (green) and extended force speed characteristic (red-cyan).

In order to introduce efficient solution methods, the time increment $d\tau$ in Equations (6) and (8) is eliminated by the angle increment $d\varphi = -vd\tau$ that leads to the time-free equations

$$v\overline{y}^{\circ} = -\overline{x}, \qquad v\overline{x}^{\circ} = (\overline{y} + 2D\overline{x}) - z_0\Omega(\overline{z} + 2Dv\overline{u}), \tag{15}$$

$$vv^{\circ} - 2D(z_0\Omega)^2\overline{u}^2 v = (z_0\Omega)^2\,\overline{zu} - z_0\Omega(\overline{uy} + 2D\overline{ux}) - f\Omega/c, \tag{16}$$

where prime in $(.)' = d(.)/d\tau$ is replaced by a circle in $(.)^{\circ} = d(.)/d\varphi$ denoting differentiation with respect to the polar angle φ. In Equations (15) and (16), the related road level and slope are sinusoidal and dependent on the polar angle φ. Accordingly, the vertical vehicle vibrations and the longitudinal speed oscillations are calculable inside one period by means of shooting methods to obtain numerical solutions for $v_p(\varphi), \overline{y}(\varphi)$ and $\overline{x}(\varphi)$. The associated characteristic multipliers ρ_i are investigated by means of the new equation system

$$v_p \begin{bmatrix} \Delta\overline{y} \\ \Delta\overline{x} \\ \Delta v \end{bmatrix}^{o} + \begin{bmatrix} 0 & -1 & \overline{y}_p^o \\ -1 & -2D & \overline{x}_p^o + z_0\Omega 2D\overline{u} \\ z_0\Omega\overline{u} & 2Dz_0\Omega\overline{u} & v_p^o - 2D(z_o\Omega\overline{u})^2 \end{bmatrix} \begin{bmatrix} \Delta\overline{y} \\ \Delta\overline{x} \\ \Delta v \end{bmatrix} = 0$$

which is obtained when the perturbations $\overline{y} = \overline{y}_p + \Delta\overline{y}$, $\overline{x} = \overline{x}_p + \Delta\overline{x}$, and $v = v_p + \Delta v$ are inserted into Equations (15) and (16) and the perturbation equations are linearized in the Δ quantities. According to the Floquet theory, the calculated periodic solutions are asymptotically stable with respect to φ if all multipliers ρ_i satisfy $|\rho_i| < 1$. For known velocity $v_p(\varphi)$, the applied polar angle can be integrated back to time by which the sequence of associated time distances is determined. Note that the four-dimensional time system (13) and (14) is not ergodic since the polar angle φ is a state variable that grows infinitely by permanent rotation. This disadvantage is avoided by eliminating the time variable by means of the polar angle, which leads to the three-dimensional angle system (15) and (16) where the polar angle now represents the independent integration variable restricted to one-periodic interval. The solutions of this new time-free equation system are ergodic, and multiplicative ergodic theorems are applicable in order to calculate characteristic numbers of the dynamic system of interest.

For the new angle Equations (15) and (16), new analytical solutions are derived by means of the introduction of the Fourier expansions

$$\overline{y}(\varphi) = y_c\cos\varphi + y_s\sin\varphi + \cdots, \qquad \overline{z}(\varphi) = \cos\varphi \tag{17}$$

$$\overline{x}(\varphi) = x_c\cos\varphi + x_s\sin\varphi + \cdots, \qquad \overline{u}(\varphi) = -\sin\varphi, \tag{18}$$

$$v(\tau) = v_o + v_c\cos 2\varphi + v_s\sin 2\varphi + \cdots\ , \tag{19}$$

where the zeroth Fourier coefficient v_o of the speed expansion in Equation (19) takes the role of the averaged speed v in Section 3. The insertion of the expansions (17) and (18) into Equation (15) and the coefficient comparison of the sinusoidal terms of $\cos\varphi$ and $\sin\varphi$ leads to the same result, already noted in Equations (10) and (11). All other terms are cutoff and can only be taken into account when higher expansions are introduced. The insertion of the expansions (17), (18), and (19) into the speed Equation (16) and the coefficient comparison leads to

$$v_s = \frac{D(z_o\Omega)^2 v_o^2\left[(z_o\Omega)^2(1+4D^2v_o^2-v_o^4)-4v_o^4\right]}{2\left[(v_o^2-1)^2+(2Dv_o)^2\right]\left[4v_o^2+(z_o\Omega)^4D^2\right]}, \tag{20}$$

$$v_c = \frac{(z_o\Omega)^2 v_o^3\left\{1+v_o^2\left[4D^2-1+D^2(z_o\Omega)^2\right]\right\}}{\left[(v_o^2-1)^2+(2Dv_o)^2\right]\left[4v_o^2+(z_o\Omega)^4D^2\right]}, \tag{21}$$

which represent new results for the amplitudes of the travel speed oscillations. In Figure 3b, the resultant speed amplitude A_v is plotted against the mean speed $v_o = \mathrm{v}\Omega/\omega_1$ by

$$A_v = \sqrt{v_c^2+v_s^2}\,, \quad \lim_{v_o\to\infty} A_v = D(z_o\Omega)^2[4+(z_o\Omega)^2]/8 \tag{22}$$

and marked by a thick green line. Obviously, the speed amplitude vanishes if the vehicle slows. It is increasing up to the resonance speed and decreases again for further growing speed, up to the asymptote given in Equation (22). In addition to the above coefficient comparison of terms with $\cos 2\varphi$ and $\sin 2\varphi$, all terms with $\cos 0$ lead to the extended speed driving force characteristic

$$f\Omega/c = \frac{D(z_o\Omega)^2 v_o^3}{(v_o^2-1)^2+(2Dv_o)^2}\left\{v_o^2+(z_o\Omega)^2\frac{v_o^2\left[1-(z_o\Omega)^2D^2-4D^2\right]-1}{2\left[4v_o^2+(z_o\Omega)^4D^2\right]}\right\} \tag{23}$$

where the first part coincides with Equation (12) and the second part gives a correction of second order. The extended speed driving force characteristic in Equation (23) is plotted in Figure 3b. The red line marks the first approximation noted in Equation (12). Its second order approximation, noted in Equation (23), is marked by a thick cyan line. It is close to the red line of the first approximation. Thin black lines represent speed driving force characteristics with negative slope where the calculated solutions are unstable and therefore not realizable.

5. Stability in Mean and Robustness with Regard to Disturbances

Note that the velocity Equation (13) is determined by the products of road level z and slope u multiplied by the two vibration coordinates y and $x = y'$. These products are called covariances [2]. They determine the dependency of the response of the system from the road excitation. The application of the product rule of differentiation to the road Equation (7) and the vehicle Equation (8) leads to the covariance equation system

$$\begin{bmatrix}\overline{zy}\\\overline{uy}\\\overline{zx}\\\overline{ux}\end{bmatrix}' + \begin{bmatrix}0 & -v & -1 & 0\\ v & 0 & 0 & -1\\ 1 & 0 & 2D & -v\\ 0 & 1 & v & 2D\end{bmatrix}\begin{bmatrix}\overline{zy}\\\overline{uy}\\\overline{zx}\\\overline{ux}\end{bmatrix} = z_0\Omega\begin{bmatrix}0\\0\\\overline{z^2}+2Dv\overline{zu}\\\overline{zu}+2Dv\overline{u^2}\end{bmatrix}, \tag{24}$$

where the related speed v of the vehicle is determined by Equation (13) as

$$v' + 2D(z_0\Omega)^2\overline{u}^2 v = z_0\Omega(\overline{uy}+2D\overline{ux}) - (z_0\Omega)^2\,\overline{zu} + f\Omega/c\,. \tag{25}$$

Level and slope of the road are $\overline{z}(\tau) = \cos(v\tau)$ and $\overline{u}(\tau) = -\sin(v\tau)$, respectively. Again, this is a DA equation system where transcendentals take the role of algebraic terms. For numerical integrations, the sinusoidal terms in Equation (3) are generated by means of the road oscillator, noted in Equation (7). The oscillator is transformed by means of the polar coordinates

$$\overline{z}(\varphi) = \cos\varphi, \qquad \overline{u}(\varphi) = -\sin\varphi, \qquad \varphi' = -v, \tag{26}$$

which are inserted into Equations (24) and (25). The remaining equation $\varphi' = -v$ determines the derivative of the polar angle and couples the polar angle back to the speed of the vehicle.

In order to derive analytical approximations, the averaging method is applied by taking the time mean values $<\overline{z}^2> = <\overline{u}^2> = 1/2$ and $<\overline{zu}> = 0$ in Equations (24) and (25) that leads to

$$\left(1 - v^2\right)\overline{uy} + 2D\overline{ux} = z_o\Omega v D$$
$$-2Dv^2\,\overline{uy} + \left(1 - v^2\right)\overline{ux} = z_o\Omega v/2,$$

when in Equation (24) for $(.)' = 0$ the first two rows are applied to eliminate $\overline{zy}$ and $\overline{zx}$. Subsequently, the covariances $\overline{uy}$ and $\overline{ux}$ are calculated and inserted into Equation (25)

$$D(z_0\Omega)^2 v = z_0\Omega(\overline{uy} + 2D\overline{ux}) + f\Omega/c,$$

which finally leads to the same speed driving force characteristic

$$f\Omega/c = (z_0\Omega)^2 D v^5/[\left(v^2 - 1\right)^2 + (2Dv)^2],$$

already noted in Equation (12). Note that the above applied time mean values are exact when the speed of the vehicle is constant. For oscillating speeds, the time mean values are approximations.

Following [2], the averaged Equation (24) are applied to investigate the stability in mean of the averaged speed by means of the perturbation equation system

$$\begin{bmatrix} \Delta v \\ \Delta\overline{zv} \\ \Delta\overline{uy} \\ \Delta\overline{zx} \\ \Delta\overline{ux} \end{bmatrix} + \begin{bmatrix} D(z_0\Omega)^2 & 0 & -z_0\Omega & 0 & -2Dz_0\Omega \\ -\overline{uy_0} & 0 & -v_0 & -1 & 0 \\ \overline{zy_0} & v_0 & 0 & 0 & -1 \\ -\overline{ux_0} & 1 & 0 & 2D & -v_0 \\ \overline{zx_0} - z_0\Omega D & 0 & 1 & v_0 & 2D \end{bmatrix} \begin{bmatrix} \Delta v \\ \Delta\overline{zv} \\ \Delta\overline{uy} \\ \Delta\overline{zx} \\ \Delta\overline{ux} \end{bmatrix} = 0 \tag{27}$$

which is obtained by means of the perturbations $v = v_0 + \Delta v$ and $\overline{\alpha\gamma} = \overline{\alpha\gamma_0} + \Delta\overline{\alpha\gamma}$ for $(\alpha, \gamma) \in \{z, u, y, x\}$. The insertion of the perturbations into the averaged Equations (24) and (25) and linearization in the Δ-terms yields the fifth order stability Equation (27) with the characteristic equation $\Delta(\lambda) = A_0\lambda^5 + A_1\lambda^4 + \cdots A_5 = 0$. The determinant of Equation (27) yields

$$A_5 = \frac{(z_0\Omega)^2 D v_0^4 [v_0^4 - 6(1 - 2D^2)v_0^2 + 5]}{(1 - v_0^2)^2 + (2Dv_0)^2}. \tag{28}$$

According to the Hurwitz criterion, $A_5 < 0$ determines divergence that gives the boundary of monotonous instability. Obviously, the stability condition $A_5 > 0$ coincides with the positive slope condition of the speed driving force characteristic calculable by differentiating Equation (12) with respect to the travel speed v that gives

$$\frac{d}{dv}\left(\frac{f\Omega}{c}\right) = (z_0\Omega)^2 v^4 D \frac{v^4 - 6(1-2D^2)v^2 + 5}{\left[\left(1-v_0^2\right)^2 + (2Dv)^2\right]^2} > 0\,.$$

This result coincides with Equation (28) except that the positive definite denominator in Equation (28) is squared. Hence, the negative slope condition and the instability in mean lead to the same instability boundary. This is plausible and physically explainable, as already done in the yellow area in Figure 2a. The instable speed range vanishes with increasing damping.

In addition to the stability behavior, the robustness of the limit cycle calculation with regard to disturbances is of interest. When disturbances are generated by brief shocks, the eigenvalues of the stability matrix in Equation (27) must be calculated. They determine the growth behavior with which the disturbances increase or decrease, respectively. In the case of stationary disturbances, noise models are introduced. Figure 4a shows a stochastic limit cycle obtained for the quarter car model in the case that the angle motion on the sinusoidal road form is perturbed by additive noise given by

$$d\phi_\tau = -V_\tau d\tau + \sigma dW_\tau, \qquad \Delta W_n = \sqrt{\Delta\tau}\, N_n. \tag{29}$$

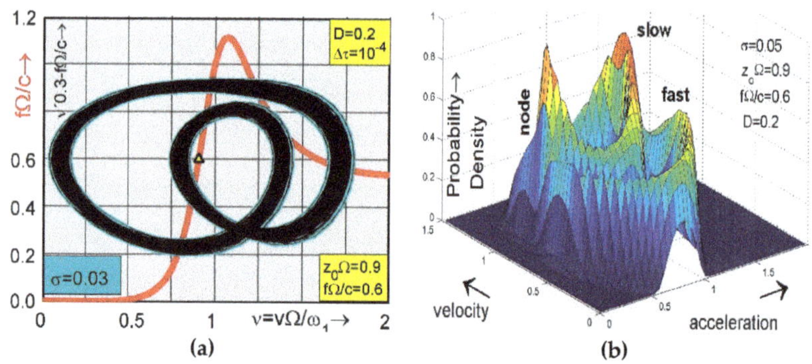

Figure 4. (**a**) Limit cycle flow of a bundle of irregular trajectories with double-periodic side limits when the acceleration is plotted against the true vehicle speed. (**b**) Double-periodic crater-like probability density of the vehicle speed and acceleration with fast and slow motions in the phase plane.

In Equation (29), capital letters with index τ denote set functions [21,22] dependent on time. Noise is generated by normally distributed numbers N_n with zero mean [23]. The stochastic angle perturbation takes into account that the road surface is no longer sinusoidal but more realistically irregular and noisy with bounded realizations. For small noise intensities σ, this leads to response realizations which are bounded, as well.

Initial results are shown in Figure 4a where trajectories of a stochastic limit cycle are plotted in the phase plane of travel velocity and acceleration scaled by 0.3 and shifted by the applied driving force. The realizations are calculated by means of Equations (24)–(26) where the polar angle in Equation (26) is replaced by Equation (29). The applied damping is given by $D = 0.2$, the driving force by $f\Omega/c = 0.6$, the road level by $z_0\Omega = 0.9$, and the noise intensity by $\sigma = 0.03$. The Euler scheme is applied with the time step $\Delta\tau = 10^{-4}$. The mean value of the shifted acceleration and true velocity is marked by a yellow triangle on the red curve of the speed driving force characteristic, indicating that the mean acceleration is vanishing and the mean travel speed coincides with Equation (12). The comparison with the double-periodic yellow limit cycle in Figure 2b shows that its sharp line is widened to a bundle of nonperiodic realizations, the boundaries of which are double periodic with two loops and one node of two crossing limit flows. Figure 4b shows the associated double-crater-like probability distribution density on the phase plane of velocity and acceleration calculated for the stronger noise intensity $\sigma = 0.05$. Clockwise rotation in the phase plane of acceleration and true speed is slow when the probability density is high and vice versa fast for low densities. Note that Figure 4a,b present new results in nonlinear stochastic vehicle dynamics, obtained by applying noise perturbations that are bounded by means of the applied sinusoidal terms. For small intensity σ, the limit flow in Figure 4a possesses periodic side limits in spite of the fact that all limit cycle realizations are random. This is a new effect presented in this paper. For growing intensity σ, the inner side limits of the limit cycle shown in Figure 4a become broader and finally disappear, such that only the outer side limits remain, and the whole phase plane in Figure 4a is covered by realizations of velocity and acceleration.

6. Stable Travel Speeds of Road Vehicle Systems

To extend the above investigations to higher order models, consider the quarter car model with two degrees of freedom shown in Figure 5a. The motions of the car body of mass M and the wheel mass m are described by the equations of motion

$$\ddot{y} + 2D\omega_1(\dot{y} - \dot{x}) + \omega_1^2(y - x) = 0, \qquad z(s) = z_0\cos(\Omega s), \tag{30}$$

$$\ddot{x} + 2B\omega_2\left[(1+\beta)\dot{x} - \dot{z} - \beta\dot{y}\right] + \omega_2^2[(1+\gamma)x - z - \gamma y] = 0, \tag{31}$$

where the system frequencies ω_1, ω_2 and the damping coefficients D, B are introduced together with the stiffness and damping ratios γ and β, and defined as follows:

$$\omega_1^2 = c/M, \qquad 2D\omega_1 = d/M, \qquad \gamma = c/k,$$

$$\omega_2^2 = k/m, \qquad 2B\omega_2 = b/m, \qquad \beta = d/b.$$

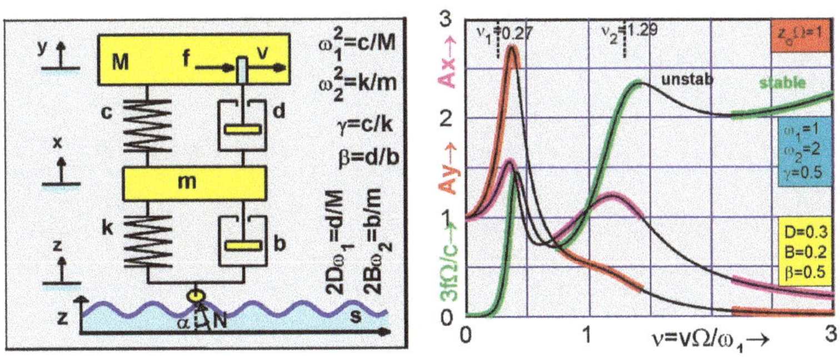

Figure 5. (**a**) Quarter car model with 2 +1/2 DOF rolling on wavy ground without losing road contact. (**b**) Vertical vibration amplitudes of car body (red) and wheel (pink). Speed driving force characteristic marked by green for stable speeds and by black lines when the travel speed is unstable.

The parameter γ denotes the stiffness ratio of the car and wheel spring c and k, respectively. Correspondingly, β is the damping ratio of the car and wheel damper d and b, respectively. The reference frequencies ω_2 and ω_1 describe the decoupled vibrations of the wheel and car body. In addition to Equations (30) and (31), the dynamic balance in horizontal direction gives a third equation of motion that determines the travel speed, as follows:

$$(M+m)\,\dot{v} = f + \left[k(x-z) + b(\dot{x} - \dot{z})\right]\tan\alpha\,, \qquad \tan\alpha = dz/ds. \tag{32}$$

Note that Equation (32) is of first order with respect to the car speed v, and s denotes the longitudinal coordinate of the travel path. Note that both masses are assumed to be concentrated in the contact point of road and vehicle so that only planar translations are considered. Rotations are excluded.

It is appropriate to introduce the dimensionless vibration and road coordinates by means of $(y,x) = (\overline{y},\overline{x})/\Omega$ and $(z,u) = z_o(\overline{z},\overline{u})$, respectively. The insertion of these coordinates into Equations (30) and (31) leads to the dimensionless equations of motion

$$\ddot{\overline{y}} + 2D\omega_1\left(\dot{\overline{y}} - \dot{\overline{x}}\right) + \omega_1^2(\overline{y} - \overline{x}) = 0, \qquad v = \mathrm{v}\Omega/\omega_1, \tag{33}$$

$$\ddot{\overline{x}} + 2\beta\omega_2\left[(1+\beta)\dot{\overline{x}} - \beta\dot{\overline{y}}\right] + \omega_2^2[(1+\gamma)\overline{x} - \gamma\overline{y}] = z_o\Omega\omega_2^2(\overline{z} + 2Bv\overline{u}), \tag{34}$$

where v is the related speed of the vehicle rolling on road with level $\overline{z} = \cos\Omega s$ and slope $\overline{u} = -\sin\Omega s$. In order to derive a first approximation, it is assumed that the oscillating speed of the vehicle can be averaged by $v(\tau) = v = \text{const}$. In this case, the travel path is $s = vt$ and the equations of motion become linear. They are solved by the set-up

$$\overline{y}(t) = y_c\cos(v\omega_1 t) + y_s\sin(v\omega_1 t), \qquad \overline{z}(t) = \cos(v\omega_1 t),$$

$$\overline{x}(t) = x_c\cos(v\omega_1 t) + x_s\sin(v\omega_1 t), \qquad \overline{u}(t) = -\sin(v\omega_1 t)$$

In the stationary case, the insertion of these set-ups into Equations (33) and (34) and the coefficient comparison leads to the linear matrix equation

$$\begin{bmatrix} 1 & 2Dv & v^2-1 & -2Dv \\ -2Dv & 1 & 2Dv & v^2-1 \\ 1+\gamma-v^2\kappa^2 & 2B(1+\beta)v\kappa & -\gamma & -2B\beta v\kappa \\ -2B(1+\beta)v\kappa & 1+\gamma-v^2\kappa^2 & 2B\beta v\kappa & -\gamma \end{bmatrix}\begin{bmatrix} x_c \\ x_s \\ y_c \\ y_s \end{bmatrix} = z_o\Omega\begin{bmatrix} 0 \\ 0 \\ 1 \\ -2Bv \end{bmatrix} \tag{35}$$

where κ is the reference frequency ratio given by $\kappa = \omega_2/\omega_1$. The coefficients x_c, x_s, y_c, y_s of the sinusoidal set-up are calculated by means of Equation (35) and inserted into

$$A_y = \sqrt{y_s^2 + y_c^2}, \qquad \text{and} \qquad A_x = \sqrt{x_s^2 + x_c^2},$$

which are plotted in Figure 5b against the related speed of the vehicle. The amplitude A_y of the car body is marked by red and the wheel amplitude A_x by pink. In unstable speed ranges, both amplitudes are marked by thin black lines.

The vertical vibration amplitudes of both masses are calculated for the reference frequencies $\omega_1 = 1/s$, $\omega_2 = 2/s$, $(\kappa = 2)$, for the stiffness-ratio $\gamma = 0.5$ and for the damping values $D = 0.3$, $B = 0.2$, and $\beta = 0.5$. The applied road wave is chosen by $z_o\Omega = 1$. This is equivalent to the height–length ratio $z_o/L = 0.159$ when the wave frequency $\Omega = 2\pi/L$ is inserted into $z_o\Omega = 1$. The frequency responses in Figure 5b possess two resonance speeds: the first at $v_1 = 0.27$ and the second at $v_1 = 1.29$. Both eigenvalues are valid for vanishing damping. In order to decide the realizability of the stationary vehicle speeds applied in Equation (35), the new speed Equation (32) is investigated in the dimensionless form

$$\frac{m+M}{m}\dot{v}/\omega_2 + 2B(z_o\Omega)^2\overline{u}^2 v = f\Omega/k + z_o\Omega(\overline{xu} + 2B\overline{u\dot{x}}/\omega_2) - (z_o\Omega)^2\overline{uz}\,. \tag{36}$$

As already shown in Section 3, Equation (36) can approximately be evaluated by averaging the oscillating speed with $\dot{v} = 0$ and introducing the time mean values $\overline{uz} = 0$, $\overline{xu} = -x_s/2$, and $\overline{u\dot{x}}/\omega_2 = \kappa v x_c/2$ into Equation (36). This gives the new approximated force characteristic

$$f\Omega/k = B(z_o\Omega)^2 v + z_o\Omega(x_s - 2B\kappa v x_c)/2\,, \tag{37}$$

where x_s and x_c are calculated by means of Equation (35). New numerical evaluations of Equations (35) and (37) are plotted in Figure 5b and marked by green for stable speeds when the slope of the speed driving force characteristic is positive and by black-thin lines when the slope is negative. In the latter case, the speed applied in Equation (35) is unstable and physically not realizable. Finally, it is noted that both instability speed ranges are vanishing for growing damping coefficients.

The instability behavior is numerically investigated by means of simulations applied to a slightly modified quarter car model with two equal masses m and a third damper B, as shown in Figure 6a. The car is rolling with velocity v on a wavy road described by $z(s) = z_o \cos \Omega s$ in dependence on the travel way coordinate s. The speed frequency $v\Omega$ is related to the reference frequency ω_1 given by $\omega_1^2 = c/m$. The vertical vibrations of this car model are described by the two nonlinear equations of motion

$$\ddot{y}_1 + 2d_1\omega_1(\dot{y}_1 - \dot{y}_2) + 2D\omega_1(\dot{y}_1 - \dot{z}) + \omega_1^2(y_1 - y_2) = 0, \tag{38}$$

$$\ddot{y}_2 + 2d_2\omega_1(\dot{y}_2 - \dot{z}) - 2d_1\omega_1(\dot{y}_1 - \dot{y}_2) + \omega_1^2(2y_2 - y_1 - z) = 0, \tag{39}$$

where $\dot{z} = v\Omega u$ couples both vertical vibrations to the horizontal velocity described by

$$2\dot{\mathrm{v}} + \omega_1[\omega_1(z - y_2) + 2d_2(\dot{z} - \dot{y}_2) + 2D(\dot{z} - \dot{y}_1)]\Omega u = f/m. \tag{40}$$

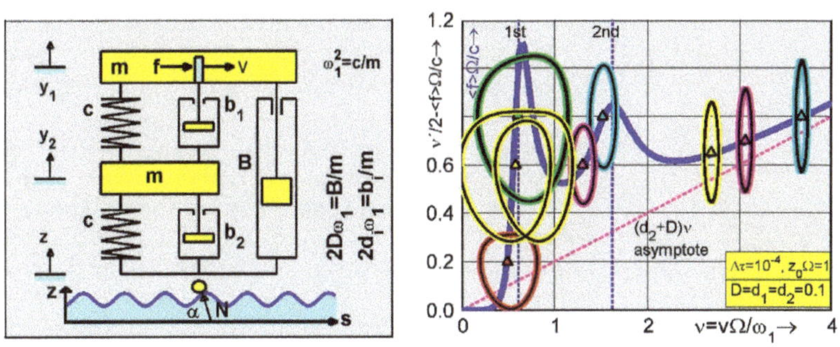

(a) (b)

Figure 6. (**a**) Extended vehicle model with wheel and car body on a wavy road without losing contact. (**b**) Bifurcation from one-periodic limit cycle (red and green) into a double-periodic one (yellow) before the first resonance. Vehicle becomes stuck before the first and second resonance with strong oscillations.

The damping coefficients are introduced by $B/m = 2D\omega_1$ and $b_i/m = 2d_i\omega_i$ for $i = 1, 2$. For numerical integration, the dimensionless time $\tau = \omega_1 t$ and the velocity $v = \mathrm{v}\Omega/\omega_1$, together with the related coordinates $(\overline{z}, \overline{u}) = (z, u)/z_o$ and $(\overline{y}_i, \overline{x}_i) = (y_i, x_i)\Omega$, are inserted into the velocity Equation (40) and vehicle Equations (38) and (39).

Figure 6b shows the average vehicle speed for a given driving force $f\Omega/c$, marked by a blue thick line. It is calculated for the road level $z_o\Omega = 1$ and the damping values $D = d_1 = d_2 = 0.1$. In the middle range of the related driving force $(0.60 < f\Omega/c < 0.82)$, there are five branches of stationary solutions: two solution branches are unstable and three are stable. The stable speeds possess positive slopes in the driving force speed characteristic; meanwhile, speeds started in negative slope ranges do not remain in this range. They migrate to higher or lower speed ranges. Figure 6b shows three limit cycles (red, yellow, green) before the first resonance, two further limit cycles (red, cyan) between the first and second resonance and the last three limit cycles for overcritical speeds near the asymptote $f\Omega/c = (d_2 + D)v$, which is marked by a dashed red line. Mean values of shifted acceleration and true speed are plotted by colored triangles on positive slopes.

7. Summary and Remarks on Main New Results

When vehicles are rolling on uneven roads and driven by a constant force, vertical vibrations of the vehicle are induced that are coupled with the horizontal motions of the vehicle, affecting the travel velocity, which fluctuates around a stationary speed with zero mean acceleration. The coupling between both planar motions is caused by the permanent direction change of the contact force to ground along the contour of the road profile. In the case of sinusoidal road profiles, the vehicle becomes stuck before the resonance speed and needs more power supply to overcome the resonance in order to reach higher speeds. The speed range immediately after the resonance where the slope of the speed driving force characteristic is negative, is not realizable. In this range, the calculated vertical vibration amplitudes are unstable. This is completely different from the results in linear road vehicle dynamics. To describe the dynamic effects obtained, the following main new results are presented:

1. Multiple Sommerfeld effects are shown before and between all resonances. Regular speeds of operation are far beyond. The periodic limit cycles in Figure 6b are new results showing how the car speed becomes stuck on positive slopes of the driving force speed characteristic.
2. In Figure 4a,b, new stochastic limit cycles are shown. For small intensities of white noise, a bundle of realizations is obtained described by a periodic probability distribution over the phase plane of the oscillating true speed and zero mean travel acceleration.
3. Fourier expansions are introduced into the nonlinear equations of motion in order to calculate new analytical approximations for the mean travel speed and the amplitude of the speed oscillations. These expansions depend on the polar angle which takes the role of time.

In future, the one wheel vehicle is extended to half car models by which additional effects are introduced by road excitations with time delays when there is a finite distance between the front and rear wheel. Analytical investigations by means of Fourier expansions are applied to higher order vehicle models. For numerical solutions, shooting methods restrict the integration range to one period of the polar angle by which unstable period limit cycles can be calculated, as well. The stability of the periodic solutions is investigated by means of the characteristic multipliers calculated by means of the Floquet theory.

Funding: This research received no external funding.

Institutional Review Board Statement: Not applicable.

Informed Consent Statement: Not applicable.

Data Availability Statement: This study does not report any data.

Conflicts of Interest: The author declares no conflict of interest.

References

1. Sommerfeld, A. Beiträge zum dynamischen Ausbau der Festigkeitslehre. *Z. Des Ver. Dtsch. Ing.* **1902**, *46*, 391–394. (In German)
2. Wedig, W.V. Turbulent Travel Speeds in Nonlinear Vehicle Road Dynamics. *J. Nonlinear Dyn.* **2020**, *100*, 147–158. [CrossRef]
3. Wedig, W.V. Velocity Jumps in Road-Vehicle Dynamics. *Procedia Eng.* **2016**, *144*, 1076–1085. Available online: https://www.sciencedirect.com/science/article/pii/S187770581630279X (accessed on 6 May 2016). [CrossRef]
4. Wedig, W.V. Jump Phenomena in Road-Vehicle Dynamics. *Int. J. Dyn. Control* **2016**, *4*, 21–28. [CrossRef]
5. Wedig, W.V. New Resonances and Velocity Jumps in Nonlinear Road-Vehicle Dynamics. *Procedia IUTAM* **2016**, *19*, 209–218. Available online: https://www.sciencedirect.com/science/article/pii/S2210983816000286?via%3Dihub (accessed on 1 January 2016). [CrossRef]
6. Wedig, W.V. Lyapunov Exponents and Rotation Numbers in Rotor- and Vehicle Dynamics. *Procedia Eng.* **2017**, *199*, 875–881. Available online: www.sciencedirect.com (accessed on 21 September 2017). [CrossRef]
7. Wedig, W.V. Velocity Turbulences in Stochastic Road-Vehicle Dynamics. In *Proceedings of the 15th International Probabilistic Workshop*; Dresden TUD Press: Dresden, Germany, 2017; Volume 1, pp. 1–13.
8. Doods, C.J.; Robson, J.D. The Description of Road Surface Roughness. *J. Sound Vib.* **1973**, *31*, 175–183. [CrossRef]

9. Sobczyk, K.; MacVean, D.; Robson, J. Response to Profile-Imposed Excitation with Randomly Varying Transversal Velocity. *J. Sound Vib.* **1977**, *52*, 37–49. [CrossRef]
10. Davis, B.R.; Thompson, A.G. Power Spectral Density of Road Profiles. *Veh. Syst. Dyn.* **2001**, *35*, 409–415. [CrossRef]
11. Popp, K.; Schiehlen, W.O. *Fahrdynamik*; Teubner-Verlag: Leipzig, Germany, 1993. (In German)
12. Wedig, W.V. *Dynamics of Cars Driving on Stochastic Roads, In Computational Stochastic Mechanics CSM-4*; Spanos, P., Deodatis, G., Eds.; Millpress: Rotterdam, The Netherlands, 2003; pp. 647–654.
13. Blekhman, I.; Kremer, E. Vertical-longitudinal dynamics of vehicles on road with uneveness. *Procedia Eng.* **2017**, *199*, 3278–3283. Available online: www.sciencedirect.com (accessed on 21 September 2017). [CrossRef]
14. Blekhman, I.; Kremer, E. Vibrational resistance to vehicle motion due to road unevenness. *J. Sound Vib.* **2017**, *405*, 306–313. [CrossRef]
15. Barti, S.K.; Sinha, A.; Samantary, A.K.; Bhattacharyya, R. Sommerfeld effect in a single DOF system with base excitation from motor driven mechanism. *Mech. Mach. Theory* **2020**, *148*, 103808.
16. Fidlin, A.; Drozdetskaya, O. On the averaging in strongly damped systems: The general approach and its application to asymptotic analysis of the Sommerfeld effect. *Procedia IUTAM* **2016**, *18*, 43–52. [CrossRef]
17. Bisoi, A.; Samantaray, A.K.; Bhattacharyya, R. Sommerfeld effect in a two-disk rotor dynamic system at various unbalance conditions. *Meccanica* **2018**, *53*, 681–701. [CrossRef]
18. Bharti, S.K.; Bisoi, A.; Sinha, A.; Samantaray, A.K.; Bhattacharyya, R. Sommerfeld effect at forward and backward critical speeds in a rigid rotor shaft system with anisotropic supports. *J. Sound Vib.* **2019**, *442*, 330–349. [CrossRef]
19. Dimentberg, M.; McGovern, L.; Norton, R.; Chapdelaine, J.; Harrison, R. Dynamics of an Unbalanced Shaft Interacting with a Limited Power Supply. *Nonlinear Dyn.* **1997**, *13*, 171–187. [CrossRef]
20. Bharti, S.K.; Sinha, A.; Samantary, A.K.; Bhattacharyya, R. The Sommerfeld effect of second kind: Passage through parametric instability in a rotor with non-circular shaft and anisotropic flexible supports. *Nonlinear Dyn.* **2020**, *100*, 3171–3179. [CrossRef]
21. Arnold, L. *Stochastic Differential Equations*; Wiley: New York, NY, USA, 1974.
22. Wedig, W.V. Digital Simulation of Road-Vehicle Systems. *Prob. Eng. Mech.* **2012**, *27*, 82–87. [CrossRef]
23. Kloeden, P.; Platen, E. *Numerical Solution of Stochastic Differential Equations: A Review*; Springer: Heidelberg, Germany, 1995.

applied sciences

MDPI

Article

Static and Dynamic Analysis of Non-Pneumatic Tires Based on Experimental and Numerical Methods

Andrea Genovese *, Dario Garofano, Aleksandr Sakhnevych , Francesco Timpone and Flavio Farroni

Department of Industrial Engineering, University of Naples Federico II, Via Claudio 21, 80125 Naples, Italy; d.garofano@studenti.unina.it (D.G.); ale.sak@unina.it (A.S.); francesco.timpone@unina.it (F.T.); flavio.farroni@unina.it (F.F.)

* Correspondence: andrea.genovese2@unina.it

Abstract: Since the beginning of their production, pneumatic tires have experienced tremendous improvements in structure and materials, becoming the dominant design in the world tires market. Nevertheless, relying upon pressurized air, they are affected by maintenance and security issues that can lead to fatal accidents. Therefore, tire-makers are investigating new tire designs, called Airless or Non-Pneumatic, with the aim of removing air-related problems. The research about such tires is still at an early stage, especially if compared to the one conducted on the pneumatic ones. In this paper, the development of a methodology capable of studying the mechanical behavior of a Non-Pneumatic Tire (NPT) by means of experimental data and numerical approach is illustrated. The experimental activities consisted of a scanner acquisition of the NPT and a footprint analysis for the calculation of the radial stiffness and contact patch pressure distribution. Moreover, the Digital Image Correlation (DIC) technique was applied to carry out a more specific study about the spoke's deformation. From the acquired 3D model, a calculation of the NPT vertical deflection with finite element analysis (FEA) was performed—validating the model and then submitting it to a steady state analysis—that allows the simulation of a steady state rolling tire with the possibility to replicate different values of slip ratio. The results of the experimental activities are in good agreement with the ones obtained with FEA, further validating the developed methodology.

Keywords: non-pneumatic tire; vehicle dynamics; finite element analysis; steady state analysis; tire characterization; footprint; contact patch; longitudinal interaction

Citation: Genovese, A.; Garofano, D.; Sakhnevych, A.; Timpone, F.; Farroni, F. Static and Dynamic Analysis of Non-Pneumatic Tires Based on Experimental and Numerical Methods. *Appl. Sci.* **2021**, *11*, 11232. https://doi.org/10.3390/app112311232

Academic Editor: Seong-Ik Han

Received: 8 October 2021
Accepted: 23 November 2021
Published: 26 November 2021

1. Introduction

The tire is one of the most important components of vehicles, being responsible for a large set of functions: connecting the transmission to the ground with the forces necessary for motion, supporting the vehicle loads, ensuring the directionality of the vehicle by enabling steering in the driver's desired trajectory, and improving passengers' comfort by contributing to the vehicle suspension [1]. The pneumatic tire design is the market leader because of four main advantages over the NPT used in the past: lower energy loss on rough surfaces, lower radial stiffness, lower contact pressure, and lower mass [2]. Nevertheless, relying upon the containment of pressurized air, it is susceptible to be punctured and eventual burst may cause fatal accidents. Moreover, the air pressure inside the tire should be kept in a proper range, representing a potential discomfort for its users: the loss of pressure—caused by the leaking around the wheel rim and by the oxygen absorption in rubber—leads the tire to flatten in the contact area, increasing wear and fuel consumption. To overcome these disadvantages, researchers developed an innovative concept, the Non-Pneumatic tire (NPT), that does not require compressed air to work [3,4]. This concept was designed to own the key characteristics of the pneumatic tire, bringing advantages such as the absence of maintenance for proper internal pressure, a wider range of application in fields where the pneumatic tire struggles—such as the military and agricultural fields, and the recyclability of its components, so its production can be environmentally friendly and more efficient in the use of resources [5]. A proper definition of the NPT is reported in the code of Federal regulations of the United States of America [6], where it is defined as a mechanical device that transmits the vertical load and tractive forces from the roadway to the vehicle, generates the tractive forces that provide the directional control of the vehicle, and does not rely on the containment of any gas or fluid for providing those functions. Although this definition does not specify a structure, an NPT is generally composed of the following elements: a rigid hub connected to an annular outer ring, called shear band or shear beam, by means of flexible polyurethane elements called spokes and a tread. Since the rigid hub and the tread explicate the exact same functions of the ones in the pneumatic tire, the peculiar elements are the shear beam and spokes. The shear beam is constituted by two inextensible membrane layers separated by a relatively low modulus elastic layer [2].

For what concerns the spokes, researchers and tire makers developed a huge variety of structures that differ from each other in geometries and material characteristics. Several comparative studies were conducted to evaluate the effects of different spokes' structures on performance parameters. Aboul-Yazid et al. [7] compared three different concepts to investigate the influence of spokes structures on the contact pressure, radial stiffness, and stresses. Each concept was also modeled with and without the shear band to highlight its influence. Results showed that the presence of the shear band has a great influence on the deformed shapes of the spokes and on the contact pressure distribution. The objects of the aforementioned work were also modeled using finite element (FE) method by Kucewicz et al. [8] to perform a static radial test with the aim of assessing the effects of different spoke structures on vertical displacement of the rim, deformation shape, stress distribution, and contact pressure. Mathew N. J. et al. [9] introduced a model of NPT with natural rubber material instead of synthetic one for the tread. Moreover, a comparison study was conducted on various structures such as the honeycomb, plate spokes, triangular, and diamond shape to identify the structure that exhibits the least deformation. The diamond shape was found to perform better. Vinay et al. [10] studied the hexagonal honeycomb structure developing three models varying the geometric dimensions of the hexagon. They noted that even small changes in honeycomb's geometry strongly affect the vertical displacement values; in particular, the increase in the spoke's thickness causes a reduction in spokes' deformation. Rugsaj et al. [11] developed a 3D FE model of a commercially available NPT for designing the proper radial spoke. Further, a test campaign on specimens taken from the NPT was developed to obtain the hyperelastic constants that describe the material behavior. The numerical simulations were aimed at selecting the optimal thickness to replicate the radial characteristics of the tire. For the same NPT, Suvanjumrat et al. [12] studied the possibility of using 3D printing technique to build the complex shape of spoke. They compared the mechanical properties of 3D printing material with the one of the actual spokes. They found that the performances of the 3D printed spoke do not differ significantly from the ones of the original NPT, so the 3D printing technique can be used in future developments. Zhao Y. et al. [13] developed a new concept named "MEW", i.e., mechanical elastic wheel, whose peculiarity was represented by the use of mechanical hinges for the spokes. In their work, they performed numerical and experimental studies with the aim of investigate the wheel's overall stiffness and its relationship with the number of hinges used. An unconventional spokes design was proposed by Zhang et al. [14]. Studying the kangaroo's lower limb structure, they proposed spokes having the same shape and proportions of the kangaroo's thighs and calves and developed numerical simulation to compare the overall stiffness of the NPT designed with a pneumatic tire showing that the former has better load-bearing characteristics, better handling performance, and better comfort.

Besides the comparative tests and innovative designs of specific elements of the NPT, researchers also focused on the overall performances. Jackowski et al. [15] estimated the energy consumption of a pneumatic and non-pneumatic tire, chosen for size and application, evaluating experimentally the rolling resistance of both. Results showed that for the smaller values of load, the NPT is characterized by lower or quite the same rolling resistance. Veeramurthy et al. [16] studied the effect of geometric and material parameters of non-pneumatic tire elements on overall performances. In particular, the effects of design variables such as the thickness of spokes, shear band, and the material of spokes on radial stiffness and rolling resistance were analyzed. Xiaochao Jin et al. [17], using the finite element analysis (FEA), investigated both static and dynamic behavior of NPTs with different honeycomb spoke structures to find the optimal geometric characteristics that minimize the rolling resistance, a key factor for the fuel consumption.

All the aforementioned works used the FEA, a powerful tool for engineers since it can be used to design NPTs [18] by predicting the performance parameters upon different testing conditions with varying sizes, geometries, and materials of the NPT itself. With FEA, tires can be analyzed in static condition, whose main purpose is the evaluation of the radial stiffness and stress–deformation distribution, and dynamic condition, in which the tire is rolling and the interaction forces between the tire and the road are evaluated. In recent years, to perform dynamic simulation, transient FEA has been extensively applied using an explicit FE code whose computational cost is very high [19]. However, a steady state rolling analysis, which uses an implicit time integration method, is becoming widely preferred due to its effectiveness and efficiency [20]. To the author's knowledge, no studies about the dynamic behavior of a non-pneumatic tire applying the steady state analysis to an NPT with plate spoke structures and detailed tread pattern have been conducted.

The aim of this work comprises the development of a methodology capable of analyzing the static and dynamic behavior of an NPT, assessing the radial stiffness and contact patch in the static step and the longitudinal interaction, along with the dynamic radial stiffness, in the dynamic one. All the FE models created in this work were developed, processed and postprocessed in Abaqus CAE, a widely used software in FE application.

As a case study, the Michelin X-Tweel UTV was chosen, which is a non-pneumatic tire suitable for side-by-side vehicles or utility vehicles. The technical data regarding the case study, taken from the datasheet provided by the manufacturer, are summarized in Table 1.

Table 1. Michelin X-Tweel UTV technical data.

MPSN	Size	Max Load	Max Speed	Mass
21709	26 x 9N14	715 lb 325 kg	62 mph 100 kph	50 lb 22.7 kg

The experimental activities conducted on the NPT consisted of a laser scanning operation and a footprint analysis. Moreover, the NPT was also prepared for the application of Digital Image Correlation (DIC), a tool that allows a more specific study of spokes' deformation.

With the information acquired by the scanner, the FE model of the NPT was built and subjected to the same test conditions of the real tire. In order to validate the FE model, the results of the FEA were compared with the experimental ones. Once the FE model was statically validated, the dynamic behavior of the tire was investigated by submitting it to a steady state transport analysis, in which the longitudinal interaction was obtained.

2. Experimental Activities

A scanning operation by means of a 3D scanner and a footprint analysis were carried out on the NPT object of the investigation. These activities have the main target of enabling the following modeling and simulation steps, allowing us to characterize the static behavior of the tire, obtaining the radial stiffness and the contact patch evaluations. In order to perform a more detailed study of the spoke's deformation, the test rig was implemented with an additional setup to allow the application of the DIC technique.

2.1. 3D Model Acquisition

The 3D model acquisition is a preliminary and useful operation for the creation of the FE model because it provides the geometric basis from which to start the realization of the mesh. As the object is very complex, using a reverse engineering approach, all the geometric dimensions were obtained using a 3D scanner. The acquisition of the NPT geometry was realized by means of a commercial laser scanner, the FARO CAM2 Edge ScanArm HD, which uses innovative blue laser technology that allows obtaining high speed, accuracy, and resolution. Despite the black color of the tire, it was not necessary to apply a coating spray thanks to the blue laser technology [21]. The laser scanner, using the principle of triangulation, acquires a cloud of points from the surface of the object and—thanks to several best-fitting operations—all the surfaces were generated in a CAD environment. Figure 1a shows the real tire whilst Figure 1b depicts the CAD model. In order to be as accurate as possible, the real tire was subjected to a sectioning operation, through which part of the section profile was established. This destructive operation allowed us to investigate the internal material distribution of the tire; geometric characteristics of the spokes, such as the thickness that varies along both the radial and axial direction of the tire; and to acquire more information related to the arrangement of the plies, as can be seen in Figure 1c.

(a) (b) (c)

Figure 1. (**a**) Michelin X-Tweel UTV. (**b**) Michelin X-Tweel UTV CAD model. (**c**) Shear band cross-section.

2.2. Radial Characterization and Footprint Analysis: Setup and Test Campaign

The tire radial characterization aims at the evaluation of radial stiffness, contact patch, and relative pressure distribution. It can be performed using a press, mounting the tire on a fixed hub that is lowered until it crashes against a fixed plate, representing the road surface, while measurement instruments, throughout the entire test, collect the imposed forces and the displacement made by the center of the hub. Eventually, the contact patch can be obtained with pressure-sensitive films placed between the tread and the fixed plate, subsequently scanned, and analyzed with suited software to calculate the pressure distribution.

The footprint analysis was performed using a hydraulic press, which allows the imposition of different values of vertical load through a manual hydraulic system and to set camber angles varying the inclination of the rigid surface. The test rig, shown in Figure 2, was constituted by a manually actuated hydraulic press (1) mounted on a fixed frame (2), allowing only the displacement in the vertical direction. The tire was mounted on a vertically mobile frame (3) that flows inside two linear guide rails placed between the fixed and mobile frame (4). The tire was crushed against the surface of a fixed plate (5), which is integral to the press-fixed structure.

The measurement instruments used in the tests were a load cell (A), to acquire the signal of the vertical force imposed; a draw-wire sensor (B), to acquire the signal related to the displacement of the center rim; and a pressure sensitive film, to acquire the pressure distribution inside the contact patch. For what concerns the pressure-sensitive films, the ones used in the tests are composed of two polyester bases. One is coated with a layer of microencapsulated color-forming material and the other with a layer of the color-developing material. The film was positioned between the tire and the fixed plate (5). When the vertical load was applied, the microcapsules broke and the color-forming material reacted with the color-developing material, printing the image of the footprint on the film. The latter was then scanned and analyzed with a software that allows from the color intensity to obtain the pressure distribution. The procedure for the contact patch acquisition is represented in Figure 3.

Figure 2. Test rig components and measurement instruments: (1) manually actuated hydraulic press, (2) fixed frame, (3) vertically mobile frame, (4) linear guide rails, (5) fixed plate, (A) load cell, (B) draw-wire sensor.

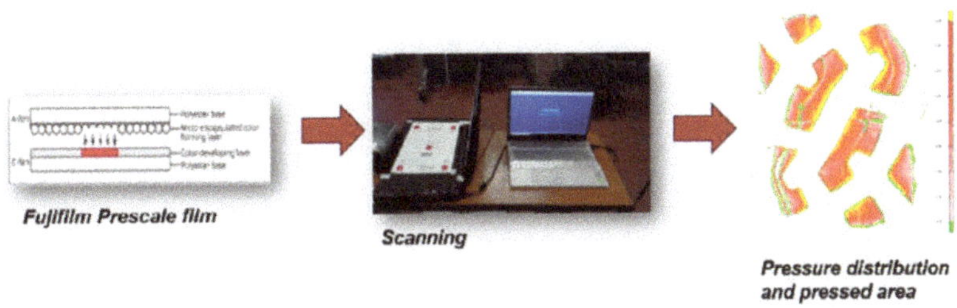

Figure 3. Contact patch acquisition procedure.

The test rig was further modified for the application of the DIC technique. It provides the field of displacements and strains by comparing the digital images of the specimen in the undeformed and deformed states. In particular, it consists of the tracking of selected points of interests, or pixels, on the images of the object recorded before, during, and after the deformation process [21,22]. In Figure 4, the preparation of the NPT is shown, in which the surfaces of the spokes were covered with chalk and a black sheet was placed in the background. Some markers were placed in precise control points to help the comparison of homologous points during the experiment. The setup was completed with a CCD camera sensor placed with its optical axis normal to the longitudinal plane of the tire.

The test campaign was composed of seven different load cases for three different values of camber angle, giving twenty-one tests to be conducted. To obtain better results, each test was repeated three times for each load and camber condition. The test procedure was the following: at the beginning of the test, the tire was brought in contact with the film placed on the fixed plate and no vertical load was applied; after twenty seconds, the vertical force was applied and kept constant for one hundred and twenty seconds; after that time, the tire was unloaded and maintained in position for other twenty seconds, when the test ended. The pressure sensitive film was then collected, scanned, and stored. The Vertical-Force–Displacement and Radial-Stiffness–Displacement experimental graphs were obtained averaging the data obtained in the three repetitions for a fixed load and camber condition. The standard deviation affecting the data was in all the cases <5%, with the highest value—found in correspondence of the highest value of Vertical Force in camber 0° condition—equal to 4.69%, giving a value of Force equal to 3563.5 ± 167 N. These graphs are shown in Figure 5.

Figure 4. NPT setup for DIC tool.

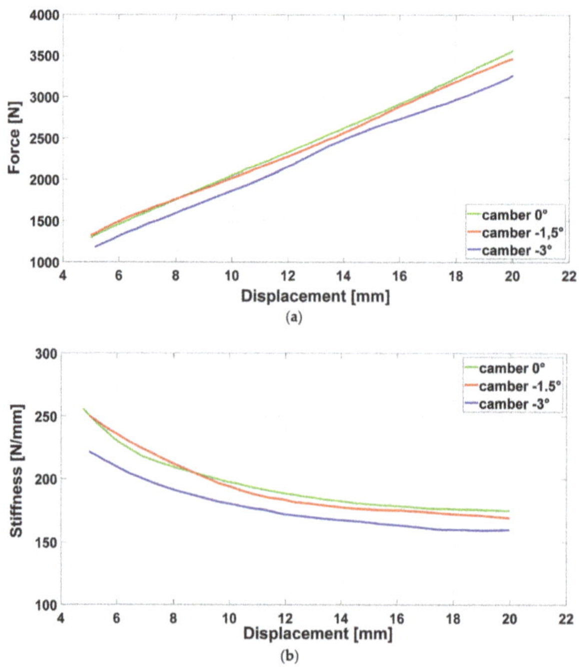

Figure 5. (**a**) Force vs. Displacement experimental diagram; (**b**) Radial Stiffness vs. Displacement experimental diagram.

For what concerns the contact patch, the pressure distribution and the pressed area are strongly related to the two variables chosen. To highlight the influence of the single variable, one was maintained fixed whilst the other one was varied. In Figure 6a, the camber angle was held fixed at 0° while the load was increased. As was expected, the pressed area increased as the vertical load increased. In Figure 6b, the influence of the camber angle can be noticed. Fixing the vertical load at 1500 N, the camber angle was changed, highlighting the differences in the contact patch; further, the pressure distribution was affected by the increasing camber, showing average higher values highlighted in yellow according to the pressure scale.

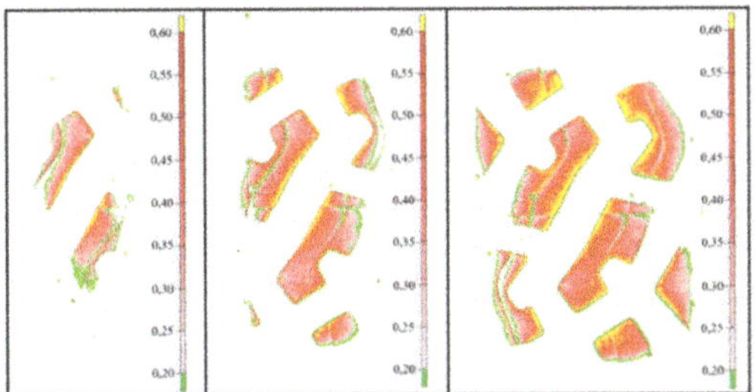

Camber [°]	0	0	0
Load [N]	500	2000	3500
Pressed Area [mm^2]	1602	3999	7215

(**a**)

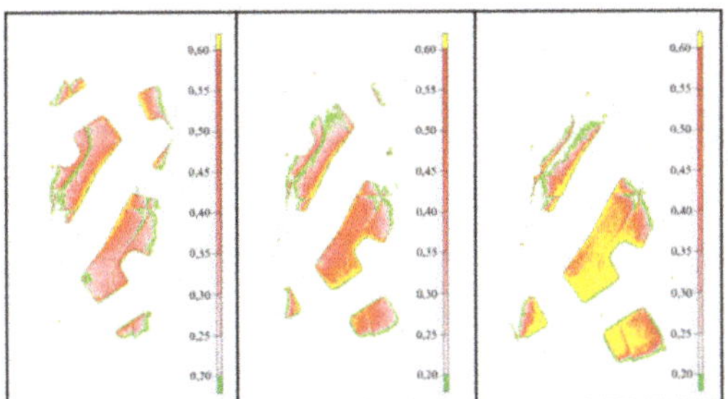

Camber [°]	0	-1,5	-3
Load [N]	1500	1500	1500
Pressed Area [mm^2]	3159	3091	2833

(**b**)

Figure 6. Contact patch analysis: (**a**) Influence of load; (**b**) influence of camber angle.

3. Static Analysis Using FE Method

With the aim to create a model that actually reproduces the mechanical behavior of the tire, matching the experimental results previously obtained, and investigating its dynamic behavior, the process developed to submit the reference tire to the footprint analysis by means of the FE method is presented.

3.1. FE Modeling

Starting from the outcome of the laser scanning operation, the FE model of the Michelin X-Tweel UTV was developed. In the modeling phase, five elements were considered: spokes, inner ring, shear band, belt plies, and tread.

The spokes modeling was a critical step, since they strongly affect both the radial behavior of the tire and the elements number. They can be modeled using solid elements starting from the geometry acquired with the laser scanner, realizing a 3D model. However, to minimize the elements number and thus, the computational time, a 2D–3D mixed model was studied. The mixed model is made of 4-node linear shell elements placed on the mid-surface of the spokes for thinner zone and 8-node linear solid elements for the thicker one, which is the interface of the spokes with the inner ring and the shear band. The connection between 3D and 2D elements was realized with shell elements located inside the solid elements, with the nodes equivalenced. This transition zone is required to ensure the continuity of the component because solid elements nodes have 3 degrees of freedom while shell elements nodes have 6 degrees of freedom. The model of the spoke is shown in Figure 7a whilst a focus on the transition zone is shown in Figure 7b. In the mentioned figure, some solid elements were hidden to allow the view of the internal shell elements.

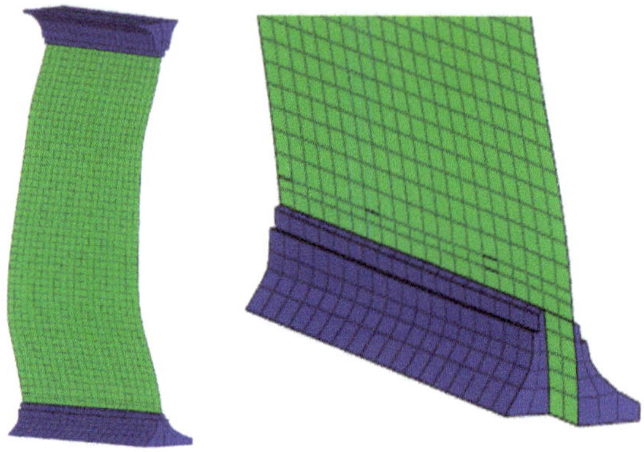

(a) (b)

Figure 7. (**a**) Spoke 2D–3D mixed model with shell elements (green) and solid elements (blue), (**b**) focus on the transition zone, with solid elements hidden.

The inner ring and the shear band were modeled with 8-node linear solid elements with the hybrid formulation or Hermann formulation, which consists of the separated integration of pressure and displacement field in order to prevent volumetric locking that may occur during compression of hyperelastic materials, which may cause convergence problems. To recreate the chemical linking existing between the top and bottom surfaces of the spokes and, respectively, the inner ring and shear band, and between the shear band and the tread, a tie surface to surface constraint was used.

The belt plies were reproduced using a simplifying technique that consists of modeling a belt ply with surface elements [23]. Each surface element is integrated into the three-dimensional structure using the embedded elements technique while the mechanical behavior of the plies is simulated through the introduction of the rebar layers. The information about the location of the surfaces that reproduce the belt plies can be deduced from the cross-section of the shear band obtained in the sectioning step previously described. The adopted rebar properties were 1.5 mm^2 as resistant area of the single fiber with a distance between two bar equal to 1.33 mm and an orientation angle equal to 0°. The orientation angle is defined as the angle between the direction of the fibers and the direction of the wheel travel.

The tread is composed by the repetition, about the axial direction of the tire, of a pattern constituted by six blocks that repeats itself every 20°. Each element was modeled with a 4-node linear tetrahedron with hybrid formulation. Since the tread elements were meshed only in the contact zone in order to not increase excessively the number of elements and because they do not work, outside the contact zone, during the contact patch simulation, a slightly smaller element size was used to better describe the contact between the tread and the road.

Finally, the road was modeled with a rigid quadrilateral element with 4 nodes, placed in a one-point contact position with the tread. The road was assumed to be rigid and modeled as a discrete rigid body. In Figure 8a, the entire FE model of the NPT is presented, whilst in Figure 8b, some elements are hidden to allow the view of the surfaces representing the belt plies embedded in the shear band.

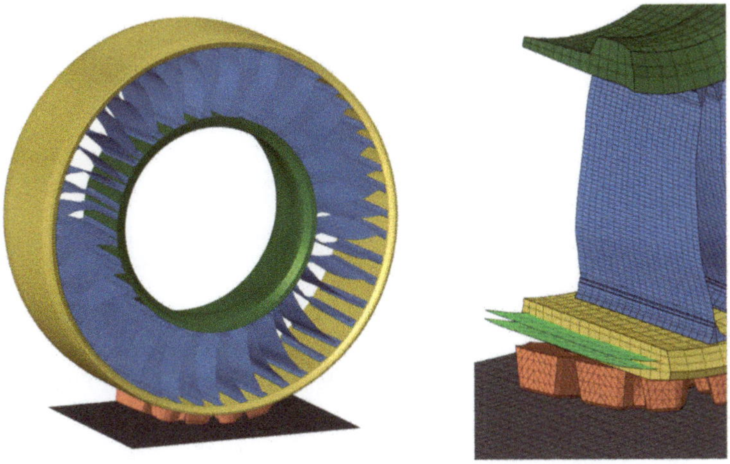

(a) (b)

Figure 8. FE model of the Michelin X-Tweel UTV: (**a**) Entire FE model, (**b**) details of belt plies.

For what concern the material properties to be assigned to the elements of the NPT, Żmuda et al. [24] provided the material coefficients needed to describe the mechanical behavior of the rubber constituting the tread. The inner ring, the spokes, and shear band are made of PU and modeled with a linear elastic material model with a Young modulus found by Rugsaj et al. [25] by means of the inverse method, whilst the belt plies are made of nylon. The material properties adopted for the analysis are summarized in Table 2.

Table 2. Material properties.

Material	Material Model	Young Modulus E [MPa]	Poisson's Ratio	Hyperelastic Coefficients
Polyurethane	Linear Elastic	62	0.48	-
Rubber	Hyperelastic Neo-Hookean	-	0.482	C_{10} = 0.833 D_1 = 0.0437
Nylon	Linear Elastic	3500	0.30	0

The Neo-Hookean hyperelastic model was chosen to describe the tread elements to provide more accuracy in portions characterized by large deformations and where the material is highly confined, especially when the loading is highly compressive. Concerning the polyurethane, the use of a linear elastic material model represents a simplification that allows for reduction of the computational time, maintaining the approximation in a small range of error. Notice that more precise results can be obtained by using more complex, nonlinear material models, without affecting the validity of the proposed procedure.

In order to reproduce the experimental test conditions, the right set of constraints and boundary conditions must be applied. Starting from the constraints, the hub of the tire was replaced by the kinematic coupling between the hub reference point and its external nodes. This constraint was applied to replace the hub since it is infinitely stiffer than the other elements of the tire.

To model the mechanical behavior of the road, a rigid body constraint was imposed between a reference point, placed outside the contact zone and the road elements. In this way, the displacements and rotations imposed to the reference point are equal to the ones of the nodes of the road.

Concerning the boundary conditions, a reference point was defined in the center rim of the tire and to that, a displacement, representing the crushing of the tire against the surface, was imposed.

To simulate the tire–road contact, the penalty method was chosen. It provides for the identification of the surfaces experiencing the contact dividing them into a master surface—in this case, the road—and a slave surface, the portion of the tread placed in contact with the road. This operation is mandatory for the individuation of the nodes that, during the crashing, could experience the interpenetration of bodies. In the hypothesis of finite sliding between the two surfaces, the friction is simulated by Coulomb's law assuming equal to 0.8 the coefficient of friction. The same friction model and value was used for the following steady state transport analysis.

3.2. Results and Model Validation

In order to validate the methodology, the results of the FEA were compared with the experimental ones. The comparison was carried out on two different planes: a qualitative comparison that involves the deformation shapes of the experimental and FE model, and a quantitative comparison that includes the following: the Force and Radial Stiffness vs. Displacement diagrams that globally analyze the mechanical behavior of the tire; the DIC results that locally study the deformation of the spokes; the contact patch comparison that focuses on the contact phenomenon of the tread elements in terms of pression.

The qualitative comparison between simulation and experimental results is shown in Figure 9, where the deformed shape of the FE model is overlapped with the real deformed shape, obtained with a 3500 N load with 0° camber.

Figure 9. Comparison of the deformed shapes.

In order to quantitative validate the model behavior, in Figure 10a, the Force vs. Displacement diagram of the FEA is shown, which is in good agreement with the experimental one. For a wide range of displacements, the reaction force is quite the same for both with a diverging behavior only present for the smallest and largest displacements. At the smallest displacements, the divergences can be explained by the errors that affect the measurement instruments when they acquire the displacement signals at low loads whilst, for the largest displacements, the differences can be explained by the highly nonlinear behavior of the material that constitutes the tire and does not allow to match the exact force trend for the entire range of deformation. For what concerns the Radial Stiffness vs. Displacement diagrams are shown in Figure 10b; further, in this case, the FE model is in good agreement, replicating the softening spring characteristic behavior of a NPT, which means that the radial stiffness of the tire decreases with an increasing displacement applied to center of the rim.

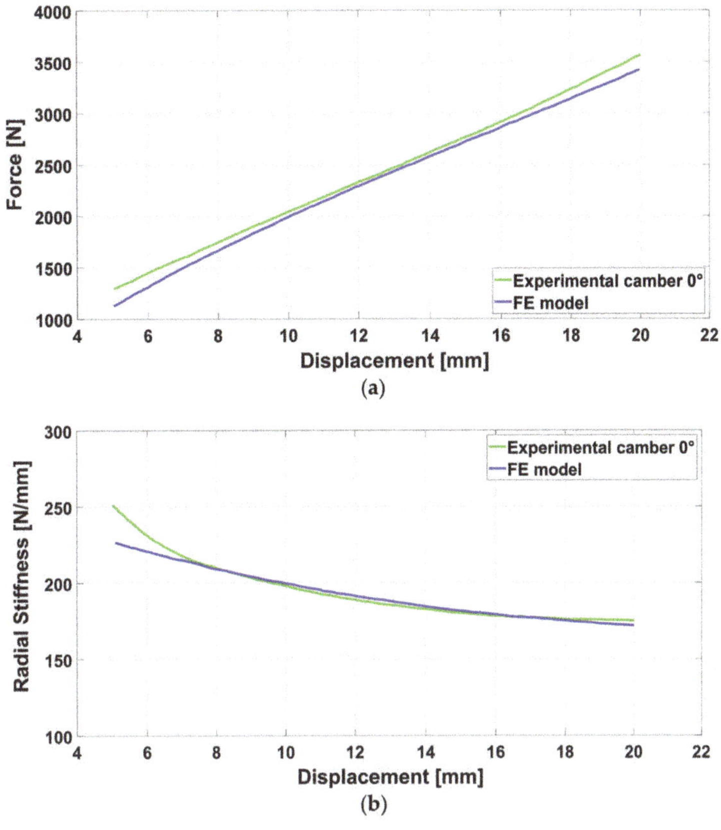

Figure 10. (**a**) Force vs. Displacement comparison diagram. (**b**) Radial Stiffness vs. Displacement comparison diagram.

The spokes deformations were compared with the ones obtained from the application of the DIC method. It was performed on three pairs of spokes but, in this paragraph, we present the results of only one to show the methodology applied. On each spoke, five reference points (I–V) were chosen to investigate the deformation in specific points. In Figure 11a, the deformed shape of the NPT is shown with the red markers representing the position of the points of interests after the deformation, whilst the green before, both on the tire and on a coordinate axis system placed on a plane normal to the optical axis of the camera sensor and with the origin corresponding to the undeformed position of the first marker. The same was done in the FE model, querying the displacement of the nodes that in the model represent the markers, as shown in Figure 11b. In Table 3, we report the values of the displacement of each marker, along the x and y direction of the coordinate axis system previously defined and shown in Figure 11a, obtained experimentally with the DIC tool and numerically with FEA and compared with the calculus of the percentage error, found as follows:

$$\Delta y_{\%error} = \frac{\Delta y_{experimental} - \Delta y_{FEM}}{\Delta y_{experimental}} \times 100, \quad \Delta x_{\%error} = \frac{\Delta x_{experimental} - \Delta x_{FEM}}{\Delta x_{experimental}} \times 100 \quad (1)$$

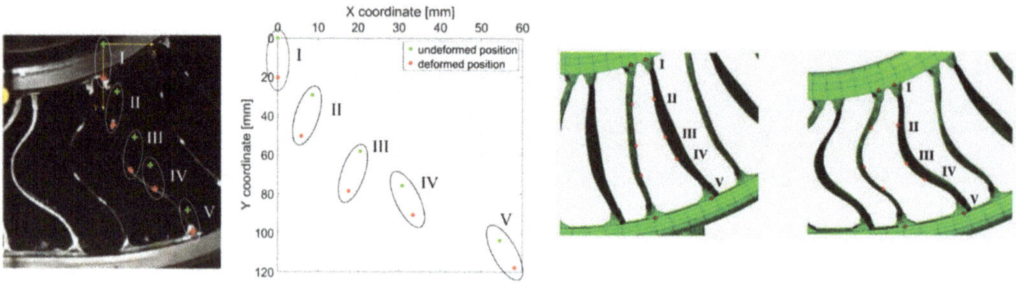

(a) **(b)**

Figure 11. DIC tool markers: (**a**) marker position in undeformed and deformed state on tire (left) and coordinate axis system (right); (**b**) marker position in undeformed (left) and deformed (right) state in FE model.

Table 3. DIC tool and FEA displacement values and percentage errors.

Spokes Marker	ΔY Experimental (mm)	ΔY Fem (mm)	ΔY % Error	ΔX Experimental (mm)	ΔX Fem (mm)	ΔX % Error
I	20.35	19.56	3.88	0	0	0
II	20.97	19.78	5.67	−2.72	−2.83	−4.04
III	20.35	19.75	2.94	−3.816	−4.19	−9.80
IV	15.15	16,475	−8.75	2.59	2.80	−8.11
V	13.91	12.64	9.13	3.56	3.28	7.86

The largest displacement in the X direction was experienced by the III point, since it is placed in the middle of the spokes where the bending is higher. Instead, the largest displacement in the Y direction was found in the first points, being directly under the load application. The DIC tool validated the FE model realized to model the spokes since the values of error found, caused by the discretization of the spokes in the FE model that did not allow to place the markers in the exact same positions occupied in the experimental phase, are within the target values set ≅ 10% for all the spokes analyzed with the DIC tool.

For what concerns the contact patch, the footprint shows the same agreement as the previous comparisons. In Figure 12a,b is represented the pressure distribution of the contact patch of both experiments, one obtained with the pressure sensitive film during the experimental step, and one obtained with FEA. The FE result shows the same distribution of the experiment, with the pressure peaks detected in the same points indicated by the yellow area in the film. In Figure 12c is shown the contact opening between the tread and road surfaces, in which the blue color is representative of the contact condition whilst the red color is the detachment. Further, in this case, the extension of the contact area obtained with FEA matches the experimental one. Particularly, the contact area obtained with the experimental activity is 5495.26 mm^2, whilst the contact area obtained with FEA is 5136.42 mm^2, with a percentage error of 6.53%.

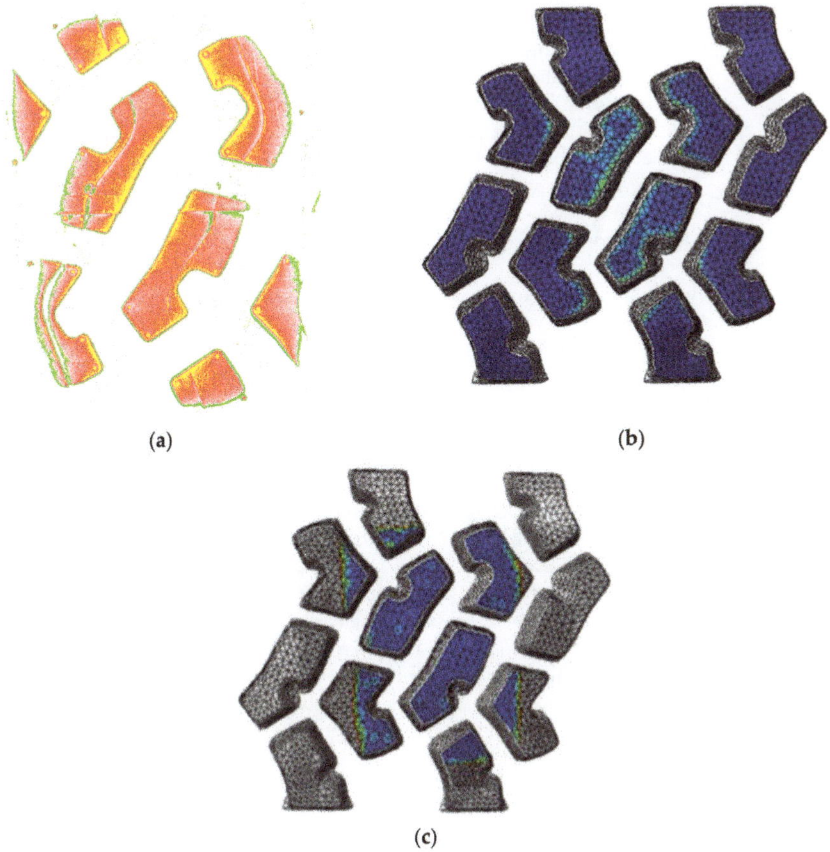

Figure 12. Contact patch comparison: experimental pressure distribution (a), FE contact pressure distribution (b), and contact opening plot (c).

4. Dynamic Study: Steady State Analysis

The phenomenon of rolling structures can be studied in a steady or transient state. Although the transient analysis allows to simulate the rolling, imposing directly the value of acceleration, the computational cost is very high. Therefore, the steady state analysis, using a specific kinematic formulation to describe the phenomenon, shows a less-expensive computational cost, though it needs some mesh requirements. In this paragraph, we introduce the kinematic description at the basis of the steady state transport analysis—a specific procedure of the Abaqus environment [26]—and the guidelines for the creation of a FE model capable of meeting its requirements explained.

Only in recent periods, in FE codes, objects with complex geometric shape were allowed to undergo the steady state simulation. Korunovic et al. [27] showed the approach to submit a pneumatic tire with a detailed tread pattern to a steady state analysis. This analysis was also applied to NPTs. Deng et al. [28] simulated the steady state rolling of the "MEW"—the NPT with a slick tread developed by Zhao Y.—obtaining a good match between the experimental and FEM analysis results, for what concerns the graph of longitudinal force vs. slip ratio and contact patch.

The mathematical formulation behind this analysis forced the construction of a FE model with a methodology different from the one shown in the static analysis and this step was also complicated by the complex geometry of the case study.

In fact, the model must be generated starting from the rotation of a bidimensional surface or three-dimensional sector along about an axis. This requirement is a consequence of the peculiar kinematic description of the phenomenon that removes the time dependence in favor of a spatially dependent analysis, using specific coordinates. The rolling movement of an axisymmetric body is described in a reference frame that moves with the ground motion of the body and, in this moving frame, the rigid body rotation is described in spatial or Eulerian manner whilst the deformation in a Lagrangian manner, introducing the requirement of continuous lines in the circumferential direction—streamlines that define a static mesh in which the material can flow. In this case, the presence in the circumferential direction of voids between the spokes and the treads can cause nonconvergence issues because, during the rotation, the elements of these components leave the mesh defined in the static step. To overcome this problem, the spokes and the treads cannot be meshed individually, as was made for the static analysis, but together, with the inner and outer ring for the spokes and the shear band for the treads. Thus, all nodes and elements at the interface of these components must be the same and equivalenced without any kind of discontinuity. In this way, the elements and nodes of the inner and outer ring can easily flow inside the streamlines thanks to their cylindrical geometry and can transmit the deformation status to the elements and nodes of the spokes. The same is for the tread blocks.

With the previous guidelines, the FE model of the NPT was made, revolving a single sector, shown in Figure 13, along the tire axis. It is recommended to choose the sector with the smallest angle to be as close as possible to the revolution of a bidimensional cross-section. In this case, the sector has a 60° extension due to the repetition pattern of its elements. In fact, the single pair of spokes has an angular extension of 15° whilst the tread pattern repeats itself every 20°. The meshing process of the components was the same as the one shown before, aside from some changes in the spokes modeling. In addition, the constraints and the boundary condition applied were the same as the static analysis, including the value of friction coefficient imposed between the road and tread surfaces equal to 0.8.

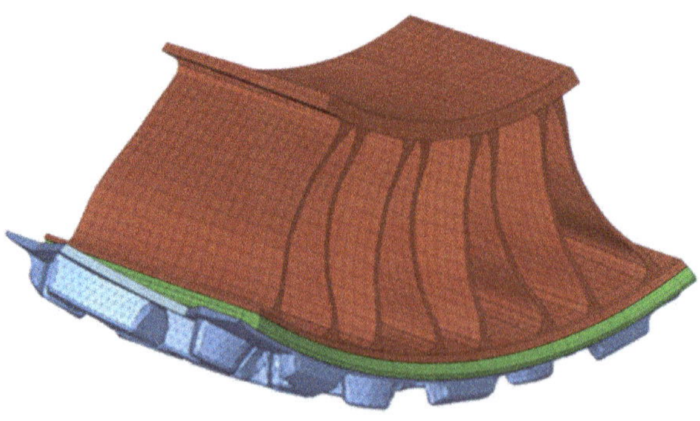

Figure 13. FE model of the 60° sector.

As was explained at the beginning of this chapter, the spokes were meshed with the inner and outer ring. Unlike the model used in the footprint analysis, the spokes here were modeled completely with 8-node linear solid elements instead of shell.

Steady State Analysis

The rolling simulation consisted of two main steps: in the first step, the FE model was built rotating the 60° sector and the footprint analysis was reproduced imposing the same vertical displacement of the static tests (20 mm). The static footprint is an essential requirement for the second step, in which the steady state analysis was performed. The first simulation step is fundamental to check if the radial stiffness of the FE model built for the dynamic simulation matches the one of the FE model built for the static characterization, verifying that it behaves the same way. This check was carried out calculating the percentage error between the values of Forces obtained for homologous values of Displacements for the two different models, finding that the maximum value of percentage error was 6%.

The computational cost of the footprint analysis for this model was higher than the one previously described due to the modeling of the tread along the entire circumference of the tire and the modeling of the spokes with 3D elements that led to a higher number of elements and nodes. In fact, the previous FE model was constituted by 139,543 nodes and 129,042 elements, whilst the model created for the steady state analysis was made up of 274,984 nodes and 805,836 elements. Once the footprint analysis was realized, the steady state analysis was performed. At first, it was mandatory to define the longitudinal velocity of the center rim and the angular velocity of the tire, the value of which is found by means of an iterative process. In the free rolling motion, the following equations are verified:

$$V = \Omega * R_e,\ F_x = 0,\ M_y = 0, \tag{2}$$

where V is the speed of the center of the rim, Ω is the angular velocity of the tire, R_e is the effective rolling radius of the tire, F_x is the longitudinal force, and M_y is the torque. When the simulation is set, both Ω and R_e are unknown and the only input is the longitudinal velocity V, set equal to 10 km/h. The iteration started with an approximated value of R_e obtained using the following equation deduced from geometric consideration [29]:

$$R_e = R - \frac{\delta}{3} \tag{3}$$

where R is the nominal radius and δ is the vertical deflection imposed in the footprint analysis. This equation gives a first attempt at effective rolling radius R_e equal to 323.4 mm and from the first equation of (2) can be derived an angular velocity Ω of 8.6 rad/s. With these inputs, the simulation was set imposing at first, an angular velocity of 7.7 rad/s, and subsequently, it was varied to a value of 9.7 rad/s. Covering this range of values of angular velocity, the one corresponding to a value of longitudinal force F_x equal to zero will be the angular velocity of free rolling condition. In Figure 14 is represented the longitudinal force F_x vs. the angular velocity Ω diagram, from which can be observed that F_x is zero for an angular velocity of 8.425 rad/s. For the first Equation (2), the effective rolling radius R_e is

$$R_e = \frac{V}{\Omega} = \frac{2777.8\ \text{mm/s}}{8.425\ \text{rad/s}} = 329.7\ \text{mm}, \tag{4}$$

where the velocity V of 10 km/h was transformed into mm/s.

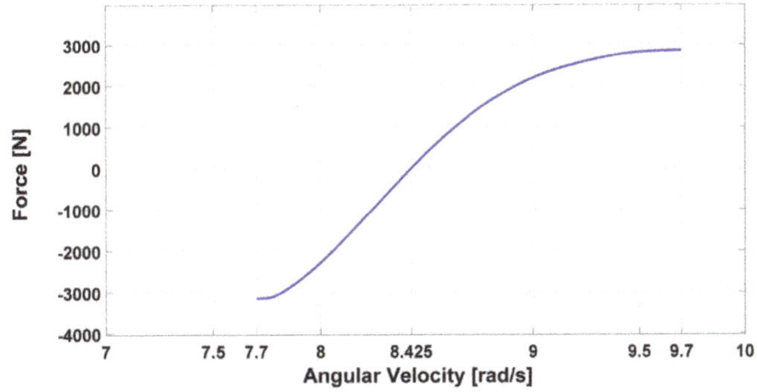

Figure 14. Longitudinal force vs. angular velocity for the calculation of the effective rolling radius.

The effective rolling radius is necessary also for the slip ratio definition:

$$S = -\frac{(V - \Omega * R_e)}{V} * 100 \quad (5)$$

with which it is possible to define the longitudinal force vs. slip diagram, shown in Figure 15 for three different values of vertical deflection. The diagrams are not symmetric with respect to the y-axis, representing a different behavior in braking and driving condition. Nevertheless, the linear trend lasts until the slip ratio reaches the value of about −5% and 5% for all the conditions, before reaching the nonlinear shape and the plateau for higher values of slip ratio.

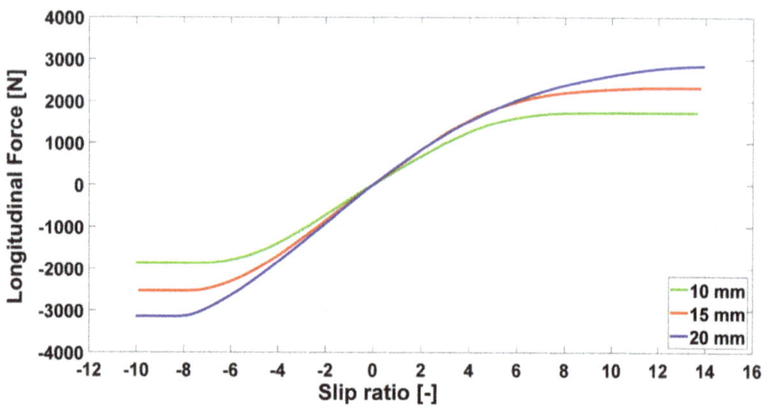

Figure 15. Longitudinal force vs. slip ratio for three different vertical displacements of the center rim.

The results of the simulation also showed a peculiar aspect regarding the vertical reaction, the index of the radial stiffness of the NPT, as a function of the slip ratio. Observing Figure 16, it is possible to note the variation of the vertical reaction with the slip ratio for a 10 km/h longitudinal speed. This diagram highlights two interesting features of the dynamic behavior of the NPT: 1—In free rolling condition (slip ratio equal to 0), the dynamic stiffness decreases as soon as the tire starts rolling, according to the Pacejka studies on pneumatic tires [30]; 2—the vertical reaction presents a lower value at zero slip than the one calculated in static condition and then it grows with the slip ratio, in a nonsymmetric manner for negative or positive value of the slip ratio. The latter result could be explained analyzing the spokes' deformation. During the nonzero slip conditions, the spokes are tensioned by the relative motion between the shear band and the hub, stiffening the vertical reaction.

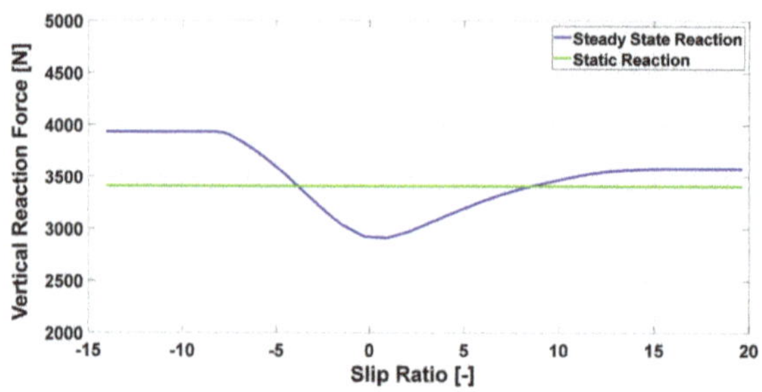

Figure 16. Vertical reaction vs. slip ratio.

Moreover, the nonsymmetric shape of the spokes in radial direction induces a different deformation for negative or positive values of slip ratio, as showed in Figure 17, where the deformed shapes corresponding to the braking and driving states are reported, overlapped to highlight the differences. During braking, the spokes placed inside the contact patch exhibited a translation to the left in the opposite direction of motion, which in these figures, is from left to right. This deformation is caused by the relative velocity (slip velocity) existing between the road and the tire. The opposite happens for the driving condition, as can be imagined, with the spokes being deformed to the right.

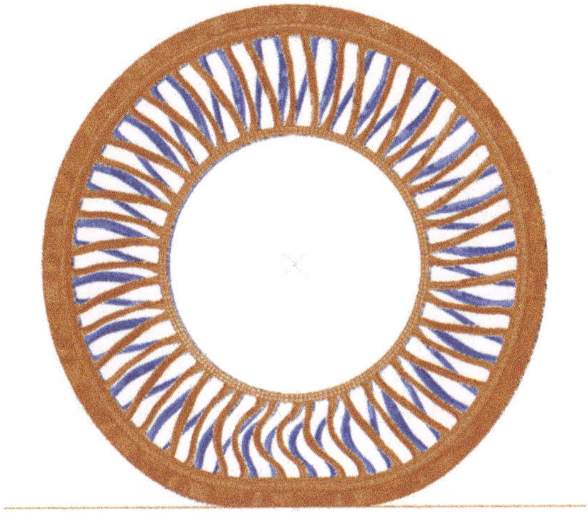

Figure 17. Deformed blue shape: braking condition. Deformed orange shape: driving condition.

5. Conclusions

The purpose of this paper is to present the development of a methodology capable of characterizing both statically and dynamically a non-pneumatic tire by means of experimental activities and numerical simulations. The methodology provides for the digital acquisition of the shape of the NPT, in this case, the Michelin X-Tweel UTV chosen as reference, realized using the 3D scanner FARO CAM2 Edge ScanArm HD that allowed us to subsequently obtain the exact geometry used to build the FE model. The footprint analysis of the NPT was realized in the DII department of Federico II University of Naples, obtaining the radial stiffness and the contact patch pressure distribution. From the output of the 3D scanning operation, a FE model of the reference tire was built and submitted to the footprint analysis using FEA. The numerical simulations showed a good agreement with the experimental data, with regard to the radial stiffness and spokes deformation, for which a study was performed with DIC technique. Once the model was statistically validated, a steady state analysis was performed on a FE model specifically built for this simulation, studying the rolling at a velocity of 10 km/h. From the simulation, the effective rolling radius, the longitudinal force vs. slip ratio graph, and the vertical reaction vs. slip graph were obtained, allowing us to perform an analysis on the longitudinal interaction and the dynamical behavior.

The methodology, composed of both an experimental and numerical method, allowed us to obtain interesting results in the static and dynamic characterization of complex objects such as NPTs. However, it is affected by some limitations. The simulation of the dynamic condition was realized through the steady state analysis, which does not allow the study of transient state in which important effects occur. Furthermore, the numerical approach needs a set of information regarding the geometry and materials of the object of the simulation that force the reverse engineering approach to be obtained if they are not possessed, leading also to destructive operations.

Further studies will be focused on the effect of different camber values on the longitudinal characteristic of the NPT, the application of the methodology to the study of lateral dynamics, and to the outdoor validation of the obtained results.

Author Contributions: Conceptualization, A.G. and F.F.; methodology, A.G., D.G. and F.F. software, A.G. and D.G; validation, A.S. and F.T.; formal analysis, A.G., F.T. and F.F.; investigation, A.G. and D.G.; resources, A.S. and F.T.; data curation, D.G. and A.S.; writing—original draft preparation, A.G., D.G. and A.S.; writing—review and editing, F.T. and F.F.; visualization, D.G., F.F. and A.S.; supervision, A.G. and F.F.; project administration, A.G., A.S., F.T. and F.F.; funding acquisition, A.G., A.S., F.T. and F.F. All authors have read and agreed to the published version of the manuscript.

Funding: This work was partly supported by the project "FASTire (Foam Airless Spoked Tire): Smart Airless Tyres for Extremely-Low Rolling Resistance and Superior Passengers Comfort" funded by the Italian MIUR "Progetti di Ricerca di Rilevante Interesse Nazionale (PRIN)"–grant no. 2017948FEN.

Acknowledgments: The authors would like to thank Giuseppe Iovino and Gennaro Stingo for their technical support.

Conflicts of Interest: The authors declare no conflict of interest.

References

1. Gent, A.N.; Walter, J.D. *The Pneumatic Tire*; The University of Akron: Akron, OH, USA, 2006; Volume 854.
2. Rhyne, T.B.; Cron, S.M. Development of a non-pneumatic wheel. *Tire Sci. Technol.* **2006**, *34*, 150–169. [CrossRef]
3. Manesh, A.; Tercha, M.J.; Ayodeji, O.; Anderson, B.; Meliska, B.J.; Ceranski, F. Tension-Based Non-Pneumatic Tire. U.S. Patent 8104524B2, 31 January 2012.
4. Manesh, A.; Tercha, M.J.; Anderson, B.; Meliska, B.J.; Ceranski, F. Tension-Based Non-Pneumatic Tire. U.S. Patent 8109308B2, 7 February 2011.
5. Bras, B.; Cobert, A. Life-Cycle Environmental Impact of Michelin Tweel® Tire for Passenger Vehicles. *SAE Int. J. Passeng. Cars Mech. Syst.* **2011**, *4*, 32–43. [CrossRef]
6. Office of the Federal Register N.A. and R.A. 49 CFR 571.129—Standard No. 129; New Non-Pneumatic Tires for Passenger Cars. Available online: https://www.govinfo.gov/app/details/CFR-2011-title49-vol6/CFR-2011-title49-vol6-sec571-129 (accessed on 3 August 2021).
7. Aboul-Yazid, A.M.; Emam, M.A.A.; Shaaban, S.; El-Nashar, M.A. Effect of spokes structures on characteristics performance of Non-Pneumatic Tires. *Int. J. Automot. Mech. Eng. ISSN* **2015**, *11*, 2212–2223. [CrossRef]
8. Kucewicz, M.; Baranowski, P.; Małachowski, J. Airless Tire Conceptions Modeling and Simulations. In Proceedings of the 13th International Scientific Conference, Wroclaw, Poland, 22–24 June 2016; Springer: Cham, Switzerland, 2016; pp. 293–301. [CrossRef]
9. Mathew, N.J.; Sahoo, D.K.; Chakravarthy, E.M. Design and Static Analysis of Airlesstyre to Reduce Deformation. *IOP Conf. Ser. Mater. Sci. Eng.* **2017**, *197*, 012042. [CrossRef]
10. Vinay, T.; Marattukalam, K.J.; Varghese, S.Z.; Samuel, S.; Sreekumar, S. Modeling and Analysis of Non-Pneumatic Tyres with Hexagonal Honeycomb Spokes. *Int. J. Recent Technol. Mech. Electr. Eng.* **2015**, *2*, 19–24.
11. Rugsaj, R.; Suvanjumrat, C. Proper Radial Spokes of Non-Pneumatic Tire for Vertical Load Supporting by Finite Element Analysis. *Int. J. Automot. Technol.* **2019**, *20*, 801–812. [CrossRef]
12. Suvanjumrat, C.; Rugsaj, R. Study of 3D printing for forming spoke of non-pneumatic tire using finite element method. *IOP Conf. Ser. Mater. Sci. Eng.* **2021**, *1137*, 012020. [CrossRef]
13. Zhao, Y.; Du, X.; Lin, F.; Wang, Q.; Fu, H. Static stiffness characteristics of a new non-pneumatic tire with different hinge structure and distribution. *J. Mech. Sci. Technol.* **2018**, *32*, 3057–3064. [CrossRef]
14. Zhang, Z.; Fu, H.; Zhao, Q.; Tan, D.; Yang, K. Pattern design and performance analysis of a flexible spoke bionic non-pneumatic tire. *J. Braz. Soc. Mech. Sci. Eng.* **2021**, *43*, 41. [CrossRef]
15. Jackowski, J.; Żmuda, M.; Wieczorek, M. Energy consumption estimation of non-pneumatic tire and pneumatic tire during rolling. *J. KONES* **2018**, *25*, 159–168. [CrossRef]
16. Veeramurthy, M.; Ju, J.; Thompson, L.L.; Summers, J.D. Optimisation of geometry and material properties of a non-pneumatic tyre for reducing rolling resistance. *Int. J. Veh. Des.* **2014**, *66*, 193–216. [CrossRef]
17. Jin, X.; Hou, C.; Fan, X.; Sun, Y.; Lv, J.; Lu, C. Investigation on the static and dynamic behaviors of non-pneumatic tires with honeycomb spokes. *Compos. Struct.* **2018**, *187*, 27–35. [CrossRef]
18. Steven, M. Cron Role of Abaqus in the Development of the Michelin Tweel Tire. In Proceedings of the 2010 Simulia Users Conference, Providence, RI, USA, 25–27 May 2010; pp. 1–27.
19. Shiraishi, M.; Yoshinaga, H.; Miyori, A.; Takahashi, E. Simulation of dynamically rolling tire. *Tire Sci. Technol.* **2000**, *28*, 264–276. [CrossRef]

20. Kim, K.W. Finite element analysis of a steady-state rolling tire taking the effect of tread pattern into account. *Int. J. Automot. Technol.* **2006**, *7*, 101–107.
21. Martorelli, M.; Speranza, D.; Ferraro, P.; Genovese, A.; Gloria, A.; Pagliarulo, V. Optical characterizations of airless radial tire. In Proceedings of the 2020 IEEE 7th International Workshop on Metrology for AeroSpace (MetroAeroSpace), Pisa, Italy, 22–24 June 2020; pp. 561–565. [CrossRef]
22. Pan, B.; Qian, K.; Xie, H.; Asundi, A. Two-dimensional digital image correlation for in-plane displacement and strain measurement: A review. *Meas. Sci. Technol.* **2009**, *20*, 062001. [CrossRef]
23. Korunović, N.; Fragassa, C.; Marinković, D.; Vitković, N.; Trajanović, M. Performance evaluation of cord material models applied to structural analysis of tires. *Compos. Struct.* **2019**, *224*, 111006. [CrossRef]
24. Zmuda, M.; Jackowski, J.; Hryciów, Z. Numerical research of selected features of the non-pneumatic tire. In Proceedings of the AIP Conference Proceedings; American Institute of Physics Inc.: New York, NY, USA, 2019; Volume 2078, p. 020027.
25. Rugsaj, R.; Suvanjumrat, C. Determination of material property for non-pneumatic tire spokes by inverse method. *Key Eng. Mater.* **2018**, *777*, 411–415. [CrossRef]
26. Abaqus Documentation 6.6. from Dassault Systems, SIMULIA Corp. Available online: https://classes.engineering.wustl.edu/2009/spring/mase5513/abaqus/docs/v6.6/index.html (accessed on 1 March 2021).
27. Korunovic, N.; Trajanovic, M.; Stojkovic, M. Detailed vs. Simplified Tread Tire Model for Steady-State Rolling Analysis. *Strojarstvo* **2012**, *54*, 153–160.
28. Deng, Y.; Zhao, Y.; Lin, F.; Xiao, Z.; Zhu, M.; Li, H. Simulation of steady-state rolling non-pneumatic mechanical elastic wheel using finite element method. *Simul. Model. Pract. Theory* **2018**, *85*, 60–79. [CrossRef]
29. Jazar, R.N. *Vehicle Dynamics: Theory and Applications*; Springer: New York, NY, USA, 2008.
30. Clark, S.K. *The Mechanics of Pneumatic Tires*; Government Printing Office: Washington, DC, USA, 1971.

MDPI
St. Alban-Anlage 66
4052 Basel
Switzerland
Tel. +41 61 683 77 34
Fax +41 61 302 89 18
www.mdpi.com

Applied Sciences Editorial Office
E-mail: applsci@mdpi.com
www.mdpi.com/journal/applsci